Biology of Parrotfishes

Biology of Parrotfishes

Editors

Andrew S. Hoey
ARC Centre of Excellence for Coral Reef Studies
James Cook University
Townsville, QLD
Australia

Roberta M. Bonaldo
Grupo de História Natural de Vertebrados Museu de Zoologia
Universidade Estadual de Campinas
Campinas, SP
Brazil

CRC Press
Taylor & Francis Group
Boca Raton London New York

CRC Press is an imprint of the
Taylor & Francis Group, an **informa** business

A SCIENCE PUBLISHERS BOOK

Cover credits
Clockwise from top left:
Bolbometopon muricatum (João Paulo Krajewski)
Chlorurus bleekeri (João Paulo Krajewski)
Scarus perrico (Kendall D. Clements)
Sparisoma amplum (Kendall D. Clements)

CRC Press
Taylor & Francis Group
6000 Broken Sound Parkway NW, Suite 300
Boca Raton, FL 33487-2742

First issued in paperback 2020

© 2018 by Taylor & Francis Group, LLC
CRC Press is an imprint of Taylor & Francis Group, an Informa business

No claim to original U.S. Government works

ISBN-13: 978-1-4822-2401-6 (hbk)
ISBN-13: 978-0-367-78140-8 (pbk)

Library of Congress Cataloging-in-Publication Data

Names: Hoey, Andrew S., editor.
Title: Biology of parrotfishes / editors, Andrew S. Hoey, ARC Centre of
 Excellence for Coral Reef Studies, James Cook University, Townsville, QLD,
 Australia, Roberta M. Bonaldo, Grupo de Historia Natural de Vertebrados
 Museu de Zoologia, Universidade Estadual de Campinas, Campinas, SP, Brazil.
Description: Boca Raton, FL : CRC Press, Taylor & Francis Group, [2018] | "A
 Science Publishers book." | Includes bibliographical references and index.
Identifiers: LCCN 2017047039 | ISBN 9781482224016 (hardback : alk. paper)
Subjects: LCSH: Parrotfishes.
Classification: LCC QL638.S3 B56 2017 | DDC 597/.7--dc23
 LC record available at https://lccn.loc.gov/2017047039
Classification: LCC HE151 .J356 2016 | DDC 388--dc23
 LC record available at https://lccn.loc.gov/2016028261

Visit the Taylor & Francis Web site at
http://www.taylorandfrancis.com

and the CRC Press Web site at
http://www.crcpress.com

Foreword

No one questions that parrotfishes have evolved from wrasses, and we show this close relationship by grouping them in the same suborder, Labroidei (Nelson, 2006). Parrotfishes were recognized as a distinct group by Aristotle who wrote, "All fishes are saw-toothed excepting the Scarus" and "of all fishes the so-called Scarus, or parrrot, is the only one known to chew the cud like a quadruped." He was, of course, referring to the unique pharyngeal mill of scarids that grinds limestone fragments ingested with turf algae into a fine sand, and at the same time reducing the algae to more digestible fragments. Another unique scarid character that facilitates digestion is the very long intestine and the lack of a stomach. Parrotfishes have evolved to utilize a new resource of nutrition that is denied other herbivores. Once the herbivorous acanthurids, siganids, and pomacentrids have grazed algae to a low stubble, the scarid fishes still have a food resource. Surely this, the morphological differences, and being recognized as a family for 215 years support recognition as a family. The divers and fishermen readily distinguish parrotfishes from wrasses. If we tell them a parrotfish belongs in the wrasse family, they will think we are joking.

Jack Randall
Honolulu

Preface

Parrotfish are found on almost every coral reef in the world. It is this ubiquity, coupled with their brilliant colouration and fused 'beak-like' jaws, that have long attracted the attention of those looking and working on tropical reefs. Parrotfishes also have an incredibly diverse and complex array of reproductive and mating strategies that vary both among and within species. However, it is their unique feeding action that has stimulated much scientific endeavour. The morphological innovations of the oral jaws allow parrotfishes to bite through reef carbonates, while the pharyngeal jaws allow them to grind ingested carbonates into sand particles. These innovations not only enable parrotfishes to access nutritional resources that are largely unavailable to other fishes, but make them one of the most important groups of fishes within coral reef ecosystems. No other group of fishes is so inextricably linked to the structural dynamics of their ecosystem. Despite their importance to reef ecosystems, the threats to parrotfish are numerous and severe: from the global effects of ocean warming and acidification to the local effects of overfishing, pollution and habitat degradation.

The aim of this book is to synthesise what is currently known about the biology of parrotfishes, and to consider why are parrotfishes so important to the ecology of coral reefs? The book provides a series of reviews that are intended to provide a firm grounding in the understanding of the morphology, diet, demography, distribution, functional ecology, and current threats of this group. Importantly, it provides new insights into their diet and food processing ability, their life-histories, and the influence of habitat and environment on parrotfish populations, and also identifies emerging research topics and future directions. We hope this book will appeal to students, early-career and established researchers, alike, and will stimulate further investigation into this fascinating and unique group of fishes.

Lastly, we wish to thank to all of those who contributed to this book. We invited the international authorities on various aspects of the biology of parrotfishes to contribute to the book and were overwhelmed by their positive and enthusiastic responses. We would also like to thank David Bellwood for initiating our interest in parrotfishes, sharing his extensive knowledge, and guiding our scientific development. We sincerely thank the reviewers of each chapter of this book for their constructive and insightful comments. Finally, we are extremely grateful for the ongoing support from our families (especially Jess, Kiara, Caelen, and João) for their ongoing support that has enabled us to undertake important and interesting scientific pursuits.

Andrew Hoey (Townsville, Australia)
Roberta Bonaldo (Campinas, Brazil)

Contents

CHAPTER
1

Cranial Specializations of Parrotfishes, Genus *Scarus* (Scarinae, Labridae) for Scraping Reef Surfaces

Kenneth W. Gobalet

Department of Biology, Emeritus, California State University, Bakersfield, California 93311
Current address: 625 Wisconsin St., San Francisco, California, USA 94107
Email: kgobalet@csub.edu

Introduction

Parrotfishes (family Labridae) forage by excavating or scraping surfaces of rocks and carbonate substrate that are encrusted with algae, bacterial mats, and detritus (Bellwood 1994, Choat et al. 2004, Rice and Westneat 2005), often leaving scratches and scars on the rock and coral surfaces (Cousteau 1952, Newell 1956, Clements and Bellwood 1988, Bellwood and Choat 1990, Bellwood 1994, 1996b). Ingested material is then ground into a slurry by their impressive pharyngeal jaws, that have been described to be "like a cement mixer in reverse" (Bellwood 1996b). Analysis of their gut contents indicates that they consume staggering quantities of inorganic residue (Randall 1967, Clements and Bellwood 1988, Bellwood 1995a, 1995b, Choat et al. 2002), accounting for over 70% of the gut volume in some cases (Gobalet 1980), and recent work has shown this residue is a major contributor to island-building sediments (Perry et al. 2015). The unique morphology of parrotfish feeding apparatus has facilitated the functional decoupling of the mandibular and pharyngeal jaws, with the mandibular jaws collecting the materials that are pulverized by the pharyngeal jaws.

Parrotfishes have distinctive modifications of their skulls associated with feeding on massive quantities of abrasive material that is scraped from resistant surfaces. Several early studies describing the anatomical features of parrotfishes largely focused on the mandibular, or oral, jaws (Cuvier and Valenciennes 1839, Boas 1879, Lubosch 1923, Gregory 1933, Monod 1951, Board 1956). In the last few decades there have been several more extensive studies of the mandibular and pharyngeal jaws, as well as the associated musculature (Tedman 1980a, b, Clements and Bellwood 1988, Bellwood 1994, Monod et al. 1994, Bullock and Monod 1997, Wainwright et al. 2004, Price et al. 2010). However, the connective tissue elements of the jaws of labroid fishes have been minimally addressed (for exceptions see van Hasselt 1978, Tedman 1980b, Bellwood and Choat 1990, Bellwood 1994). In this chapter the specializations of the bones, joints and ligaments of the mandibular jaws of parrotfishes, that allow them to withstand the stress generated during frequent contact with hard surfaces, are described and interpreted along with other elements of the head. The investigators cited above have also noted many of the features described here, but what makes this study noteworthy is the detail of the study and the elaboration of the connective tissue features. In particular, I provide detailed anatomical descriptions of five parrotfish species that reside in the southern Gulf of California (Thomson et al. 1979): the azure parrotfish *Scarus compressus,* bluechin parrotfish *Sc. ghobban*, bumphead parrotfish *Sc. perrico*, bicolor parrotfish *Sc. rubroviolaceus,* and loosetooth parrotfish *Nicholsina denticulata.*

The study of these species complement Clements and Bellwood (1988) and Bellwood (1994) who included one or more of these species in their authoritative studies. The descriptions presented here are a refinement and substantial update of Gobalet (1980). I fully agree with Clements and Bellwood (1988) that in the absence of any data from electromyography, cine radiology, or readings from force transducers, much of the interpretation made here is logical but speculative. It is hoped that this chapter stimulates additional investigations on this unique group of fishes.

Materials

The specimens examined in this study were collected while spear fishing from the coast of the Baja Peninsula, Mexico. Most of the specimens were collected near Danzante Island (just south of Loreto and east of Puerto Escondido, Baja California, Sur). Additional specimens were collected from Pulmo Reef located between La Paz and Cabo San Lucas just north of Punta Los Frailes. For the study, 19 *Sc. compressus* (Standard Length (SL) range: 206-559 mm), 25 *Sc. ghobban* (SL 206-482 mm), 18 *Sc. perrico* (SL 263-540 mm), 10 *Sc. rubroviolaceus* (SL 206-394 mm), a single *Nicholsina denticulata* (SL 291mm), 17 *Mycteroperea rosacea* (Epinephelidae, SL 349-610 mm) and small numbers of several other labrids, and epinephelids were collected (see Gobalet 1980 for details). Dissections were completed on fresh material and specimens preserved for later study. Skeletonized material supplemented the dissections, most of which are now housed at the Ichthyology Department, California Academy of Sciences, San Francisco. The skeletons were prepared by maceration, enzyme digestion, or with the use of dermestid beetles. Identifications follow Rosenblatt and Hobson (1969) and the nomenclature follows Page et al. (2013). The terminology for skeletal elements generally follows Rognes (1973) or Patterson (1977). The features described below are for *Scarus* except where indicated otherwise. The anatomical differences between these four *Scarus* species are subtle at best.

Results and Discussion

Detailed and technical descriptions of the hard and soft connective tissue elements of the cranium of parrotfishes are present in the appendix to this chapter, as is a table of abbreviations used in the figures. Parrotfishes are not delicate nibblers, but feed by forceful scraping or excavating chunks of algae-bearing substrate. Their feeding requires a coordinated action of the locomotor, sensory, and mandibular jaws (Rice and Westneat 2005). When their open jaws come in contact with rock surfaces, often the whole body thrashes to maintain contact with what is often an irregular substrate. Though they propel themselves toward the substrate with their pectoral fins in typical labriform motion, they break prior to contact. Rice and Westneat (2005: p 3512) provide a classic description of parrotfish feeding: "During many *Scarus* bites, it appears as though the fish is slamming its head into the rock". Grooves may actually be left on the rocks (Cousteau 1952, Newell 1956, Bellwood and Choat 1990) depending upon whether or not the species is a browser, excavator, or a scraper (Bellwood 1994). Chunks were missing from the scraping edges of the jaws of many specimens in this study and a large *Sc. compressus* had a longitudinal fracture across the palatine-ectopterygoid suture and ventral palatine. Bonaldo et al. (2007) quantified the dental damage to three species of *Sparisoma* off the coast of northeastern Brazil, and suggested the frequency of damage was related to the harder composition of the basaltic rock substratum at this marginal reef environment. Irrespective, these injuries testify to the hazards of this feeding behavior.

There are numerous connective tissue elements that encircle and tightly interconnect the bones surrounding the comparatively small mouth of parrotfishes. Ligaments and connective tissue bands encircle the snout within the lips and attach to the mass of connective tissue between the broad posterolateral surface of the coronoid process and the maxilla. These findings are consistent with Board's (1956) assessment that these bands collectively serve to resist distortion of the jaws during contact with the substrate and during jaw closing. They apparently help to prevent the dorsal displacement of one premaxilla (upper jaw) relative to the other during feeding and complement the interpremaxillary cruciate ligament (Fig. 1A) in this function. Further, the maxillary-dentary ligaments that attach to elements of the upper and lower jaws (Fig. 1C) are too substantial to serve only for mandibular-maxillary coupling that leads to upper jaw protrusion in actinoperygian fishes (Schaeffer and Rosen 1961). Alfaro and Westneat (1999) have documented upper jaw protrusion in *Sc. iseri* despite the inferences of Bellwood (1994) and Wainwright et al. (2004) that it is limited in parrotfishes.

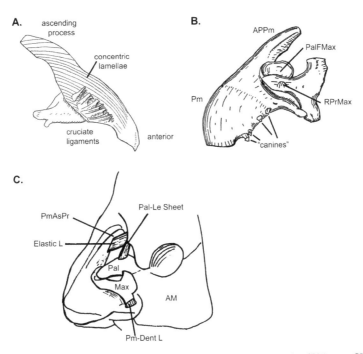

Fig. 1. Mandibular jaws and connective tissues of parrotfishes. A. *Scarus perrico* (530 mm SL): medial view of the left premaxilla. B. *Scarus ghobban* (482 mm SL): maxilla and premaxilla in lateral view. C. *Scarus ghobban*: dorsolateral view of the rostrum showing the elastic ligament and maxillary-dentary ligament (Pm-Dent L).

Trabeculae of bone are laid down along lines of stress (Murray 1936) and the concentric laminae of the medial premaxillae (Fig. 1A) look like a diagrammatic representation of the stress lines one would expect if a load were applied by the premaxillary tip (e.g. see images in Kardong 2006: p 151). The most superficial laminae arch almost the complete length of the bone and the laminae of the posterior portion of the robust ascending process are oriented almost perpendicular to the rostrum so they contact the rostral cartilage when the upper jaw is abducted with the laminae and cartilage dampening the forces. The premaxillary-frontal elastic ligament (Fig. 1C: Elastic L) apparently stretches during abduction and protrusion and could help dampen the dorsal deflection of the anterior tip of the premaxilla while the ascending process is anteroventrally positioned. It may also recoil to retract the upper jaw across the substrate.

Though many parrotfishes scrape flat or convex surfaces (Choat and Bellwood 1985, Konow and Bellwood 2005), shearing forces resulting from feeding on heterogeneous surfaces might tend to dislocate the premaxillae or dentaries (i.e., upper and lower oral jaws) relative to each other. Cruciate ligaments are positioned to resist shearing forces (Beecher 1979) and the cruciate ligaments between the premaxillae (Fig. 1A) are radially arranged and probably can resist shearing forces over a range of positions. The symphysis between the dentaries is broad and bears a series of long interdigitating ridges and grooves (for illustrations, see Bellwood 1994: p 16). The ridges are perpendicular to the radius of curvature of the outer edge of the beak, an orientation that increases the area of contact and thus the surface for transmission of forces from one bone to the other (Herring 1972). Stresses would thus be minimized through the serrate joint and the cruciate ligaments.

A forward thrust with abducted jaws against an unyielding substrate will force the premaxilla against the premaxillary condyle of the maxilla; the maxilla against the palatine; and the ascending process of the premaxilla against the rostrum. Menisci are present between maxilla and premaxilla, maxilla and vomer, and the rostral cartilage between the premaxilla and rostrum are positioned to provide cushioning. Consistent with the findings of Clements and Bellwood (1988) there is no synovial connection between the neurocranium and anterior suspensorium as exists in the less derived epinephelids. The lateral ethmoid-palatine ligament and bands (Fig. 1C: Pal-Le Sheet), and the endopterygoid-lateral ethmoid ligament restrict free motion of the anterodorsal portion of the suspensorium. These connections also would transmit forces from the palatine to the neurocranium as well as limit suspensorial abduction consistent with the reduced suction feeding (Clements and Bellwood 1988, Alfaro and Westneat 1999, Wainwright et al. 2004). Therefore, there appears to have been an evolutionary tradeoff between the selective forces encouraging reinforcement of the skull versus the generation of suction (Alfaro and Westneat 1999).

The palatine must withstand the forces transmitted to it. Longitudinal forces from the upper jaws will also be directly transmitted to the neurocranium because the posterior palatine fits in a notch on the lateral ethmoid. This is noteworthy in large specimens of *Sc. compressus* and *Sc. perrico*, which have a high posterior edge of the palatine. The maxillary condyle of the palatine is a particularly conspicuous and robust feature in large specimens (Fig. 2A: PaMax). Trabeculae within the anterior palatine generally have an orientation that reflects the application of longitudinal forces (Hildebrand and Goslow 2001). The lachrymal (Fig. 2B: La) is tightly bound to the preorbital process by the lachrymal-lateral ethmoid bands and ligaments. Anteriorly the tough lachrymal-palatine ligament connects the lachrymal with the lateral surface of the palatine. Stresses may also be dissipated along the track from the palatine to the preorbital process of the neurocranium via these bones and ligaments.

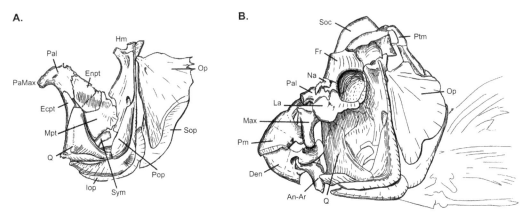

Fig. 2. A. Lateral view of left suspensorium and opercular series of *Scarus compressus* (477 mm SL); B. Lateral view of skull of *Scarus perrico* (510 mm SL).

Attention has deservedly been given to the intramandibular joint of derived percimorphs (Konow and Bellwood 2005, Konow et al. 2008, Price et al. 2010). In pomacanthids this novel joint between the dentary and anguloarticular allows 35 degrees of motion and permits gape closure when the mandibles are fully protruded (Konow et al. 2008). More derived parrotfishes (*Hipposcarus+Chlorurus+Scarus*) possess this intramandibular joint (Streelman et al. 2002) and it has been treated as a key portion of a unique four-bar linkage (Bühler 1977, Wainwright et al. 2004). Price et al. (2010) pose that the modulation of this joint may allow parrotfishes to maintain a consistent orientation with a wide gape on the surface throughout the scraping bite. The parrotfish innovations of the intramandibular joint and the pharyngeal jaws together led to rapid diversification of the oral jaws (Price et al. 2010). Parrotfishes have higher jaw-closing lever ratios than the wrasses, reflecting the greater force required to scrape hard substrata (Bellwood 2003, Wainwright et al. 2004, Westneat et al. 2005). These previous evaluations were made on the mechanics of the entire mandible with a pivot between the quadrate and mandible. I suggest that the mechanics is even more complicated because it is a double lever. Wainwright et al. (2004) hinted at this. Of particular interest are modifications of the mandible that enhance force applied at the dentary tip. A distinctive syndesmosis between the dentary and anguloarticular is present along with a shift in the insertion of the A2 of the adductor mandibulae to the coronoid process from the typical actinopterygian insertion on the ascending process of the anguloarticular (Winterbottom 1973). The consequence is a shortened out-lever of the mandible with the intramandibular joint as the pivot from that seen in generalized percimorphs like *Mycteroperca* (Fig. 3). The quadrate-mandibular articulation is the other joint. The A3 subdivision of the adductor mandibulae attaches to the medial anguloarticular (Fig. 4B, C) and is in a position to effect adduction around the quadrate-mandibular joint but it likely has only a minor role because it is quite thin. The A2 subdivision of the adductor mandibulae, on the other hand, is in a position to adduct the dentary on its pivot at the intramandibular joint. The A2 thus would be an important adductor of the dentary as previously noted by Lubosch (1923). In generalized percimorphs like *Mycteroperca* the A2 inserts on the ascending process of the anguloarticular, close to the quadratomandibular joint, which is thus the fulcrum of a third class lever and being close to the pivot is positioned to enhance speed rather than force. In *Scarus* the insertion of A2 is on the coronoid process of the dentary, and the fiber direction is almost parallel with the anterodorsal ramus of the anguloarticular (Figs. 3B, 4A). With this orientation it can generate little force that would cause mandibular rotation around the quadrate-mandibular joint and being roughly perpendicular to the coronoid process has a mechanically optimal orientation at least during limited rotation. Therefore, this is a first class lever with a shortened out-lever arm (Lo in Fig. 3B). Its in-lever of the dentary is also lengthened as a result of the elongation of the coronoid process. For a given in-force generated by the adductor mandibulae, the out-force at the tip of the dentary will be three times that of the generalist which feeds by inertial suction (Fig. 3). The Aw muscle is also in a position to abduct (Fig. 4C: Aw ab) or adduct (Fig. 4C: Aw ad) the dentary around the intramandibular joint. The muscle is delicate, however, and likely functions to modulate the position of the dentary rather than generate much force.

Considering the presence of only subtle anatomical differences among the members of the genus *Scarus* studied here, one can speculate on how these sympatric species divide the resources because it does not appear to be on the basis of their feeding. The gut content of *Scarus* spp. is composed primarily of fine particles (Hoey and Bellwood 2008, Bonaldo et al. 2014), with over 70% of the gut contents of the *Scarus* species in this study passing through a 630 µm mesh (Gobalet 1980). This small particle size makes it extremely difficult to evaluate what they are targeting, and it would take a creative, perhaps molecular, approach to discriminate what exactly has been pulverized and resides in the intestines (see Clements and Choat, Chapter 3). Considering that parrotfishes have been estimated to spend in excess of 84-91% of the daylight hours feeding (*Chlorurus* spp: Bellwood 1995a) and their impact on reefs can be bioerosion in excess of 5,000 kg per individual per year (Bellwood et al. 2003, 2012) it is logical that they are going to possess anatomical features consistent with the forceful cropping of chunks of inorganic materials. Collectively, the numerous structural adaptations in parrotfishes described above contribute to a spectacular eating machine.

A1A2

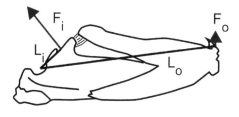

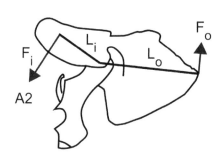

A. *Mycteroperca rosacea*

$$F_i \times L_i = F_o \times L_o$$
$$F_o = 0.16 \, F_i$$

B. *Scarus compressus*

$$F_i \times L_i = F_o \times L_o$$
$$F_o = 0.47 \, F_i$$

Fig. 3. Comparison of the lower oral jaw of a generalized percimorph and *Scarus*. A. Lateral view of the right mandible of *Mycteroperca rosacea* (610 mm SL); B. Lateral view of the right mandible of *Scarus compressus* (457 mm SL). F_i = in-force generated by the adductor mandibulae; F_o = out-force at the tip of the dentary, L_i = in-lever (distance from fulcrum to the point of application of the in-force); L_o = out-lever (distance from the fulcrum to the point of application of the out-force). *Scarus* demonstrates three times the mechanical advantage as in the generalist, *Mycteroperca*.

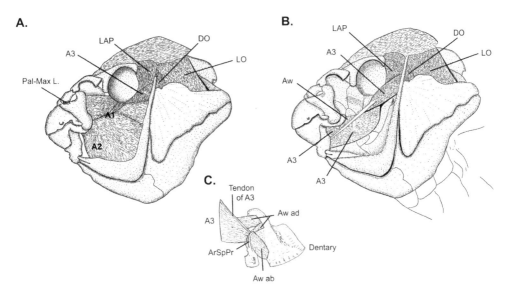

Fig. 4. Muscles of the head of *Scarus*. A. Lateral view of the muscles of the head of *Scarus ghobban* (460 mm SL). Connective tissue of the A1A2 to the premaxillary symphysis has been removed. B. Lateral muscles of the head of *Scarus ghobban* (460 mm SL) with the A1A2 complex of the adductor mandibulae removed. C. Medial complex of the left adductor mandibulae of *Scarus perrico* (540 mm SL) in medial view. Anterior is to the right.

Summary

In this chapter I have presented the details of the anatomy of parrotfishes of the genus *Scarus* that emphasize the features of the head that resist the forces applied during the scraping of rocky substrates that are encrusted with the organic materials they ingest. These descriptions and interpretations complement the growing literature on this monophyletic group of almost 100 species (Parenti and Randall 2011). The connective tissue elements (ligaments, menisci, fascia, joints) were emphasized with the following specializations being of particular interest: within the tissues surrounding the mouth, and likely deeper, are connective tissue bands and ligaments that encircle the snout; serrate joints between the dentaries and cruciate ligaments between the ascending processes of the premaxillae are positioned to resist dislocations; the intramandibular joint between the dentary and anguloarticular is a syndesmosis that likely functions to dampen forces generated during the scraping of the rigid surfaces upon which parrotfishes feed; this joint also enhances the leverage of the system powered by the A2 portion of the adductor mandibulae acting on the enlarged coronoid process; menisci are present between the premaxillae and maxillae, maxillae and vomers, and between the ascending processes of the premaxillae and the rostrum; the boney structure of the ascending processes of the premaxillae are concentrically laminar to resist compressive forces; there are unique elastic ligaments interconnecting the frontals with the ventral surfaces of the premaxillary ascending processes. Along with a highly derived pharyngeal grinding apparatus, the impressive mandibular jaws and their associated connective tissues have contributed to the parrotfishes having a substantial influence on the turnover of substrate in tropical and subtropical reefs.

Acknowledgements

Without the assistance of the following individuals this project would not have been possible: Milton Hildebrand, Karel Liem, Steve Strand, Sandy Tanaka, Peter Moyle, Aida Parkinson, Jim Deacon, Harold Silverman, Bob Ensminger, Mike Zumwalt, Wayne Schrader, Walter Hill, Dale Stevenson, Robert Daniels, Don Baltz, Jim Broadway, Terry Hansen, Glenn Douglass, John M. Hash, Michael Schimmel, Jeffrey Schimmel and my wife Kay Schimmel-Gobalet.

References Cited

Alfaro, M. and M.W. Westneat. 1999. Motor patterns of herbivorous feeding: electromyographic analysis of biting in the parrotfishes *Cetoscarus bicolor* and *Scarus iseri*. Brain Behav. Evol. 54: 205–222.

Anker, G.C. 1977. The morphology of the head-muscles of a generalized *Haplochromis* species: *H. elegans* Trewavas 1933 (Pisces, Cichlidae). Neth. J. Zool. 28: 234–271.

Beecher, R.M. 1979. Functional significance of the mandibular symphysis. J. Morphol. 159: 117–130.

Bellwood, D.R. 1994. A phylogenetic study of the parrotfishes family Scaridae (Pisces: Labroidei), with a revision of genera. Rec. Aust. Mus. Suppl. 20: 1–86.

Bellwood, D.R. 1995a. Direct estimate of bioerosion by two parrotfish species, *Chlorurus gibbus* and *C. sordidus*, on the Great Barrier Reef, Australia. Mar. Biol. 121: 419–429.

Bellwood, D.R. 1995b. Carbonate transport and within-reef patterns of bioerosion and sediment release by parrotfishes (family Scaridae) on the Great Barrier Reef. Mar. Ecol. Prog. Ser. 117: 127–136.

Bellwood, D.R. 1996a. Production and reworking of sediment by parrotfishes (family Scaridae) on the Great Barrier Reef, Australia. Mar. Biol. 125: 795–800.

Bellwood, D.R. 1996b. Coral reef crunchers. Nature Australia 25: 48–55.

Bellwood, D.R. 2003. Origins and escalation of herbivory in fishes: a functional perspective. Paleobiology 29: 71–83.

Bellwood, D.R. and J.H. Choat. 1990. A functional analysis of grazing in parrotfishes (family Scaridae): the ecological implications. Environ. Biol. Fishes 28: 189–214.

Bellwood, D.R., A.S. Hoey and J.H. Choat. 2003. Limited functional redundancy in high diversity systems: resilience and ecosystem function on coral reefs. Ecol. Lett. 6: 281–285.

Bellwood, D.R., A.S. Hoey and T.P. Hughes. 2012. Human activity selectively impacts the ecosystem roles of parrotfishes on coral reefs. Proc. R. Soc. B. 16: rspb20111906.

Board, P.A. 1956. The feeding mechanism of the parrotfish *Sparisoma cretense* (Linne). Proc. Zool. Soc. Lond. 127: 59–77.

Boas, J.E.V. 1879. Die zähne der Scaroiden. Zeitschrift für wissenschaftliche Zoologie 32: 189–215.

Bonaldo, R.M., J.P. Krajewski, C. Sazima and I. Sazima. 2007. Dentition damage in parrotfishes feeding on hard surfaces at Fernando de Noronha Archipelago, southwest Atlantic Ocean. Mar. Ecol. Prog. Ser. 342: 249–254.

Bonaldo, R.M., A.S. Hoey and D.R. Bellwood. 2014. The ecosystem roles of parrotfishes on tropical reefs. Oceanogr. Mar. Biol. Annu. Rev. 52: 81–132.

Bühler, P. 1977. Comparative kinematics of the vertebrate jaw frame. Fortschr. Zool. 24: 123–138.

Bullock, A.E. and T. Monod. 1997. Myologie céphalique de deux poissons perroquets (Teleostei: Scaridae). Cybium 21: 173–199.

Choat, J.H. and D.R. Bellwood. 1985. Interactions amongst herbivorous fishes on a coral reef: influence of spatial variation. Mar. Biol. 89: 221–234.

Choat, J.H., K.D. Clements and W.D. Robbins. 2002. The trophic status of herbivorous fishes on coral reefs. I: Dietary analyses. Mar. Biol. 140: 613–623.

Choat, J.H., W.D. Robbins and K.D. Clements. 2004. The trophic status of herbivorous fishes on coral reefs. II: Food processing modes and trophodynamics. Mar. Biol. 145: 445–454.

Clements, K.D. and D.R. Bellwood. 1988. A comparison of the feeding mechanisms of two herbivorous labroid fishes, the temperate *Odax pullus* and the tropical *Scarus rubroviolaceus*. Mar. Freshw. Res. 39: 87–107.

Cousteau, J.-Y. 1952. Fish men explore a new world undersea. Natl. Geogr. Mag. 102: 431–472.

Cuvier, G. and A. Valenciennes. 1839. Histoire naturelle des poissons: Tome quatorzième. Berger-Levrault, Paris, France.

Elshoud-Oldenhave, M.J.W. and J.W.M. Osse. 1976. Functional morphology of the feeding system in the ruff – *Gymnocephalus cernua* (L. 1758) – (Teleostei, Percidae). J. Morphol. 150: 399–422.

Gobalet, K.W. 1980. Functional morphology of the head of parrotfishes of the genus *Scarus*. PhD dissertation. University of California, Davis, USA.

Gobalet, K.W. 1989. Morphology of the parrotfish pharyngeal jaw apparatus. Am. Zool. 29: 319–331.

Gosline, W.A. 1968. The suborders of perciform fishes. Proc. U. S. Natl. Mus. 124: 1–78.

Gregory, W.K. 1933. Fish skulls; a study of the evolution of natural mechanisms. Trans. Am. Philos. Soc. 23: 75–481.

Herring, S.W. 1972. Sutures–a tool in functional cranial analysis. Cells Tissues Organs 83: 222–247.

Hildebrand, M. and G. Goslow. 2001. Analysis of Vertebrate Structure. John Wiley & Sons, Inc., New York, USA.

Hoey, A.S. and D.R. Bellwood. 2008. Cross-shelf variation in the role of parrotfishes on the Great Barrier Reef. Coral Reefs. 27: 37–47.

Kardong, K.V. 2006. Vertebrates: Comparative Anatomy, Function, Evolution. McGraw-Hill, Boston, USA.

Konow, N. and D.R. Bellwood. 2005. Prey-capture in *Pomacanthus semicirculatus* (Teleostei, Pomacanthidae): functional implications of intramandibular joints in marine angelfishes. J. Exp. Biol. 208: 1421–1433.

Konow, N., D.R. Bellwood, P.C. Wainwright and A.M. Kerr. 2008. Evolution of novel jaw joints promote trophic diversity in coral reef fishes. Biol. J. Linn. Soc. 93: 545–555.

Liem, K.F. 1970. Comparative functional anatomy of the Nandidae (Pisces: Teleostei). Fieldiana Zool. 56: 1–166.

Lubosch, W. 1923. Die kieferapparat der Scariden und die frage der Streptognathie. Verh. Anat. Ges. 32: 10–29.

Monod, T. 1951. Notes sur le squelette viscéral des Scaridae. Bull. Soc. Hist. Nat. Toulouse 86: 191–194.

Monod, T., J.C. Hureau and A.E. Bullock. 1994. Ostéologie céphalique de deux poissons perroquets (Scaridae: Teleostei). Cybium 18: 135–168.

Murray, P.D.F. 1936. Bones, a Study of the Development and Structure of the Vertebrate Skeleton. Cambridge University Press, Cambridge, UK.

Newell, N.D. 1956. Geological reconnaissance of Raroia (Kon Tiki) Atoll, Tuamotu Archipelago. Bull. Am. Mus. Nat. Hist. 109: 311–372.

Osse, J.W.M. 1968. Functional morphology of the head of the perch (*Perca fluviatilis* L.): an electromyographic study. Neth. J. Zool. 19: 289–392.

Page, L.M., H. Espinosa-Pérez, L.T. Findley, C.R. Gilbert, R.N. Lea, N.E. Mandrak, R.L. Mayden and J.S. Nelson. 2013. Common and Scientific Names of Fishes from the United States, Canada and Mexico, 7th edition. American Fisheries Society, Bethesda, USA.

Parenti P. and J.E. Randall. 2011. Checklist of the species of the families Labridae and Scaridae: an update. Smithiana Bulletin 13: 29–44.

Patterson, C. 1975. The braincase of pholidophorid and leptolepid fishes, with a review of the actinopterygian braincase. Philos. Trans. R. Soc. B Biol. Sci. 269: 275–579.

Patterson, C. 1977. Cartilage bones, dermal bones and membrane bones, or the exoskeleton versus the endoskeleton. pp. 77–121 *In*: S.M. Andrews, R.S. Miles and A.D. Walker (eds.). Problems in Vertebrate Evolution. Linnean Society Symposium Series. No. 4. Academic Press, New York, USA.

Perry, C.T., P.S. Kench, M.J. O'Leary, K.M. Morgan and F.A. Januchowski-Hartley. 2015. Linking reef ecology to island building: parrotfish identified as major producers of island-building sediment in the Maldives. Geology 43: 503–506.

Price, S.A., P.C. Wainwright, D.R. Bellwood, E. Kazancioglu, D.C. Collar and T.J. Near. 2010. Functional innovations and morphological diversification in parrotfish. Evolution 64: 3057–3068.

Randall, J.E. 1967. Food habits of reef fishes of the West Indies. Stud. Trop. Oceanogr. 5: 665–847.

Rice, A.N. and M.W. Westneat. 2005. Coordination of feeding, locomotor and visual systems in parrotfishes (Teleostei: Labridae). J. Exp. Biol. 208: 3503–3518.

Rognes, K. 1973. Head skeleton and jaw mechanism in Labrinae (Teleostei labridae) from Norwegian waters. Arb. Univ. Bergen Mat. Nat. Ser. 4: 1–149.

Rosenblatt, R.H. and E.S. Hobson. 1969. Parrotfishes (Scaridae) of the eastern Pacific, with a generic rearrangement of the Scarinae. Copeia 3: 434–453.

Schaeffer, B. and D.E. Rosen. 1961. Major adaptive levels in the evolution of the actinopterygian feeding mechanism. Am. Zool. 1: 187–204.

Schultz, L.P. 1958. Review of the parrotfishes family Scaridae. Bull. U.S. Nat. Mus. 214: 1–143.

Starks, E.C. 1926. Bones of the Ethmoid Region of the Fish Skull. Stanford University Publications, California, USA.

Streelman, J.T., M. Alfaro, M.W. Westneat, D.R. Bellwood and S.A. Karl. 2002. Evolutionary history of the parrotfishes: biogeography, ecomorphology, and comparative diversity. Evolution 56: 961–971.

Tedman, R.A. 1980a. Comparative study of the cranial morphology of the labrids *Choeroden venustus* and *Labroides dimidiatus* and the scarid *Scarus fasciatus* (Pisces : Perciformes) I. Head skeleton. Mar. Freshw. Res. 31: 337–349.

Tedman, R.A. 1980b. Comparative study of the cranial morphology of the labrids *Choeroden venustus* and *Labroides dimidiatus* and the scarid *Scarus fasciatus* (Pisces: Perciformes) II. Cranial myology and feeding mechanisms. Mar. Freshw. Res. 31: 351–372.

Thomson, D.A., L.T. Findley and A.N. Kerstitch. 1979. Reef Fishes of the Sea of Cortez. John Wiley & Sons, Inc., New York, USA.

van Hasselt, M.J.F.M. 1978. Morphology and movements of the jaw apparatus in some Labrinae (Pisces, Perciformes). Neth. J. Zool. 29: 52–108d.

Wainwright, P.C., D.R. Bellwood, M.W. Westneat, J.R. Grubich and A.S. Hoey. 2004. A functional morphospace for the skull of labrid fishes: patterns of diversity in a complex biomechanical system. Biol. J. Linn. Soc. 82: 1–25.

Westneat, M.W., M.E. Alfaro, P.C. Wainwright, D.R. Bellwood, J.R. Grubich, J.L. Fessler, K.D. Clements and L.L. Smith. 2005. Local phylogenetic divergence and global evolutionary convergence of skull function in reef fishes of the family Labridae. Proc. R. Soc. B Biol. Sci. 272: 993–1000.

Winterbottom, R. 1973. A descriptive synonymy of the striated muscles of the Teleostei. Proc. Acad. Nat. Sci. Phila. 125: 225–317.

APPENDIX

The abbreviations used in the figures are as follows:

Am	adductor mandibulae
A1	portion of adductor mandibulae
A2	portion of adductor mandibulae
A3	portion of adductor mandibulae
An-Ar	anguloarticular
APPm	ascending process of premaxilla
Ar	articular
ArSpPr	splint process of anguloarticular
AStF	anterior subtemporal fossa
Aw	portion of adductor mandibulae
Aw ab	abducting portion of Aw
Aw ad	adducting portion of Aw
BH	basihyal (glossohyal)
BPhGr	basipharyngeal groove of neurocranium
BrStg	branchiostegal rays
CCMax	cranial condyle of maxilla
CH	ceratohyal
Den	dentary
DenCorPr	coronoid process of dentary
DHPPR	dorsal hypohyal posterior process
D.Intrahyoid Lig.	dorsal intrahyoid ligament
DO	dilator operculi
DOF	dilator operculi fossa
DPFr	dorsal process of frontal
DStF	deep subtemporal fossa
Ecpt	ectopterygoid
EH	epihyal
Enpt	endopterygoid
EoPtfa	postemporal facet of epiotic
Epo	epiotic
Eth	ethmoid
Exs	extrascapular
Fm	foramen magnum
Fr	frontal
Hm	hyomandibula
HmSoc	hyomandibular sockets
HyoHyAbd 1&2	hyohyoideus abductores
Ic	intercalcar
IH	interhyal
IOF	infraorbital foramen
Iop	interopercle
La	lachrymal
LAP	levator arcus palatini

LEth   lateral ethmoid
LO   levator operculi
LOcF   lateral occipital fossa
Max   maxilla
MaxCr   maxillary crest
MOcF   medial occipital fossa
Mpt   metapterygoid
Na   nasal
OccCon   occipital condyle
Op   opercle
Pal   palatine
PalFMax   palatine fossa of maxilla
Pal-Le sheet   sheet of connective tissue between palatine and lateral ethmoid
Pal-Max L.   palatine maxillary ligament
PaMax   maxillary condyle of palatine
Par   parietal
ParCr   parietal crest
Pm   premaxilla
PmAsPr   ascending process of premaxilla
PmCMax   premaxillary condyle of maxilla
Pm-Dent L   premaxillary-dentary ligament
Pop   preopercle
PPtoF   fossa of the posterior face of the pterotic
ProSp   ventral spike of prootic
Psp   parasphenoid
PtF   posttemporal fossa
Ptm   posttemporal
Pto   pterotic
PtoPopPr   preopercular process of pterotic
PtoCr   pterotic crest
PtF   posttemporal fossa
Q   quadrate
QMJLF   quadrato-mandibular joint lateral fossa
QMJMF   quadrato-mandibular joint medial fossa
Rar   retroarticular
RoF   rostral fossa
RPr Max   rostral process of maxilla
Soc   supraoccipital
SoCr   supraoccipital crest
Sop   subopercle
Spo   sphenotic
StF   supratemporal fossa
Sym   symplectic
VHH   ventral hypohyal
VOcF   ventral occipital fossa
Vo   vomer
VoF   ventral fossa of vomer
VoMaxFa   maxillary facet of the vomer

Detailed Anatomy of the Parrotfish Head

Skeleton

Neurocranium. The neurocranium is highly sculptured with ridges and concavities that provide for the attachment and muscle mass of the cranial and trunk musculature (Fig. 5). A rostral fossa extends back to the level of the orbit and is divided by a low midline crest (Fig. 5A: RoF). A posterodorsally asymmetrically expanded supraoccipital crest extends the length of the supraoccipital (Fig. 5B: SoCr). In *Scarus compressus* and *Sc. perrico*, the crest is high in association with their prominently enlarged foreheads. The anterior edge of the crest is nearly vertical in *Sc. perrico*, which has the most developed hump. The crest is quite low in *Sc. rubroviolaceus* and of intermediate height in *Sc. ghobban*. The parietal crest (Fig. 5A, B: ParCr) angles laterally in *Sc. rubroviolaceus*, and dorsolaterally in *Sc. perrico*, *Sc. ghobban*, and *Sc. compressus*. In *Sc. perrico*, the parietal crest is higher, extends more anteriorly, and curves to the midline along the anterior edge of the frontal. The pterotic crest angles dorsolaterally and terminates at the posterior edge of the orbit, except in *Sc. perrico*, where it curves medially on the frontal to join the parietal crest. The posterior ends of the pterotic and parietal crests are joined superficially by an extrascapular that covers

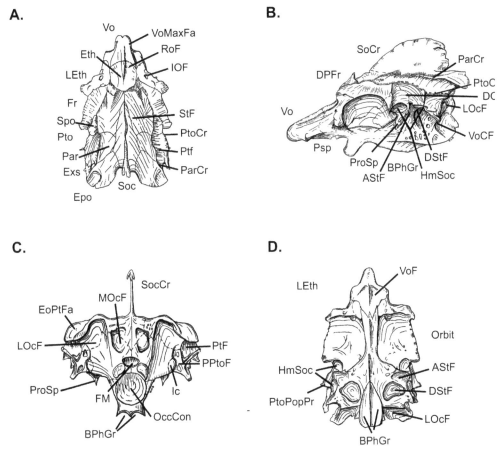

Fig. 5. Neurocranium of *Scarus compressus* (478 mm SL): A. dorsal view; B. lateral view; C. posterior view; D. *Scarus rubroviolaceus* (342 mm SL) ventral view.

the posterior end of the posttemporal fossa between the two crests (Fig. 5A: Exs). The supratemporal fossa lies between the parietal and the supraoccipital crests (Fig. 5A: StF).

Three prominent fossae are present in the postorbital region of the ventral neurocranium. Between the sockets of the neurocranial-hyomandibular joint is the anterior subtemporal fossa (Fig. 5D: AStF). Posterior and medial to this fossa is the deep subtemporal fossa (Fig. 5D: DStF). Bellwood (1994) considers these fossae unique to scarines, but dramatic subtemporal fossae are present in such distantly related groups including cyprinids and elopids. Thin laminae of the supraoccipital, epiotic, parietal, and pterotic separate the deep subtemporal fossa from the posttemporal and supratemporal fossae. Most of the greatly enlarged fourth levator externus muscle originates from the anterior and deep subtemporal fossae (Gobalet 1989, Bellwood 1994). The lateral occipital fossa (Fig. 5C: LOcF) is as deep as the deep subtemporal fossa and is the dominant feature of the posterior aspect of the neurocranium. The medial portion of the elevator posterior muscle originates from the lateral occipital fossa (Gobalet 1989). A small fossa on the posterior face of the pterotic, and lateral to the opisthotic, is the site of the fleshy origin of the lateral portion of the levator posterior muscle (Gobalet 1989). In *Nicholsina denticulata* this fossa is triangular and deep. It is well defined only in large specimens of *Scarus*.

The medial occipital fossae, visible in posterior view (Fig. 5C: MOcF), are limited to the region dorsal to the foramen magnum and are separated in the midline by a ventral extension of the supraoccipital crest. Cranial to the first vertebra on the posterolateral neurocranium is the ventral occipital fossa (Fig. 5B: VOcF). The supratemporal, posttemporal, ventral occipital and medial occipital fossae are points of attachment for trunk muscles. A dilator fossa is present posterior to the postorbital process (Fig. 5B: DOF). The posterolateral pterotics bear fossae of the levator operculi. The preopercular process of the lateral pterotic (Fig. 5D: PtoProPr) separates the dilator fossa from the levator fossa.

The anterolateral vomer has a broad convex surface against which the cranial condyle of the maxilla abuts. Posterior to this surface is a lateral expansion to which the vomeropalatine and vomeroendopterygoid ligaments attach. In anterior view, the vomer of *Sc. rubroviolaceus* is the shape of an inverted "V." It is more rounded in the other three species of *Scarus*. The ventral vomer has a sharp midline ridge that is the anterior extension of the keel of the parasphenoid (Fig. 5B: Psp) to which the anterior fibers of the adductor arcus palatini muscle attach.

A cartilaginous interspace noted by Starks (1926) in two parrotfish species separates the posterior edge of the ethmoid from the frontal in *Sc. compressus*, *Sc. perrico* and *Sc. rubroviolaceus*. This cartilaginous interspace is lacking in *Sc. ghobban*. The ethmoid is outwardly convex in *N. denticulata* and there is no cartilaginous interspace.

In *Scarus*, the ethmoid appears to be a bone of multiple origins. At least three ossification centers are indicated, a subcircular dorsal plate [the rostrodermethmoid plus supraethmoid (Patterson 1975)] which covers the ethmoid cartilage and paired cones of cartilage bone which form the medial portion of the anterior myodome. In large specimens these centers grow together. Starks (1926) describes the endochondral components of the ethmoid as remaining separate in three species of *Callyodon,* a genus since subsumed within *Scarus* (Bellwood 1994). This pattern of formation of the ethmoid, and the dominance of the endochondral ethmoid as a major bone of the anterior myodome, may be taxonomically and phylogenetically important features. The condition of the ethmoid in *Mycteroperca rosacea* (Epinephelidae) and *Morone saxatilis* (Moronidae) is that of a single rostrodermethmoid-supraethmoid that doesn't form the anterior myodome and does not separate the lateral ethmoids. In representative labrids (*Labrus, Symphodus, Ctenolabrus* and *Centrolabrus*; Rognes 1973), and in *Halichoeres nicholsi, Bodianus diplotaenia* and *Semicossyphus pulcher,* the anterior myodomal component of the ethmoid forms as a pair of posteroventrally-directed growths from the dorsal plate and the separation of the lateral ethmoids is not as complete as in *Scarus*. A third endochondral ossification between the lateral ethmoids is present in *N. denticulata* in the midline ventral to the two described for *Scarus*.

The ventrolateral portion of the preorbital process of the lateral ethmoid bears a rough lachrymal facet and anterior and lateral to the olfactory foramen the lateral ethmoid is notched for the posterodorsal portion of the palatine. Anterior to this notch is a flat facet for the medial palatine.

Posterior and medial to the anterior-most point of the frontal is a dorsal process that is of a form unique to each of the five parrotfishes studied (Fig. 5B: DPFr). In *Sc. perrico* the process is part of the parietal crest. In *Sc. compressus* the process is high and stands alone. A parasagittal ridge rises anteriorly in *Sc. ghobban* to abruptly terminate at a high point. In *Sc. rubroviolaceus* it is a nondescript bump on the transversely flattened frontal. The frontal slopes antero-ventrally at this level in *N. denticulata*.

The supraorbital region of the frontal is sculptured with low outwardly directed ridges. The frontal is robust and cancellous anterior to the dorsal process in *Sc. perrico* and *Sc. compressus*. This region is flatter in *Sc. ghobban* and *Sc. rubroviolaceus*. Ventrally directed laminae of the frontals form part of the medial wall of the orbit. Anteriorly these laminae meet the ethmoid. These laminae angle ventromedially but remain separated across the midline except in *N. denticulata* where they meet at the midorbital level. They also meet in four species of parrotfishes studied by Starks (1926).

Cartilage-fills the cavity between the ventral wings of the frontals in *Calotomus* (Starks 1926), *Scarus* and *N. denticulata*. Transverse, ventrally directed laminae of the frontals below the supraoccipital separate this cartilage from the braincase.

The "Y"-shaped basisphenoid of the midline splits the entrance to the posterior myodome and forms the base of the orbital opening to the cranial chamber. The prootics meet in the midline ventral to the dorsal forks of the basisphenoid where they form the roof of the anterior portion of the myodome. Anteriorly directed laminae of the basioccipital are the roof of the posterior portion of the myodome. The lateral walls of the myodome are formed by the prootics and parasphenoid. The prootic also forms the posterior wall and part of the ventral portion of the incomplete pituitary capsule.

In the posterior-most corner of the neurocranium, as seen in dorsal view, is the slightly convex and spatulate epiotic. Its flattened dorsal surface has a facet for the ventral surface of the dorsal ramus of the posttemporal. The splinter-like intercalar limb of the posttemporal is bound by ligaments to the intercalar. The complex exoccipital contributes to the lateral, medial, and ventral occipital fossae, the deep subtemporal fossa, occipital condyle, and walls of the foramen magnum.

The sphenotic forms most of the postorbital process, contributes to the posterior wall of the orbit, to the round anterior socket of the hyomandibula, to the dilator fossa, and contributes to the roof of the anterior subtemporal fossa. I was unable to distinguish a sphenotic distinct from a dermosphenotic as Patterson (1977) stated occurs in the majority of teleosts.

The prootic contributes to the anterior subtemporal fossa, the deep subtemporal fossa, and to the myodome. A prominent spike from which branchial levators originate (Gobalet 1989) projects ventrally from the lateral commissure lateral to the posterior opening of the par jugularis of the trigeminofacialis chamber. The spike is absent from *N. denticulata*.

The medial edge of the pterosphenoid bears a small preotic wing described by Rognes (1973) in some labrids. The preotic wing in *Scarus* is an extension of the pterosphenoid within the connective tissue membrane covering part of the opening of the cranial chamber.

When present, these thin laminae are rarely symmetrical. The preotic wing is absent from *Sc. perrico*. It is a small equilateral triangle and is present in seven of twelve specimens of *Sc. compressus* and only on the left side. Seven *Sc. rubroviolaceus* observed have some bilateral representation of the preotic wing in which it is usually broad and square. In *Sc. ghobban* the preotic wing is typically squared-off on the left side and round on the right. In a 206 mm specimen it is two tiny splinters. It was absent from only one of 13 *Sc. ghobban*. The preotic wing is triangular in the single specimen of *N. denticulata*.

The longest bone in the neurocranium is the parasphenoid, which ventrally has a sharp keel that caudally separates the neurocranial grooves of the synovial basipharyngeal joint (Fig. 5B: Psp, BPhGr). Anteriorly the keel bares a characteristic rudder-shaped process ventrally. The scar of Baudelot's ligament is found dorsal to the neurocranial grooves that receive the upper pharyngeal condyles and in line with the ridge separating the deep subtemporal fossa from the ventral occipital fossa. The neurocranium is highly trabecular and suggestive of considerable reinforcement due to forces applied to it from the action of both the mandibular and pharyngeal jaws.

Mandibular jaws. The distinctive jaws of *Scarus* show extreme modifications for their habit of scraping rock and calcareous surfaces. The jaws are short, robust, and the quadratomandibular joint is well anterior of the orbit. The exposed surfaces of the dentary and premaxilla are composed of numerous tiny denticles cemented together into thick outwardly convex beaks with tapered and squared-off cutting edges (Bellwood 1994). Worn denticles are constantly replaced from internal germinative tissues.

The anguloarticular is quite distinctive (Figs. 2, 3B, 6, also see Bellwood and Choat 1990: p 196, Bellwood 1994: p 18). There is a syndesmosis between it and the dentary and a diarthrosis between it and the quadrate. In *Scarus* the vertically oriented anguloarticular consists of two spatulate parts. One medially concave portion is directed ventromedially from the quadrate fossa. A ridge is found on the anterior edge of the medial face. The interopercular-mandibular ligament attaches posterior to this ridge. The posteroventral portion of this arm is ankylosed to the small retroarticular. The anterodorsomedially-directed arm is at an oblique angle to the ventral arm and is twisted forty-five degrees relative to it. This ramus fits in the notch on the lateral dentary and it is called the anterior articular ascending process by Bellwood (1994). I should note what I am naming the anguloarticular is based on Patterson (1977) whereas other investigators (e.g. Bellwood 1994, 2003, Wainwright 2004) use the name articular.

The quadrate fossa of the anguloarticular has medial and lateral facets (Fig. 6: QMJLF). In smaller specimens of all four species the lateral facet is a continuous almost semicircular surface. In large specimens of *Sc. compressus* and *Sc. perrico*, the lateral facet has both posterior-facing and dorsal-facing portions. A raised transverse ridge meets a similar ridge on the lateral mandibular condyle of the quadrate during adduction. The medial facet is directed posterodorsally and is offset posteroventrally relative to the lateral facet. A dorsally projecting process, located just posterior to the fossa, is the attachment point of a quadratomandibular

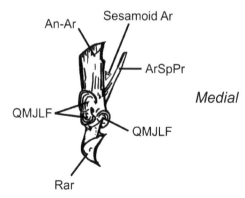

Fig. 6. *Scarus compressus*: posterior view of anguloarticular.

ligament. A unique splint process of the anguloarticular projects dorsally from above the medial facet (Fig. 4C, 6: ArSpPr). This is the articular medial spine of Bellwood (1994). The small pyramidal sesamoid articular tightly adheres to the anguloarticular dorsal to the splint process. No Mechelian groove or cartilage is found in *Scarus*, but is present in the less derived *N. denticulata*.

The dentary symphysis is a long, broad serrate joint (beautifully illustrated in Bellwood 1994: p 16). The dentary is concave inside the tapered scraping edge. Its massive coronoid process projects posteriorly from the denticled portion of the bone and its broad dorsolateral surface is concave, rough, and rounded-off posteriorly. A ventrolateral flange covers the articular notch.

The hemispherical dental surfaces of the premaxillae are similar to those of the dentaries (Fig. 1B, 2: PM). The posterior edge of the premaxilla is squared-off except for the medially positioned ascending process and the notched posterolateral corner. The maxilla is tightly bound to the premaxilla by the maxillary-premaxillary posterior ligament in this notch. The anterior end of the ventral portion of the ascending process bears two sharp parallel ridges. The lateral ridge is expanded at the base into a small maxillary condyle. The medial ridges broaden the contact surfaces between the premaxillae, surfaces tightly bound by cruciate ligaments (see below). Concentric arches of boney laminae are visible on the medial face (Fig. 1A). Large specimens of *Sc. ghobban* and *Sc. rubroviolaceus* bear "canines" on the lateral surfaces of the premaxillae above the corner of the mouth (Fig. 1B). *Sc. perrico* lacks these canines and some *Sc. compressus* have tiny raised denticles in this position.

Only limited motion is possible between the anterior maxilla and the premaxilla (Figs. 1B, 4: Max, Pm). The lateral surface of the maxilla is flattened, and has a thin anteriorly recurved dorsal crest. A rostral process is located lateral to the palatine fossa (Fig. 1B: RPrMax). The dorso-medial edge of the head of the maxilla forms the medial wall of an elongate palatine fossa. The large scar of the adductor mandibulae tendon is located on the posterior face ventral and lateral to the palatine fossa.

The anterior-most portion of the maxilla bears a premaxillary condyle. This condyle has synovial joints anterolaterally with the maxillary facet of the premaxilla, and anteromedially with the broad maxillary-premaxillary anterior ligament. A small flat cranial condyle that glides on the vomer is found on the posterior portion of the medial aspect of the head. The medial surface of the ventrolateral maxilla is grooved for the tough maxillary-dentary and maxillary-premaxillary posterior ligaments.

Circumorbitals. The dorsal portion of the medial surface of the broad lachrymal (Fig. 2B: La) is notched for the tight joint with the preorbital process of the neurocranium. The dorsomedial portion of the lachrymal is concave and the dorsal edge is slightly dished in for the olfactory pit. The anterior portion of the medial surface is thickened and scarred from the tough lachrymal-palatine ligament. A thin anteroventral extension covers the dorsal crest of the maxilla.

There are usually three infraorbitals in addition to the lachrymal though in each *Scarus* species an individual was found with four on one side and three on the other. One specimen of *Sc. compressus* had two on each side. Gosline (1968) indicated that this reduction in circumorbital number from the perciform total of six (including the lachrymal) is a specialization.

Nasal. The lateral portion of the fan-shaped posterior portion of the nasal is attached to the frontal-lateral ethmoid suture and to the lateral ethmoid, dorsolateral to the olfactory foramen. The nasal is laterally notched around the olfactory pit and is laterally recurved over the dorsal portion of the maxillary process of the palatine.

Hyoid (Fig. 7). The hyoid bar is thin, broad, ventrally convex, dorsally deeply recessed and anterodorsally recurved medially. Of the four bones that comprise it, the ventral hypohyal (Fig. 7A: VHH) is unusual in that it underlies 60% of the length of the ceratohyal (in both *Scarus* and *N. denticulata*) and anteromedially meets its opposite at an oval condyle. The dorsal hypohyal bears a posteriorly recurved hook (Fig. 9A: DHPPr) that forms the anterior border of the dorsal recess. Cartilage separates the anterior ceratohyal from the ventral hypohyal. The epihyal (Fig. 7A: IH) is tightly bound to the interopercular thus preventing pivoting around the now vestigal interhyal. This apparent immobility is consistent with the proposal of Bellwood (1994) that scraping parrotfishes have limited suspensorial abduction and gular depression. Five flattened branchiostegal rays attach ventrolaterally to the hyoid bar (Fig. 7A: BrStg).

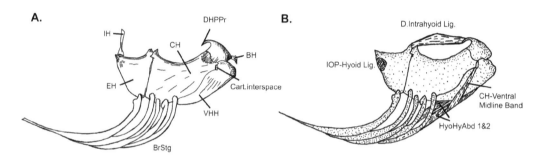

Fig. 7. A. *Scarus perrico* (349 mm S.L.): lateral view of right hyoid, anterior is to the right.
B. *Scarus* sp.: lateral view of the ligaments of the right hyoid, anterior is to the right.

Suspensorium and opercular series. The suspensorium is roughly rectangular in shape with a deeply notched dorsal border. The quadratomandibular joint (Fig. 2A: Q) is positioned nearly as far rostral as the maxillary process of the palatine (Fig. 2A: PaMax). Flanges on the mandibular condyle of the quadrate meet those of the anguloarticular and apparently serve to limit adduction (see description of the anguloarticular fossa above). The hyomandibular articulation with the neurocranium is through synovial joints on two condyles. The posterodorsal hyomandibula projects posteriorly in line with the condyles. A sharp ridge of the hyomandibular arises from the pillar of the anterior condyle runs to the opercular condyle, and defines the ventral boundary of the dilator fossa.

The extensive fossa of the adductor mandibulae muscle is deepest dorsal to the quadrate at the junction of the endopterygoid and metapterygoid. A dorsally concave horizontal ridge of the quadrate defines the ventral edge of the fossa and the angle of the preopercle defines the posteroventral border. The dorsal portions of the convex medial surfaces of the endopterygoid (Fig. 2A: Enpt) and metapterygoid (Fig. 2A: Mpt) are the insertion point of the adductor arcus palatini muscle.

The palatine and endopterygoid are bound at a serrate suture. The stout maxillary condyle of the palatine (Fig. 2A: PaMax) projects at an acute angle to the anterior border of the suspensorium.

The lateral surface of the palatine (Fig. 2A: Pal) bears a scar from the lachrymal-palatine ligament, and the medial surface bears a prominent scar of the vomero-palatine ligament in line with the ventral edge of the maxillary process. The posterior edge of the palatine fits in a notch on the lateral ethmoid anteroventral to the olfactory foramen and a concavity of the posterior portion of the medial surface reinforces the contact.

The opercle (Fig. 2A: Op) is thickened ventral to the hyomandibular socket and a sharp ridge extends posteroventrally from the socket half the breadth of the bone. The adductor operculi and levator operculi attach dorsal to this ridge. The subopercle (Fig. 2A: Sop) has a dorsal process that projects anterior to the ventral portion of the opercle and on the medial interopercle (Fig. 2A: Iop) is a facet against which the epihyal (Fig. 7A: EH) abuts. The ligament to the epihyal attaches to the fossa anterior to the facet.

Ligaments

The *nasal-frontal ligament* is a band of connective tissue interconnecting the posterior end of the nasal with the anterior edge of the frontal lateral to its anterior-most point and continues laterally onto the lateral ethmoid. The dorsal surfaces of the nasals are interconnected across the midline by tough connective tissue over the ascending processes of the premaxillae. The anterior attachment of these *internasal and interpalatine bands* is on the dorsal edge of the palatine or on the maxillary-nasal ligament. Ventral to the internasal bands, a tough independent sheet interconnects the posterodorsal edges of the palatines.

Connective tissue arises on the medial surfaces of the rostral processes of the maxillae and spreads out medially as its fibers cross the ascending process of the premaxillae. With the interpalatine bands, this *intermandibular cross ligament* forms the dorsal cover of the rostral fossa. Connective tissue interconnects the posterodorsal lachrymal with the anterior lateral ethmoid, the *lachrymal-lateral ethmoid ligaments*. Ligaments also join the notched portion of the lachrymal with the preorbital process, immobilizing the union.

The *nasal-lachrymal bands* are continuous with the subcutaneous connective tissue of the snout that binds the nasal to the lachrymal anterior and posterior to the olfactory pit. A short, tough, *lachrymal-palatine ligament* interconnects the medial surface of the anterodorsal part of the lachrymal with the lateral surface of the palatine, posterior to, and in line with, the ventral edge of the maxillary process. This distinctive ligament was also observed in epinephelids.

Considerable connective tissue attaches to the symphysis of the ascending processes of the premaxillae. The deep layer of this *premaxillae associated connective tissue* arises from between the dentary and the ventral maxilla. The outer layer is a superficial tendon of the adductor mandibulae. It arises from the A1A2 complex posterior to the coronoid process of the dentary. The deep and superficial parts fuse before their attachment to the premaxillary symphysis.

Strong *maxillary-nasal and maxillary-palatine ligaments* join the rostral process of the maxilla to the anterior tip of the nasal, and to the anterolateral surface of the maxillary process of the palatine anterior to the lachrymal-palatine ligament (Fig. 4A: Pal-MaxL). The maxillary-nasal ligament, the nasal bone, and the nasal-frontal ligament form a track that interconnects the maxilla and the neurocranium. It probably functions to limit the forward rotation of the maxilla. The narrow and delicate *anguloarticular-maxillary ligament* (Primordial Ligament) attaches to the ridge of the anguloarticular dorsal to the quadrate fossa. It passes medial to the A2 and lateral to the A3 portions of the adductor mandibulae and joins the massive tendon of A1 that inserts on the posterior face of the maxilla. A distinct band interconnects the ventral portion of the groove of the ventromedial maxilla and the anterodorsal portion of the coronoid process of the dentary. A thick cuff of connective tissue attaches to the posterior border of the lateral face of the coronoid process and covers the posterior and posterolateral portions of the ventral maxilla. This is the *maxillary-dentary ligament and cuff*. The tough *posterior maxillary-premaxillary ligamentous* strap interconnects the medial surface of the notched posterolateral portion of the premaxilla with the dorsal portion of the groove in the medial surface of the ventral maxilla. This tight connection severely limits independent motion of these elements.

A tough, broad band of parallel fibers connects the medial portion of the head of the maxilla, ventral to the cranial condyle, with the ventral portion of the groove in the posterior aspect of the ascending process of the premaxilla. This is the *anterior maxillary-premaxillary ligament*. The attachment to the ascending process is broader than the attachment to the maxilla. The medial facet of the premaxillary condyle of the maxilla glides on the lateral surface of this band in a synovial cavity. A *maxillary-rostral cartilage ligament* connects the anterolateral portion of the rostral cartilage (described below) with the medial maxilla dorsal to its cranial condyle. The ascending processes of the premaxillae are bound together by a continuous superficial band of *interpremaxillary cruciate ligaments* (Fig. 1A). The broad premaxillary symphysis contains approximately nine pairs of tough crisscrossing ligaments. They are radially arranged and attach ventrally on one element and dorsally on the opposite bone. The cruciate design of ligaments is optimal for resisting shearing forces (Beecher 1979) and Bellwood (1994) also noted them in parrotfishes. These ligaments are also cruciate in *N. denticulata*.

The *premaxillary-frontal elastic ligament* is a cylindrical bundle of elastic fibers that connects the lateral surface of the posterior half of the ascending process of the premaxilla (and, at times, the anteroventral portion of the ascending process) with the anterior tip of the frontal and with the lateral ethmoid ventral to it (Fig. 1C). This extraordinary ligament stretches and recoils with manipulation. It is also found in *N. denticulata*.

The tough *vomero-palatine ligament* is directed dorsally from the edge of the laterally expanded portion of the vomer to the scar on the medial palatine in line with the ventral edge of the maxillary process. This ligament has a transverse orientation in *N. denticulata*. The tough, strap-like *vomero-endopterygoid ligament* attaches to the same ridge as the vomero-palatine ligament on the lateral expansion of the vomer, and is directed posteroventrally to the anterior edge of the medial surface of the endopterygoid. The posterodorsal and posterior edge of the palatine connect to the anterior tip of the frontal, and to the lateral ethmoid by the *palatine-lateral ethmoid ligament and sheet* (Fig. 1C: Pal-Le Sheet). The nasal may attach to the dorsal edge of the connective tissue sheet. The ventrolateral edge of this sheet is a short, tough ligament that connects the ventral portion of the posterior edge of the palatine with the anterior face of the lateral ethmoid, lateral to the palatine notch. The joint between the anterior suspensorium and the preorbital process in this location is not synovial.

The tough, short *lateral ethmoid-endopterygoid ligament* connects the ventral preorbital process, medial to the lachrymal facet, with the dorsal edge of the anterior endopterygoid. This is a ligament that limits suspensorial abduction. The joint capsules positioned between the hyomandibula and sphenotic are surrounded by the *hyomandibulo-neurocranial ligaments* that restrict motion to the medial-lateral plane. The posterior ligaments extend caudally behind the synovial capsule. The *anguloarticular-dentary sheet of connective tissue* connects the anterior edge of the anguloarticular to the rough superficial surface of the bony portion of the posterior dentary. An *interanguloarticular sheet of connective tissue* lies deep within the lower lip and is continuous with the connective tissue mass between the dentary and maxilla. This band interconnects the anterior faces of the anguloarticulars across the ventral midline.

A tough, thick *anguloarticular-dentary ligament* connects the medial face of the anteriodorsal spatulate ramus of the anguloarticular with a rough triangular pedicel on the lateral wall of the dentary. The dorsolateral surface of the anguloarticular is connected to the medial surface of the laterally expanded flange of the coronoid process by a ligament.

Anterior-posterior and medial-lateral motions are possible at this syndesmosis. The tough, wide *interopercular-mandibular ligament* narrows from its attachment to the anterior end of the interopercle to its attachment to the posterior and medial faces of the retroarticular, and the ventromedial anguloarticular. A *quadrato-mandibular ligament* connects the anguloarticular, posterior to the quadrate fossa, with the quadrate concavity ventral to the mandibular condyles. Fine connective tissue also interconnects the quadrate and anguloarticular on the lateral surface of the joint. The medial connection across the joint is a tough band. Additional connective tissue interconnects the anterior preopercle with the anguloarticular.

The *interopercular-hyoid ligament* is an extensive mass of short fibers interconnecting the posterolateral surface of the epihyal with a concave facet on the medial interopercle (Fig. 7B: IOP-Hyoid Lig.). It limits rotation between these elements. These tough, short, circular *urohyal-hypohyal ligaments* interconnect the anterolateral facets of the urohyal to the medial surfaces of the ventral hypohyals. The tough *posterior interhyoid cross ligament* interconnects the dorsomedial processes of the dorsal hypohyals over the joint between the first and second basibranchials. The *anterior interhyoid cross ligament* is a small band that interconnects the cartilages between the ceratohyals and ventral hypohyals across the midline ventral to the urohyal. A tough band forms the dorsal edge of the hyoid bar and interconnects the dorsal processes of the dorsal hypohyal, ceratohyal and epihyal. This is the *dorsal intrahyoid ligament* (Fig. 7B: D. Intrahyoid Lig.). Board (1956) identified this in *Sparisoma* and it is also present in *N. denticulata*. Osse (1968) identified (probably incorrectly) the hyohyoideus proprius muscle in this position in *Perca fluviatilis*.

The *ceratohyal-ventral midline band* is transparent and extends ventromedially from the ceratohyal above the cartilaginous interspace between the ceratohyal and ventral hypohyal (Fig. 7B). It is lateral to the first hyohyoideus adductoris and meets its counterpart in the midline and may extend onto the first branchiostegal rays and ventral urohyal. This ligament is probably the tendon of the hyohyoideus ventralis par caudalis in *Haplochromis* described by Anker (1977). The fibers of the *urohyal-first basibranchial ligament* interconnect the flat ventral surface of the first basibranchial with the dorsal spine of the urohyal. Except for Liem's (1970) description of the Nandidae, the *opercular-interopercular ligament* is rarely mentioned. It probably exists in all bony fishes that open the mouth through the levator operculi-opercular series coupling. The ligament arises from most of the anterior edge of the opercle and attaches to the posterolateral surface of the interopercle and may include the dorsal process of the subopercle. The thin *hyomandibular-opercular ligament* encircles this synovial ball-and-socket joint between these two bones. The unusually tough *Baudelot's ligament* interconnects the parasphenoid ventral to the ridge separating the deep subtemporal and ventral occipital fossae with the anterior cleithrum at the base of the elongate dorsal spike.

Menisci

A thin *maxillary-premaxillary meniscus* is positioned within the synovial joint between the anterolateral face of the premaxillary condyle of the maxilla and the facet on the base of the lateral ridge of the posterior aspect of the ascending process of the premaxilla. As a result, the maxilla and premaxilla also appear to move as a single unit that is also the case in *N. denticulata* even though this meniscus is absent.

A *maxillary-vomerine meniscus* is located between the cranial condyle of the maxilla and the anterolateral facet of the vomer. Connective tissue from the posterior edge of this meniscus attaches to the vomer and connective tissue from the dorsal part of the meniscus attaches within the groove in the ventral portion of the ascending process of the premaxilla. In *N. denticulata* the maxillary cranial condyle is tiny and the meniscus is absent.

The *rostral cartilage* is located between the ascending processes of the premaxillae and the vomer. A synovial capsule for the ascending process of each premaxilla is located on the dorsal surface of the cartilage and the grooved ventral surface of the rostral cartilage slides in a synovial joint on the midline ridge of the rostral fossa (5A: RoF). This is effectively a large meniscus. A thin *interhyoid meniscus* is found in the synovial joint between the ventromedial condyles of the ventral hypohyals in the midline.

Muscles

The A1 and A2 portions of the *adductor mandibulae* (Fig. 4) are distinguished on the basis of their points of insertion because there are not obvious superficial subdivisions (Fig. 4A). None of the superficial adductor mandibulae inserts on the anguloarticular as described by Board (1956) or Tedman (1980b) nor does the internal subdivision described by Lubosch (1923). The A3, which inserts on the anguloarticular (Fig. 4B: A3), and the subdivided Aw (Figs. 4B, C: Aw) are the "deep" portions of the adductor mandibulae. The details described here are consistent with Clements and Bellwood (1988) and Bellwood (1994).

The massive *A1* (Fig. 4A: A1) has a fleshy origin from most of the adductor fossa of the suspensorium and from the surface of A3. The A1 inserts by a thick tendon on the posterior maxilla and to the medial portion of the coronoid process of the dentary. Superficial attachments to connective tissue probably also directly influence the premaxillae. There doesn't appear to be a discreet dorsal portion as determined by Clements and Bellwood (1988) in *Sc. rubroviolaceus* and no aponeurotic connection to the anguloarticular was identified.

The fleshy origin of *A2* (Fig. 4A: A2) is from the posterior portion of the horizontal ridge of the quadrate. A2 is parallel-fibered and inserts on the medial surface of the posterior end of the coronoid process of the dentary. This is consistent with Lubosch (1923), Clements and Bellwood (1988) and Bellwood (1994). Dorsally its short, flat tendon is continuous with the tendon of A1. This apparently is the A1 alpha muscle of Tedman (1980b). In *Leptoscarus vaigiensis* (Schultz 1958), *Sparisoma cretense* (Board 1956), and *N. denticulata*, the A2 is separate from A1 and inserts along the posterior border of the ascending process of the anguloarticular and on the tip of the coronoid process of the dentary.

Most of the mass of *A3* (Fig. 4B) is located posterodorsal to A1 and originates from the lateral surface of the ventral portion of the levator arcus palatini and from the hyomandibula ventral to it. These fibers attach to a long, narrow, and flat tendon that inserts on the sesamoid articular on the medial face of the ascending process of the anguloarticular. A broad, but very thin portion of A3 originates from the metapterygoid and quadrate and attaches to the medial surface of the tendon. A few fibers of A3 that have origins on the ventral quadrate insert directly on the anguloarticular posteroventral to the sesamoid articular, not on the dentary as suggested by Clements and Bellwood (1988) for *Sc. rubroviolaceus*. The insertion of A3 is not on the splint process of the anguloarticular as described by Lubosch (1923). A3 has two points of insertion to the medial surface of the body of the anguloarticular in *N. denticulata*.

The small *Aw* is subdivided into a ventral abductor (Fig. 4C: Aw ab) and a pair of dorsal adductors (Fig. 4C: Aw ad) and is a more complicated muscle than described by Clements and Bellwood (1988). The abductor portion (Aw beta of Bellwood 1994: p 38) originates along the anterior edge of the splint process of the anguloarticular (Fig. 4C: ArSpPr) and from the bone ventral to the splint. The fleshy insertion is on the posterior edge of the dentary. One part of the adductor portion of Aw arises from the tendon of A3 and inserts on the coronoid process medial to A2. This is the muscle tendo-dentary of Lubosch (1923) and part of the deep portion of A3 of Board (1956) and possibly Aw gamma of Bellwood (1994: p 38). The other part of Aw arises from the end of the splint process and inserts on the dentary anterior to, and in line with, the fibers that originate on the tendon of A3. This is the muscle artic-dentary of Lubosch (1923) who considers this subdivision of the medial adductor to be a scarid character. Though the anguloarticular of *N. denticulata* is not highly modified (ascending process present; splint process rudimentary; no *Scarus*-like anguloarticular-dentary joint), the Aw subdivision is very close to that of *Scarus*. An additional portion of the medial adductor of *N. denticulata* arises from the medial surfaces of the endopterygoid and quadrate and inserts on the ventral portion of the medial anguloarticular. This is the adductor mandibulae medialis of Lubosch (1923) for *Leptoscarus vaigiensis* (then called *Sc. coeruleopunctatus*) and is present in *Sparisoma cretense* (Board 1956).

The *levator arcus palatini* (Fig. 4A, B: LAP) originates from the postorbital process of the neurocranium and fans out to insert on the hyomandibula dorsal to the transverse ridge. A thin layer of its fibers originate along the superficial edge of the dilator fossa caudal to the orbit, cover the dilator operculi and insert on the caudal end of the transverse ridge. The *dilator operculi* (Fig. 4A, B: DO) originates from the dilator fossa of the neurocranium medial to the adductor arcus palatini and from the hyomandibula dorsal to the transverse ridge. This conical muscle comes together medial to the dorsal process of the preopercle and inserts on the anterodorsal corner of the opercule. It isn't always distinct from the levator operculi caudal to it.

The *levator operculi* (Figs. 4A, B: LO) originates from the levator fossa of the pterotic and from the posterodorsal process of the hyomandibula. There is extensive pinnation within this muscle and the fiber direction is posteroventral toward the insertion on the dorsomedial opercule.

The *adductor arcus palatini* originates from the ventral fossa of the vomer (Fig. 5D: VoF), from the keel of the parasphenoid (Fig. 5B: Psp) and midline anterior to the keel, and from the lateral parasphenoid and prootic dorsal to the keel. It inserts on the dorsomedially-directed portion of the endopterygoid and metapterygoid and on the anterior portion of the medial hyomandibula. The fiber direction is generally posterolateral to posteroventrolateral from the origin.

The pinnate *adductor operculi* has a fleshy origin from the ventral-facing surface of the neurocranium between the lateral occipital fossa and the deep subtemporal fossa and an aponeurotic origin anterodorsal to the ridge separating the ventral occipital fossa from the media wall of the deep subtemporal fossa. The fiber direction is ventrolateral from the origin to its fleshy insertion on the medial opercule. The fascia of the ventral edge of the ovoid insertion is on the crest of the ridge of the medial opercule.

The small *intermandibularis* interconnects the medial surfaces of the dentaries dorsal to the posterior portion of the symphysis. The fleshy attachment is posteroventral to a large foramen, lateral to the insertion of the geniohyoideus, and dorsal to the insertion of the abductor portion of the Aw. It is ovoid in cross section. The *geniohyoideus* originates from all but the dorsal portion of the lateral surface of the posterior ceratohyal, the anterior epihyal, and the heads of the branchiostegal rays. It fuses with its counterpart in the midline anterior to the first branchiostegal ray and inserts on the dentary lateral to the posterior end of the symphysis. The insertion is by a mass of connective tissue medially and by a tendinous band dorsolaterally. This muscle is subdivided in *Sparisoma cretense* (Board 1956).

The *hyohyoideus inferioris* originates from the medial surfaces of the arched ventral portion of the first two branchiostegal rays. The sheet of fibers angles anteromedially and meets its counterpart ventral to the urohyal to which there is a connection through fascia. Fibers may also insert either on the ventrolateral urohyal or on the lateral surface of the hypaxial muscle that is continuous with the sternohyoid muscle. The hyohyoideus inferioris and the first hyohyoideus abductoris define the anteroventral opening to the opercular chamber.

There are five distinct branchiostegal ray abductors (*hyohyoidei abductores*) that interconnect each ray and the hyoid (Fig. 7B: HyoHyAbd). The first abductor has a fleshy attachment to the anterior edge on the medial surface of the first branchiostegal ray. The attachment on the hyoid bar is tendinous on the ventral hypohyal posterior to the urohyal-ventral hypohyal ligament. There is an attachment to the lateral surface of the anteroventral urohyal. The second abductor originates by a flat tendon from the ventral edge of the ventral hypohyal and its fleshy insertion is on the medial surface of the second branchiostegal ray ventral to its head. The abductors of the three caudal branchiostegal rays have fleshy attachments to the proximal portions of the medial surfaces of their respective rays and tendons of origin from the ventral portion of the medial surface of the hyoid covering the suture between the ventral hypohyal and the ceratohyal. The three thin tendons are slightly staggered anterior to posterior.

The *hyohyoidei adductores* are thin muscles between the distal portions of the branchiostegal rays. The dorsal-most adductor interconnects the dorsal portion of the fifth branchiostegal ray with much of the medial surface of the opercular.

From a tendinous origin from the more ventro-posterior process of the pterotic the *protractor pectoralis* fans out to a long, fleshy insertion on the anterior face of the dorsal elongation of the cleithrum dorsal and lateral to Baudelot's ligament. In *N. denticulata* the protractor pectoralis is parallel fibered and inserts on the lateral edge of the cleithrum posterodorsal to the pharyngeal facet. The protractor pectoralis may have a role in parrotfish pharyngeal jaw stabilization through the pharyngocleithral joint (Gobalet 1989).

The thin *levator pectoralis* is continuous with the epaxial musculature medial to it. Its origin from the posterodorsal process of the hyomandibula may be either fleshy or aponeurotic. Additional fibers originate from the levator fossa ventromedial to the levator operculi. The insertion is on the anterior edge of the dorsal portion of the supracleithrum, and on the lateral surface of the posttemporal.

The *sternohyoideus* muscle is a ventromedial mass of complex fibers that originate from the anterior surface of the ventral portion of the cleithrum. Fibers from each side meet in the midline on the posterior edge of the urohyal and on the ventral surface of the more dorsal of the two posterior processes of the urohyal. Dorsal to this mass, and also originating from the anteroventral cleithrum and from the connective tissue cover of the medial mass, are anteriorly directed fibers. They also insert on the more dorso-posterior processes of the urohyal. The fiber direction of this sheet is almost parallel to the dorsal edge of the urohyal. The medial mass fans out anteriorly and dorsally. Originating from the lateral surface of the connective tissue cover of the ventromedial mass is a thin sheet of fibers that are continuous with the hypaxial musculature ventral to them. Together they insert in the lateral groove of the urohyal. A narrow tendon extends dorsally from their myocomma covering and inserts on the ventral end of the third hypobranchial. A thin sheet of muscle fibers is found lateral to the myocomma posterior and lateral to the tendon to the third hypobranchial. The tendon is also present in *Sparisoma cretense* arising from the "isthmus muscle" (Board 1956). Winterbottom (1973) gives an independent name, the sternobranchialis, to a muscle with this connection. Elshoud-Oldenhave and Osse (1976) have called this tendon to the third hypobranchial a ligamentous projection in *Gymnocephalus cernua*. The small tendon to the third hypobranchial in *N. denticulata* arises from the superficial fascia of the sternohyoideus muscle. This connection probably is a modification of the ligamentum urohyal caudale of Anker (1977). Wainwright et al. (2004) consider the sternohyoideus to be reduced in parrotfishes in contrast with the wrasses.

CHAPTER

2

Innovation and Diversity of the Feeding Mechanism in Parrotfishes

Peter C. Wainwright[1] and Samantha A. Price[2]

[1] Department of Evolution & Ecology, Center for Population Biology,
University of California, Davis, California-95616
Email: pcwainwright@ucdavis.edu

[2] Department of Biological Sciences, Clemson University, Clemson, South Carolina-29634, U.S.A.
Email: saprice@ucdavis.edu

Introduction

The feeding activities of parrotfishes are one of the fundamental ecological processes in coral reef ecosystems. These activities involve scraping hard rocky surfaces to remove turf algae, detritus, bacteria, and a wide range of encrusting invertebrates. This mixture of dead coral skeletons, the invertebrate and microbial organisms that colonize these surfaces and the detritus of organic debris is then passed to the pharyngeal jaw apparatus of parrotfish where it is mixed with mucous and ground to a fine slurry before being passed to the intestines (Bellwood and Choat 1990, Choat 1991, Choat et al. 2002). Here, nutrients are extracted from the slurry (Crossman et al. 2005) and fine sand is excreted back into the environment (Frydle and Stearn 1978, Bellwood 1995a, 1995b, Bruggenmann et al. 1996). The grazing activities of parrotfishes play a major role in disturbing benthic communities (Burkepile and Hay 2011, Brandl et al. 2014), preventing large algae from getting established and allowing corals and a more diverse community of encrusting organisms to become established and persist. The excretion of sand and concomitant bioerosion of the reef by parrotfishes occurs on a profound level as well, with accounts concluding that parrotfish are the major biological producers of sand in many reef systems (Bellwood 1995a, 1995b, Malella and Fox Chapter 8). Many groups of reef fishes are herbivores, microbiotivores or detritivores but the singular impact of parrotfishes is because they are the only major group that removes the calcareous surface layers of the reef as they graze.

The unique ability of parrotfish to feed in this way is closely linked to the presence of several evolutionary novelties in the feeding mechanism that facilitate their ability to scrape rocky substrates and pulverize these scrapings. In this chapter we will focus on three of these innovations: the parrotfish pharyngeal mill apparatus, the cutting edge of the oral dentition, and the intramandibular joint in the oral jaws. We describe each of the three innovations, review their evolutionary history, their impact on parrotfish feeding abilities, and the impact that each has had on the evolutionary diversification of parrotfishes.

Major Innovations in the Parrotfish Feeding Mechanism

The Pharyngeal Mill Apparatus

Parrotfishes are phylogenetically nested within the Labridae (Westneat and Alfaro 2005). Herbivory appears to have evolved at least three times within Labridae: once in *Pseudodax*, at least once in the odacines (Clements et al. 2004), and once in parrotfish. Parrotfish (Figs 1 and 2) are by far the largest radiation of herbivorous labrids with about 100 described species. All parrotfish share a derived condition of the pharyngeal jaw apparatus, a pharyngeal mill (Fig. 2) that appears to be crucial to their abilities as herbivores (Gobalet 1989, Bullock and Monod 1997). This system is built on a suite of already existing modifications of the pharyngeal jaw system that are shared by labrid fishes (Kaufman and Liem 1982, Bellwood

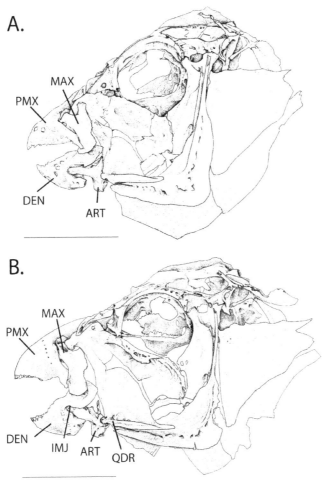

Fig. 1. Diagrams of the skull of parrotfishes prepared by hand from cleared and stained specimens, A. *Cetoscarus bicolor* and B. *Chlorurus sordidus*. Note that while both of these species possess cutting edge dentition on the jaws, *Cetoscarus* lacks an intramandibular joint between the dentary and articular bones while *Chlorurus* has this novel joint. Scale bars = 10 mm. Abbreviations: ART, articular; DEN, dentary; IMJ, intramandibular joint; MAX, maxilla; PMX, premaxilla; QDR, quadrate. Diagrams prepared by Ian Hart.

1994, Wainwright et al. 2012). The labrid condition, termed pharyngognathy, involves three derived features. (1) Fused left and right lower pharyngeal jaw bones (5th ceratobranchials) into a single structural lower jaw that is stronger and able to withstand higher forces. (2) Well developed joints between the underside of the neurocranium and the dorsal surface of the upper pharyngeal jaws that stabilize the upper jaws when the lower jaw is pulled up against them in biting actions. (3) The presence of a direct muscular connection between the neurocranium and the lower pharyngeal jaw that results in a powerful bite (Kaufman and Liem 1982, Stiassny and Jensen 1987).

The modifications in parrotfish are substantial and include extensive elaboration of the paired fourth epibibranchial bones that sit lateral to the pharnygobranchials (the upper jaw bones that bear tooth plates), holding the upper jaws in a medial position while biting occurs, thus stabilizing them and guiding them during anterior-posterior movements of the upper jaw (Gobalet 1989, Chapter 1). The joints between the upper pharyngeal jaws and the neurocranium are extended anterior-posteriorly and are convex, allowing the upper jaws a long scope as they slide forward and backward while the muscular sling generates a biting action (Fig. 2). It is suspected that the characteristic milling action of parrotfishes is produced by an anterior-posterior motion of the upper jaws while the lower jaw bites against it (Gobalet 1989, Wainwright 2005). The teeth on both the upper jaws

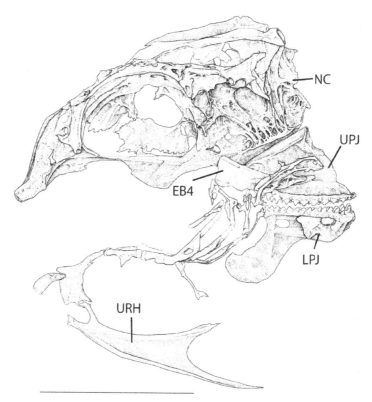

Fig. 2. Diagram of the pharyngeal mill apparatus of the parrotfish *Chlorurus sordidus*, prepared by hand from a cleared and stained specimen. Note the anterior-posterior elongation of the joint between the neurocranium and upper pharyngeal jaw, and the teeth on both the upper and lower pharyngeal mill. Scale bar = 10 mm. Abbreviations for bone names: EB4, fourth epibranchial; LPJ, lower pharyngeal jaw (5th ceratobranchials); NC, neurocranium; UPJ, upper pharyngeal jaw (pharyngobranchial); URH, urohyal. Diagram prepared by Ian Hart.

and lower jaw are organized into anterior-posterior rows with the newest teeth coming in anteriorly and older teeth being moved posteriorly until they are worn away. This characteristic structure of the pharyngeal jaws is found in all parrotfish (Bellwood 1994) though it varies somewhat among genera and species in terms of the extent of the anterior-posterior elongation of the joint with the neurocranium and the size of the grinding surface on the lower pharyngeal jaw (see figure 12 in Bellwood 1994). The major trend within these features is that the grinding surface and the joint with the neurocranium are more elongate in the anterior-posterior direction in the reef-associated group that includes *Bolbometopon*, *Cetoscarus*, *Hipposcarus*, *Chlorurus* and *Scarus*. Functionally, the key consequence of this large suite of derived traits characteristic of the parrotfish pharyngeal mill is that the system is specialized for milling or grinding actions, as opposed to the crushing and winnowing actions that are more typical of pharyngeal jaw function in other labrids (Liem and Sanderson 1986, Wainwright 1988).

Cutting Edge on Oral Dentition

Teeth on the upper and lower oral jaws are coalesced into a cutting edge in *Leptoscarus*, *Sparisoma*, *Cetoscarus*, *Bolbometopon*, *Hipposcarus*, *Chlorurus* and *Scarus* (Fig. 1, Bellwood and Choat 1990, Bellwood 1994). Oral jaw teeth in the remaining parrotfish, *Cryptotomus*, *Nicholsina*, and *Calotomus*, are individual, caniniform teeth as in wrasses, though *Calotomus* has somewhat flattened teeth (Bellwood 1994). Referring to the distribution of this trait on the parrotfish phylogeny (Fig. 3), it is somewhat ambiguous whether the absence of the cutting edge in these taxa is a retained primitive trait or a secondary reversal to this condition. However, a maximum likelihood reconstruction upon the phylogeny favors the interpretation that the cutting edge dentition evolved once and has been lost twice. All parrotfish that lack the cutting edge dentition are occupants of seagrass habitats and all taxa with the cutting edge except some *Sparisoma* are reef-dwellers, suggesting that there is a strong relationship between feeding on rocky substrates and the evolution and use of the cutting edge. The cutting edge gives a distinctive beak-like appearance to the jaws that is the basis of the common name 'parrotfish'. This structure is key to the ability of parrotfish to scrape the surface of rock or dead coral, removing the characteristic assemblage of coral skeleton, algae, microbes, detritus and encrusting invertebrates that they feed upon. Whether scraping or excavating, the feeding activities of parrotfish on reefs depend critically on this modified dental arrangement (Clements and Bellwood 1988, Bellwood and Choat 1990).

Intramandibular Joint

Parrotfish in the genera *Hipposcarus*, *Chlorurus* and *Scarus* have a well-developed joint between the dentary and articular bones of the lower jaw (Fig. 1b). In these taxa, the large section two of the adductor mandibulae muscle has the derived condition of inserting on the dentary rather than the articular bone and thus has the unusual property of crossing two joints, both the quadrate-articular joint and the articular-dentary joint. In other parrotfish and in wrasses the mandible is a single rigid structure formed by a dentary and articular that are held tightly together by many short ligaments (Fig. 1a). The intramandibular joint permits motion at the quadrate-articular joint, as in other teleosts, as well as the joint between the dentary and articular. The introduction of this joint alters the linkage mechanics of the oral jaw system, resulting in a novel four-bar linkage that transmits motion of the lower jaw to the upper jaws (Wainwright et al. 2004). Exactly how the intramandibular joint functions during feeding is not known, although one inferred

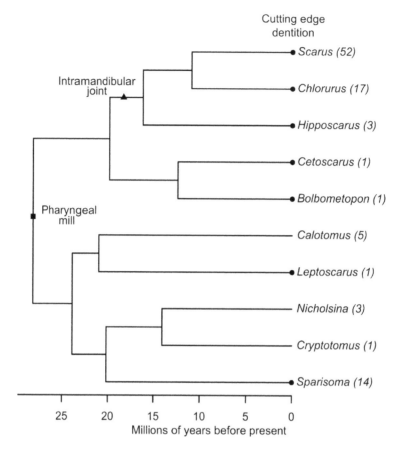

Fig. 3. Time calibrated molecular phylogeny of parrotfish genera (Kazancioğlu et al. 2009). Inferred origins of the parrotfish pharyngeal mill and the intramandibular joint are indicated by a square and triangle respectively. The distribution of the cutting edge dentition among genera is indicated at the tips with dark circle. A likelihood reconstruction of the history of the cutting edge dentition favored a single origin of the trait and two losses.

consequence of the joint and modified attachment of the adductor mandibulae is that the mechanical advantage, or force transmission, of the adductor mandibulae during oral jaw biting is almost twice as high as it is in taxa that lack the joint (Bellwood 1994, Wainwright et al. 2004, see also Gobalet Chapter 1). At present it is also not known if movement occurs at both joints during normal feeding behavior. If movement occurs at both joints during feeding this could allow fish to modulate the orientation of the dentary during biting actions, maintaining a favorable orientation throughout the scrape. Such a function could result in enlargement of the region of contact between the teeth and substrate during scraping. A better understanding of the function of the intra-mandibular joint will be an important goal in future research.

Phylogenetic Distribution of Feeding Innovations

Some lineages of parrotfishes have acquired all three of the innovations described above (Fig. 3). Parrotfish in the group made up by *Scarus, Chlorurus* and *Hipposcarus* have the modified pharyngeal mill, cutting edge dentition on their oral jaws, and an intramandibular joint. The possession of all three of these innovations appears to be unique among reef fishes and even among teleost fishes, although none of these traits are unique to parrotfishes.

Remarkably, many of the complex modifications found in the parrotfish pharyngeal mill are also found in the herbivorous wrasse *Pseudodax* (Bellwood 1994). Although a labrid, *Pseudodax* is not closely related to parrotfishes and the presence of the pharyngeal mill has evolved independently in this lineage. This is particularly noteworthy as the pharyngeal mill configuration of the pharyngeal jaws is not known to occur in any teleost outside of Labridae. Cutting edge dentition involving a cement layer around coalesced dentition in the oral jaws is found in another labrid, *Odax* (Clements and Bellwood 1988). *Odax* is a temperate herbivore that feeds mostly on large fucoids and laminarian macroalgae (Clements and Bellwood 1988). The pharyngeal jaws of *Odax* show the typical wrasse condition, lacking the modifications characteristic of the pharyngeal mill seen in parrotfishes and *Pseudodax* (Clements and Bellwood 1988, Bellwood 1994). Similarly, *Pseudodax* lacks the cutting edge dentition made of smaller coalesced teeth, although it does have large, flattened incisiform teeth that provide a different type of cutting edge in the oral jaws (Bellwood 1994). A few other teleost lineages have cutting edge dentition formed by coalesced or fused teeth, including members of Oplegnathidae and Tetraodontidae.

Finally, an intramandibular joint has evolved several times in reef fishes, including some members of Acanthuridae, Pomacanthidae, Chaetodontidae, Blenniidae, Girellidae and Siganidae (Vial and Ojeda 1990, Purcell and Bellwood 1993, Bellwood 2003, Konow et al. 2008, Konow and Bellwood 2005, Ferry-Graham and Konow 2010), and some non-reef lineages: *Helostoma* and some Poeciliidae (Gibb et al. 2008, Ferry et al. 2012). In all cases this trait is associated with feeding by biting the benthos (Konow et al. 2008). Bellwood (2003) noted that these reef lineages, together with parrotfishes, make up the major herbivorous fishes on modern reefs. Given that intramandibular joints have apparently evolved numerous times in benthic feeding reef fishes, there is a need to better understand the functional benefits of this modification in benthic feeding fishes (Konow et al. 2008) and whether the function of the extra joint is similar in each case. Some possible advantages of the additional joint are that it permits (1) a greater angular sweep of the lower jaw, although this trait is normally associated with overall shortening of the lower jaw (Purcell and Bellwood 1993), (2) Modulation of the orientation of the toothed surface of the lower jaw through the sweep of the bite (Price et al. 2010), (3) Effective biting while the upper jaws are protruded (Konow et al. 2008) or (4) that the flexibility and associated complexity in muscular attachments result in greater dexterity in movements of the lower jaw during feeding.

Although all parrotfish possess the grinding pharyngeal mill, many seagrass-dwelling lineages lack the cutting edge on the oral jaws that is essential for scraping hard surfaces on reefs. Only *Scarus*, *Chlorurus* and *Hipposcarus*, a lineage nested inside a larger clade of reef-dwelling parrotfishes, have the mobile intramandibular joint. As discussed in Bonaldo et al. (2014), the phylogeny suggests that parrotfish may have invaded reef habitats twice, once along the branch below the node uniting *Bolbometopon* and *Scarus*, and a second time within *Sparisoma*. Most parrotfish living in seagrass feed in a different manner from those taxa on reefs because of the absence of the ubiquitous hard substrata that promotes scraping behavior. In seagrass, parrotfish feed on blades of seagrass, epiphytes that live on seagrass and large algal plants. These are taken by a browsing behavior in which they are removed from their holdfast or separated from the rest of the plant by cropping or biting and tearing.

Morphological and Functional Diversity of Parrotfish

The complex phylogenetic distribution of the three innovations discussed above implies the presence of functional diversity among parrotfishes (Fig. 3). There are considerable differences between taxa in their feeding biology, with *Bolbometopon, Cetoscarus* and *Chlorurus* digging deep gashes in the reef during forays (termed excavators by Bellwood and Choat 1990) while *Hipposcarus* and *Scarus* feed with much more superficial scrapes of rock or sometimes sandy surfaces, taking less carbonate while they primarily remove epilithic organisms (Bellwood and Choat 1990). The reef-dwelling *Sparisoma* species appear to be superficial scrapers with *Sparisoma viride* and its sister species, *Sp. amplum*, being informally described as an excavator (e.g. Bellwood 1994) and some authors also categorizing *Sp. chrysopterum* and *Sp. rubripinne* as excavators (Bernardi et al. 2000). Within the lineage that possesses the intramandibular joint there is wide diversity in feeding mode, from superficial scraping to excavating. This suggests that the intramandibular joint may have a general benefit to scraping hard substrate that is not specific to either extreme on the axis from deep excavating to superficial scraping.

A functional morphospace for the parrotfish feeding system can be produced by a principal components analysis summarizing eight functional traits of the feeding mechanism in 34 species, including representatives of all genera except *Nicholsina* (Figs 4 and 5). The traits used in this analysis are described in detail elsewhere (Wainwright

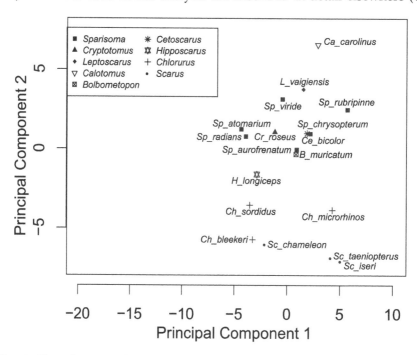

Fig. 4. Plot of Principal Component 1 vs 2 from a phylogenetic PCA run on nine morphological variables associated with the functional morphology of the parrotfish feeding mechanism. The variables included were mechanical advantage of jaw opening and jaw closing, transmission coefficients of the oral jaws and hyoid 4-bar linkages, oral jaw gape distance, maximum upper jaw protrusion distance, and masses of the adductor mandibulae, sternohyoideus, and levator posterior muscles. Average values of each trait for several specimens per species were corrected for body while accounting for phylogenetic relationships where necessary. Data are from Wainwright et al. 2004 and Price et al. 2010.

et al. 2004) but include the horizontal width of the oral gape, maximum premaxillary protrusion distance, the mechanical advantage of jaw opening and closing muscles, and the transmission coefficient of the four-bar linkage that operates the oral jaws (transmission coefficients are the inverse of mechanical advantage), as well as the mass of three major muscles; the adductor mandibulae complex, the sternohyoideus and the levator posterior. The adductor mandibulae is a complex of muscles that function to adduct the oral jaws during biting, the sternohyoideus is involved in ventral depression of the hyoid bar during suction, which is poorly developed in parrotfish, and the levator posterior muscle is a major biting muscle from the pharyngeal jaw system that pulls the lower pharyngeal jaw up against the upper jaw.

After size-correcting traits by calculating residuals of species means from Log-Log regressions on the cube root of body mass the position of 34 species in principal component space reveals major features of the morphological diversity (Figs 4 and 5). Principal component one (PC1) is negatively correlated with all morphological traits and represents an axis that captures species at one extreme with relatively large muscles, a large mouth, high protrusion distance, and high values of four-bar transmission coefficients and jaw lever mechanical advantage, and species at the other extreme with small values of these traits (Table 1; Fig. 4). Principal component two (PC2) primarily involves a trade-off between jaw lever mechanical advantage and the gape width. In bivariate plots of PC1 vs 2 and PC3 vs 4 a group composed of species of *Scarus* is apparent, and a second group made up of all other parrotfishes with the excavators *Chlorurus* and *Bolbometopon* somewhat is set apart from this group (Figs 4 and 5). Two interesting points are that *Sp. viride* is

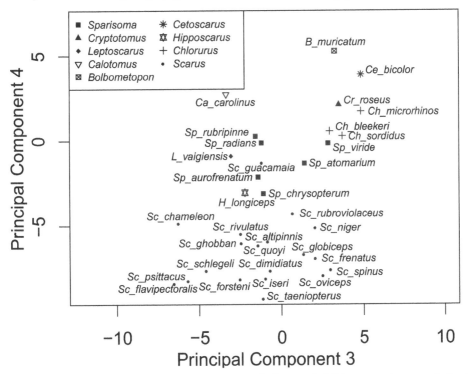

Fig. 5. Plot of Principal Component 3 vs 4 from a phylogenetic PCA run on nine morphological variables associated with the functional morphology of the parrotfish feeding mechanism. See Fig. 4 legend for member variables and Table 1 for loadings and variance explained. Data are from Wainwright et al. 2004 and Price et al. 2010.

intermediate between the excavator group and other *Sparisoma*, and *Hipposcarus* does not fall with *Scarus* or *Chlorurus*, but is intermediate between *Scarus* and *Sparisoma*. This last observation again points to the fact that the intramandibular joint, present in *Scarus*, *Chlorurus* and *Hipposcarus*, is not associated with a narrow range of functional morphology, but instead supports considerable diversity.

Table 1. Principal Component loadings from a phylogenetic PCA run on morphological traits of parrotfishes.

	PC1	PC2	PC3	PC4
Gape Distance	-0.354	0.652	-0.435	0.152
Protrusion Distance	-0.705	0.182	-0.031	-0.435
Adductor Mass	-0.772	0.068	0.539	0.015
Sternohyoideus Mass	-0.924	0.049	0.087	0.148
Levator Posterior Mass	-0.572	0.013	-0.320	0.649
Jaw Closing Lever	-0.186	-0.764	-0.347	-0.170
Jaw Opening Lever	-0.368	-0.767	-0.067	0.204
Oral Jaw KT	-0.620	0.046	-0.592	-0.391
Hyoid KT	-0.766	-0.089	0.416	-0.089
Cumulative Variance Explained	0.393	0.576	0.714	0.813

Bivariate plots of individual variables are also revealing. A plot of jaw closing lever ratio against mass of the levator posterior muscle sets *Scarus* apart from all other taxa (Fig. 6). Species of *Scarus* have extremely high jaw closing mechanical advantage of the oral

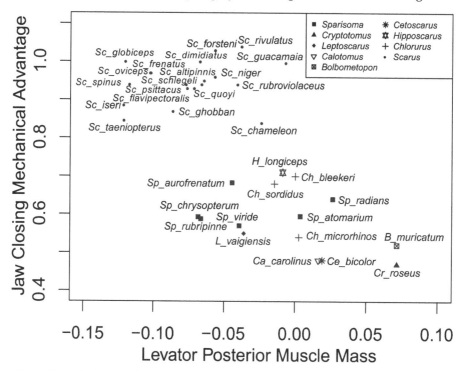

Fig. 6. Plot of jaw closing mechanical advantage versus levator posterior muscle mass in 34 species of parrotfishes. Levator posterior muscle mass is expressed as a residual from a body size correction regression. Data are from Wainwright et al. 2004 and Price et al. 2010.

jaw mandible but extensive diversity in the size of the levator posterior muscle from the pharyngeal jaws. The excavators, *Chlorurus*, *Bolobometopon* and *Cetoscarus* have very large levator posterior muscles and relatively low jaw closing mechanical advantage, placing them in the lower right region of this plot with *Calotomus*. The large levator posterior muscle may reflect that excavators often remove solid pieces of reef carbonate that must be reduced in the pharyngeal mill before being swallowed. The remaining species are intermediate in the two traits, except *Cryptotomus* which has the largest levator posterior muscle and lower jaw closing mechanical advantage of all parrotfish. This is interesting because *Cryptotomus* is the smallest parrotfish, about 75 mm adult body size.

Curiously, although *Scarus* species all have very high mechanical advantage of jaw closing, they have moderate to very high transmission coefficient in the oral jaw four-bar linkage (Fig. 7). This indicates that while *Scarus* transfer a large amount of adductor muscle force to the cutting edge of the lower jaw during biting, many of them also generate a large amount of movement in the upper jaw for a given amount of rotation of the lower jaw. This appears to be an unusual case where a 'force modified' linkage system operates in series with a second system that is 'displacement modified'.

When the transmission coefficient of the oral jaw 4-bar linkage is plotted against mechanical advantage of jaw opening (Fig. 7), the major pattern is once again the high diversity found in *Scarus* and *Chlorurus*. The mechanical advantage of jaw opening ranges from 0.19 to 0.38 among species of *Scarus* and the most extreme values of 4-bar transmission coefficient are found in *Scarus* or *Chlorurus*. *Sp. viride* has a very low 4-bar transmission

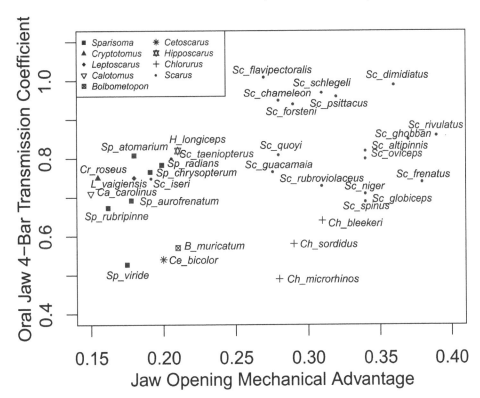

Fig. 7. Plot of transmission coefficient of the 4-bar linkage of the oral jaws versus jaw opening mechanical advantage in 34 species of parrotfish. Data are from Wainwright et al. 2004 and Price et al. 2010.

coefficient, like all other excavators (*Chlorurus*, *Bolbometopon* and *Cetoscarus*). Most of the species of *Sparisoma*, *Calotomus*, *Leptoscarus*, *Hipposcarus* and *Cryptotomus* are clumped in a region with intermediate values of 4-bar transmission coefficient and low jaw opening mechanical advantage.

A plot of adductor mandibulae mass against width of the oral jaw gape continues the trend of separating *Scarus*, excavators and the other taxa (Fig. 8). Once again, *Scarus* shows high diversity in adductor mass, with species spanning most of the range seen across all parrotfishes. As expected, the largest adductor muscles are found in the excavators, *Chlorurus*, *Bolbometopon* and *Cetoscarus*, but they are joined by three species of *Scarus* that also have large adductors. The smallest adductor muscles are found in some *Sparisoma*, *Leptoscarus*, *Calotomus* and some *Scarus*. The width of the oral gape is highest in *Sparisoma*, *Calotomus* and *Leptoscarus* and smallest in some *Scarus*.

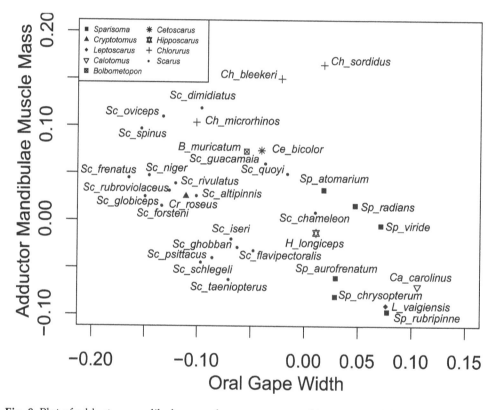

Fig. 8. Plot of adductor mandibulae muscle mass versus oral jaw gape width in 34 species of parrotfishes. Data are from Wainwright et al. 2004 and Price et al. 2010.

Evolutionary Implications of Feeding Innovations

The evolutionary history of the three feeding innovations as seen in Fig. 3 suggests a structure for posing questions about the effect that they have had on parrotfish macroevolution. Two of the innovations, the modified pharyngeal mill and the intramandibular joint, have apparently each evolved only once in parrotfish, while the cutting edge of the oral jaws has a more ambiguous history, with the strongest support for a single origin and two independent losses. Both cutting edge dentition and the intramandibular joint are distributed such that lineages with and without the innovation can be compared, but because all parrotfish have the pharyngeal mill, studies of its effect on diversification must be made between parrotfish and other labrids.

Previous studies have explored macroevolutionary impacts of the pharyngeal mill and the intramandibular joint. A study of lineage diversification rates in labrids revealed a significant correlation between possession of the parrotfish pharyngeal mill and increased rates of speciation (Alfaro et al. 2009) although the influence of co-distributed characters cannot be ruled out. More detailed explorations have revealed that elevated rates of diversification are found only in the clade that includes *Scarus* and *Chlorurus* (Alfaro et al. 2009, Kazancioğlu et al. 2009, Cowman and Bellwood 2011) and the higher rate seen across parrotfish as compared to wrasses appears to be due to a trickle-down effect of the high rate in *Scarus* and *Chlorurus*. Thus, while the highest diversification rates across Labridae are found in the *Scarus/Chlorurus* clade, even moderately elevated rates of diversification are not seen in other parrotfishes (Alfaro et al. 2009, Cowman and Bellwood 2011). Interestingly, this shift in diversification on the branch leading to *Scarus/Chlorurus* is very close to the inferred origin of the intra-mandibular joint on the branch leading to *Scarus/Chlorurus/Hipposcarus* (Fig. 3). Most authors have concluded that the higher diversification rate seen in *Scarus/Chlorurus* is more likely related to strong sexual selection in this clade, as reflected by strong dichromatism, than to functional innovations of the feeding mechanism (Streelman et al. 2002, Alfaro et al. 2009, Kazancioğlu et al. 2009, Cowman and Bellwood 2011). Sexual selection, or change in the strength of sexual selection, is one of the most commonly found factors that influences diversification rate (Coyne and Orr 2005). It is important to emphasize, however, that in spite of the popularity of this hypothesis, formal analyses of the relationship between sexual selection and diversification rate in parrotfishes, or more broadly in labrids, have not yet been conducted.

What about functional and ecological diversity? Is there a relationship between the three parrotfish innovations and diversity in the functional morphology of the feeding system, the food they eat, and where and how they eat it? Unfortunately, there may not be enough detailed information about the micro-habitat feeding locations and diet in individual species of parrotfish to evaluate diversity in these traits. But, when the diversity of morphological traits is viewed in the context of the time-calibrated molecular phylogeny of parrotfishes a very interesting pattern becomes immediately apparent. *Scarus*, although very diverse in terms of feeding functional morphology (Figs 4-8), is a young lineage, roughly 5-10 million years old crown age (Smith et al. 2008, Kazancioğlu et al. 2009, Cowman and Bellwood 2011, Choat et al. 2012). Furthermore, the lineage that also includes *Chlorurus* and *Hipposcarus* is even more morphologically diverse (Figs 4-8). High diversity evolving over a relatively short period of time implies that the rate of evolution has been high. Indeed, this intuition was confirmed in a model-fitting study where the estimated rate of evolution of feeding traits in the *Scarus/Chlorurus/Hipposcarus* clade was found to be much higher relative to other parrotfishes (Fig. 8; Price et al. 2010). The jaw closing and opening mechanical advantage, oral jaw 4-bar transmission coefficient and mass of the adductor mandibulae muscle have all evolved between 4 and 23 times faster in this clade than in other parrotfish (Fig. 9).

Parrotfish with the intramandibular joint show high rates of evolution in the functional morphology of the oral jaw feeding apparatus. It is interesting that these elevated rates are restricted to aspects of the oral jaws, where the intramandibular joint occurs. Neither mass of the sternohyoideus nor the levator posterior muscle evolve at different rates in the two groups of parrotfish (Fig. 9; Price et al. 2010). These two muscles are not directly associated with the oral jaw system. Price et al. (2010) suggested that the elevated rates

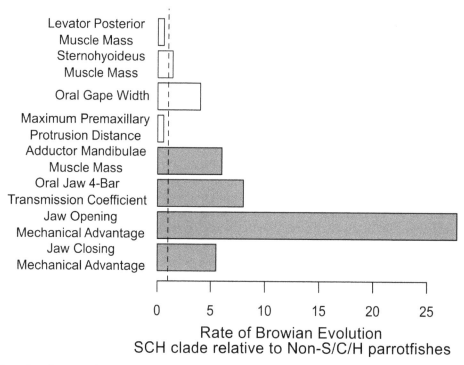

Fig. 9. Bar diagram depicting the rates of evolution of functional morphological traits of the parrotfish feeding apparatus. Rates of evolution of traits are expressed as a ratio of rate of its evolution in the clade that possesses the intramandibular joint (*Scarus*, *Chlorurus* and *Hipposcarus*) and the rate in all other parrotfishes. The dashed line indicates equal rates in both clades (which occurs at a relative rate of 1). Color indicates whether the difference in rates between the clades is significant (grey) or not (white) according to the p-value derived from parametric bootstrapping. Jaw closing and opening mechanical advantage, oral jaw transmission coefficient, and mass of the adductor mandibulae all evolve faster in the intramandibular joint clade, while the sternohyoideus and levator posterior muscles show no difference between groups.

of evolution found in oral jaw traits in the *Scarus*/*Chlorurus*/*Hipposcarus* clade may have come about because of the increased mechanical complexity of the jaws that is produced by the introduction of the second jaw joint. The introduction of the novel joint may increase the range of configurations that can function effectively, thus removing a constraint on diversification. One difficulty with this interpretation is that we would expect functional diversity to reflect variation in feeding ecology, and yet, it is thought that this clade exhibits relatively little diversity in feeding ecology beyond the well-documented differences between the excavating *Chlorurus* and scraping *Scarus* (Bellwood and Choat 1990). Nevertheless, some evidence of ecological diversity and community complementarity has been described (e.g. Rasher et al. 2013). It may be worth future studies generating more detailed data on the microhabitat grazing locations on reefs, where each species feeds and the way in which they scrape the substrate, to determine if the high functional morphological diversity, seen particularly in *Scarus*, is associated with ecological variation. This sort of ecological variation in the substrate that species feed on has been found in surgeonfishes (Brandl et al. 2014, Brandl and Bellwood 2014) where there is considerable variation among taxa in the surface topography of the hard substrate that is grazed and the diversity of substrates grazed by single species. Parrotfish typically occur in high diversity systems with many other species of parrotfish. Species of *Scarus* and *Chlorurus* in the Indo-Pacific can be found feeding in groups with other species and *Scarus* frequently occupy microhabitats with more than six congeners (Russ 1984, Hoey and Bellwood 2008).

It is possible that functional diversity among species of parrotfish on reefs, beyond the difference between excavators and scrapers, may result in ecological diversity (Choat et al. 2002), such as that which facilitates complementarity (Burkepile and Hay 2008, 2011) and the coexistence of so many species. However, no such axis of ecological diversity has yet been identified. The presence of high rates of evolution in oral jaw functional morphology in the *Scarus/Chlorurus/Hipposcarus* clade is therefore something of a conundrum. One interesting possibility is that both the high rate of functional morphological evolution and the high speciation rate in this group are both a secondary result of strong sexual selection. The high rates of morphological evolution may be tied to sexual selection for aspects of head shape. However, it is difficult to imagine that the variation in size of the adductor mandibulae muscle seen among species of *Scarus* (Fig. 4), which spans most of the range found in parrotfishes, does not have consequences for the feeding ecology of these species. The interplay between sexual selection, functional variation and ecological diversification in parrotfishes remains an area deserving of continued investigation.

References Cited

Alfaro, M.E., C.D. Brock, B. Banbury and P.C. Wainwright. 2009. Does evolutionary innovation in pharyngeal jaws lead to adaptive radiation in labrid fishes? BMC Evol. Biol. 9: 255.

Bellwood, D.R. 1994. A phylogenetic study of the parrotfish Family Scaridae (Pisces: Labroidei), with a revision of genera. Rec. Aust. Mus. Suppl. 20: 1–86.

Bellwood, D.R. 1995a. Direct estimates of bioerosion by two parrotfish species, *Chlorurus gibbus* and *C. sordidus*, on the Great Barrier Reef, Australia. Mar. Biol. 121: 419–429.

Bellwood, D.R. 1995b. Carbonate transfer and intra-reefal patterns of bio-erosion and sediment release by parrotfishes (family Scaridae) on the Great Barrier Reef. Mar. Ecol. Prog. Ser. 117: 127–136.

Bellwood, D.R. 2003. Origins and escalation of herbivory in fishes: a functional perspective. Paleobiology 29: 71–83.

Bellwood, D.R. and J.H. Choat. 1990. A functional analysis of grazing in parrotfishes (family Scaridae): the ecological implications. Environ. Biol. Fish. 28: 189–214.

Bernardi, G., D.R. Robertson, K.E. Clifton and E. Azzuro. 2000. Molecular systematics, zoogeography, and evolutionary ecology of the Atlantic parrotfish genus *Sparisoma*. Mol. Phylogen. Evol. 15: 292–300.

Bonaldo, R.M., A.S. Hoey and D.R. Bellwood. 2014. The ecosystem roles of parrotfishes on tropical reefs. Oceanogr. Mar. Biol. Annu. Rev. 52: 81–132.

Brandl, S.J. and D.R. Bellwood. 2014. Individual-based analyses reveal limited functional overlap in a coral reef fish community. J. Anim. Ecol. 83: 661– 670.

Brandl, S.J., A.S. Hoey and D.R. Bellwood. 2014. Micro-topography mediates interactions between coral, algae, and herbivorous fishes on coral reefs. Coral Reefs 33: 421–430.

Bruggemann, J.H., A.M. van Kessel, J.M. van Rooij and A.M. Breeman 1996. Bioerosion and sediment ingestion by the Caribbean parrotfish *Scarus vetula* and *Sparisoma viride*: Implications of fish size, feeding mode and habitat use. Mar. Ecol. Prog. Ser. 134: 59–71.

Bullock, A.E. and T. Monod. 1997. Myologie céphalique de deux poisons perroquets (Teleost: Scaridae). Cybium. 21: 173–199.

Burkepile, D.E and M.E. Hay. 2008. Herbivore species richness and feeding complementarity affect community structure and function on a coral reef. Proc. Natl. Acad. Sci. USA. 105: 16201–16206.

Burkepile, D.E. and M.E. Hay. 2011. Feeding complementarity versus redundancy among herbivorous fishes on a Caribbean reef. Coral Reefs. 30: 351–362.

Choat, J.H. 1991. The biology of herbivorous fishes on coral reefs. pp. 120–155. *In*: P.F. Sale (ed.). The Ecology of Fishes on Coral Reefs. Academic Press, San Diego.

Choat, J.H., K.D. Clements and W.D. Robbins. 2002. The trophic status of herbivorous fishes on coral reefs. 1: Dietary analyses. Mar. Biol. 140: 613–623.

Choat, J.H., O.S. Klanten, L. Van Herwerden, D.R. Robertson and K.D. Clements. 2012. Patterns and processes in the evolutionary history of parrotfishes (Family Labridae). Biol. J. Linnean Soc. 107: 529–557.

Clements, K.D. and D.R. Bellwood. 1988. A comparison of the feeding mechanisms of two herbivorous labroid fishes, the temperate *Odax pullus* and the tropical *Scarus rubroviolaceus*. Aust. J. Mar. Freshwater Res. 39: 87–107.

Clements, K.D., M.E. Alfaro, J.L. Fessler and M.W. Westneat. 2004. Relationships of the temperate Australasian labrid fish tribe Odacini (Perciformes; Teleostei). Mol. Phylogen. Evol. 32: 575–587.

Cowman, P.F. and D.R. Bellwood. 2011. Coral reefs as drivers of cladogenesis: expanding coral reefs, cryptic extinction events, and the development of biodiversity hotspots. J. Evol. Biol. 24: 2543–2562.

Coyne, J.A. and H.A. Orr. 2005. Speciation. Sinauer Associates, Sunderland Mass., USA.

Crossman, D.J., J.H. Choat and K.D. Clements. 2005. Nutritional ecology of nominally herbivorous fishes on coral reefs. Mar. Ecol. Prog. Ser. 296: 129–142.

Ferry-Graham, L.A. and N. Konow. 2010. The intramandibular joint in *Girella*: a mechanism for increased force production? J. Morphology 271: 271–279.

Ferry, L.A., N. Konow and A.C. Gibb. 2012 Are kissing gourami specialized for substrate-feeding? Prey capture kinematics of *Helostoma temmickii* and other anabantoid fishes. J. Exp. Zool, Part A, Ecol. Genet. Physiol. 317A: 571–579.

Frydl, P. and C.W. Stearn. 1978. Rate of bioerosion by parrotfish in Barbados reef environments. J. Sediment. Petrol. 48: 1149–1157.

Gibb, A.C., L.A. Ferry-Graham and L.P. Hernandez 2008. Functional significance of intramandibular bending in poeciliid fishes. Environ. Biol. Fish. 83: 473–485.

Gobalet, K.W. 1989. Morphology of the parrotfish pharyngeal jaw apparatus. Am. Zool. 29: 319–331.

Hoey, A.S. and D.R. Bellwood. 2008. Cross-shelf variation in the role of parrotfishes on the Great Barrier Reef. Coral Reefs. 27: 37–47.

Kaufman L.H. and K.F. Liem. 1982. Fishes of the suborder Labroidei (Pisces: Perciformes): phylogeny, ecology, and evolutionary significance. Breviora 472: 1–19.

Kazancioğlu, E., T.J. Near, R. Hanel and P.C. Wainwright. 2009. Influence of feeding functional morphology and sexual selection on diversification rate of parrotfishes (Scaridae). Proc. R. Soc., B. 276: 3439–3446.

Konow, N. and D.R. Bellwood 2005. Prey-capture in *Pomacanthus semicirculatus* (Teleostei, Pomacanthidae): functional implications of intramandibular joints in marine angelfishes. J. Exp. Biol. 208: 1421–1433.

Konow, N., D.R. Bellwood, P.C. Wainwright and A.M. Kerr. 2008. Evolution of novel jaw joints promote trophic diversity in coral reef fishes. Biol. J. Linnean Soc. 93: 545–555.

Liem, K.F. and S.L. Sanderson. 1986. The pharyngeal jaw apparatus of labrid fishes – a functional morphological perspective. J. Morph. 187: 143–158.

Price, S.A., P.C. Wainwright, D.R. Bellwood, E. Kazancioglu, D.C. Collar and T.J. Near. 2010. Functional innovations and morphological diversification in parrotfishes. Evolution. 64: 3057–3068.

Purcell, S.W. and D.R. Bellwood. 1993. A functional analysis of food procurement in two surgeonfish species, *Acanthurus nigrofuscus* and *Ctenochaetus striatus* (Acanthuridae). Environ. Biol. Fish. 37: 139–159.

Rasher, D.B., A.S. Hoey and M.E. Hay. 2013. Consumer diversity interacts with prey diversity to drive ecosystem function. Ecology. 94: 1347–1358.

Russ, G.R. 1984. Distribution and abundance of herbivorous grazing fishes in the central Great Barrier Reef. II. Patterns of zonation of mid-shelf and outershelf reefs. Mar. Ecol. Prog. Ser. 20: 35–44.

Smith, L.L., J.L. Fessler, M.E. Alfaro, J.T. Streelman and M.W. Westneat. 2008. Phylogenetic relationships and the evolution of regulatory gene sequences in the parrotfishes. Mol. Phylogen. Evol. 49: 136–152.

Stiassny M.L.J. and J. Jensen. 1987. Labroid interrelationships revisited: morphological complexity, key innovations, and the study of comparative diversity. Bull. Mus. Comp. Zool. 151: 269–319.

Streelman, J.T., M.E. Alfaro, M.W. Westneat, D.R. Bellwood and S.A. Karl. 2002. Evolutionary history of the parrotfishes: biogeography, ecomorphology, and comparative diversity. Evolution 56: 961–971.

Vial, C.I. and F.P. Ojeda . 1990. Cephalic anatomy of the herbivorous fish *Girella laevifrons* (Osteichthyes: Kyphosidae): mechanical considerations of its trophic function. Revista Chilena de Historia Natural. 63: 247–260.

Wainwright, P.C. 1988. Morphology and ecology: the functional basis of feeding constraints in Caribbean labrid fishes. Ecology. 69: 635–645.

Wainwright, P.C. 2005. Functional morphology of the pharyngeal jaw apparatus. pp. 77–101. *In*: R. Shadwick and G.V. Lauder (eds.). Biomechanics of Fishes. Academic Press, Amsterdam.

Wainwright, P.C., D.R. Bellwood, M.W. Westneat, J.R. Grubich and A.S. Hoey. 2004. A functional morphospace for the skull of labrid fishes: patterns of diversity in a complex biomechanical system. Biol. J. Linnean Soc. 82: 1–25.

Wainwright, P.C., W.L. Smith, S.A. Price, K.L. Tang, L.A. Ferry, J.S. Sparks and T.J. Near. 2012. The evolution of pharyngognathy: a phylogenetic and functional appraisal of the pharyngeal jaw key innovation in labroid fishes and beyond. Syst. Biol. 61: 1001–1027.

Westneat, M.W. and M.E. Alfaro. 2005. Phylogenetic relationships and evolutionary history of the reef fish family Labridae. Mol. Phylogen. Evol. 36: 370–390.

CHAPTER 3

Nutritional Ecology of Parrotfishes (Scarinae, Labridae)

Kendall D. Clements[1] and J. Howard Choat[2]

[1] School of Biological Sciences, University of Auckland, Private Bag 92019,
   Auckland 1142, New Zealand
   Email: k.clements@auckland.ac.nz
[2] College of Marine and Environmental Sciences, James Cook University, Townsville,
   QLD 4811, Australia
   Email: John.Choat@jcu.edu.au

Introduction

G.E. Hutchinson's classic "Homage to Santa Rosalia" paper (Hutchinson 1959) has been very influential in shaping ecological thinking on biodiversity. As pointed out by Brown (1981), Hutchinson's main thesis is that an understanding of biodiversity requires us to determine how energy, i.e. nutrients, are acquired by and partitioned among species. Hutchinson (1959) could thus be seen as an early argument for the importance of nutritional ecology in biodiversity research. Foley and Cork (1992) define nutritional ecology as the study of "the relationships between the foods that animals eat, how those foods are obtained, and how they are processed." Although nutritional ecology is sometimes misrepresented as involving purely post-ingestive processes, Foley and Cork's definition shows that it is broader than that, and examines the full gamut of nutritional relationships between animals and their environment. Nutritional ecology thus includes three main elements: (i) the biochemical composition of foods, (ii) food detection and ingestion, and (iii) the extraction and digestion of nutrients from these foods.

The view that nutrition lies at the heart of feeding biology seems self-evident, since feeding is the process by which animals obtain nutrients for growth, maintenance, energy and reproduction. Examining nutritional requirements involves understanding the causes of feeding, i.e. to obtain nutrition. A focus on nutrition should thus be central to the study of trophodynamics and resource partitioning, and this is indeed the case in well-studied examples of trophic diversification such as ruminants (Van Soest 1994) and marsupials (Hume 1999). It is therefore noteworthy that the study of trophic diversification in herbivorous fishes on coral reefs has developed very differently, with a focus on the effects of feeding rather than its causes (Clements et al. 2009). Work on feeding and diet in these fishes has concentrated on pre-ingestive factors, and contemporary studies predominantly involve feeding observations and habitat use (e.g. Cheal et al. 2013, Brandl and Bellwood

2014, Adam et al. 2015a) or cage experiments that measure the effects of feeding on the benthic biota (e.g. Burkepile and Hay 2008, 2011, Cernohorsky et al. 2015). Nutrition is thus subsidiary to a consideration of ecological role, leading to a viewpoint where the causes and effects of feeding are inverted, as best illustrated by the concepts of feeding complementarity and redundancy (e.g. Burkepile and Hay 2011, Nash et al. 2015). Nowhere is this conceptual viewpoint more evident than in parrotfishes (Labridae, Scarinae), whose feeding biology is generally categorised in terms of its effects on the substratum, i.e. grazing, excavating, browsing and coral predation (Bonaldo et al. 2014, Adam et al. 2015b). The dietary substrata of parrotfishes as a focus of investigation have in effect become where parrotfishes feed, not what they actually ingest. As a result of this, parrotfishes are arguably the least understood group of coral reef fishes in terms of nutritional ecology, i.e. their use of nutritional resources from the environment (Foley and Cork 1992). Indeed, the recent and comprehensive review of parrotfish ecology by Bonaldo et al. (2014) suggests that "detailed studies are still necessary to verify the food items responsible for providing nutrition to different species in the group."

In the remainder of this chapter we will take the framework of parrotfish feeding substrata provided by recent reviews (Bonaldo et al. 2014, Adam et al. 2015b), and use this to generate a series of predictions about what the nutritional targets of parrotfish feeding may be. This requires removing the ambiguity around the use of the word 'substrata', i.e. biochemical substrates versus feeding surfaces, and exploring the differences between potential food resources in terms of macronutrient composition. Next, we will evaluate these nutritional predictions in terms of what we know about macronutrient intake in parrotfishes, given the high growth rates of parrotfishes (Choat and Robertson 2002, Taylor and Choat 2014) and thus their likely high demand for protein (Bowen et al. 1995). We will then examine feeding and digestive mechanisms in parrotfishes: how do these match the predictions of dietary targets based on nutritional parameters? Finally, we will examine two lines of data on assimilation of nutrients by parrotfishes: tissue fatty acid analysis and stable isotope analysis. We will conclude by showing that a nutritional approach can generate considerable new insight into parrotfish trophic biology and resource partitioning, and for the first time provide a unified hypothesis that explains the apparently disparate feeding ecologies of this diverse, abundant and ecologically important group of reef fishes.

Nutritional Resources

Early studies on parrotfishes indicated a diet of coral (e.g. Al-Hussaini 1945, Gohar and Latif 1959), but most subsequent workers consider these fishes as herbivores while acknowledging that some species also predate sessile invertebrates including corals (Bonaldo et al. 2014, Adam et al. 2015b) and sponges (McAfee and Morgan 1996, Dunlap and Pawlik 1998). Bonaldo et al. (2014) define five categories of ecological roles for parrotfishes: (i) grazing (i.e. "scraping algae-covered substratum"), (ii) browsing (i.e. "biting off parts of erect macroalgae" or seagrass), (iii) coral predation, (iv) bioerosion, and (v) production, reworking and transport of sediment. Bonaldo et al. (2014) identify the primary feeding substratum of parrotfish worldwide as the "epilithic algal matrix" (EAM), which they define as "a conglomeration of short, turf-forming filamentous algae (< one cm high), macroalgal propagules, microalgae, sediment, detritus, and associated fauna." These authors go on to suggest that detritivory may be widespread among parrotfishes depending on the level of selectivity involved when feeding on the EAM, but do not clearly define detritus beyond non-algal components of the EAM. Bioerosion is discussed at length, but the dietary targets of the bioeroding species, i.e. predominantly excavators (Bellwood and Choat 1990), are not defined.

Adam et al. (2015b) broadly describe Caribbean parrotfishes in terms of the first four of the ecological categories used by Bonaldo et al. (2014). They define parrotfishes as selective herbivores, and make the generalization that *Sparisoma* species tend to be browsers of macroalgae, while *Scarus* species mainly target turf algae. In addition, Adam et al. (2015b) identify several parrotfish species as coral predators, including the bioeroding *Sp. viride* that also targets endolithic algae using an excavating feeding mode. Endolithic or boring algae are phototrophic organisms including cyanobacteria, rhodophytes and chlorophytes that actively penetrate hard substrata such as the calcareous substratum of coral reefs, resulting in microbioerosion (Tribollet 2008, Verbruggen and Tribollet 2011). Although endolithic phototrophs inhabit dead coral rock, live corals and calcareous algae, communities tend to be less well developed under dense turf or calcareous algae (Vooren 1981, Bruggemann et al. 1994a, Tribollet 2008).

It is clear from the above that parrotfishes are considered to target a very wide range of dietary resources, including EAM, macroalgae, seagrasses, corals, sponges and endolithic algae. However, assessing the differences between these food resources in terms of macronutrient composition requires clarifying their biotic make-up. EAM is particularly problematical in this regard, as both 'turf algae' and 'detritus' are highly heterogeneous. Algal turfs vary greatly in taxonomic composition, consisting of diverse mixtures of rhodophytes, chlorophytes and phaeophytes as well as cyanobacteria (Connell et al. 2014, Harris et al. 2015). The latter can make up a considerable proportion of turf biomass and substratum cover, but this can be highly patchy over small spatial scales (Williams and Carpenter 1997, Adey 1998, Harris et al. 2015). Other food components of EAM include microalgae such as diatoms and dinoflagellates, meiofauna, heterotrophic bacteria and organic detritus. The relationships between algae, cyanobacteria and diatoms on reef surfaces will be complex both spatially and temporally. Cyanobacteria are generally the pioneer phototrophic colonists of bare substrata, but the precise successional trajectories of epilithic community development from microbial biofilms to algal turfs and macroalgae depend greatly on environmental factors such as sedimentation and nutrient availability and biotic factors such as the intensity and type of grazing (Williams et al. 2000, Taylor et al. 2006, Skov et al. 2010). Heavily grazed epilithic communities on coral reefs tend to remain in an early successional stage where they are dominated by cyanobacteria (Tsuda and Kami 1973, Sammarco 1983, Wilkinson and Sammarco 1983). It should be noted in this context that epilithic and endolithic cyanobacteria are major contributors of both fixed carbon and fixed nitrogen to coral reefs (Wilkinson and Sammarco 1983, Williams and Carpenter 1997, Tribollet et al. 2006, Casareto et al. 2008).

The dietary categories of macroalgae and seagrasses are also heterogeneous in the sense that both can be densely colonized by epibionts. Macroalgae including *Dictyota*, *Lobophora*, *Padina* and *Halimeda* can be heavily colonized by cyanobacteria (Penhale and Capone 1981, Barott et al. 2011, Fricke et al. 2011, Hensley et al. 2013). The senescent tips of seagrass blades are also heavily epiphytized by a range of organisms including rhodophytes and cyanobacteria (Van Montfrans et al. 1984, Bologna and Heck 1999, Yamamuro 1999, Holzer et al. 2013). Finally, the composition of endolithic communities varies with successional stage, but typically communities are dominated by boring cyanobacteria such as *Mastigocoleus testarum* and *Plectonema terebrans* and microscopic, siphonaceous chlorophytes such as *Ostreobium quekettii* (Tribollet 2008, Grange et al. 2015). *Ostreobium* spp. are more tolerant of low light than euendolithic cyanobacteria, and thus penetrate deeper into the substratum (Grange et al. 2015).

The major food resources of parrotfishes thus appear to lie among the following: chlorophytes, rhodophytes, phaeophytes, seagrasses, cyanobacteria, heterotrophic bacteria, diatoms, coral, sponges and organic detritus. The first four of these groupings are predominantly composed of carbohydrate, although the composition of this varies considerably (Choat and Clements 1998, Jung et al. 2013). Chlorophytes, rhodophytes and seagrasses all store forms of starch, whereas phaeophytes predominantly store energy as laminarin and the sugar alcohol mannitol (Goldberg 2013, Jung et al. 2013). The lipid content of macroalgae and seagrasses is generally <10 percent dry weight, although in some species of Dictyotalean algae it can be up to 20 percent dry weight (Table 1). Protein content of most tropical macroalgae and seagrasses is in the range of 3-20 percent and 2.5-9 percent dry weight, respectively (Table 1), but in some rapidly growing rhodophytes and chlorophytes it can be up to ca. 30 percent dry weight (Angell et al. 2015). Overall, the protein content of chlorophytes and rhodophytes is about 33-45 percent higher than phaeophytes (Angell et al. 2015), and protein levels among proposed parrotfish dietary sources are lowest in seagrasses (Table 1). The lipid content of cyanobacteria is at the high end of the range for macroalgae, and protein content is much higher, with most cyanobacterial species being >40 percent dry weight protein (Table 1). Lipid content of corals is ca. 11 percent dry weight, while protein content is about 20-26 percent dry weight (Table 1). Mean lipid and protein content of 71 species of Caribbean sponges was 11.4 percent and 20.7 percent dry weight, respectively (Chanas and Pawlik 1995; Table 1). The macronutrient content of heterotrophic bacteria is likely to be similar to that of cyanobacteria, while that of

Table 1. Protein and lipid content (as % dry weight) of potential dietary resources of parrotfishes.

Diet category	Lipid	Protein	Reference
Chlorophytes	-	11.4-19.5	Barbarino and Lourenço (2005)
	-	12.9-23.2	Lourenço et al. (2002)
	3.6-7.2	7.0-12.3	McDermid and Stuercke (2003)
	2.6-11.4	4.0-13.9	McDermid et al. (2007)
Rhodophytes	-	15.5-22.4	Barbarino and Lourenço (2005)
	-	12.5-27.2	Lourenço et al. (2002)
	1.3-4.4	2.7-21.2	McDermid and Stuercke (2003)
	1.5-3.9	3.2-12.3	McDermid et al. (2007)
Phaeophytes	-	11.1-14.8	Barbarino and Lourenço (2005)
	-	12.4-18.4	Lourenço et al. (2002)
	2.6-20.2	6.4-13.0	McDermid and Stuercke (2003)
	3.8-16.1	10.3-12.0	McDermid et al. (2007)
Seagrasses	2.9-5.1	2.5-9.2	McDermid et al. (2007)
Cyanobacteria	-	30-55	López et al. (2010)
	-	18.9-70.8	Nagarkar et al. (2004)
	13.4-19.2	-	Pushparaj et al. (2008)
	7.9-12.6	37.3-52.2	Vargas et al. (1998)
Diatoms	-	15.4	Barbarino and Lourenço (2005)
	7.2-19	12-34	Brown (1981)
Coral	-	20.6-26.1	Edmunds and Davies (1986)
	8.6-34.8	-	Harland et al. (1992)
Sponges	4-30	8-58	Chanas and Pawlik (1995)
EAM algae	-	8.9-13.9	Crossman et al. (2001)
EAM detritus	-	17.8-24.8	Crossman et al. (2001)

EAM detritus is around 20 percent (Crossman et al. 2001), although this estimate is based on samples that probably contained a mixture of heterotrophic bacteria, cyanobacteria, diatoms and particulate organic detritus. Protein content of EAM algae falls in the range of chlorophyte and rhodophyte algae (Crossman et al. 2001; Table 1). Overall, cyanobacteria and animal material have the highest levels of protein among the food resources that parrotfishes are thought to target, and phaeophytes and seagrasses the lowest.

Macronutrient Intake in Parrotfishes

Surprisingly little is known about macronutrient intake in parrotfishes. Furthermore, most of the studies that contain relevant data were conducted over two decades ago. Lobel and Ogden (1981) examined the basis for food selection in the Caribbean sparisomatine *Sparisoma radians*, and concluded that this species targeted older, epiphytized blade tips of the seagrass *Thalassia testudinum*. Blades of *T. testudinum* with and without epiphytes did not differ in caloric value, but the former diet was 559 percent more digestible to *Sp. radians* (Lobel and Ogden 1981). The lack of difference in caloric value between blades with and without epiphytes is unsurprising, because much of this will have involved cellulosic cell wall content, which will be indigestible to parrotfish. The same parrotfish species was later shown to be able to maintain daily energy intake by increasing gut throughput rate even when fed low energy foods such as the calcareous chlorophyte *Halimeda incrassata* (Targett and Targett 1990).

More recently, Goecker et al. (2005) showed that high nitrogen (i.e. protein) content was a consistent and significant predictor of preference by *Sp. radians* when selecting *Thalassia testudinum* tissues for consumption. The mechanism for choice was thought to be gustation, since experimental fish appeared to actively sample different food items prior to preferentially consuming the most protein-rich on offer. This result is interesting in the broader context of parrotfish feeding, as it indicates that the preference for high protein items may involve a feeding pattern that includes trying different nutritional sources. Other papers that have sought to test feeding selectivity in seagrass-feeding parrotfishes removed epiphytes from seagrass prior to using it in choice experiments (e.g. Kirsch et al. 2002), and so do not really test the hypothesis of selectivity as it pertains in the wild.

Harmelin-Vivien et al. (1992) compared the biochemical composition of ingesta from the foregut of the excavating scarinine *Chlorurus spilurus* (then called *Scarus sordidus*) with that of freshly collected food, which they considered to consist of epilithic algal turf and endolithic algae from the upper 2 mm of coral rock. The proportions of total carbohydrate and chlorophyll pigments to total organic matter (i.e. once calcareous material was removed) were both significantly lower in foregut contents (18 percent and 0.17 percent dry weight, respectively) than in fresh food (32.1 percent and 0.37 percent, respectively). The authors explained this discrepancy in terms of dilution by high concentrations of mucus in the foregut of *Ch. spilurus* (Harmelin-Vivien et al. 1992). This explanation seems unlikely, as (a) mucus contains glycosidic residues (Varute and Jirge 1971, Holley et al. 2015) that would contribute to, rather than dilute, total sugar content, and (b) the foregut and hindgut of *Ch. spilurus* did not differ in the percentage of total organic matter made up by total sugars (19.5 and 19.2 percent, respectively). Another explanation is that the parrotfish ingested much less algal turf than expected, i.e. the diet was not rich in total carbohydrate. Protein was not measured in this study.

Low levels of ingested carbohydrate were also reported along with high levels of ingested protein in parrotfish in a more recent study conducted on the Great Barrier Reef (Crossman et al. 2005). Macronutrient intake was investigated in several herbivorous fishes including the excavator *Ch. microrhinos* and the scrapers *Sc. frenatus* and *Sc. schlegeli* by comparing the biochemical composition of contents from the foregut (Crossman et al. 2005). Material ingested by the parrotfishes contained a much higher proportion of protein and lipid, and a much lower proportion of carbohydrate, than that ingested by acanthurid and kyphosid species that feed on macroalgae or epilithic turf algae (Crossman et al. 2005; Table 2). The material ingested by these parrotfishes is thus very different in macronutrient composition from EAM algae and detritus (Table 1) and material ingested by algivores. Overall, a principal components analysis of these data and those from posterior gut contents (Crossman et al. 2005; Fig. 1) shows parrotfishes to be distinct from co-occurring algivorous acanthurids and kyphosids. These differences appear to be largely driven by carbohydrate, protein and lipid, not ash. Although one would have expected the ash content of parrotfishes to be much higher than the algivorous fish species due to their excavating and scraping feeding modes, this is not the case in *Sc. frenatus*, which appears to ingest little inorganic material. This work strongly indicates that these parrotfish species target nutritional resources that are very distinct from total EAM, including the turf algae on which *Acanthurus lineatus* predominantly feeds (Crossman et al. 2005).

Table 2. Macronutrient and ash composition of foregut contents from seven species of herbivorous fishes from the Great Barrier Reef. Values are mean ± SE % dry weight. Data from Crossman et al. (2005)

Fish species	Ash	Protein	Carbohydrate	Lipid
Chlorurus microrhinos	48.3 ± 13.7	16.8 ± 7.7	4.0 ± 2.2	8.2 ± 3.4
Scarus frenatus	19.7 ± 5.4	44.9 ± 4.3	4.6 ± 0.2	17.5 ± 1.3
Scarus schlegeli	55.9 ± 6.8	22.1 ± 3.8	4.0 ± 0.2	7.5 ± 1.1
Acanthurus lineatus	41.4 ± 2.1	11.0 ± 1.1	28.8 ± 2.5	5.9 ± 0.7
Kyphosus cinerascens	21.6 ± 2.9	8.2 ± 1.2	42.1 ± 3.6	4.7 ± 0.6
Kyphosus vaigiensis	24.1 ± 2.2	7.7 ± 0.8	37.9 ± 2.1	4.5 ± 0.5
Naso unicornis	32.2 ± 4.4	8.3 ± 1.1	30.3 ± 2.2	6.7 ± 0.7

It should be noted that the feeding mechanism of suspension-feeding freshwater fishes such as cyprinids and clupeids involves the secretion of mucus produced by epibranchial organs into the oesophagus (Holley et al. 2015). Mucus was shown to contribute about 10 percent of the foregut contents in gizzard shad (*Dorosoma cepedianum*), and the glycoprotein content of this mucus increases foregut N levels by about 18 percent (Holley et al. 2015). However, a comparable level of endogenous input by mucus in parrotfishes could not explain the macronutrient composition that we see in parrotfish foregut contents, especially since mucus inputs would also elevate levels of carbohydrate in the foregut, i.e. making the ingesta composition seem more like algae, not less.

Arguably the most informative body of work on parrotfish nutritional ecology is that of Henrich Bruggemann and colleagues in the mid-1990 (Bruggemann et al. 1994a-c, 1996). This work was conducted in the Caribbean and focused on the excavator *Sparisoma viride* and the scraper *Scarus vetula*. Feeding was examined in the context of five main food resources: macroalgae, large (or dense) turfs, sparse turfs on endolithic algae, sparse turfs on crustose corallines, and crustose corallines. Both parrotfish species tended to avoid coralline algae and fed mainly on sparse turfs on endolithic algae, which along with large turfs contained the highest levels of protein (Bruggemann et al. 1994a). The biotic composition of large and sparse turfs was not determined, although different types of dead

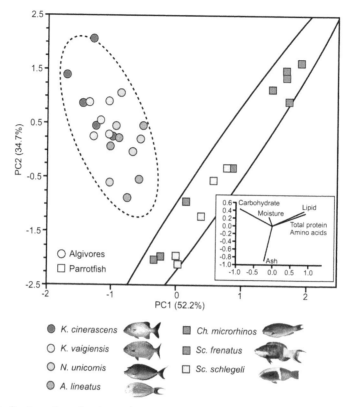

Fig. 1. Scatter plot of principal components analysis of nutritional components from foregut and hindgut contents of seven species of herbivorous fishes from the Great Barrier Reef. Ellipses show 95% confidence intervals around the two dietary groups: algivores and parrotfishes. Color coded symbols denote study species. Modified from Crossman et al. (2005).

coral substrata with turfs were compared (Bruggemann et al. 1994c). It was concluded that the excavating *Sp. viride* predominantly ingests substrate-bound algae (Bruggemann et al. 1994c), i.e. euendoliths, while *Sc. vetula* ingests mainly epilithic algal turfs. Feeding by *Sc. vetula* on macroalgae was never observed, while *Sp. viride* took occasional bites on macroalgae including *Dictyota* and *Lobophora* (Bruggemann et al. 1994a). Both of these phaeophytes can be heavily epiphytized by rhodophytes and cyanobacteria, although the extent and biotic composition of epiphytic communities is variable (Barott et al. 2011, Fricke et al. 2011).

The conclusion that the main dietary target of *Sc. vetula* is epilithic turfs seems surprising given that all size classes of this species preferred sparse turfs growing over endolithic algae (Bruggemann et al. 1994c). Why is sparse turf preferred over large turfs if the dietary target is the epilithic turf itself? The largest size classes of *Sc. vetula* ingested significant quantities of endolithic algae (Bruggemann et al. 1994c), and the differences in biochemical composition between endolithic algae and the algal components of epilithic turf raises the question of what epilithic resources are actually ingested. In a similar vein, an alternative to the idea that feeding on macroalgae by *Sp. viride* may supplement protein intake through incidental ingestion of invertebrates (Bruggemann et al. 1994a) is that the target of ingestion when feeding on macroalgae is protein-rich epiphytes such as cyanobacteria.

The excellent work of Bruggemann and colleagues highlights the complexity of defining and sampling resource targets for herbivorous fishes on coral reefs, even when nutritional parameters are considered.

Finally, the nutritional resources targeted by parrotfishes are also illuminated by a consideration of the factors causing selectivity when feeding on corals. Numerous studies have shown that some parrotfish species target live corals in both the Indo-Pacific and the Caribbean (Rotjan and Dimond 2010, Bonaldo et al. 2012, 2014). Some corallivorous butterflyfishes remove individual polyps, causing adjacent polyps to retract, and thus prefer to feed on ungrazed corals (Rotjan and Dimond 2010). In contrast, parrotfishes repeatedly graze the same coral colonies and remove large areas of tissue, leaving large feeding scars (Rotjan and Dimond 2010, Bonaldo et al. 2012). Corallivory by parrotfishes is highly variable both spatially and temporally, and on the Great Barrier Reef coral predation by the excavator *Ch. microrhinos* appears to be more prevalent on inshore reefs than mid-shelf reefs (Bonaldo et al. 2012). It was suggested that this pattern may be caused by the lower nutritional value of the more heavily-sedimented EAM resources on inshore reefs, thus possibly causing excavating parrotfishes such as *Ch. microrhinos* to switch to corallivory to maintain nutrient intake (Bonaldo et al. 2012). However, even if this is correct it does not explain why parrotfishes repetitively target certain coral colonies while avoiding others (Rotjan and Dimond 2010).

Numerous experimental studies have tested selectivity in coral predation by parrotfishes on the basis of several parameters including C:N ratio, skeletal hardness, density of macroboring invertebrates, nematocysts and zooxanthellae, and development of reproductive structures including density of gametes (Rotjan and Dimond 2010). Grazed corals appeared to have lower nutritional quality (i.e. using C:N ratio as a proxy), but this appeared to be an effect rather than a cause of feeding as it did not predict the likelihood of grazing (Rotjan and Dimond 2010). Similarly, zooxanthellae cell density, nematocysts and coral color failed to predict selectivity of feeding. Seasonal differences in coral predation by parrotfishes in both the Indo-Pacific and Caribbean suggest that grazing may be most intense immediately prior to coral spawning, and grazing appears to be heavier on shallow corals with increased density of macroboring invertebrates (Rotjan and Dimond 2010, Bonaldo et al. 2012). However, these correlative factors do not explain the causes behind selective feeding at other places and times (Rotjan and Dimond 2010). Interestingly, the density of euendolithic algae in live corals has never been examined as a cause of parrotfish predation, despite the fact that (a) euendoliths are important constituents of live corals (Le Campion-Alsumard et al. 1995, Titlyanov et al. 2008, Gutiérrez-Isaza et al. 2015), and (b) parrotfishes are known to selectively graze dead coral surfaces with dense euendolith communities (Bruggemann et al. 1994a, Chazottes et al. 1995, Tribollet 2008, Carreiro-Silva et al. 2009, Grange et al. 2015).

To summarize this section, the over-riding feature of macronutrient intake in parrotfishes appears to be a strong feeding preference for protein that is reflected in high protein content of ingesta. Material ingested by parrotfishes is much lower in carbohydrate content than would be expected if algae were the dietary target. In saying this we acknowledge the numerous observational, caging and smorgasbord studies that indicate feeding on algae, especially by Caribbean sparisomatines (Adam et al. 2015b). However, we stress that feeding observations in the wild can rarely determine the actual items ingested, especially when feeding is on highly heterogeneous assemblages such as EAM. Few studies have examined the real targets of selectivity in macroalgal feeding by parrotfishes, but a recent study suggests that feeding on *Sargassum* by *Sc. schlegeli* targets epibionts rather than algal thallus itself (Vergés et al. 2012). Caging enclosure studies (e.g.

Burkepile and Hay 2008, 2011) are problematical in determining selectivity as fishes are prohibited from foraging normally, and because of this may resort to feeding on food items that are not a major part of the normal diet. Unfortunately, the condition of fishes at the end of these cage experiments was not compared to that of free-feeding wild counterparts.

Feeding and Digestive Mechanisms in Parrotfishes

We now move on to examine how predictions of dietary targets based on nutritional parameters stack up against what we know of feeding and digestive mechanisms in parrotfishes. Does a consideration of the latter mechanisms help us narrow down the range of potential food resources that parrotfishes could be targeting for nutrition? In this section we focus mainly on pharyngeal anatomy, gut anatomy, digestive enzymes, microbiota composition, and short-chain fatty acid (SCFA) levels in the posterior intestine.

Pharyngeal Anatomy

The osteology and myology of the trophic apparatus of parrotfishes has been extensively studied (e.g. Gobalet 1989, Bellwood and Choat 1990, Bellwood 1994, Wainwright et al. 2004, Carr et al. 2006, Gobalet Chapter 1, Wainwright and Price Chapter 2). The characteristic fused-beak morphology of the premaxilla and dentary enables parrotfishes to feed using a scraping or excavating feeding mode on hard, calcareous substrata, leaving characteristic feeding scars (Bellwood and Choat 1990, Bonaldo and Bellwood 2009). It is noteworthy that such feeding on non-calcareous, volcanic substrata can result in damage to the oral jaws of sparisomatines in the Atlantic (Bonaldo et al. 2007).

The osteology and myology of the pharyngeal apparatus has been described elsewhere (Clements and Bellwood 1988, Gobalet 1989). Gobalet (1989) summarizes the function of this complex apparatus as follows: "The slurry of fine material that reaches the gut of parrotfishes is the product of the massive pharyngeal apparatus which acts on the ingested organic and inorganic material. Extraordinary dentition and bone-muscle systems are required to accomplish a task for which we would require a mortar and pestle. The adaptation may allow the parrotfish to exploit not only macroalgae, but endolithic algae and bacteria as well." Interestingly, however, accelerated rates of evolution in the genera *Scarus*, *Chlorurus* and *Hipposcarus* appear to be associated with the intramandibular joint characteristic of this scarinine clade, not the generalized scarine pharyngeal apparatus itself, which is also present in sparisomatines (Price et al. 2010). We note in this context that excavating sparisomatines (i.e. *Sp. amplum* and *Sp. viride*) are nevertheless somewhat convergent with excavating scarinines in the morphology of the pharyngeal bones, having finer dentition and an expanded toothed surface on the lower pharyngeal bone (Fig. 2).

The amount of information available on the osteology and myology of the pharyngeal apparatus stands in stark contrast to what is known of pharyngeal soft anatomy in parrotfishes, including gill rakers. Detailed descriptions of pharyngeal anatomy are limited to a handful of elderly papers (Al-Hussaini 1945, Board 1956, Gohar and Latif 1959), although ranges of gill rakers for genera are presented by Bellwood (1994). Board (1956) describes the feeding apparatus of *Sp. cretense*, and although this is probably the most detailed study of parrotfish feeding morphology to date, it has only been cited a few times (Google Scholar). This under-appreciated study shows that parrotfish pharyngeal anatomy is both highly specialized and distinctive compared to that of other herbivorous coral reef fishes such as acanthurids, siganids and kyphosids. In *Sp. cretense*, gill rakers of the third arch interdigitate with posterior gill-rakers of the second arch, and those on

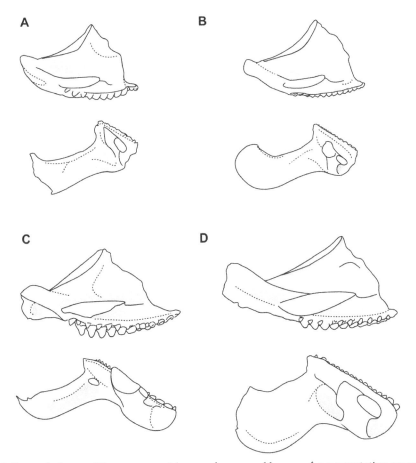

Fig. 2. Lateral views of the upper and lower pharyngeal bones of representative scraping and excavating sparisomatine and scarinine parrotfishes. Key: A. The scraper *Sparisoma axillare*; B. The excavator *Sparisoma amplum*; C. The scraper *Scarus zelindae*; D. The excavator *Chlorurus microrhinos*.

the lower half of the second arch interdigitate with posterior rakers of the first gill arch (Board 1956). The resulting structure clearly prevents loss of fine particulate ingesta through the gills.

A complex pharyngeal valve sits on the roof of the pharynx immediately behind the muscular roof of the mouth in *Sp. cretense* (Board 1956). The anterior portion of this valve contains both mucus-secreting goblet cells and taste buds, while the submucosa of the posterior section immediately anterior to the upper pharyngeal bones contains numerous mucus-secreting cells. Board (1956) summarizes the function of the gill rakers and pharyngeal valve as follows: "It is suggested that the fish carries on a type of filter feeding, the filter consisting of the posterior gill-rakers of the second and third gill arches. Food particles which collect on the filter are scraped off by the anterior portion of the pharyngeal valve. Particles on the floor of the pharynx behind the last gill slit are pushed backward in a film of mucus by the posterior portion of the valve."

Al-Hussaini (1945) describes pharyngeal anatomy and histology in the excavator *Ch. sordidus*, although the function of this is not explored in detail. The anterior and posterior gill rakers in *Ch. sordidus* are even more interdigitated than in *Sp. cretense* (Board 1956), and the two species display a similar pharyngeal valve. Gohar and Latif (1959) also report a similar pharyngeal valve structure in *Sc. ghobban* and *Hipposcarus harid*, and note that they did not observe this structure in non-scarine labrids. It thus appears that this characteristic and highly specialized pharyngeal anatomy is present in the three most diverse genera of parrotfishes, i.e. *Scarus*, *Sparisoma* and *Chlorurus*. Our re-examination of these papers led us to look more generally at pharyngeal anatomy in parrotfishes, and the results of this are presented in Figs 3-5.

Within the sparisomatines, *Cryptotomus*, *Nicholsina* and *Calotomus* have shorter and more fleshy gill rakers than *Sparisoma* (Fig. 3). *Cryptotomus* and *Nicholsina* have fewer gill rakers than *Calotomus* and *Sparisoma*, but these differences are small compared to the differences in gill raker number between sparisomatines and scarinines in general (Figs 4-5).

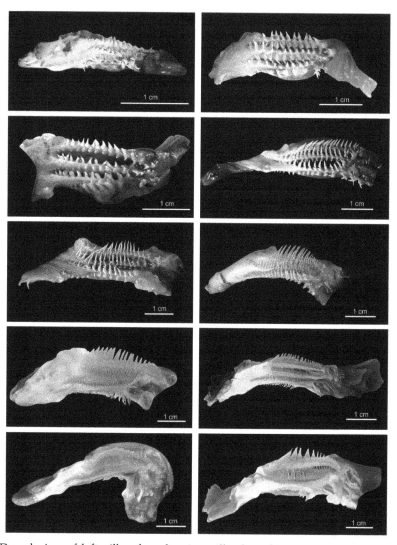

Fig. 3. Dorsal view of left gill arches showing gill rakers from representative parrotfish species. Species: A. *Cryptotomus roseus*; B. *Nicholsina usta*; C. *Calotomus carolinus*; D. *Sp. axillare*; E. *Sp. frondosum*; F. *Sp. amplum*; G. *Sc. ghobban*; H. *Sc. rivulatus*; I. *Sc. frenatus*; J. *Ch. microrhinos*. Scale = 1 cm.

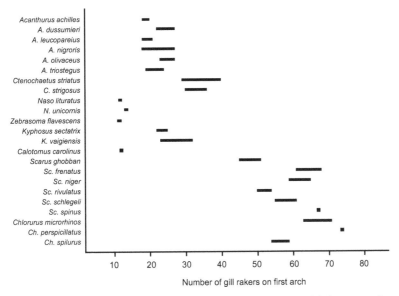

Fig. 4. Range of gill raker counts on first arch of herbivorous reef fish species from the Indo-Pacific. Sources: *Acanthurus* species, Randall (1966); *Ctenochaetus* species, Randall and Clements (2001); *Naso* species, Smith (1966); *Zebrasoma*, Randall (1955); *Kyphosus* species, Knudsen and Clements (2013); *Calotomus, Scarus* and *Chlorurus*, this study.

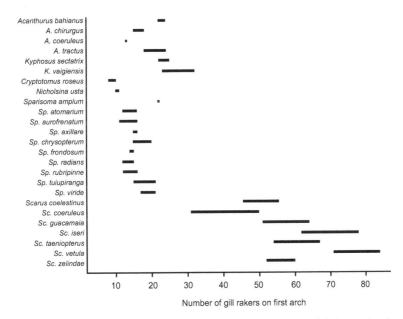

Fig. 5. Range of gill raker counts on first arch of herbivorous reef fish species from the Atlantic. Sources: *Acanthurus bahianus, A. chirurgus* and *A. coeruleus*, Randall (1956); *A. tractus*, Bernal and Rocha (2011); *Kyphosus* species, Knudsen and Clements (2013); *Sp. atomarium*, Gasparini et al. (2003); *Sp. aurofrenatum, Sp. chrysopterum, Sp. rubripinne, Sp. viride, Sc. coelestinus, Sc. coeruleus, Sc. guacamaia, Sc. iseri, Sc. taeniopterus* and *Sc. vetula*, Westneat (2002); *Cryptotomus roseus, Nicholsina usta, Sp. amplum, Sp. axillare, Sp. frondosum, Sp. radians, Sp. tuiupiranga* and *Sc. zelindae*, this study.

Counts of gill rakers on the first gill arch in scarinines are much higher than those in algal-feeding acanthurids and kyphosids (Figs 4-5), and even the detritivorous *Ctenochaetus* (Fig. 4). The gill rakers are very closely spaced in the scraping and excavating scarinine parrotfish genera *Scarus* and *Chlorurus* (Fig. 3), with interdigitation of anterior (outer) and posterior (inner) rakers (Al-Hussaini 1945, Gohar and Latif 1959). The interdigitation of gill rakers on the upper and lower limbs of the gill arches allows even the relatively widely-spaced gill rakers in sparisomatine parrotfishes to act as an effective filter for fine particles (Board 1956), as in tilapia (Northcutt and Beveridge 1988). Within *Sparisoma*, the excavating species (*Sp. amplum* and *Sp. viride*) have higher counts than scraping and browsing species (Fig. 5). The overall arrangement in *Sparisoma* and scarinines resembles that seen in suspension-feeding fishes such as gizzard shad (Smoot and Findlay 2010, Holley et al. 2015), detritivorous tilapia (Northcott and Beveridge 1988, Sanderson et al. 1996) and mullet (Guinea and Fernandez 1992). In these microphytophagous and detritivorous fishes, low-density organic particles such as bacteria, microalgae and organic detritus are separated from high-density inorganic particles such as sand using crossflow filtration involving the gill rakers and mucus entrapment (Sanderson et al. 1996, Smoot and Findlay 2010). This is similar to the system proposed for *Sp. cretense* by Board (1956). In short, the pharyngeal anatomy of *Sparisoma* and scarinines appears designed to retain fine organic particles such as microorganisms or detritus.

Gut Anatomy

Scraping and excavating species in the genera *Sparisoma*, *Chlorurus* and *Scarus* have a characteristic intestinal morphology with an anterior dorsal bulb, a smooth duodenal region, and ileal sacculations which form imperfect pouches that have been suggested to slow the passage of fine digesta (Al-Hussaini 1945, Board 1956, Gohar and Latif 1959, Clements and Raubenheimer 2006; Fig. 6). The sparisomatine genera *Cryptotomus*, *Nicholsina* and *Calotomus*, which are considered to be browsers (Bonaldo et al. 2014), lack these sacculae (Fig. 6). While sacculation of the intestine was considered a unique, unreversed synapomorphy of *Sparisoma*, *Cetoscarus*, *Bolbometopon*, *Chlorurus*, *Hipposcarus* and *Scarus* by Bellwood (1994), more recent phylogenetic hypotheses challenge this view (Wainwright and Price 2016). Intestinal sacculation in parrotfishes thus appears to be associated with genera that ingest inorganic calcareous sediment in the course of feeding.

The pharyngeal anatomy of parrotfishes described above indicates that *Sparisoma* and scarinines have the capacity to retain microscopic particles within the pharynx using mucus and gill rakers, and perhaps some selective filtration also occurs in the sacculated gut of these species. It is possible that the larger particles of organic and inorganic material might tend to pass through the central lumen, while microscopic organic material and fluid gets squeezed into the sacculae (see Gohar and Latif 1959), where it may be retained among the villi and microvilli for digestion.

Digestive Enzymes, Gastrointestinal Microbiota and Hindgut Fermentation

Very little work has been done on digestive enzymes in herbivorous coral reef fish taxa, and we are aware of only a single paper on parrotfishes. Papoutsoglou and Lyndon (2006) examined carbohydrases, proteases and lipases in *Sp. cretense*, and found that levels of carbohydrase activity were comparable to those seen in omnivorous and herbivorous freshwater fishes such as cichlids and carp. In contrast, proteolytic activity was surprisingly high, and was more comparable to that seen in carnivorous than herbivorous fishes (Papoutsoglou and Lyndon 2006).

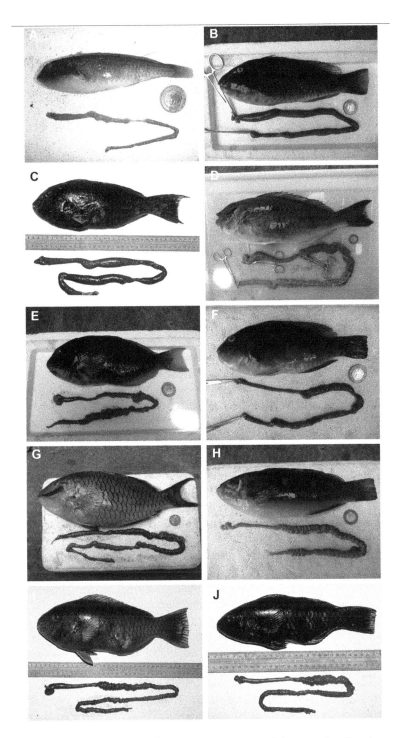

Fig. 6. Alimentary tract of representative parrotfishes species. Species: A. *Cryptotomus roseus*; B. *Nicholsina usta*; C. *Calotomus carolinus*; D. *Sp. axillare*; E. *Sp. frondosum*; F. *Sp. radians*; G. *Sp. amplum*; H. *Sc. zelindae*; I. *Sc. rivulatus*; J. *Sc. schlegeli*.

The taxonomic composition of hindgut microbial communities in parrotfishes resembles that in detritivorous and omnivorous acanthurids, rather than in acanthurids that feed on epilithic algae or macroalgae (Smriga et al. 2010, Miyake et al. 2015). Similarly, the low concentration and pattern of microbial fermentation products (SCFA) in the posterior intestine resembles that seen in detritivorous and planktivorous acanthurids (Clements and Choat 1995, Choat and Clements 1998, Crossman et al. 2005), not algivorous fish species. This is unsurprising, given that gut transit times are rapid in parrotfishes (Smith and Paulson 1974, Choat et al. 2004), and hindgut fermentation requires lengthy gut retention times (Van Soest 1994). Parrotfishes and detritivorous acanthurids are characterized by low levels of acetate and high levels of the branch-chain fatty acids isobutyrate and isovalerate (Clements and Choat 1995, Crossman et al. 2005). The branched-chain SCFA isobutyrate and isovalerate are produced by fermentation of the amino acids valine and leucine, respectively, and serve as markers of protein fermentation, which tends to be reduced when carbohydrate is prevalent as in herbivorous diets (Davila et al. 2013). The high proportion of branched-chain SCFA in parrotfishes is indicative of protein fermentation, rather than the carbohydrate fermentation characteristic of algivorous taxa (Clements and Choat 1995, Crossman et al. 2005). This is fully consistent with the high levels of protein and low levels of carbohydrate in parrotfishes mentioned previously.

Assimilation of Nutrients

The previous section showed that trophic anatomy, digestive physiology and gut microflora in parrotfishes are distinct from that seen in algivorous taxa such as acanthurids and kyphosids. Pharyngeal anatomy in parrotfishes resembles that of microphagous freshwater suspension feeders and mullet, while what little is known of digestive physiology suggests a diet rich in protein with the capacity to digest soluble carbohydrates. We now explore two lines of evidence that bear on nutrient uptake by parrotfishes: tissue fatty acid composition and stable isotope analysis.

Tissue Fatty Acid Analysis

Tissue fatty acid analysis is widely used to determine trophic relationships between consumer taxa and available food resources, and individual fatty acids can serve as biomarkers for particular dietary taxa (Kelly and Scheibling 2012). A study conducted in the Northwestern Hawaiian Islands showed that parrotfishes displayed a very distinct tissue fatty acid signature compared to macroalgae-feeding *Kyphosus*, macroalgae-feeding *Naso* and a range of EAM-feeding acanthurid species (Piché et al. 2010). The EAM-feeding acanthurid taxa examined included species that feed on the broad spectrum of EAM resources, i.e. epilithic algal fractions (e.g. *Acanthurus achilles* and *Zebrasoma flavescens*), mixtures of algae and diatoms (e.g. *A. nigroris* and *A. triostegus*), and detritus and microalgae (e.g. *Ctenochaetus strigosus* and *A. olivaceus*) (Jones 1968). The very distinctive tissue fatty acid pattern seen in the parrotfish species examined (*Ch. perspicillatus*, *Ch. spilurus* and *Sc. dubius*) indicates that these fishes are assimilating fatty acids distinct from those assimilated by both EAM-feeding acanthurids and macroalgal-feeding *Naso* and *Kyphosus*, as the similarities to other fish taxa show that the results are not driven by phylogenetic relationship.

Parrotfishes were most distinct from the kyphosid and acanthurid taxa examined in having high levels of the monounsaturated vaccenic acid, 18:1n-7 (Piché et al. 2010 supplement; Fig. 7), known to be a biomarker for bacteria in marine benthic food webs (Kelly and Scheibling 2012), and more specifically considered a tracer for cyanobacteria

(Yang et al. 2016). The three parrotfish species examined were distinct from detritivorous surgeonfishes such as *A. olivaceus* and *C. strigosus* in having lower levels of palmitoleic acid, 16:1n-7 (Piché et al. 2010 supplement; Fig. 8), a biomarker for diatoms (Kelly and Scheibling 2012). The scraper *Sc. dubius* was nonetheless richer in 16:1n-7 than the two

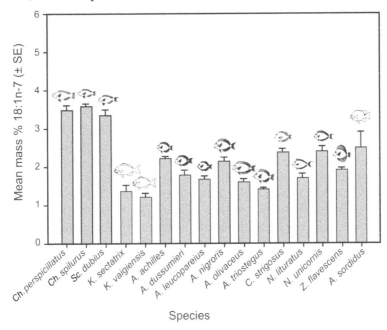

Fig. 7. Mean ± SE mass % of vaccenic acid 18:1n-7 in tissue from herbivorous and detritivorous fish species from the Northwestern Hawaiian Islands. Data from Piché et al. (2010) supplement.

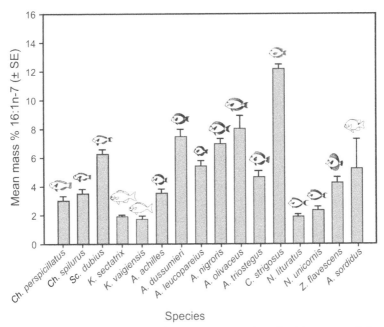

Fig. 8. Mean ± SE mass % of palmitoleic acid 16:1n-7 in tissue from herbivorous and detritivorous fish species from the Northwestern Hawaiian Islands. Data from Piché et al. (2010) supplement.

excavating *Chlorurus* species (Fig. 8), possibly indicating a greater proportion of epilithic diatom material in the former. Overall, this fatty acid analysis suggests that parrotfishes are using food resources distinct from those of EAM-feeding acanthurids, and thus if parrotfishes are ingesting EAM they are targeting different components to the acanthurids examined by Piché et al. (2010). More specifically, the results suggest that the parrotfishes are assimilating a higher proportion of bacterial or cyanobacterial fatty acids rather than fatty acids characteristic of EAM detritus or diatoms.

Stable Isotope Analysis

Although stable isotope analysis is now a standard tool in trophic studies, this type of data has not been examined in recent reviews of parrotfish ecology (Bonaldo et al. 2014, Adam et al. 2015b). Several studies using stable isotope analysis are now available that present data bearing on the nutritional resources used by parrotfishes (e.g. Nagelkerken et al. 2006, Carassou et al. 2008, Lamb et al. 2012, Page et al. 2013, Plass-Johnson et al. 2013, Dromard et al. 2015). Plass-Johnson et al. (2013) examines only parrotfishes (see below), but all of the others examine other herbivorous fish taxa including acanthurids and kyphosids.

What is most notable about the results of these studies is the consistent pattern whereby reef and seagrass-inhabiting parrotfishes are consistently depleted in $\delta^{15}N$ compared to other co-occurring herbivorous and detritivorous fish species such as acanthurids, siganids and kyphosids. Such a relatively depleted $\delta^{15}N$ signal indicates either (a) a trophic source rich in diazotrophs, i.e. nitrogen fixing bacteria such as cyanobacteria, or (b) the other fish species are assimilating a higher proportion of enriched $\delta^{15}N$. The latter explanation would be consistent with an omnivorous diet containing higher trophic levels, i.e. animals. This seems unlikely given that many of the acanthurid species examined ingest little animal material (see Carassou et al. 2008 and Page et al. 2013 in particular). A similarly depleted $\delta^{15}N$ signal to that seen in parrotfishes is also seen in herbivorous invertebrates that feed on nitrogen-fixing cyanobacteria in rocky shore (Pinnegar and Polunin 2000) and seagrass communities (Yamamuro 1999). Carassou et al. (2008) interpreted the distinctive isotopic signal of parrotfishes in their study as indicating a diet of microalgae, although they did not specifically mention diazotrophs.

Two of these stable isotope studies conclude that parrotfishes feed on a wide range of epilithic resources including macroalgae and EAM (e.g. Dromard et al. 2015, Plass-Johnson et al. 2013). These studies used statistical isotopic mixing models that incorporate priors based on dietary proportions either assumed from the literature or derived from their own gut content analysis. However, gut contents of parrotfishes typically include a large proportion of unidentifiable organic material that is usually interpreted as detritus (e.g. Choat et al. 2002, Ferreira and Gonçalves 2006), although some workers conclude dietary proportions based on the identifiable components of the ingesta alone (e.g. Randall 1967, Dromard et al. 2015). Such a practice will not reflect the proportional intake of food items overall, and is thus potentially misleading. Moreover, Dromard et al. (2015) and Plass-Johnson et al. (2013) interpreted the results of their stable isotope mixing models (a) without considering the level of statistical resolution required to discriminate among food sources with similar isotopic signals, and (b) assuming that their diet source samples included the total range of possible food resources used. Such a practice makes the results dependent on the initial assumptions concerning diet composition (Fry 2013).

Interestingly, Plass-Johnson et al. (2013) show that most scraping parrotfishes have more depleted $\delta^{15}N$ signals than excavating species, an exception being the scraper *Sc. tricolor*. The authors interpreted this difference as being the result of scraping and excavating species feeding on different proportions of macroalgal and EAM resources, a conclusion that we consider problematical given the analytical assumptions discussed above. We interpret this result as being consistent with excavating species utilising one or both of two euendolithic resources that are enriched in $\delta^{15}N$ relative to the predominantly epilithic resources used by scrapers. The first resource is substrate-bound, deeper-living euendolithic components which are not diazotrophic and thus less depleted in $\delta^{15}N$, e.g. the siphonaceous chlorophyte *Ostreobium* sp. (Grange et al. 2015). The second is macroboring invertebrates such as boring sponges or serpulid worms. In contrast, scrapers only access epilithic biota and the upper couple of mm of euendolithic communities (Bonaldo and Bellwood 2009), i.e. the components that are richer in cyanobacteria than deeper-boring *Ostreobium* sp. (Grange et al. 2015).

Overall, both tissue fatty acid analysis and stable isotope analysis are consistent with parrotfishes ingesting a diet rich in cyanobacteria, i.e. nitrogen-fixing microscopic phototrophs.

Conclusions

Our examination of a very broad literature bearing on parrotfish nutritional ecology is unequivocal, at least for *Sparisoma* and scarinine taxa for which data are available. These parrotfish taxa, which comprise the vast bulk of scarine diversity, have a specialised feeding mechanism that appears designed to harvest and selectively retain microscopic particles in the pharynx through a combination of (a) a branchial sieve formed by interdigitating gill rakers, and (b) mucus secretion from the pharyngeal valve that traps fine particles and transports them from the gill rakers to the pharyngeal mill. This structure occurs in scraping and excavating scarinines and *Sparisoma* (Board 1956), a genus generally thought to mainly browse macroalgae (Adam et al. 2015b) but which also includes scraping and excavating species (Bonaldo et al. 2014). All parrotfish for which macronutrient data are available appear to target and ingest a diet that contains a higher proportion of protein and lipid and a lower proportion of carbohydrate than epilithic turf algae or macroalgae. The community composition of hindgut microbiota and the pattern of microbial fermentation products in the hindgut are indicative of rapid processing of a largely protein-based diet. Tissue fatty acid analysis indicates a diet rich in lipid biomarkers for bacteria and/or cyanobacteria that is distinct from that used by EAM-feeding acanthurids and macroalgal browsing *Naso* and kyphosids (Piché et al. 2010). Finally, stable isotope analysis of browsing, scraping and excavating parrotfishes indicates a diet depleted in $\delta^{15}N$ relative to that used by EAM-feeding and macroalgal browsing herbivorous taxa, suggesting a high proportion of diazotrophs.

Collectively, these lines of evidence all point towards a dietary target of protein-rich microorganisms, i.e. largely cyanobacteria. These microscopic phototrophs are associated with all of the feeding substrata used by parrotfishes, i.e. as important constituents of both EAM (Cruz-Rivera and Paul 2006, Den Haan et al. 2014, Harris et al. 2015) and euendolithic communities (Tribollet 2008, Grange et al. 2015), dominant epiphytes of both macroalgae (Capone et al. 1977, Barott et al. 2011) and seagrass (Capone and Taylor 1977, Bologna and Heck 1999, Yamamuro 1999), and endosymbionts of both corals (Lesser et al. 2004, Gutiérrez-Isaza et al. 2015) and sponges (Gillan et al. 1988, Weisz et al. 2007). The hypothesis that parrotfishes are microphages that target protein-rich epilithic, endolithic and epiphytic microscopic phototrophs, i.e. predominantly cyanobacteria, thus provides a synthetic, unified explanation for all major parrotfish feeding substrata that is consistent with all of the available nutritional evidence (Clements et al. 2017). We are not suggesting that parrotfishes do not derive any nutrition from ingesting epilithic algae or macroalgae, but rather that their feeding and digestive mechanism is largely geared towards processing protein-rich microbial food resources.

Contradicting our hypothesis are numerous papers suggesting that herbivorous fishes, including parrotfishes, are deterred by toxic cyanobacterial secondary metabolites (e.g. Thacker et al. 1997, Nagle and Paul 1999). We question the generality of this statement with respect to parrotfish feeding, since (a) at least some of the work on parrotfish deterrence is equivocal (e.g. Thacker et al. 1997); (b) experiments always involve mat-forming or tufting cyanobacteria such as *Lyngbya*, and the palatability of diverse and abundant non-bloom forming epilithic and euendolithic cyanobacteria remains untested; (c) many fishes including cichlids (Reinthal 1990), pomacentrids (Lobel 1980, Sammarco 1983), acanthurids (Robertson and Gaines 1986, Montgomery et al. 1989) and siganids (Hoey et al. 2013) eat significant quantities of epilithic cyanobacteria; and (d) parrotfishes have been shown to accumulate cyanobacterial toxins (Laurent et al. 2008). The idea that parrotfishes evolved the capacity to harvest substrate-bound microorganisms in itself partly answers this question, as few herbivores can access these organisms in hard substrata such as calcareous rock and live coral (Tribollet and Golubic 2005), and as such these are likely to be resources that are not well defended chemically.

The finding that parrotfishes are microphages targeting protein-rich microorganisms such as cyanobacteria also explains the accelerated rates of evolution in scarinine clades with an intramandibular joint (Price et al. 2010). This scarinine diversity is largely an Indo-Pacific phenomenon, and the differences with the Atlantic are reflected in the lower diversity of consolidated reef habitats and geomorphologies there (Choat 2012). In this sense, the ability to harvest substrate-bound microorganisms more efficiently was probably associated with the evolution of the intramandibular joint, although the two excavating sparisomatine sister species *Sp. amplum* and *Sp. viride* appear to be successful without this innovation in the absence of competition from scarinine excavators. The diverse sparisomatine fauna of the Atlantic uses a wide variety of feeding substrata including EAM, macroalgae, seagrasses, sponges and live coral (Adam et al. 2015b). Although these fishes are thought of as mainly browsers of macroalgae and seagrass (e.g. Bonaldo et al. 2014, Adam et al. 2015b), we see this feeding as similar to the 'peanut butter and nutritionally unsuitable crackers' analogy used to explain the way aquatic insects harvest microbial biomass from leaf litter in streams (Cummins 1974, Smoot and Findlay 2010). The crackers in this sense are seagrasses and predominantly phaeophyte macroalgae that in most cases contain carbohydrates that will be largely refractory to the digestive system of parrotfishes (a possible exception is *Dictyota*, see above). The peanut butter is the protein-rich epiphytes including cyanobacteria. A similar sort of feeding is seen in wood-eating catfishes, which harvest and assimilate biofilm from submerged wood, but are unable to use the cellulose in the wood (German and Miles 2010).

Given the availability of the evidence in the relevant literature, much of which is elderly, and the fact that a dietary target of protein-rich microorganisms explains all the feeding substrata of parrotfishes, why has this hypothesis not been suggested previously? We suggest two reasons. First, the use of terms such as EAM and algal turf for what are very heterogeneous resources on coral reefs has impeded the search for the sort of fine-scale resource partitioning seen in some freshwater fish communities, e.g. Malawi cichlids (Reinthal 1990). As soon as resources are defined in nutritional terms it rapidly becomes clear what is a potential food for consumers with the ability to harvest, digest and assimilate it. We therefore endorse calls for heterogeneous resources such as EAM and algal turfs to be replaced by more realistic biological classifications of trophic resources on reefs (Connell et al. 2014, Harris et al. 2015). We note that our microphage hypothesis is entirely consistent with very recent data demonstrating that populations of both scraping and excavating scarinine species in the Philippines responded positively to increases in cover of dead calcareous substrata and declines in live coral cover (Russ et al. 2015). Both of these changes are associated with environmental disturbances such as cyclones and bleaching events that create habitat for pioneer phototrophs, i.e. epilithic and euendolithic microorganisms.

Second, studies on feeding in herbivorous reef fishes must involve testing hypotheses based around these nutritional resources, not ecological roles. An example here is the failure by many studies to examine bioerosion as a nutritional process (cf. Bruggemann et al. 1994a-c). For example, Bonaldo et al. (2014) discuss bioerosion in parrotfishes at length, yet their comprehensive review contains neither the word 'endolith' nor 'cyanobacteria.' The purpose of feeding, i.e. nutrition, had been lost. We argue that the focus on resilience and the confusion between 'ecological roles' and resource partitioning has led to a situation where functional groups are more-or-less defined in terms of how or where parrotfishes feed, not what they actually ingest. The ultimate expression of this viewpoint in coral reef fishes is the definition of herbivory in terms of algal removal, even where algae is not necessarily the target of ingestion or assimilation (e.g. Hoey and Bellwood 2008, Smith 2008, Marshell and Mumby 2012). This "cyclones as herbivores" viewpoint (Clements et al. 2009) subverts the Hutchinsonian concept of resource partitioning and rapidly becomes problematical in evolutionary studies. Trophic diversification through resource partitioning is driven by natural selection on individual fitness (Schluter 2000), and thus a focus on the role of species in reef health or resilience, e.g. complementarity or redundancy, is uninformative. A focus on nutrition clarifies the distinction between ecological role and resource partitioning, and feeding mechanism can then be placed in a functional context of nutritional acquisition from a particular set of dietary resources. Parrotfishes are a very diverse group, and the future study of their use of nutritional resources in reef environments and how this has diversified since the Pliocene (Choat et al. 2012) promises to be hugely rewarding.

Acknowledgments

We are grateful to the editors of this volume for the opportunity to contribute this chapter. We thank Roberta Bonaldo, Thiago Costa Mendes, Cadu Ferreira, Donovan German, Philip Harris, Michel Kulbicki, Jacinthe Piché, Tony Roberton, Ross Robertson, Garry Russ, Mary Sewell, Sue Taei, Brett Taylor, Aline Tribollet and Lindsey White for helpful comments and other input on this work. We thank Michael Berumen, Paul Caiger, Linda Eggertsen, Beatrice Ferreira, Gabriel Ferreira, Rossana Freitas, Renato Morais and Tane Sinclair-Taylor for help obtaining material and examining specimens, and Viv Ward for preparing figures.

References Cited

Adam, T.C., M. Kelley, B.I. Ruttenberg and D.E. Burkepile. 2015a. Resource partitioning along multiple niche axes drives functional diversity in parrotfishes on Caribbean coral reefs. Oecologia 179: 1173–1185.

Adam, T.C., D.E. Burkepile, B.I. Ruttenberg and M.J. Paddack. 2015b. Herbivory and the resilience of Caribbean coral reefs: knowledge gaps and implications for management. Mar. Ecol. Prog. Ser. 520: 1–20.

Adey, W.H. 1998. Coral reefs: algal structured and mediated ecosystems in shallow, turbulent, alkaline waters. J. Phycol. 34: 393–406.

Al-Hussaini, A.H. 1945. The anatomy and histology of the alimentary tract of the coral feeding fish *Scarus sordidus* (Klunz.). Bull. Inst. Egypte 27: 349–377.

Angell, A.R., L. Mata, R. de Nys and N.A. Paul. 2015. The protein content of seaweeds: a universal nitrogen-to-protein conversion factor of five. J. Appl. Phycol. 28: 511–524.

Barbarino, E. and S.O. Lourenço. 2005. An evaluation of methods for extraction and quantification of protein from marine macro- and microalgae. J. Appl. Phycol. 17: 447–460.

Barott, K.L., B. Rodriguez-Brito, J. Janouškovec, K.L. Marhaver, J.E. Smith, P. Keeling and F.L. Rohwer. 2011. Microbial diversity associated with four functional groups of benthic reef algae and the reef-building coral *Montastraea annularis*. Environ. Microbiol. 13: 1192–1204.

Bellwood, D.R. 1994. A phylogenetic study of the parrotfishes Family Scaridae (Pisces: Labroidei), with a revision of genera. Rec. Aust. Mus. Suppl. 20: 86.

Bellwood, D.R. and J.H. Choat. 1990. A functional analysis of grazing in parrotfishes (Family Scaridae): the ecological implications. Environ. Biol. Fishes 28: 189–214.

Bernal, M.A. and L.A. Rocha. 2011. *Acanthurus tractus* Poey, 1860, a valid western Atlantic species of surgeonfish (Teleostei, Acanthuridae), distinct from *Acanthurus bahianus* Castelnau, 1855. Zootaxa 2905: 63–68.

Board, P.A. 1956. The feeding mechanism of the fish *Sparisoma cretense* (Linné). Proc. Zool. Soc. Lond. 127: 59–77.

Bologna, P.A.X. and K.L. Heck Jr. 1999. Macrofaunal associations with seagrass epiphytes. Relative importance of trophic and structural characteristics. J. Exp. Mar. Biol. Ecol. 242: 21–39.

Bonaldo, R.M. and D.R. Bellwood. 2009. Dynamics of parrotfish grazing scars. Mar. Biol. 156: 771–777.

Bonaldo, R.M., A.S. Hoey and D.R. Bellwood. 2014. The ecosystem roles of parrotfishes on tropical reefs. Oceanogr. Mar. Biol. Annu. Rev. 52: 81–132.

Bonaldo, R.M., J.P. Krajewski, C. Sazima and I. Sazima. 2007. Dentition damage in parrotfishes feeding on hard surfaces at Fernando de Noronha Archipelago, southwest Atlantic Ocean. Mar. Ecol. Prog. Ser. 342: 249–254.

Bonaldo, R.M., J.Q. Welsh and D.R. Bellwood. 2012. Spatial and temporal variation in coral predation by parrotfishes on the GBR: evidence from an inshore reef. Coral Reefs 31: 263–272.

Bowen, S.H., E.V. Lutz and M.O. Ahlgren. 1995. Dietary protein and energy as determinants of food quality: trophic strategies compared. Ecology 76: 899–907.

Brandl, S.J. and D.R. Bellwood. 2014. Individual-based analyses reveal limited functional overlap in a coral reef fish community. J. Anim. Ecol. 83: 661–670.

Brown, J.H. 1981. Two decades of homage to Santa Rosalia: towards a general theory of diversity. Amer. Zool. 21: 877–888.

Brown, M.R. 1991. The amino-acid and sugar composition of 16 species of microalgae used in mariculture. J. Exp. Mar. Biol. Ecol. 145: 79–99.

Bruggemann, J.H., M.J.H. van Oppen and A.M. Breeman. 1994a. Foraging by the stoplight parrotfish *Sparisoma viride*. I. Food selection in different, socially determined habitats. Mar. Ecol. Prog. Ser. 106: 41–55.

Bruggemann, J.H., J. Begeman, E.M. Bosma, P. Verburg and A.M. Breeman. 1994b. Foraging by the stoplight parrotfish *Sparisoma viride*. II. Intake and assimilation of food, protein and energy. Mar. Ecol. Prog. Ser. 106: 57–71.

Bruggemann, J.H., M.W.M. Kuyper and A.M. Breeman. 1994c. Comparative analysis of foraging and habitat use by the sympatric Caribbean parrotfish *Scarus vetula* and *Sparisoma viride* (Scaridae). Mar. Ecol. Prog. Ser. 112: 51–66.

Bruggemann, J.H., A.M. van Kessel, J.M. van Rooij and A.M. Breeman. 1996. Bioerosion and sediment ingestion by the Caribbean parrotfish *Scarus vetula* and *Sparisoma viride*: implications of fish size, feeding mode and habitat use. Mar. Ecol. Prog. Ser. 134: 59–71.

Burkepile, D.E. and M.E. Hay. 2008. Herbivore species richness and feeding complementarity affect community structure and function on a coral reef. Proc. Natl. Acad. Sci. USA 105: 16201–16206.

Burkepile, D.E. and M.E. Hay. 2011. Feeding complementarity versus redundancy among herbivorous fishes on a Caribbean reef. Coral Reefs 30: 351–362.

Capone, D.G. and B.F. Taylor. 1977. Nitrogen fixation (acetylene reduction) in the phyllosphere of *Thalassia testudinum*. Mar. Biol. 40: 19–28.

Capone, D.G., D.L. Taylor and B.F. Taylor. 1977. Nitrogen fixation (acetylene reduction) associated with macroalgae in a coral-reef community in the Bahamas. Mar. Biol. 40: 29–32.

Carr, A., I.R. Tibbetts, A. Kemp, R. Truss and J. Drennan. 2006. Inferring parrotfish (Teleostei: Scaridae) pharyngeal mill function from dental morphology, wear, and microstructure. J. Morph. 267: 1147–1156.

Carassou, L., M. Kulbicki, T.J.R. Nicola and N.V.C. Polunin. 2008. Assessment of fish trophic status and relationships by stable isotope data in the coral reef lagoon of New Caledonia, southwest Pacific. Aquat. Living Resour. 21: 1–12.

Carreiro-Silva, M., T.R. McClanahan and W.E. Kiene. 2009. Effects of inorganic nutrients and organic matter on microbial euendolithic community composition and microbioerosion rates. Mar. Ecol. Prog. Ser. 392: 1–15.

Casareto, B.E., L. Charpy, M.J. Langlade, T. Suzuki, H. Ohba, M. Niraula and Y. Suzuki. 2008. Nitrogen fixation in coral reef environments. Proc. 11th Int. Coral Reef Symp., Session No. 19: 890–894.

Cernohorsky, N.H., T.R. McClanahan, I. Babu and M. Horsák. 2015. Small herbivores suppress algal accumulation on Agatti atoll, Indian Ocean. Coral Reefs 34: 1023–1035.

Chanas, B. and J.R. Pawlik. 1995. Defenses of Caribbean sponges against predatory reef fish. II. Spicules, tissue toughness, and nutritional quality. Mar. Ecol. Prog. Ser. 127: 195–211.

Chazottes, V., T. Le Campion-Alsumard and M. Peyrot-Clausade. 1995. Bioerosion rates on coral reefs: interactions between macroborers, microborers and grazers (Moorea, French Polynesia). Palaeogeogr. Palaeocl. 113: 189–198.

Cheal, A.J., M. Emslie, M.A. MacNeil, I. Miller and H. Sweatman. 2013. Spatial variation in the functional characteristics of herbivorous fish communities and the resilience of coral reefs. Ecol. Appl. 23: 174–188.

Choat, J.H. 2012. Spawning aggregations in ref fishes: ecological and evolutionary processes. pp. 85–116. *In*: Y. Sadovy de Mitcheson and P.L. Colin (eds.). Reef Fish Spawning Aggregations: Biology, Research and Management. Fish & Fisheries Series Vol. 35. **??**

Choat, J.H. and K.D. Clements. 1998. Vertebrate herbivores in marine and terrestrial environments: a nutritional ecology perspective. Annu. Rev. Ecol. Syst. 29: 375–403.

Choat, J.H., K.D. Clements and W.D. Robbins. 2002. The trophic status of herbivorous fishes on coral reefs. I: Dietary analyses. Mar. Biol. 140: 613–623.

Choat, J.H., O.S. Klanten, L. van Herwerden, D.R. Robertson and K.D. Clements. 2012. Patterns and processes in the evolutionary history of parrotfishes (Family Labridae). Biol. J. Linn. Soc. 107: 529–557.

Choat, J.H., W.D. Robbins and K.D. Clements. 2004. The trophic status of herbivorous fishes on coral reefs. II: Food processing modes and trophodynamics. Mar. Biol. 145: 445–454.

Choat, J.H. and D.R. Robertson. 2002. Age-based studies. pp. 57–80. *In*: P.F. Sale (ed.). Coral Reef Fishes: Dynaomics and Diversity in a Complex Ecosystem. Academic Press, San Diego.

Clements, K.D. and D.R. Bellwood. 1988. A comparison of the feeding mechanisms of two herbivorous labroid fishes, the temperate *Odax pullus* and the tropical *Scarus rubroviolaceus*. Aust. J. Mar. Freshwat. Res. 39: 87–107.

Clements, K.D. and J.H. Choat. 1995. Fermentation in tropical marine herbivorous fishes. Physiol. Zool. 68: 355–378.

Clements, K.D., D.P. German, J. Piché, A. Tribollet and J.H. Choat. 2017. Integrating ecological roles and trophic diversification on coral reefs: multiple lines of evidence identify parrotfishes as microphages. Biol. J. Linn. Soc. 120: 729–751.

Clements, K.D. and D. Raubenheimer. 2006. Feeding and nutrition. pp. 47–82. *In*: D.H. Evans and J.B. Claiborne (eds.). The Physiology of Fishes, 3rd ed. CRC Press, Gainesville.

Clements, K.D., D. Raubenheimer and J.H. Choat. 2009. Nutritional ecology of marine herbivorous fishes: ten years on. Func. Ecol. 23: 79–92.

Connell, S.D., M.S. Foster and L. Airoldi. 2014. What are algal turfs? Towards a better description of turfs. Mar. Ecol. Prog. Ser. 495: 299–307.

Crossman, D.J., J.H. Choat, K.D. Clements, T. Hardy and J. McConochie. 2001. Detritus as food for grazing fishes on coral reefs. Limnol. Oceanogr. 46: 1596–1605.

Crossman, D.J., J.H. Choat and K.D. Clements. 2005. Nutritional ecology of nominally herbivorous fishes on coral reefs. Mar. Ecol. Prog. Ser. 296: 129–142.

Cruz-Rivera, E. and V.J. Paul. 2006. Feeding by coral reef mesograzers: algae or cyanobacteria. Coral Reefs 25: 617–627.

Cummins, K.W. 1974. Structure and function of stream ecosystems. BioScience 24: 631–641.

Davila, A.-M., F. Blachier, M. Gotteland, M. Andriamihaja, P.-H. Benetti, Y. Sanz and D. Tomé. 2013. Intestinal luminal nitrogen metabolism: role of the gut microbiota and consequences for the host. Pharmacol. Res. 68: 95–107.

Den Haan, J., P.M. Visser, A.E. Ganase, E.E. Gooren, L.J. Stal, F.C. van Duyl, M.J.A. Vermeij and J. Huisman. 2014. Nitrogen fixation rates in algal turf communities of a degraded versus less degraded coral reef. Coral Reefs 33: 1003–1015.

Dromard, C.R., Y. Bouchon-Navaro, M. Harmelin-Vivien and C. Bouchon. 2015. Diversity of trophic niches among herbivorous fishes on a Caribbean reef (Guadeloupe, Lesser Antilles), evidenced by stable isotope and gut content analyses. J. Sea Res. 95: 124–131.

Dunlap, M. and J.R. Pawlik. 1998. Spongivory by parrotfish in Florida mangrove and reef habitats. Mar. Ecol. 19: 325–337.

Edmunds, P.J. and P.S. Davies. 1986. An energy budget for *Porites porites* (Scleractinia). Mar. Biol. 92: 339–347.

Ferreira, C.E.L. and J.E.A. Gonçalves. 2006. Community structure and diet of roving herbivorous reef fishes in the Abrolhos Archipelago, south-western Atlantic. J. Fish Biol. 69: 1–19.

Foley, W.J. and S.J. Cork. 1992. Use of fibrous diets by small herbivores: How far can the rules be bent? Trends Ecol. Evol. 7: 159–162.

Fricke, A., T.V. Titlyanova, M.M. Nugues and K. Bischof. 2011. Depth-related variation in epiphytic communities growing on the brown alga *Lobophora variegata* in a Caribbean coral reef. Coral Reefs 30: 967–973.

Fry, B. 2013. Alternative approaches for solving underdetermined isotope mixing problems. Mar. Ecol. Prog. Ser. 472: 1–13.

Gasparini, J.L., J.-C. Joyeux and S.R. Floeter. 2003. *Sparisoma tuiupiranga*, a new species of parrotfish (Perciformes: Labroidei: Scaridae) from Brazil, with comments on the evolution of the genus. Zootaxa 384: 1–14.

German, D.P. and R.D. Miles. 2010. Stable carbon and nitrogen incorporation in blood and fin tissue of the catfish *Pterygoplichthys disjunctivus* (Siluriformes, Loricariidae). Environ. Biol. Fish. 89: 117–133.

Gillan, F.T., I.L. Stoilov, J.E. Thompson, R.W. Hogg, C.R. Wilkinson and C. Djerassi. 1988. Fatty acids as biological markers for bacterial symbionts in sponges. Lipids 23: 1139–1145.

Gobalet, K.W. 1989. Morphology of the parrotfish pharyngeal jaw apparatus. Amer. Zool. 29: 319–331.

Goecker, M.E., K.L. Heck Jr. and J.F. Valentine. 2005. Effects of nitrogen concentrations in turtlegrass *Thalassia testudinum* on consumption by the bucktooth parrotfish *Sparisoma radians*. Mar. Ecol. Prog. Ser. 286: 239–248.

Gohar, H.A.F. and A.F.A. Latif. 1959. Morphological studies on the gut of some scarid and labrid fishes. Publ. Mar. Biol. Stn. Al-Ghardaqa (Red Sea) 10: 145–189.

Goldberg, W.M. 2013. The biology of reefs and reef organisms. University of Chicago Press, Chicago.

Grange, J.S., H. Rybarczyk and A. Tribollet. 2015. The three steps of the carbonate biogenic dissolution process by microborers in coral reefs. Environ. Sci. Pollut. Res. 22: 13625–13637.

Guinea, J. and F. Fernandez. 1992. Morphological and biometrical study of the gill rakers in four species of mullet. J. Fish Biol. 41: 381–397.

Gutiérrez-Isaza, N., J. Espinoza-Avalos, H.P. León-Tejera and D. González-Solís. 2015. Endolithic community composition of *Orbicella faveolata* (Scleractinia) underneath the interface between coral tissue and turf algae. Coral Reefs 34: 625–630.

Harland, A.D., P.S. Davies and L.M. Fixter. 1992. Lipid content of some Caribbean corals in relation to depth and light. Mar. Biol. 113: 357–361.

Harmelin-Vivien, M.L., M. Peyrot-Clausade and J.-C. Romano. 1992. Transformation of algal turf by echinoids and scarid fishes on French Polynesian coral reefs. Coral Reefs 11: 45–50.

Harris, J.L., L.S. Lewis and J.E. Smith. 2015. Quantifying scales of spatial variability in algal turf assemblages on coral reefs. Mar. Ecol. Prog. Ser. 532: 41–57.

Hensley, N.M., O.L. Elmasri, E.I. Slaughter, S. Kappus and P. Fong. 2013. Two species of *Halimeda*, a calcifying genus of tropical macroalgae, are robust to epiphytism by cyanobacteria. Aquat. Ecol. 47: 433–440.

Hoey, A.S. and D.R. Bellwood. 2008. Cross-shelf variation in the role of parrotfishes on the Great Barrier Reef. Coral Reefs 27: 37–47.

Hoey, A.S., S.J. Brandl and D.R. Bellwood. 2013. Diet and cross-shelf distribution of rabbitfishes (f. Siganidae) on the northern Great Barrier Reef: implications for ecosystem management. Coral Reefs 32: 973–984.

Holley, L.L., M.K. Heidman, R.M. Chambers and S.L. Sanderson. 2015. Mucous contribution to gut nutrient content in American gizzard shad *Dorosoma cepedianum*. J. Fish Biol. 86: 1457–1470.

Holzer, K.K., D.A. Sekell and K.J. McGlathery. 2013. Bucktooth parrotfish *Sparisoma radians* grazing on *Thalassia* in Bermuda varies seasonally and with background nitrogen content. J. Exp. Mar. Biol. Ecol. 443: 27–32.

Hume, I.D. 1999. Marsupial nutrition. Cambridge University Press, Cambridge.

Hutchinson, G.E. 1959. Homage to Santa Rosalia or Why are there so many kinds of animals? Amer. Nat. 93: 145–159.

Jones, R.S. 1968. Ecological relationships in Hawaiian and Johnston Island Acanthuridae (Surgeonfishes). Micronesica 4: 309–361.

Jung, K.A., S.-R. Lim, Y. Kim and J.M. Park. 2013. Potentials of macroalgae as feedstocks for biorefinery. Biores. Technol. 135: 182–190.

Kelly, J.R. and R.E. Scheibling. 2012. Fatty acids as dietary tracers in benthic food webs. Mar. Ecol. Prog. Ser. 446: 1–22.

Kirsch, K.D., J.F. Valentine and K.L. Heck Jr. 2002. Parrotfish grazing on turtlegrass *Thalassia testudinum*: evidence for the importance of seagrass consumption in food web dynamics of the Florida Keys National Marine Sanctuary. Mar. Ecol. Prog. Ser. 227: 71–85.

Knudsen, S.W. and K.D. Clements. 2013. Revision of the fish family Kyphosidae (Teleostei: Perciformes). Zootaxa 3751: 1–101.

Lamb, K., P.K. Swart and M.A. Altabet. 2012. Nitrogen and carbon isotopic systematics of the Florida reef tract. Bull. Mar. Sci. 88: 119–146.

Laurent, D., A.-S. Kerbrat, H.T. Darius, E. Girard, S. Golubic, E. Benoit, M.-P. Sauviat, M. Chinain, J. Molgo and S. Pauillac. 2008. Are cyanobacteria involved in Ciguatera fish poisoning-like outbreaks in New Caledonia? Harmful Algae 7: 827–838.

Le Campion-Alsumard, T., S. Golubic and P. Hutchings. 1995. Microbial endoliths in skeletons of live and dead corals: *Porites lobata* (Moorea, French Polynesia). Mar. Ecol. Prog. Ser. 117: 149–157.

Lesser, M.P., C.H. Mazel, M.Y. Gorbunov and P.G. Falkowski. 2004. Discovery of symbiotic nitrogen-fixing cyanobacteria in corals. Science 305: 997–1000.

Lobel, P.S. 1980. Herbivory by damselfishes and their role in coral reef community ecology. Bull. Mar. Sci. 30: 273–289.

Lobel, P.S. and J.C. Ogden. 1981. Foraging by the herbivorous parrotfish *Sparisoma radians*. Mar. Biol. 64: 173–183.

López, C.V.G., M.D.C.C. García, F.G.A. Fernández, C.S. Bustos, Y. Chisti and J.M.F. Sevilla. 2010. Protein measurements of microalgal and cyanobacterial biomass. Biores. Technol. 101: 7587–7591.

Lourenço, S.O., E. Barbarino, J.C. De-Paula, L. Otávio da S. Pereira and U.M. Lanfer Marquez. 2002. Amino acid composition, protein content and calculation of nitrogen-to-protein conversion factors for 19 tropical seaweeds. Phycol. Res. 50: 233–241.

Marshell, A. and P.J. Mumby. 2012. Revisiting the functional roles of the surgeonfish *Acanthurus nigrofuscus* and *Ctenochaetus striatus*. Coral Reefs 31: 1093–1101.

McAfee, S.T. and S.G. Morgan. 1996. Resource use by five sympatric parrotfishes in the San Blas Archipelago, Panama. Mar. Biol. 125: 427–437.

McDermid, K.J. and B. Stuercke. 2003. Nutritional composition of edible Hawaiian seaweeds. J. Appl. Phycol. 15: 513–524.

McDermid, K.J., B. Stuercke and G.H. Balazas. 2007. Nutritional composition of marine plants in the diet of the green sea turtle (*Chelonia mydas*) in the Hawaiian Islands. Bull. Mar. Sci. 81: 55–71.

Miyake, S., D. Kamanda Ngugi and U. Stingl. 2015. Diet strongly influences the gut microbiota of surgeonfishes. Mol. Ecol. 24: 656–672.

Montgomery, W.L., A.A. Myrberg Jr. and L. Fishelson. 1989. Feeding ecology of surgeonfishes (Acanthuridae) in the northern Red Sea, with particular reference to *Acanthurus nigrofuscus* (Forsskål). J. Exp. Mar. Biol. Ecol. 132: 179–207.

Nagarkar, S., G.A. Williams, G. Subramanian and S.K. Saha. 2004. Cyanobacteria-dominated biofilms: a high quality food resource for intertidal grazers. Hydrobiologia 512: 89–95.

Nagelkerken, I., G. van der Velde, W.C.E.P. Verberk and M. Dorenbosch. 2006. Segregation along multiple resource axes in a tropical seagrass fish community. Mar. Ecol. Prog. Ser. 308: 79–89.

Nagle, D.G. and V.J. Paul. 1999. Production of secondary metabolites by filamentous tropical marine cyanobacteria: ecological functions of the compounds. J. Phycol. 35: 1412–1421.

Nash, K.L., N.A.J. Graham, S. Jennings, S.K. Wilson and D.R. Bellwood. 2015. Herbivore cross-scale redundancy supports response diversity and promotes coral reef resilience. J. Appl. Ecol. 53: 646–655.

Northcott, M.E. and M.C.M. Beveridge. 1988. The development and structure of pharyngeal apparatus associated with filter feeding in tilapias (*Oreochromis niloticus*). J. Zool., Lond. 215: 133–149.

Page, H.M., A.J. Brooks, M. Kulbicki, R. Galzin, R.J. Miller, D.C. Reed, R.J. Schmitt, S.J. Holbrook and C. Koenigs. 2013. Stable isotopes reveal trophic relationships and diet of consumers in temperate kelp forest and coral reef ecosystems. Oceanogr. 26: 180–189.

Papoutsoglou, E.S. and A.R. Lyndon. 2006. Digestive enzymes along the alimentary tract of the parrotfish *Sparisoma cretense*. J. Fish Biol. 69: 130–140.

Penhale, P.A. and D.G. Capone. 1981. Primary productivity and nitrogen fixation in two macroalgae-cyanobacteria associations. Bull. Mar. Sci. 3: 164–169.

Piché, J., S.J. Iverson, F.A. Parrish and R. Dollar. 2010. Characterization of forage fish and invertebrates in the Northwestern Hawaiian Islands using fatty acid signatures: species and ecological groups. Mar. Ecol. Prog. Ser. 418: 1–15.

Pinnegar, J.K. and N.V.C. Polunin. 2000. Contributions of stable-isotope data to elucidating food webs of Mediterranean rocky littoral fishes. Oecologia 122: 399–409.

Plass-Johnson, J.G., C.D. McQuaid and J.M. Hill. 2013. Stable isotope analysis indicates a lack of inter- and intra-specific dietary redundancy among ecologically important coral reef fishes. Coral Reefs 32: 429–440.

Price, S.A., P.C. Wainwright, D.R. Bellwood, E. Kazancioglu, D.C. Collar and T.J. Near. 2010. Functional innovations and morphological diversification in parrotfish. Evolution 64: 3057–3068.

Pushparaj, B., A. Buccioni, R. Paperi, R. Piccardi, A. Ena, P. Carlozzi and C. Sili. 2008. Fatty acid composition of Antarctic cyanobacteria. Phycologia 47: 430–434.

Randall, J.E. 1955. A revision of the surgeon fish genera *Zebrasoma* and *Paracanthurus*. Pacific Sci. 9: 396–412.

Randall, J.E. 1956. A revision of the surgeon fish genus *Acanthurus*. Pacific Sci. 10: 159–235.

Randall, J.E. 1967. Food habits of reef fishes of the West Indies. Stud. Trop. Oceanogr. 5: 665–847.

Randall, J.E. and K.D. Clements. 2001. Second revision of the surgeonfish genus *Ctenochaetus* (Perciformes: Acanthuridae), with descriptions of two new species. Indo-Pacific Fishes 32: 33.

Reinthal, P.N. 1990. The feeding habits of a group of herbivorous rock-dwelling cichlid fishes (Cichlidae: Perciformes) from Lake Malawi, Africa. Envir. Biol. Fish. 27: 215–233.

Robertson, D.R. and S.D. Gaines. 1986. Interference competition structures habitat use in a local assemblage of coral reef surgeonfishes. Ecology 67: 1372–1383.

Rotjan, R.D. and J.L. Dimond. 2010. Discriminating causes from consequences of persistent parrotfish corallivory. J. Exp. Mar. Biol. Ecol. 390: 188–195.

Russ, G.R., S.-L.A. Questel, J.R. Rizzari and A.C. Alcala. 2015. The parrotfish-coral relationship: refuting the ubiquity of a prevailing paradigm. Mar. Biol. 162: 2029–2045.

Sammarco, P.W. 1983. Effects of fish grazing and demselfish territoriality on coral reef algae. I. Algal community structure. Mar. Ecol. Prog. Ser. 13: 1–14.

Sanderson, S.L., M.C. Stebar, K.L. Ackermann, S.H. Jones, I.E. Batjakas and L. Kaufman. 1996. Mucus entrapment of particles by a suspension-feeding tilapia (Pisces: Cichlidae). J. Exp. Biol. 199: 1743–1756.

Schluter, D. 2000. The ecology of adaptive radiation. Oxford University Press, Oxford.

Skov, M.W., M. Volkelt-Igoe, S.J. Hawkins, B. Jesus, R.C. Thompson and C.P. Doncaster. 2010. Past and present grazing boosts the photo-autotrophic biomass of biofilms. Mar. Ecol. Prog. Ser. 401: 101–111.

Smith, J.L.B. 1966. Fishes of the sub-family Nasinae with a synopsis of the Prionurinae. Ichthyol. Bull. 32: 635–681.

Smith, R.L. and A.C. Paulson. 1974. Food transit times and gut pH in two Pacific parrotfishes. Copeia 1974: 796–799.

Smith, T.B. 2008. Temperature effects on herbivory for an Indo-Pacific parrotfish in Panamá: implications for coral-algal competition. Coral Reefs 27: 397–405.

Smoot, J.C. and R.H. Findlay. 2010. Microbes as food for sediment-ingesting detritivores: low-density particles confer a nutritional advantage. Aquat. Microb. Ecol. 59: 103–109.

Smriga, S., S.A. Sandin and F. Azam. 2010. Abundance, diversity, and activity of microbial assemblages associated with coral reef fish guts and feces. FEMS Microbiol. Ecol. 73: 31–42.

Targett, T.E. and N.M. Targett. 1990. Energetics of food selection by the herbivorous parrotfish *Sparisoma radians*: roles of assimilation efficiency, gut evacuation rate, and algal secondary metabolites. Mar. Ecol. Prog. Ser. 66: 13–21.

Taylor, B.M. and J.H. Choat. 2014. Comparative demography of commercially important parrotfish species from Micronesia. J. Fish Biol. 84: 383–402.

Taylor, B.W., A.S. Flecker and R.O. Hall Jr. 2006. Loss of a harvested fish species disrupts carbon flow in a diverse tropical river. Science 313: 833–836.

Thacker, R.W., D.G. Nagle and V.J. Paul. 1997. Effects of repeated exposures to marine cyanobacterial secondary metabolites on feeding by juvenile rabbitfish and parrotfish. Mar. Ecol. Prog. Ser. 147: 21–29.

Titlyanov, E.A., S.I. Kiyashko, T.V. Titlyanova, T.L. Kalita and J.A. Raven. 2008. $\delta^{13}C$ and $\delta^{15}N$ values in reef corals *Porites lutea* and *P. cylindrica* and in their epilithic and endolithic algae. Mar. Biol. 155: 353–361.

Tribollet, A. 2008. The boring micro flora in modern coral reef ecosystems: a review of its roles. pp. 67–94. *In*: M. Wisshak and L. Tapanila (eds.). Current Developments in Bioerosion. Erlangen Earth Conference Series, Springer-Verlag, Berlin.

Tribollet, A. and S. Golubic. 2005. Cross-shelf differences in the pattern and pace of bioerosion of experimental carbonate substrates exposed for 3 years on the northern Great Barrier Reef, Australia. Coral Reefs 24: 422–434.

Tribollet, A., C. Langdon, S. Golubic and M. Atkinson. 2006. Endolithic microflora are major primary producers in dead carbonate substrates of Hawaiian coral reefs. J. Phycol. 42: 292–303.

Tsuda, R.T. and H.T. Kami. 1973. Algal succession on artificial reefs in a marine lagoon environment in Guam. J. Phycol. 9: 260–264.

Van Montfrans, J., R.L. Wetzel and R.J. Orth. 1984. Epiphyte-grazer relationships in seagrass meadows: consequences for seagrass growth and production. Estuaries 7: 289–309.

Van Soest, P.J. 1994. Nutritional ecology of the ruminant. Cornell University Press, Ithaca.

Vargas, M.A., H. Rodríguez, J. Moreno, H. Olivares, J.A. Del Campo, J. Rivas and M.G. Guerrero. 1998. Biochemical composition and fatty acid content of filamentous nitrogen-fixing bacteria. J. Phycol. 34: 812–817.

Varute, A.T. and S.K. Jirge. 1971. Histochemical analysis of mucosubstances in oral mucosa of mouth breeding cichlid fish and seasonal variations in them. Histochemie 25: 91–102.

Verbruggen, H. and A. Tribollet. 2011. Boring algae. Curr. Biol. 21: R876.

Vergés, A., S. Bennett and D.R. Bellwood. 2012. Diversity among macroalgae-consuming fishes on coral reefs: a transcontinental comparison. PLoS One 7: e45543.

Vooren, C.M. 1981. Photosynthetic rates of benthic algae from the deep coral reef of Curaçao. Aquat. Bot. 10: 143–154.

Wainwright, P.C., D.R. Bellwood, M.W. Westneat, J.R. Grubich and A.S. Hoey. 2004. A functional morphospace for the skull of labrid fishes: patterns of diversity in a complex biomechanical system. Biol. J. Linn. Soc. 82: 1–25.

Wainwright, P.C. and S.A. Price. 2016. The impact of organismal innovation on functional and ecological diversification. Integr. Comp. Biol. 56: 479–488.

Weisz, J.B., U. Hentschel, N. Lindquist and C.S. Martens. 2007. Linking abundance and diversity of sponge-associated microbial communities to metabolic differences in sponges. Mar. Biol. 152: 475–483.

Westneat, M.W. 2002. Scaridae. Parrotfishes. pp. 1723–1739. *In*: K.E. Carpenter (ed.). The Western Central Atlantic: FAO Species Identification Sheets for Fishery Purposes. Food and Agriculture Organization of the United Nations, Vol. 3.

Wilkinson, C.R. and P.W. Sammarco. 1983. Effects of grazing and damselfish territoriality on coral reef algae. II. Nitrogen fixation. Mar. Ecol. Prog. Ser. 13: 15–19.

Williams, G.A., M.S. Davies and S. Nagarkar. 2000. Primary succession on a seasonal tropical rocky shore: the relative roles of spatial heterogeneity and herbivory. Mar. Ecol. Prog. Ser. 203: 81–94.

Williams, S.L. and R.C. Carpenter. 1997. Grazing effects on nitrogen fixation in coral reef algal turfs. Mar. Biol. 130: 223–231.

Yamamuro, M. 1999. Importance of epiphytic cyanobacteria as food sources for heterotrophs in a tropical seagrass bed. Coral Reefs 18: 263–271.

Yang, D., S. Nam, S.-J. Hwang, K.-G. An, Y.-S. Park, K.-H. Shin and S. Park. 2016. Fatty acid biomarkers to verify cyanobacteria feeding abilities of herbivorous consumers. J. Freshwat. Ecol. 31: 77–91.

Dynamic Demography: Investigations of Life-History Variation in the Parrotfishes

Brett M. Taylor[1], Elizabeth D.L. Trip[2] and J. Howard Choat[3]

[1] Joint Institute for Marine and Atmospheric Research, University of Hawaii and NOAA Fisheries, Pacific Islands Fisheries Science Center, 1845 Wasp Boulevard, Building 176, Honolulu, Hawaii 96818, U.S.A.
Email: brett.taylor@noaa.gov
[2] Nelson Marlborough Institute of Technology, 322 Hardy Street, Private Bag 19, Nelson 7042, New Zealand
Email: elizabeth.lamantrip@gmail.com
[3] College of Science and Engineering, James Cook University, Townsville QLD 4811, Australia
Email: john.choat@jcu.edu.au

Introduction

Over the last decade, a large number of publications have focused on the impacts of parrotfishes (Scarinae, Labridae) feeding on the sessile biota of coral reefs as well as the impacts of fishing on parrotfish assemblages, both regionally and globally. Despite this, over the same time period there have been relatively few studies on the demography and population dynamics of the group. An ever-increasing literature focusing on their functional and trophic ecology as scrapers and bioeroders of reef substrate emphasizes this disparity. However, the relatively few existing demographic studies establish how much more there is to learn about variation in age-based demography through both space and time. Life-history traits and their spatial and temporal dynamics have fundamental importance to both trophic ecology and fisheries management, the latter becoming increasingly important with the widespread exploitation of parrotfishes in commercial and artisanal fisheries.

The demographic configuration of biotic communities is influenced by evolutionary and regional processes, as well as biophysical features of the surrounding environment (Ricklefs 1987). Life-history traits are important components that broadly reflect the ecological diversification distinguishing species within assemblages (Winemiller and Rose 1992). For parrotfishes, there exists a high level of ecological convergence whereby phylogenetically diverse species manifest similar sizes and foraging modes (Choat et al. 2012). As a result, we often observe a high number of species co-occurring in similar habitats. Despite this, parrotfish assemblages from various regions display considerable inter-specific demographic variability in terms of life-history traits, with maximum adult body sizes ranging from approximately 15 cm to greater than 1 m and maximum life spans ranging from less than 3 yr to greater than 40 yr. The study of demographic pattern and process is critical to both ecologists and resource managers.

Parrotfish abundance has been found to be the most influential factor structuring some reef fish assemblages (i.e., they are comparatively speciose and abundant; Campbell and Pardede 2006) and perhaps no group of marine fishes exhibits more influence on the benthic biota through feeding processes (Carpenter 1986, Choat 1991, Mumby et al. 2006). They are considered to be ecologically important elements of the global coral reef ecosystem as they have the capacity to modify the benthic biota of coral reefs either by scraping surfaces or excavating calcareous structure, thus contributing to bioerosion (Bellwood and Choat 1990, Bellwood 1995a, b). Evidence suggests that body size has a considerable effect on rates of grazing and bioerosion (Bellwood 1995b, Alwany et al. 2009, Hoey Chapter 6) and the functional performance of individual parrotfishes increases non-linearly with increasing body size (Lokrantz et al. 2008). Species abundance and composition are also key variables; for example, one species in particular, *Bolbometopon muricatum*, can account for approximately 85 percent of annual parrotfish bioerosion on unexploited outer shelf reefs of the Great Barrier Reef (GBR) and this discrepancy likely exists elsewhere (Bellwood et al. 2003, Hoey and Bellwood 2008). Additionally, parrotfishes have historically been subject to fisheries exploitation in most regions circumtropically, and evidence suggests their prevalence in reef-associated harvests is increasing in many areas where other reef fish families have declined from overexploitation (Dalzell et al. 1996, Houk et al. 2012). Hence, growth rate, maximum body size, age at maturation and life span are important characteristics of populations that relate directly to yield and sustainability (Beverton and Holt 1957). Presently, these data are scarce or non-existent for most regions where parrotfishes are harvested.

Parrotfishes have dynamic and distinctive life histories. In fact, many of their general traits and characteristics make them highly suitable study species for addressing research questions regarding population or assemblage dynamics. Parrotfishes exhibit complex sexual ontogenies whereby most species are protogynous hermaphrodites (female-to-male sex change; Reinboth 1968, Choat and Robertson 1975), representing at least two sexual pathways across species (diandry and monandry; Robertson and Warner 1978). Two phylogenetically distinct species (*B. muricatum* and *Leptoscarus vaigiensis*) are identified as functional gonochores, whereby for *B. muricatum* at least, males pass through an immature female phase (Robertson et al. 1982, Hamilton et al. 2008). Nearly all species display sexual dimorphism in body size, in which males on average are larger at a given age than females (Choat et al. 1996). Many species also phenotypically express sexual dichromatism through distinct color phases (Randall 1963), allowing rapid and non-invasive sex determination *in situ*. Further, because of their often vibrant body coloration and foraging behavior, parrotfishes are highly conspicuous members of the reef fauna, thus facilitating accurate surveys of abundance using standard underwater visual survey techniques (Watson et al. 2010). The study of parrotfish demography is enhanced by complementing these traits with length-at-age information for individuals.

Assessment of the capacity of parrotfishes to respond to both anthropogenic disturbance and natural variation is no simple task. Despite their morphological similarity they are a phylogenetically complex group showing high levels of evolutionarily recent diversification (Robertson et al. 2006; Smith et al. 2008; Alfaro et al. 2009; Choat et al. 2012). There is also evidence of substantial clade-specific demographic variation (Choat and Robertson 2002). The two major groups of parrotfishes (Sparisomatinine and Scarinine) support a number of abundant and very widespread species, especially in the Indo-Pacific, that occupy a wide variety of shallow water habitats and reef systems. At present, a disproportionate amount of the demographic data on Indo-Pacific parrotfishes has been obtained from reefs associated with continental and high-island margins of the western Pacific (primarily the GBR; Choat et al. 1996; Choat and Robertson 2002). It is unclear whether parrotfishes from the more isolated atolls and islands of the central Pacific have similar demographic profiles as evidence to date suggests that taxa from the GBR may show substantially greater life spans than conspecifics from isolated oceanic islands (Taylor and Choat 2014). In addition, although parrotfishes are harvested over most of their geographical range, the fishing methods and intensity vary widely both within and between ocean basins.

This chapter aims to facilitate future age-based research on the demography and life-history of parrotfishes. Herein, we outline procedures for the classification and analysis of demographic variation and review what has been learned to date. Over 40 years have passed since the first age-based work on this group, yet the current primary literature reflects an incipient understanding at best regarding the broad-scale patterns of demographic variation across space. Hence, this generates a series of proposed investigative problems with associated preliminary hypotheses for future research. Specifically, we address issues pertaining to phylogenetic diversity, biogeography, and habitat effects as well as their associated interactions. Finally, we discuss the importance of previous and future findings to fishery management and biodiversity conservation, given the probable increase in anthropogenic pressure in years to come.

Classifying and Analyzing Demographic Information

Deriving Age-based Information

The ability to age teleost species using calcareous structures (reviewed in Fowler 2009, Moltschaniwskyj and Cappo 2009) underpins our knowledge of life-history theory in marine fishes and provides a unique advantage over many other taxa for which such information is not easily accessible. A variety of structures (e.g., otoliths, spines, scales, vertebrae, various other bones) chronologically records ontogenetic information for individuals throughout their life span. A review of fish biology literature from the last century overwhelmingly demonstrates that these techniques have predominantly been focused on temperate species, especially those constituting highly valued fisheries (Beamish 1992, Fowler 2009). Work on tropical species has lagged tremendously behind that of temperate species for a number of reasons resulting from both logistical (lack of funding and scientific manpower in most developing tropical nations) and theoretical constraints (prior notion that tropical fishes do not deposit annual increments in calcareous structures based on the comparative lack of climatic variation; Munro 1983, Longhurst and Pauly 1987, Fowler 2009). To date, most ageing work on tropical fishes has focused on valuable fishery species. However, the initial accumulation of age-based data has taught us that life histories of tropical fishes were collectively much different than had been previously speculated, with considerable longevity in many reef fish families dispelling the idea that coral reefs are comprised of short-lived high turnover species (Pannella 1974, Sale 1980).

Sagittal otoliths (or 'ear bones') have become the primary medium for the estimation of annual ages in teleost fishes, given their superior reliability and interpretability for a wide range of taxa. Their use necessitates euthanizing individual fish, which is not an ethical issue where samples can be derived from fishery-caught specimens. However, fishery-independent sampling is an important endeavor, especially when specific research objectives (e.g., groundtruthing fishery-dependent data, comparisons across habitats or regions, phenotypic responses of populations within and outside of marine reserves) are desired, and such studies can be designed with negligible impact to populations (Kritzer et al. 2001). In some cases, the use of non-lethal ageing techniques can be a valuable tool, particularly for overfished species and those of high conservation value (e.g. Hobbs et al. 2014).

The first study to estimate age-based traits in a parrotfish examined banding patterns, presumed to represent annual increments, in various bones (opercula, cleithra and hyomandibula) of *Scarus iseri* from the southern Caribbean (Warner and Downs 1977). More recent otolith-based work on the species suggests their age designations were likely accurate and their study provided the first insights into the age-based dynamics between sexes and between primary (males with female coloration that never changed sex) and terminal phase males (vibrantly colored males that underwent metamorphosis from female coloration). The first validation of age estimates from calcareous structures came from the GBR 15 years later for *Scarus schlegeli* (Lou 1992). Since then, there have been just over 20 studies detailing age-based information for parrotfishes based on calcareous structures. Fourteen of these appear in the primary peer-reviewed literature (Warner and Downs 1977, Lou 1992, Choat et al. 1996, Grandcourt 2002, Gust et al. 2002, Choat et al. 2003, Kume et al. 2009, Paddack et al. 2009, El-Sayed Ali et al. 2011, Taylor and Choat 2014, Ebisawa et al. 2016a, b, Lessa et al. 2016, Taylor and Pardee 2017), two in book chapters (Choat and Robertson 2002, Hamilton and Choat 2012) and several more in unpublished academic theses or grey literature. Most of this work has been carried out in the Indo-Pacific region. However, despite the sparse nature of demographic parrotfish studies, a considerable diversity of inter- and intra-specific dynamics has been uncovered, which we will highlight herein.

Throughout this chapter we primarily refer to age-based information derived from annual increment patterns (annuli) in sectioned sagittal otoliths. Parrotfishes, like many teleost fishes, deposit consistent opaque and translucent zones within the otolith structure that correspond to annual patterns of growth and can be best identified via thin transverse sections through the otolith primordium (i.e., core) using low power microscopy with either reflected or transmitted light (Fig. 1). Structural properties of otoliths, for whole otoliths and banding patterns across transverse sections, broadly reflect phylogenetic differences among species, with greatest similarity among closely related species (Taylor and Choat 2014). To date, annual periodicity in otolith increments has been validated in seven species of parrotfish (*B. muricatum, Chlorurus spilurus, Scarus frenatus, Sc. niger, Sc. schlegeli, Sc. rivulatus, Sparisoma viride*), all but one are from the GBR (Lou 1992, Choat et al. 1996, Choat et al. 2003, Andrews et al. 2015; all validations performed using tetracycline tagging with the exception of Andrews et al. [bomb radiocarbon dating]). There generally exists a gradient in optical clarity within species where ease of interpretability increases from low to high latitudes (Choat et al. 2009), but consistent annual incremental structures are present at all latitudes. Further, a strong and generally linear relationship exists between otolith weight and annual age within species (Choat et al. 1996). The predictive capacity of this relationship (Lou et al. 2005, Wakefield et al. 2014) is underutilized at present among tropical fishery managers.

Classifying Demographic Traits

An initial task is to develop the appropriate suite of metrics for quantifying demographic and life-history patterns. Here, we present comparative information using mean maximum length and age. We quantify these traits as the mean length and age of the largest and

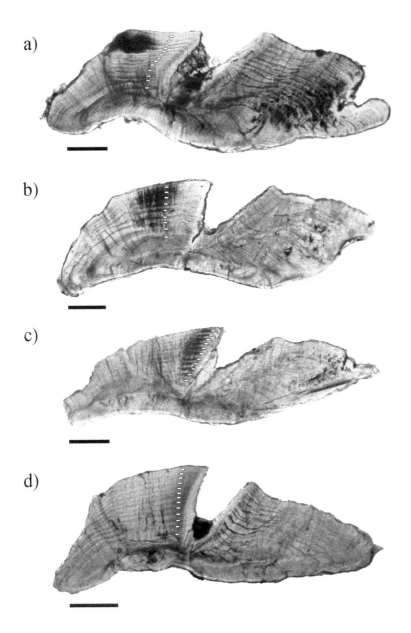

Fig. 1. Examples of transverse sections of sagittal parrotfish otoliths displaying annual increment patterns (denoted by white marks). Species are (a) *Cetoscarus bicolor* (Red Sea, Saudi Arabia), (b) *Chlorurus microrhinos* (Great Barrier Reef, Australia), (c) *Hipposcarus harid* (Red Sea, Saudi Arabia) and (d) *Scarus frenatus* (Yap, Micronesia). Scale bars are 500 μm.

oldest quartile of a sampled population (Choat and Robertson 2002, Taylor and Choat 2014). They correlate strongly with absolute maximum length and age but vary less from random population samples when exposed to intense fishing pressure. Other important traits include the mean length at age at different ages across the life span, the mean length and age at female sexual maturation and the mean length at sex change. Age at sex change in parrotfishes can be tenuous because sex change appears to be a highly length-influenced process and studies have found that not all females will change sex during their life span (Choat et al. 1996). Later in this chapter, we discuss the analytical tools that will be necessary to generate these demographic metrics and establish the patterns and processes underlying demographic variability in reef fishes.

Sexual Ontogeny and Dichromatism

Parrotfishes are frequently characterized as territorial species, displaying strong sexual dichromatism with protogyny as the dominant mode of male recruitment (Kazancıoglu et al. 2009, Streelman et al. 2002). However these generalizations obscure a more complex picture especially with respect to male recruitment and sexual dichromatism. While the majority of species are either exclusively (monandric) or partially protogynous (diandric with some males recruiting prior to sex change), at least two appear to be either primary or secondary gonochorists. Robertson et al. (1982) classified *Leptoscarus vaigiensis* as a gonochorist based on size and color phase distributions and histological analysis. Hamilton et al. (2008) used histological analysis of several Solomon Islands populations to classify *B. muricatum* as a secondary gonochorist with all male recruitment occurring through pre-maturational sex change. Although pre-maturational sex change occurs in other species of parrotfish (Robertson and Warner 1978), no evidence of post-maturational sex change (a prerequisite of protogyny; Sadovy de Mitcheson and Liu 2008) was found in this species. However it is possible that protogyny occurs in other populations of this widespread species. Additional investigation based on histological analysis may reveal other examples of secondary gonochorism although this seems unlikely in dichromatic species. All species of tropical Atlantic Sparisomatinines are monandric whereas an Indo-Pacific Sparisomatinine, *Calotomus spinidens*, possesses primary males (Robertson et al. 1982). It is unclear if the absence of diandry from tropical Atlantic species reflects trait differences derived through phylogenetic or biogeographic means.

The distribution of sexual dichromatism within parrotfishes is more complex. The majority of species are indeed dichromatic with clearly defined initial (IP) and terminal (TP) phases (*sensu* Warner and Robertson 1978). However, a conspicuous minority of species are monochromatic (Bellwood 2001). This includes large species with a schooling mode of foraging and includes some of the largest parrotfishes (*B. muricatum, Chlorurus enneacanthus* [Indo-Pacific]; *Sc. guacamaia, Sc. coelestinus, Sc. coeruleus* [tropical Atlantic]). Many of these species display specific markings, especially on the head region during spawning episodes but do not have defined IP and TP phases (Muñoz et al. 2014). The biogeography of color phase distribution is more complex. The large excavating clade parrotfishes comprising *Ch. microrhinos* (Pacific), *Ch. strongylocephalus* (Indian Ocean) and *Ch. gibbus* (Red Sea) have a predominantly red IP phase and a green TP phase. For *Ch. strongylocephalus* and *Ch. gibbus* the size distribution and relative abundance patterns conform to the usual monandric pattern with the red IP phase, which numerically dominates the female portions of the populations. For *Ch. microrhinos* the red IP phase represents less than three percent of the population with the majority of females displaying the green TP color phase (Choat unpublished data). *Ch. microrhinos* appears to be losing the IP color phase. A similar pattern is seen in *Sc. niger* in which the Red Sea and Indian Ocean populations have a distinct IP phase which is absent from the Pacific with all individuals displaying the TP color phase. Clearly the behavioral and evolutionary factors driving the development of dichromatism require further study.

Emerging Patterns in Parrotfish Demography

Spatial Heterogeneity within Ocean Basins: Geological History, Regional Habitat Structures and More

Spatial heterogeneity within and across ecosystems is a fundamentally important characteristic (Levin 2000), yet one that ecologists often attempt to control for in order to address other research questions. However, the value in studying this inherent variability is increasingly recognized (Hawkins 2012, Legendre and Legendre 2012). Regarding reproductive processes specifically, Petersen and Warner (2002) highlighted that knowledge of natural geographic variation is central to resource management, but this information is virtually non-existent for all coral reef species. This remains true for most demographic and community-level processes on coral reefs. Much of the research on coral reef fishes is conducted at very small spatial scales (Sale 1998). Regarding demographic variation across space, efforts have been focused on measuring the identifiable effects of human extraction (i.e., fishing effects). However, the selective forces associated with fishing pressure and natural geographic variability are highly dissimilar (Bohnsack 1990), thus a critical issue is the need to disentangle the variation associated with both environmental and anthropogenic factors. Stemming from this is the concept of spatial scale, particularly at which scale(s) different factors emerge as recognizably important (Wiens 1989).

Continental Margins versus Oceanic Reef Structures

Coral reefs occur in a number of geologically distinct environments. The most obvious distinction exists between continental margins and oceanic islands. Continental structures are typically characterized by broad, highly interconnected fringing and barrier reef systems. These are often comparatively eutrophic coral reef environments with considerable nutrient input from the adjacent mainland. Examples include the GBR in the western Pacific and the Mesoamerican Barrier Reef System in the western Caribbean. Conversely, oceanic coral reef systems are characterized by sparsely distributed small islands and submerged reef structures. Much of the Pacific Ocean is comprised of oceanic island systems that are characteristic of Micronesia, Melanesia and Polynesia. These reefs occur in oligotrophic regions with various levels of biotic larval interconnectivity facilitated through prevailing ocean currents. Regardless of geological history, coral-dominated environments are often made up of abundant and diverse parrotfish assemblages.

Evidence to date suggests that species associated with continental and high-island margins are larger-bodied and have longer life spans compared with conspecifics from oligotrophic oceanic reefs. Most of the examples stem from species comparisons between the GBR (continental margin) and Micronesia or the Line Islands (oceanic reef systems). Exceptions do exist, and the effects of latitude and variable fishing pressure complicate our ability to decipher the effect of geological history on demography. However, the pattern is quite consistent among conspecifics for which comparative data is available.

The most comprehensive data set for addressing this question exists for the parrotfishes, for which length-at-age data has been derived for a number of conspecifics from both the GBR and Micronesia; many of these were sampled at a similar latitude (~13-14°; Choat and Robertson 2002, Taylor and Choat 2014). Figure 2 compares length-at-age growth trajectories between regions for 10 species spanning four genera. The form of growth curves relative to region varies, but parrotfishes from the GBR consistently display larger size-at-age in the older age classes and in some cases considerably longer life spans. This

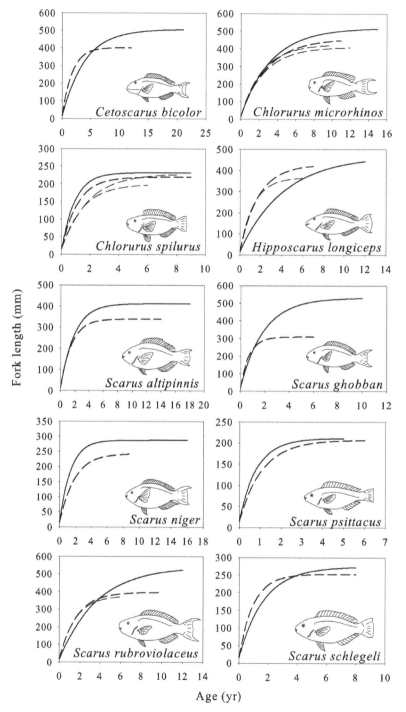

Fig. 2. Comparison of growth trajectories between conspecific parrotfish populations from the Great Barrier Reef (solid lines) and Micronesia (dashed lines). Multiple dashed lines for some species represent Micronesian islands of varying fishing intensity from lightly- to heavily-fished regions. Specimens were collected by fishery-independent sampling at each locality and age information was derived from sagittal otoliths. Von Bertalanffy growth functions were used to derive growth curves.

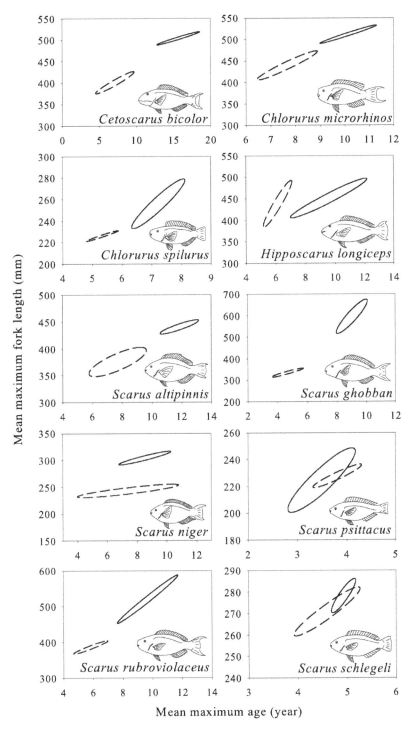

Fig. 3. Comparisons of 95% confidence ellipses surrounding estimates of mean maximum length and mean maximum age (parameters defined within text) for conspecific parrotfish populations from the Great Barrier Reef (black lines) and Micronesia (dashed lines). Specimens were collected by fishery-independent sampling at each locality and age information was derived from sagittal otoliths.

is reflected in patterns of mean maximum length and age (Fig. 3). However, the effects of fishing are potentially a major confounding factor between these data sets; parrotfishes represent a major fishery target in Micronesia whereas they are not harvested on the GBR. Could the observed differences potentially reflect truncated life spans and selection (driven by fishing pressure) for smaller-bodied fishes at Micronesian localities? This is certainly a legitimate concern and this effect has likely occurred at least to some degree. However, Micronesian samples collected across lightly and heavily exploited outer reefs of different islands displayed growth profiles more similar to each other than to the GBR and, in some cases, lightly exploited areas had shorter life spans (e.g., *Ch. microrhinos, Ch. spilurus, Sc. rubroviolaceus*; Fig. 2). At a smaller spatial scale, the same was true for species systematically sampled from marine reserves (by special permit) on Guam compared with heavily fished sites (*Ch. spilurus, Sc. psittacus*; Barba 2010). Further, this data is complemented with high-resolution *in situ* length estimates from stereo-video surveys spanning uninhabited to heavily exploited reefs of Micronesia and protected reefs of the northern GBR, where parrotfishes are not harvested. Using *Ch. microrhinos* as an example (a widespread, relatively abundant and highly targeted species throughout the Indo-Pacific), mean length decreased with increasing human density in Micronesia (Taylor et al. 2015). However, the largest length classes at all Micronesian islands (including uninhabited areas) were similar to each other and yet much smaller than those from the northern GBR, where maximum recorded lengths were nearly 20 percent greater. This suggests that the observed differences in length-at-age are indeed unique to each region. Unfortunately, this sampling represents only two distinct areas and much more sophisticated analyses and sampling programs are necessary to clarify within-region patterns. Limited information on exploited fish populations from the high-latitude, isolated oceanic islands of Hawaii has revealed even greater longevity than on the GBR, not only for parrotfishes (Howard 2008), but also for acanthurids (Claisse et al. 2009, Andrews et al. 2016).

Regional Habitat Structures and Fishing Pressure

Most work on the demography of parrotfishes has been carried out at intermediate spatial scales (10's-100's km) and highlights the influence of regional habitat, density dependence, predation and human exploitation on demographic traits. The magnitude of intra-specific variability in parrotfish life-history traits was first demonstrated by Gust et al. (2002) among mid- and outer-shelf reef systems of the GBR. Consistent differences were found for three parrotfish species (and one acanthurid) whereby considerably smaller length-at-age and shorter life spans corresponded with higher natural mortality rates on outer shelf reefs. Observed differences were attributed to density-dependence given that parrotfish densities at the outer shelf reefs were on average four times greater than those of mid shelf reefs. Early maturation and sex change at small body sizes as well as high proportions of initial phase primary males were also characteristic of the high-density outer shelf populations (Gust 2004), and suggest that considerable variability in reproductive dynamics can occur over relatively small spatial scales (~20 km).

In contrast, age-based analysis of the Caribbean stoplight parrotfish *Sp. viride* suggested limited variability in life-history traits over a broad spatial scale (Choat et al. 2003). However, latitude (proxy for water temperature) and human population density (proxy for fishing pressure) had identifiable influences on the mean maximum sizes and mortality rates of populations, but not on maximum ages. No relationship existed among fishing pressure, abundance or growth rate. At smaller spatial scales (< 10 km), life span doubled from inshore to offshore reefs (Paddack et al. 2009).

As previously mentioned, sexual dimorphism in color phases for many parrotfish species facilitates rapid and non-invasive sex determination from visual surveys. The presence of initial phase primary males is a confounding factor, as functional males displaying initial phase coloration will be classified as females. For the estimation of sex change schedules, however, this appears to be a negligible issue. Histological examination of gonads was performed on sexually dimorphic species from Micronesia, and a comparison of estimates of lengths at 50 percent sex change across species and geographic locations using histology (initial phase primary males excluded) and color-phase ratios by length class (initial phase primary males scored as females) from biological specimens demonstrates that even when the proportion of primary males is high (> 20 percent of IP individuals, > 60 percent of functional males), estimates using color phases only differ by less than six percent (Fig. 4). This is because the mean length of initial phase primary males is consistently much smaller than the length at sex change; hence, the presence of primary males has only a small influence on estimates of the length at 50 percent sex change, even when they are in high proportions.

Several studies have used this technique to examine variability in life-history traits across environmental gradients from visual survey data. Among remote atolls of the NW Hawaiian Islands and the northern Line Islands, DeMartini et al. (2005, 2008) documented shifts in the length at sex change for several parrotfish species along a gradient of predator biomass. In both locations, parrotfish length at sex change consistently decreased with increasing biomass of apex predators, implying that higher rates of mortality associated with increased predation yields a compensatory shift in life-history processes whereby sex change occurs earlier. Across seven islands of Micronesia, Taylor (2014) identified strong scale-dependence in covariates of length at sex change for *Ch. spilurus*. Fishing pressure influenced this trait considerably at the within-island scale, with higher fishing pressure yielding smaller lengths at sex change. Across islands, however, broad-scale features of island geomorphology overwhelmingly predicted patterns of length at sex change. This was consistent for all dichromatic parrotfish species for which resolution was adequate from survey data (*n* = 4). The proximal mechanism driving this pattern is unknown, but likely relates to inherent differences in assemblage structure and the influence of reef geomorphology on reproductive behavior (i.e., how reef morphology affects population size, travel distance to spawning sites and reproductive mode). Collectively, these studies demonstrate high adaptive flexibility in demographic traits of parrotfishes and contradict the model of invariant length at sex change within species (Allsop and West 2003). Further, they demonstrate that demographic variability can be driven by a range of external biotic and abiotic factors and therefore patterns will depend on both the measurable influence (i.e., value range) of factors and the spatial scale of observation.

To this point we have dealt primarily with habitat structure and the configuration of reef systems within an ocean basin and the interaction between the structure of reef ecosystems and fishing. This reveals the critical nature of a spatial and habitat context when evaluating demographic patterns in reef fishes. Without this consideration, there is a high probability of confounding the influence of habitat structure and fishing pressure on parrotfish abundances and demographic patterns.

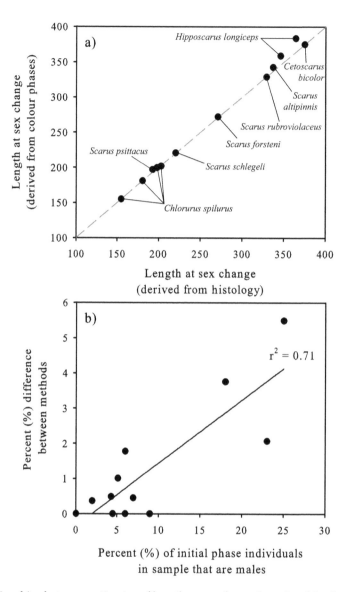

Fig. 4. (a) Relationships between estimates of length at sex change based on histological examination of gonad stages (initial phase primary males excluded; *x*-axis) and color-phase ratios (initial phase primary males scored as females; *y*-axis) by length class from biological specimens across parrotfish species and different populations within species in Micronesia. A dashed line represents a one-to-one ratio. (b) Relationship between the percent of initial phase males represented in the male portion of each population and the difference in length-at-sex-change estimates between the two above methods.

Biogeographic and Phylogenetic Signatures in Demographic Traits

Parrotfishes have undergone a relatively recent evolutionary diversification with expansion into the tropical ocean basins of the world. These basins have highly dissimilar geological and climatic histories. For instance, total area, coral reef habitat area and the number of island habitats increases by three, one and three orders of magnitude, respectively. This occurs from the Red Sea to the Caribbean and tropical Atlantic, to the Indian Ocean and to the Pacific Ocean (Table 1). Concordantly, the mean range size of parrotfish species increases by an order of magnitude across these ocean basins (Table 1). Given the level of variation, we must evaluate whether growth rates, life spans and associated turnover times of species respond to environmental disturbance on ecological and evolutionary time scales and if this is reflected among ocean basins. To assess this we consider three sources of demographic variation: (i) Environmental factors, especially changes in temperature occurring over latitudinal gradients, indicative of demographic variation at ecological time scales; (ii) biogeographic factors which reflect the differing geological histories of ocean basins and have a largely evolutionary signature (biogeographic factors are appropriately sampled over longitudinal geographic gradients that cover different ocean systems); and (iii) the influence of phylogeny and evolutionary relationships amongst the different clades of parrotfishes based on a comprehensive phylogenetic reconstruction of the taxa in question. To help put this into perspective we firstly review the main features of parrotfish phylogeny and evolutionary history.

Table 1. Broad-scale habitat and parrotfish range size metrics from major ocean basins containing coral reef habitat

	Red Sea	*Caribbean*	*Indian*	*Pacific*
Area (10^6 km²)	0.4	2.7	67.4	152.6
Habitat area (10^6 km²)	0.017	0.026	0.032	0.201
Island habitats	24	1,242	7,561	29,959
Mean parrotfish species range area (10^6 km²)	0.34	2.1	2	7.5
Pliocene/Pleistocene disturbance regime	Regionally severe, locally high	Regionally high, locally high	Regionally moderate, locally moderate	Regionally moderate, locally high

Phylogenies confirm the placement of parrotfishes as a tribe nested within the family Labridae (Westneat and Alfaro 2005). The major divergences in parrotfish evolution have been relatively recent. Phylogenies identify two major clades, the Sparisomatinines (sometimes identified as the 'seagrass clade') and the Scarinines (the 'reef clade'), although this is open to question. Although both clades appear to have undergone an initial divergence within the Tethyan region, the geography of subsequent diversification is different. Fossil evidence places the initial diversification of Sparisomatinines within the Tethys Sea with most of the subsequent diversification and reef colonization occurring within the Atlantic Ocean now dominated by the genus *Sparisoma* with 15 extant species mainly in the western Atlantic. Diversification in *Sparisoma* occurred primarily during the Pliocene and Pleistocene. A less-diverse clade comprising six species characterised by the genera *Calotomus* and *Leptoscarus* colonized the Indo-Pacific.

The Scarinines (a more-diverse clade) also appear to have originated in the Tethyan region but with the main episodes of reef colonization and diversification tending eastwards which established this clade throughout the Indo-Pacific dominated by two genera, *Scarus* (currently 52 species) and *Chlorurus* (18 species). Diversification in both genera was mainly a Pliocene/Pleistocene event with *Chlorurus* being more recent than *Scarus*. While *Chlorurus* distributions extend from the Red Sea to the central Pacific, *Scarus* is more cosmopolitan with distributions extending from the Red Sea to the east Pacific. The genus is also present in the tropical Atlantic with the possibility that colonization of this ocean was via southern Africa. More recently, a single Lessepsian migrant (*Sc. ghobban*) has colonized the eastern Mediterranean. Three additional Scarinine genera *Bolbometopon*, *Cetoscarus* and *Hipposcarus* form distinct clades with their origination preceding *Scarus* and *Chlorurus*. These genera are confined to the Indo-Pacific.

A number of trends emerge when length distributions of parrotfishes are mapped onto a complete phylogeny. Firstly, there are clear phylogenetic differences in size structure with the mean size of the older clades (the Sparisomatinine parrotfishes) being significantly lower than that of the more recent clade (the Scarinines). Sizes within the Scarinines are variable with the differences being driven by a few very large species ranging from 65 to 130 cm maximum fork length. There are also geographic correlations with body size. The Atlantic parrotfishes have marginally smaller mean lengths than those of the Indo-Pacific but length distribution is complex. Overall differences among ocean basins largely reflect the domination of the Atlantic fauna by Sparisomatinine parrotfishes.

Demographic Variation over Geographic Gradients

Variation in demographic processes, primarily growth rates and longevity reflect the three factors (environmental, biogeographic, and phylogenetic) identified at the start of this section. The following analyses are designed to distinguish between these sources of variation, especially the alternatives of common geological and oceanic histories (historical biogeography) versus common evolutionary relationships (phylogeny).

The geography of demographic processes is approached hierarchically, first by analysing trends along latitudinal gradients and secondly along a longitudinal axis that incorporates different ocean basins, each with its unique geological and oceanographic history. Sampling latitudinal gradients provides a means of estimating the influence of systematic changes in environmental variables, primarily temperature, on growth rates and longevities often over small spatial scales. Previous work shows that ectotherms respond predictably in terms of growth rates and size structure to changes in temperature in accordance with the temperature-size rule (Atkinson 1994, Trip et al. 2008). As ectotherms, we expect parrotfish body size to be inversely related to water temperature; i.e., maximum body size will increase with increasing latitude as mean water temperature decreases (Atkinson 1994, Atkinson and Sibly 1997). Evidence from other reef fish families suggests maximum age will also conform to this pattern (Robertson et al. 2005, Trip et al. 2008). Sampling over longitudinal gradients across ocean basins may reflect changes that have occurred over evolutionary time scales including differences in life span (Trip et al. 2008).

Latitudinal Gradients: The Impact of Temperature

The predictable nature of latitudinal variation is illustrated by the following data set that incorporates estimates of mean maximum length and mean maximum age of nine populations of the widespread parrotfish *Ch. spilurus* over a gradient of 22° of latitude. Mean maximum length and age estimates were highly correlated. Individuals at low latitude sites were smaller than those at high latitudes and, on the average, had relatively short life spans (Fig. 5). The largest and longest-lived individuals occurred in the colder and more productive waters in higher latitudes which is similar to the latitudinal distribution of life span seen in acanthurids (Robertson et al. 2005, Trip et al. 2008) and other species (e.g., Cappo et al. 2013). Estimates for the different populations show a great deal of variation within the central-southern part of the sampling gradient. This reflects the complexity of reef habitats in this region and the fact that the sampled populations are drawn from both the western Pacific and the eastern Indian Oceans. This indicates that longitudinal variation in reef habitats and the surrounding ocean environment may reflect the influence of long-term historical factors and should be incorporated into future sampling designs.

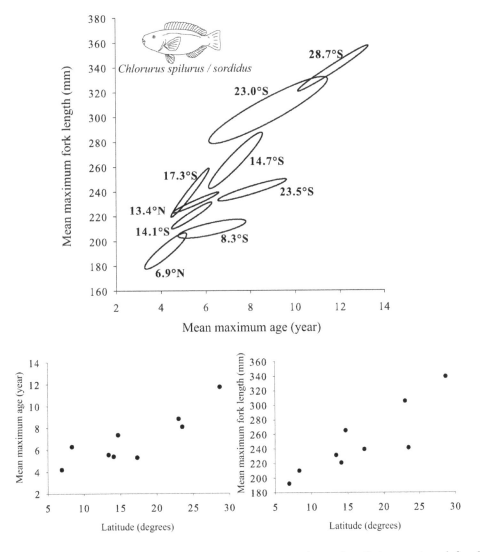

Fig. 5. Relationships between mean maximum age, mean maximum length (parameters defined in text) and latitude for the *Chlorurus spilurus/sordidus* clade. Ellipses represent 95% confidence regions. Sample sites are as follows: 6.9°N: Pohnpei, Micronesia; 8.3°S: Solomon Islands; 13.4°N: Guam, Micronesia; 14.1°S: Scott Reef, Western Australia; 14.7°S: Lizard Island, Great Barrier Reef; 17.3°S: Rowley Shoals, Western Australia; 23.0°S: Ningaloo, Western Australia; 23.5°S: One Tree, Great Barrier Reef; 28.7°S: Abrolhos, Western Australia.

Longitudinal Gradients and Life Span

Previous studies of longitudinal variation of the acanthurid *Ctenochaetus striatus* (Trip et al. 2008) identified a trend in longevity, with maximum ages achieved by western Pacific populations being significantly greater than those from localities within the Indian Ocean. Differences in growth rate emerged at more localized scales. To determine if the trend in longevity occurred in other groups, we obtained age and size estimates of five species or clades of parrotfishes by sampling over a gradient of 105° of longitude extending from the central Red Sea to the western Pacific Ocean. This gradient encompassed three ocean basins of varying size, habitat availability and the extent and severity of disturbances over the Pliocene to Pleistocene epochs. These range from the western Pacific contiguous with the Pacific Ocean proper characterized by a variety of reef habitats covering a wide area of the coastal and oceanic ecosystem to the Red Sea, relatively limited in geographic extent and the amount of reef habitat available. Moreover, the impacts of major environmental disturbances will be regionally moderate in larger ocean basins and severe in smaller systems with limited reef habitats as in the Red Sea (Table 1).

For all five species groups there was a consistent trend of a reduction in life span from the western Pacific to the Red Sea. Populations from Indian Ocean localities had lower longevities than those of the Pacific and generally were higher than those of the Red Sea (Fig. 6). Trends in size were less consistent. Red Sea populations displayed smaller mean maximum sizes than those of the western Pacific. However, for populations of *Ch. sordidus/ spilurus*, *Sc. niger* and *Sc. frenatus*, the maximum sizes were greatest in the central and western Indian Ocean while the mean ages were relatively low, especially in the Seychelles. This confirms that growth patterns are likely to be sensitive to conditions prevailing at particular reef sites while longevities appear to be consistent over whole ocean basins.

Separating Phylogenetic vs Geographic Effects

At a global scale parrotfishes show phylogenetic partitioning. The Indo-Pacific region is dominated by a high diversity of Scarinine parrotfishes with relatively fewer Sparisomatinines (five species of the genus *Calotomus* and one *Leptoscarus*). Only three Indo-Pacific Sparisomatinines are widespread (*L. vaigiensis*, *Ca. carolinus* and *Ca. spinidens*). In contrast, the tropical Atlantic supports only nine Scarinines of which six species occur within the greater Caribbean (Robertson and Cramer 2014). Sparisomatinine parrotfishes display a greater diversity within the tropical Atlantic (15 species) with one species, *Sparisoma cretense*, colonizing temperate reef environments. The 15 currently recognized species of *Sparisoma* are widely distributed over the tropical Atlantic with seven in the Greater Caribbean (*sensu* Robertson and Cramer 2014), five endemic to Brazil and its offshore islands and three endemic to the central and eastern Atlantic.

Comparative demographic analysis of the Sparisomatinine versus the Scarinine parrotfishes sampled from the Greater Caribbean and western Pacific Ocean regions reveals that the former manifests smaller mean sizes and reduced life spans (Choat and Robertson 2002). In general, Sparisomatinine parrotfishes comprised populations of relatively small fast growing species with potentially high turnover rates. Specifically, the tropical Atlantic and Indo-Pacific species display very short life spans (< five years), rapid growth and relatively small sizes. Only one tropical Atlantic species, *Sp. viride*, achieved a mean maximum age in excess of five years and the maximum sizes achieved were less than 40 cm fork length. For Scarinines the greatest mean maximum ages occurred in Pacific species. The three oldest species (*B. muricatum*, *Ce. bicolor* and *Sc. frenatus*) achieved mean

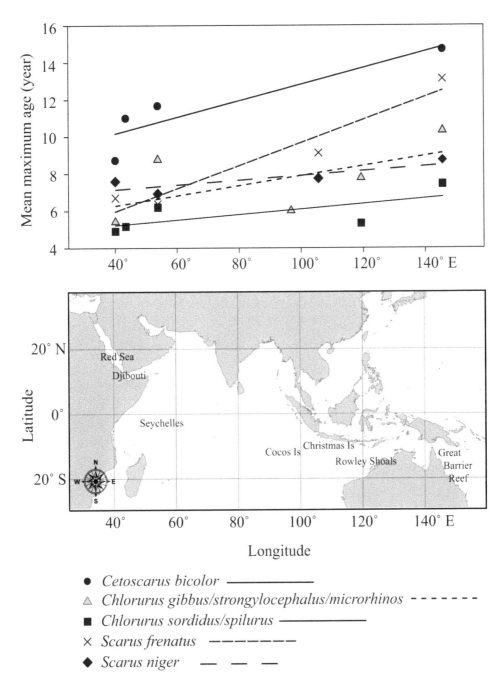

Fig. 6. Consistent patterns of increase in mean maximum age (defined in text) for species or species clades along a longitudinal gradient reflecting an increase in ocean basin size from the Red Sea to the western Pacific Ocean.

maximum ages in excess of 15 yr. The three oldest tropical Atlantic species *Sc. guacamaia*, *Sc. coelestinus* and *Sc. coeruleus* achieved mean maximum ages of 12.3, 12.8 and 10.2 yr, respectively, despite the fact that *Sc. guacamaia* is the second largest parrotfish species recorded anywhere (Fig. 7). For Scarinines, the tropical Atlantic species have shorter life

spans and grow rapidly compared to Pacific species. A further distinguishing feature of the
Pacific species were the variable growth patterns, most notably asymptotic growth with
extended life spans in the relatively small species *Ch. spilurus*, *Sc. frenatus* and *Sc. niger*.

In summary, Sparisomatinine species have small sizes and short life spans regardless
of their oceanic and biogeographic history suggesting that phylogeny was the dominant
influence on demography. Scarinines generally have larger mean sizes and greater life
spans than Sparisomatines regardless of their biogeographic history and present location.
However, there are also between-ocean differences, with species in the tropical Atlantic
being shorter lived at a given length than those of the Pacific despite the presence of some
very large species endemic to the Atlantic (Fig. 7). Given the combined demographic data
set it appears there may be some feature of the tropical Atlantic that selects for relatively
fast growth, short life spans and higher population turnover in parrotfishes relative to the
Pacific. The underlying reasons for these differences are unclear. However, the reduced
life span of Red Sea and Caribbean species compared to those of the Indo-Pacific suggests
that Ocean basin size and the frequency and magnitude of disturbance to the reef habitat
would be appropriate starting points.

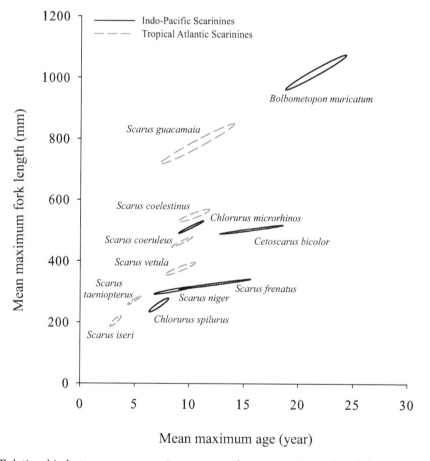

Fig. 7. Relationship between mean maximum age and mean maximum length (parameters defined
in text) across Indo-Pacific (black lines in 95% confidence ellipses) and tropical Atlantic (grey dashed
lines) Scarinine parrotfishes. Species from the Indo-Pacific tend to have longer life spans for a given
maximum body size.

What are the Main Questions Raised by these Preliminary Analyses and How Might they be Resolved?

We identify four issues that have emerged from recent ecological studies on parrotfishes: (1) the importance of carrying out demographic analyses at multiple spatial scales; (2) identifying ecological and evolutionary processes in the response of populations to environmental variation; (3) the impact of demographic studies on the management of fisheries; and (4) the need for more comprehensive demographic analyses.

The Importance of Spatial Scale in Demographic Analysis

Both the abundance and demographic traits of parrotfishes may be modified by environmental disturbances, either anthropogenic or natural. The critical questions facing reef scientists and especially managers concern the spatial extent of these effects. Do they constitute a footprint that extends over whole ocean basins or are they restricted to specific reef systems? Previously, these questions have been answered through integrative analyses that tested for the effects of environmental disturbance on the demographic traits in question. In the case of anthropogenic factors, the spatial boundaries of the effect are not often defined and in many cases it is assumed that they were regional or even global in extent (Legendre and Legendre 2012). In many instances, spatial structuring in data sets is either controlled for or the possibility ignored with the consequence that anthropogenic effects emerged as significant factors in species abundance and distributions. However, it is now clear that the influence of environmental disturbance and variation on ecological traits may be highly scale-dependent, especially with specific regards to parrotfish assemblages (Taylor 2014, Taylor et al. 2015).

The dominant message arising is that both the scale of observation and the spatial pattern of reef configuration and habitat structure are critical in interpreting parrotfish life-history information and their response to fishing pressure (Taylor et al. 2015). Hierarchical sampling designs that integrate measures of oceanographic history, reef configuration and habitat structure with estimates of fishing mortality are a priority. In these circumstances, multi-scale sampling of demographic and life-history traits, independent of anthropogenic factors becomes valuable, especially age-based reproductive data. Reproductive data such as age at maturity is important to fisheries management and related generation times are also critical in estimating evolutionary responses to natural or anthropogenic disturbances.

Ecological vs Evolutionary Demographic Responses

Demographic traits in ectotherms, especially fishes, are highly sensitive to environmental variation. Short-term responses represent phenotypic plasticity and involve rapid shifts in physiological and demographic traits, especially growth rates. This enables individuals to cope with environmental variation and is particularly important in marine fish where dispersive larvae may lead to marked habitat changes between generations (Warner 1997). For reef fishes, marked differences in growth rates can be observed among populations occupying different habitats separated by small linear distances (Gust et al. 2002), suggesting that this variation is best interpreted as phenotypic plasticity (Dudgeon et al. 2000). In multi-scale demographic studies, differences in growth rates often emerge at localized scales (Robertson et al. 2005, Trip et al. 2008, Paddack et al. 2009).

In contrast, consistent differences in life span were observed above at regional scales and reflected differences between ocean basins. This was interpreted as a longer-term evolutionary response to historical patterns in ocean environments seen at geographic scales. However, Gust et al. (2002) and Gust (2004) identified significant variation in growth rates and life span in parrotfish populations at very local scales. The study sites did not manifest the levels of temperature change associated with variation over latitudinal gradients. Demographic variation was interpreted as a consequence of density-dependent processes as recruitment strength varied significantly among the sites tested. Moreover, reductions in life span were associated with the larger number of piscivores at some sites. It is clear that both phenotypic plasticity and adaptation to prevailing oceanic regimes moulded by the environmental history of an area can result in scale-associated demographic variation, especially for life span. Research priorities in this area include investigation of trade-offs between initial growth rates and life span at local and geographic scales and the plasticity of development, growth and maturation with respect to temperature and feeding regime. To date, the common-garden experiments necessary to resolve these issues have been restricted to particular taxa, primarily pomacentrids. However, advances in reef fish culture including groups such as labrids coupled with advances in genomics now establish the possibility of resolution through common-garden experiments and genomic analysis in a number of species.

Demographic Studies and Their Impact on Management and Conservation

Parrotfishes are increasingly important fishery resources throughout the majority of their geographic range. Hence, the impact of human exploitation on parrotfish communities has emerged as a critical management and conservation issue (Jackson et al. 2014). Knowledge of life-history traits is the cornerstone to understanding and predicting population dynamics and this information underpins historical fishery management (Thorson et al. 2014). However, traditional species-level management is impractical in coral reef fisheries where several hundreds of species are harvested at various rates from the same reef system. Rather, ecosystem-based management approaches are advocated for tropical coastal environments. Ecosystem-based management of parrotfish fisheries is highly regarded because of the ecological role of parrotfishes and their influence on benthic biota through processes of grazing and bioerosion (Mumby 2006, 2014). But age-based demographic information remains an imperative component contributing to our ability to successfully manage and conserve parrotfishes as both fishery resources and functional grazers of reef substrate.

Vulnerability of a species to overexploitation is largely dependent on the inherent sensitivity (the biological response to increased mortality) of a population to harvest. Sensitivity, in turn, is primarily driven by life-history traits, especially those influencing population growth rate and turnover time. Parrotfish fisheries target a large number of species (Rhodes et al. 2008), yet the magnitude of demographic responses to fishing pressure are often highly species-specific (Dulvy and Polunin 2004, Clua and Legendre 2008, Bellwood et al. 2012, Valles and Oxenford 2014, Taylor et al. 2014). These responses are predicted by inter-specific variability in life-history trait values, whereby large-bodied and late-maturing species are most sensitive (Taylor et al. 2014). Length-based traits such as mean maximum length are reliable predictors (Clua and Legendre 2008), and maximum length metrics are easily derived from fishery surveys or underwater visual census. However, age at female maturation has been found to be the optimal predictor of vulnerability to overexploitation among parrotfish species (Taylor et al. 2014), thus encouraging the collection of age-based information for harvested species. The spatial variability in life-history traits demonstrated in this chapter suggests that trait values cannot be extrapolated across space for any given species. This creates a problem for tropical fisheries science given the dearth of age-based information for most species. However, the relationships among species (i.e., species' trait values relative to those of others) will likely remain consistent across space, and thus established relationships provide a foundation for predicting responses elsewhere.

Human exploitation has had a profound effect on parrotfish communities, both directly (through harvesting of parrotfishes; Hawkins and Roberts 2003, Bellwood et al. 2012) and indirectly (through removal of predators; DeMartini et al. 2005, 2008). However, given the small- and broad-scale variability in demographic trends of parrotfishes, a key message that emerges is that fishing pressure should not be the default hypothesis for explaining observed differences in parrotfish life history, abundance or community structure. Although fishing has clearly demonstrable impacts, these emerge at relatively local scales and are highly context-dependent. Therefore, fisheries and biodiversity management should be crafted to particular localities and reef systems. It would be unwise, for example, to extrapolate findings from the western tropical Atlantic or the east coast of Africa as guides for management of central Pacific reefs other than in very general terms. What is clear, however, is that latitude (temperature), predation pressure and geomorphological factors such as reef size and habitat diversity may have profound effects on the demography and diversity of target taxa in the absence of fishing. Parrotfishes appear to be a model group for understanding some of these effects across broad spatial scales, but more research is required.

The information we have presented is largely biased towards observations across space. Temporal observations, however, are of greater importance to regional fisheries management and have a higher capacity to elucidate fishery-induced demographic responses. The data-poor condition of most coral reef associated fisheries means that temporal data on length structure, composition and age-based demography of harvested parrotfishes are rare. Simple monitoring programs can alleviate many of these gaps, but such programs require sustained funding, resources and capacity. Questel and Russ (Chapter 14) provide a review of the responses of parrotfish communities to sustained protection in marine reserves.

From a life-history perspective, a potential concern is the long-term 'Darwinian' effects of fishing on parrotfish traits (Conover 2000, Law 2007). Selective mortality is amplified in many parrotfish fisheries because of the consistent sexual dimorphism within species. As a result, males are often targeted at much higher rates. As protogynous hermaphrodites for which sexual transition is driven by social structure (Muñoz and Warner 2003), alterations of size-based sex ratios have notable effects on the reproductive dynamics of populations. This can include changes in population fecundity (Ratner and Lande 2001), mean length and age at maturation and sex change (Law 2000, Hamilton et al. 2007, Hamilton and Caselle 2015), proportional prevalence of primary males (Munday et al. 2006), and mating strategies (Rowe and Hutchings 2003). Such phenotypic responses have been observed in many taxa but the magnitude and consequences of evolutionary changes are poorly understood. Because of the low level of heritability of life-history traits, Law (2007) suggested that evolutionary changes caused by sustained fishery selection would occur on decadal time scales. Archaeological evidence from the Hawaiian Islands demonstrates that body size distributions were similar for hundreds of years until modern times (Longenecker et al. 2014). Present-day body size distributions have decreased substantially amid concurrent increases in fishing pressure. Continuous management of important life-history traits, such as mean age at maturity, has been advocated for incorporating fishery-induced evolution into future management of fish stocks (Kuparinen and Merila 2007). This necessitates the systematic collection of age-based life-history information from locations over time. Few research or management programs focused on coral reef fisheries exercise the requisite foresight, but this approach would provide valuable insights to fishery-induced changes over the long-term, especially in areas where parrotfish fisheries are expanding.

Demographic Analysis

Length-at-age data is incorporated into demographic analyses primarily through growth models. A majority of demographic studies involving reef fishes have used the von Bertalanffy Growth Function (VBGF) to model the relationship between length and age, generate estimates of rates and compare growth performance across populations. For parrotfishes, this has provided information of spatial patterns in their demographic characteristics (e.g. Gust et al. 2002). However a number of authors have questioned the biological interpretation of the VBGF parameters, especially K (Knight 1968, Roff 1980, Craig 1999, Lester et al. 2004, Trip et al. 2008, 2014a).

A misconception is that K is a measure of growth rate, even though K is a reciprocal of time with units of time^{-1} (not length.time^{-1} as would be expected for a growth rate). Rather, K varies with the age at which the growth trajectory reaches (or would reach) asymptotic length L_{∞}, and is a measure of the curvature of a growth trajectory. Asymptotic growth curves generate high K values, whereas indeterminate curves generate lower values. While this may be correlated with differences in growth rate, this interpretation requires the following caveats. First, growth rate varies with age over the length of the life span and is thus a dynamic demographic trait that would best be represented using growth rate-at-age as opposed to a single value such as parameter K. Second, parameter K does not adequately reflect differences in growth rate between groups of fishes (e.g. between sexes). To illustrate this point: we use the following examples, the bumphead parrotfish *B. muricatum* (Hamilton and Choat 2012) and the humphead wrasse *Cheilinus undulatus* (Choat et al. 2006). In both species, males and females grow at a similar rate until sexual maturity is achieved. In *B. muricatum*, males continue to grow at similar rates whereas female growth decelerates considerably (Hamilton and Choat 2012). The differences are even more striking in *C. undulatus* where male growth rate accelerates and becomes linear. If K was a measure of growth rate, then we would expect to find K values that either (a) are similar between males and females reflecting the absence of sex-specific growth over the first years of life, or (b) are greater in males reflecting faster growth in males following sexual maturity. Yet, fitting VBGF growth trajectories to male and female *B. muricatum* generates a greater K value in females (0.15) than males (0.10). This is because females reach asymptotic size earlier. If, however, K values were used to establish the patterns of differences in growth rate between the sexes in this species, the conclusions would lead to a diagnosis of faster growth in females.

Alternative solutions to the traditional VBGF exist and are being increasingly used in demographic studies of reef fishes, including in parrotfishes (e.g. Trip et al. 2008, Claisse et al. 2009, Ruttenberg et al. 2011, Donovan et al. 2013, Taylor and Choat 2014). A model that has received increasing attention in the recent years is the re-parameterised VBGF (rVBGF, Francis 1988). This model describes the relationship between length and age in the same way as the traditional VBGF (they are two equations of the same function), but generates model parameters that differ significantly from the traditional VBGF parameters. The rVBGF produces parameter estimates of mean length at three ages across the life span, which is inherently biologically relevant and interpretable and allows the comparison of mean length at a given age (i.e. growth rate up to that age) across populations (or sexes). Trip et al. (2014a) demonstrated two other benefits in using the rVBGF parameters (over the traditional VBGF parameters). First, the rVBGF parameters are significantly more stable than the VBGF ones, indicating that they are better suited for comparative purposes. And second, the rVBGF parameters produce estimates of mean length at age in the ascending part of the growth trajectory (e.g. parameter of mean length at age one, L_1) that are closer to the most likely 'true' value (the value found when modelling growth using large numbers of new recruits) than VBGF parameter K. We propose that the rVBGF be used and presented alongside the VBGF, especially when estimating growth rate (mean length at age) and comparing growth among populations.

Female maturation has proven to be a reliable predictor of vulnerability to exploitation among parrotfishes, and thus represents a valuable demographic metric (Taylor et al. 2014). However, collecting and analysing reproductive data is usually costly and time-consuming. An alternative analytical tool was proposed by Scott and Heikkonen (2012), who use a Broken-Stick (BS) model to estimate age at sexual maturation from length-at-age information. The BS models length-at-age data using two linear regressions that describe, respectively, the ascending and asymptotic (or would-be asymptotic) arms of the growth trajectory, and that smoothly join at the point where growth slows down or reaches the asymptote (the change-point). Scott and Heikkonen (2012) successfully demonstrated that the change-point may be used as a proxy for age at sexual maturation in north Atlantic plaice. While the BS model represents a potentially powerful tool for the estimation of age at sexual maturation in future demographic studies, it requires field-testing and calibration in reef species manifesting a variety of growth curves.

Discussion

The necessity of a more comprehensive database on parrotfish population dynamics is underlined by two factors. First, with the serial depletion of many groups of reef fishes by fishing, those groups lower in the trophic pyramid, primarily parrotfishes, are now targeted at higher rates by reef fisheries on a global scale (Pauly et al. 1998). Secondly, predictions of environmental change in tropical ocean environments including temperature cover a range within which significant demographic changes would be expected (Munday et al. 2008). This review confirms the findings of previous studies on the demography of parrotfishes: age-based demographic traits are highly variable in their expression, both intra- and inter-specifically. Difficulties in determining the factors responsible for demographic variation reflect the fact that diversity in growth rates, the timing of maturation and sex change and longevity interact to provide a variety of solutions to the problems posed by environmental variation and anthropogenic disturbance. Clarification of the underlying processes requires that the investigation of demographic traits must incorporate spatial analyses that identify the scales at which the main effects emerge. This mandates a hierarchical sampling design incorporating spatial scale from the level of individual reef systems to biogeographic regions covering whole ocean basins. Such investigations are likely to be complex.

Systematic sampling over latitudinal and longitudinal gradients confirms established patterns of demographic variation as a consequence of environmental variation and also retrieves novel results that reflect the influence of ocean basin history. Demographic trends over a comprehensive latitudinal gradient confirm the temperature-size rule with mean size and age positively correlated with latitude as seen in temperate water labrids (Trip et al. 2014b). Longitudinal trends appear to reflect the evolutionary history of reef systems in terms of tectonic processes. Within the Indo-Pacific, ocean basin size and the diversity and extent of reef habitat appears to be associated with extended life spans. At the western margins of the Indian Ocean, the Gulf of Aden and the Red Sea, life spans of equivalent clades are approximately half of those of the western Pacific. More importantly, life spans of parrotfishes in the tropical Atlantic are also reduced compared with those of the Pacific Ocean but with the added complication of a strong phylogenetic signal that reflects the evolutionarily-distinct nature of the faunas in each ocean. The combination of latitudinal and longitudinal sampling also identifies the need to distinguish between demographic changes that occur on ecological time scales including variation in growth rates and those that appear to reflect evolutionary processes such as biogeographic trends in life span.

The use of demographic proxy values derived from related species or by extrapolating results from a single locality to other localities may be seen as a solution to the complexity and species diversity of coral reef fisheries. The emerging picture of species and locality-specific variation in parrotfishes and also in many other groups suggests that caution is needed in developing such proxies to compensate for the data-poor status of many tropical fisheries. While there is no substitute for species and locality-specific sampling, the continued development of new analytical methods including estimation of age-and size-specific maturity from growth curves is worth pursuing. This reflects the importance of this parameter in fisheries science and the demanding nature of sampling additional reproductive data for coral reef fishes.

Conclusion

The historical harvest of parrotfishes and variable harvest rates through both space and time complicate our ability to uncover natural demographic trends. Without historical baselines, ecologists must retrospectively interpret present patterns amid a geographic matrix muddled by a range of human influences. Disentangling the effects and identifying the scales and magnitudes of natural and human-induced variability is a management priority. The proposed spatial scale of many of the hypotheses presented demands enhanced coordination among fish biologists from various geographic regions. Regional 'age and growth' life-history assessments of common species are a necessary initial step and have great utility to fisheries management in the respective regions. Yet, simple comparisons of length-at-age and maturation or sex change schedules among regions do not facilitate an understanding of the pattern-driving processes or the implications for fishery and conservation management at broader scales. Rather, future assessments should employ spatially-explicit, mixed-effects modeling over several spatial scales and comprehensive age-based data sets to tease apart interactive effects influencing demographic processes.

Acknowledgments

We thank J. Ackerman, M. Berumen, E. DeMartini, R. Hamilton, J. Kritzer, S. Newman, J.M. Posada, W. Robbins, D.R. Robertson, J. Robinson and G. Russ for the discussion of ideas in this paper and/or logistical field support. Reviews of previous drafts by two anonymous reviewers were greatly appreciated. Logistical support was provided by Caribbean Marine Research Centre, Conservation Society of Pohnpei, Department of Fisheries Bahamas, Department of Fisheries Western Australia, Fundación Científica Los Roques, Instituto Nacional de Parques, James Cook University, Khaled Bin Sultan Living Oceans Foundation, King Abdullah University of Science and Technology, Lizard Island

Research Station, National Geographic Society, Seychelles Fishing Authority, The Nature Conservancy, Universidad Simón Bolívar, University of Guam Marine Laboratory, and Yap Marine Resources Management Division. Financial support for this work was provided by the Australian Research Council, the Pacific Islands Fisheries Science Center and the Smithsonian Tropical Research Institute.

References Cited

Alfaro, M.E., C.D. Brock, B.L. Banbury and P.C. Wainwright. 2009. Does evolutionary innovation in pharyngeal jaws lead to rapid lineage diversification in labrid fishes? BMC Evol Biol 9: 255.

Allsop, D.J. and S.A. West. 2003. Life history: changing sex at the same relative body size. Nature 425: 783–784.

Alwany, M.A., E. Thaler and M. Stachowitsch. 2009. Parrotfish bioerosion on Egyptian Red Sea reefs. J. Exp. Mar. Biol. Ecol. 371: 170–176.

Andrews, A.H., J.H. Choat, R. Hamilton and E.E. DeMartini. 2015. Refined bomb radiocarbon dating of two iconic fishes of the Great Barrier Reef. Mar. Freshwater Res. 66: 305–316.

Andrews, A.H., E.E. DeMartini, J.A. Eble, B.M. Taylor, D.C. Lou and R.L. Humphreys. 2016. Age and growth of bluespine unicornfish (*Naso unicornis*): a half-century life-span for a keystone browser, with a novel approach to bomb radiocarbon dating in the Hawaiian Islands. Can. J. Fish. Aquat. Sci. 73: 1575–1586.

Atkinson, D. 1994. Temperature and organism size—a biological law for ectotherms? Adv. Ecol. Res. 25: 1–58.

Atkinson, D. and R.M. Sibly. 1997. Why are organisms usually bigger in colder environments? Making sense of a life history puzzle. Trends. Ecol. Evol. 12: 235–239.

Barba, J. 2010. Demography of parrotfish: age, size and reproductive variables. M.S. Thesis, James Cook University, Townsville, Queensland.

Beamish, R.J. 1992. The importance of accurate ages in fisheries science. pp. 8–22. *In*: D.A. Hancock (ed.). Proceedings of the Australian Society for Fish Biology workshop on the Measurement of Age and Growth in Fish and Shellfish. Bureau of Rural Resources, Australian Government Publishing Service. Canberra, Australia.

Bellwood, D.R. 2001. Scaridae. Parrotfishes. pp: 3468–3492. *In*: K.E. Carpenter and V. Niem (eds.). The Living Marine Resources of the Western Central Pacific. Bony Fishes Part 4 (Labridae to Latmeridae), Estuarine Crocodiles, Sea Turtles, Sea Snakes, and Marine Mammals. FAO.

Bellwood, D.R. 1995a. Direct estimate of bioerosion by two parrotfish species, *Chlorurus gibbus* and *C. sordidus*, on the Great Barrier Reef, Australia. Mar. Biol. 121: 419–429.

Bellwood, D.R. 1995b. Carbonate transport and within-reef patterns of bioerosion and sediment release by parrotfishes (family Scaridae) on the Great Barrier Reef. Mar. Ecol-Prog. Ser. 117: 127–136.

Bellwood, D.R. and J.H. Choat. 1990. A functional analysis of grazing in parrotfishes (family Scaridae): the ecological implications. Environ. Biol. Fish. 28: 189–214.

Bellwood, D.R., A.S. Hoey and J.H. Choat. 2003. Limited functional redundancy in high diversity systems: resilience and ecosystem function on coral reefs. Ecol. Lett. 6: 281–285.

Bellwood, D.R., A.S. Hoey and T.P. Hughes. 2012. Human activity selectively impacts the ecosystem roles of parrotfishes on coral reefs. P. Roy. Soc. Lond. B. Bio. 279: 1621–1629.

Beverton, R.J.H. and S.J. Holt. 1957. On the dynamics of exploited fish populations. UK Ministry of Agriculture and Fisheries, London.

Bohnsack, J.A. 1990. The potential of marine fishery reserves for reef fish management in the U.S. Southern Atlantic. NOAA Technical Report NMFS-SEFC 261, 40 pp.

Campbell, S.J. and S.T. Pardede. 2006. Reef fish structure and cascading effects in response to artisanal fishing pressure. Fish. Res. 79: 75–83.

Cappo, M., R.J. Marriott and S.J. Newman. 2013. James's rule and causes and consequences of a latitudinal cline in the demography of John's Snapper (*Lutjanus johnii*) in coastal waters of Australia. Fish. B-NOAA. 111: 309–324.

Carpenter, R.C. 1986. Partitioning herbivory and its effects on coral reef algal communities. Ecol. Monogr. 56: 345–364.

Choat, J.H. 1991. The biology of herbivorous fishes on coral reefs. pp. 120–155. *In*: P.F. Sale (ed.). The Ecology of Fishes on Coral Reefs. Academic Press, San Diego.

Choat, J.H. and D.R. Robertson. 1975. Protogynous hermaphroditism in fishes of the family Scaridae. pp: 263–283. *In*: R. Reinboth (ed.). Intersexuality in the animal kingdom. Springer-Verlag, Heidelberg.

Choat, J.H. and D.R. Robertson. 2002. Age-based studies. pp: 57–80. *In*: P.F. Sale (ed.). Coral Reef Fishes: Dynamics and Diversity in a Complex Ecosystem. Academic Press, San Diego.

Choat, J.H., L.M. Axe and D.C. Lou. 1996. Growth and longevity in fishes of the family Scaridae. Mar. Ecol-Prog. Ser. 145: 33–41.

Choat, J.H., D.R. Robertson, J.L. Ackerman and J.M. Posada. 2003. An age-based demographic analysis of the Caribbean stoplight parrotfish *Sparisoma viride*. Mar. Ecol-Prog. Ser. 246: 265–277.

Choat, J.H., C.R. Davies, J.L. Ackerman and B.D. Mapstone. 2006. Age structure and growth in a large teleost, *Cheilinus undulatus*, with a review of size distribution in labrid fishes. Mar. Ecol-Prog. Ser. 318: 237–246.

Choat, J.H., J.P. Kritzer and J.L. Ackerman. 2009. Ageing in coral reef fishes: do we need to validate the periodicity of increment formation for every species of fish for which we collect age-based demographic data. pp. 23–54. *In*: B.S. Green, B.D. Mapstone, G. Carlos and G.A. Begg (eds.). Tropical Fish Otoliths: Information for Assessment, Management and Ecology. Springer, Netherlands.

Choat, J.H., O.S. Klanten, L. Van Herwerden, D.R. Robertson and K.D. Clements. 2012. Patterns and processes in the evolutionary history of parrotfishes (Family Labridae). Biol. J. Linn. Soc. 107: 529–557.

Claisse, J.T., M. Kienzle, M.E. Bushnell, D.J. Shafer and J.D. Parrish. 2009. Habitat- and sex-specific life history patterns of yellow tang *Zebrasoma flavescens* in Hawaii, USA. Mar. Ecol-Prog. Ser. 389: 245–255.

Clua, E. and P. Legendre. 2008. Shifting dominance among Scarid species on reefs representing a gradient of fishing pressure. Aquat. Living. Resour. 21: 339–348.

Conover, D.O. 2000. Darwinian fishery science. Mar. Ecol-Prog. Ser. 208: 303–307.

Craig, P.C. 1999. The von Bertalanffy growth curve: when a good fit is not enough. Naga 22: 28–29.

Dalzell, P., T.J.H. Adams and N.V.C. Polunin. 1996. Coastal fisheries in the Pacific islands. Oceanog. Mar. Biol. 34: 395–531.

DeMartini, E.E., A.M. Friedlander and S.R. Holzwarth. 2005. Size at sex change in protogynous labroids, prey body size distributions, and apex predator densities at NW Hawaiian atolls. Mar. Ecol-Prog. Ser. 297: 259–271.

DeMartini, E.E., A.M. Friedlander, S.A. Sandin and E. Sala. 2008. Differences in fish-assemblage structure between fished and unfished atolls in the northern Line Islands, central Pacific. Mar. Ecol-Prog. Ser. 365: 199–215.

Donovan, M.K., A.M. Friedlander, E.E. DeMartini, M.J. Donahue and I.D. Williams. 2013. Demographic patterns in the peacock grouper (*Cephalopholis argus*), an introduced Hawaiian reef fish. Environ. Biol. Fish. 96: 981–994.

Dudgeon, C.L., N. Gust and D. Blair. 2000. No apparent genetic basis to demographic differences in scarid fishes across continental shelf of the Great Barrier Reef. Mar. Biol. 137: 1059–1066.

Dulvy, N.K. and N.V.C. Polunin. 2004. Using informal knowledge to infer human-induced rarity of a conspicuous reef fish. Anim. Conserv. 7: 365–374.

Ebisawa, A., K. Kanashiro, I. Ohta, M. Uehara and H. Nakamura. 2016a. Changes of group construction accompanying with growth and maturity in blue-barred parrotfish (*Scarus ghobban*), and influences of the fishing targeting the immature group to the stock. Reg. Stud. Mar. Sci. 7: 32–42.

Ebisawa, A., I. Ohta, M. Uehara, H. Nakamura and K. Kanashiro. 2016b. Life history variables, annual change in sex ratios with age, and total mortality observed on commercial catch on Pacific steephead parrotfish, *Chlorurus microrhinos* in waters off the Okinawa Island, southwestern Japan. Reg. Stud. Mar. Sci. 8: 65–76.

El-Sayed Ali, T., A.M. Osman, S.H. Abdel-Aziz and F.A. Bawazeer. 2011. Growth and longevity of the protogynous parrotfish, *Hipposcarus harid*, *Scarus ferrugineus* and *Chlorurus sordidus* (Teleostei, Scaridae), off the eastern coast of the Red Sea. J. Appl. Ichthyol. 27: 840–846.

Fowler, A.J. 2009. Age in years from otoliths of adult tropical fish. pp: 55–92. *In*: B.S. Green, B.D. Mapstone, G. Carlos and G.A. Begg (eds.). Tropical Fish Otoliths: Information for Assessment, Management and Ecology. Springer, Netherlands.

Francis, R.I.C.C. 1988. Are growth parameters estimated from tagging and age-length data comparable? Can. J. Fish. Aquat. Sci. 45: 936–942.

Grandcourt, E.M. 2002. Demographic characteristics of a selection of exploited reef fish from the Seychelles: preliminary study. Mar. Freshwater. Res. 53: 122–130.

Gust, N. 2004. Variation in the population biology of protogynous coral reef fishes over tens of kilometers. Can. J. Fish. Aquat. Sci. 61: 205–218.

Gust, N., J.H. Choat and J.L. Ackerman. 2002. Demographic plasticity in tropical reef fishes. Mar. Biol. 140: 1039–1051.

Hamilton, S.L., J.E. Caselle, J.D. Standish, D.M. Schroeder, M.S. Love, J.A. Rosales-Casian, and O. Sosa-Nishizaki. 2007. Size-selective harvesting alters life histories of a temperate sex-changing fish. Ecol. Appl. 17: 2268–2280.

Hamilton, R.J., S. Adams and J.H. Choat. 2008. Sexual development and reproductive demography of the green humphead parrotfish (*Bolbometopon muricatum*) in the Solomon Islands. Coral Reefs 27: 153–163.

Hamilton, R.J. and J.H. Choat. 2012. Bumphead Parrotfish – *Bolbometopon muricatum*. pp. 490–496. *In*: Y. Sadovy de Mitcheson and P.L. Colin (eds.). Reef Fish Spawning Aggregations: Biology, Research and Management. Springer, Dordrecht.

Hamilton, S.L. and J.E. Caselle. 2015. Exploitation and recovery of a sea urchin predator has implications for the resilience of southern California kelp forests. P. Roy. Soc. Lond. B. Bio 282: 20141817.

Hawkins, B.A. 2012. Eight (and a half) deadly sins of spatial analysis. J. Biogeogr. 39: 1–9.

Hawkins, J.P. and C.M. Roberts. 2003. Effects of fishing on sex-changing Caribbean parrotfishes. Biol. Conserv. 115: 213–226.

Hobbs, J.-P.A., A.J. Frisch, S. Mutz and B.M. Ford. 2014. Evaluating the effectiveness of teeth and dorsal fin spines for non-lethal age estimation of a tropical reef fish, coral trout *Plectropomus leopardus*. J. Fish. Biol. 84: 328–338.

Hoey, A.S. and D.R. Bellwood. 2008. Cross-shelf variation in the role of parrotfishes on the Great Barrier Reef. Coral Reefs 27: 37–47.

Houk, P., K. Rhodes, J. Cuetos-Bueno, S. Lindfield, V. Fread and J.L. McIlwain. 2012. Commercial coral-reef fisheries across Micronesia: a need for improving management. Coral Reefs 31: 13–26.

Howard, K.G. 2008. Community structure, life history, and movement patterns of parrotfishes: large protogynous fishery species. PhD Thesis. University of Hawaii, Honolulu.

Jackson, J.B.C., M.K. Donovan, K.L. Cramer and V.V. Lam. 2014. Status and trends of Caribbean coral reefs: 1970–2012. Global Coral Reef Monitoring Network, IUCN, Gland, Switzerland.

Kazancioglu, E., T.J. Near, R. Hanel and P.C. Wainwright. 2009. Influence of sexual selection and feeding functional morphology on diversification rate of parrotfishes (Scaridae). Proc. R. Soc. Lond. B. Bio. 276: 3439–3446.

Knight, W. 1968. Asymptotic growth: an example of nonsense disguised as mathematics. J. Fish. Res. Board. Can. 25: 1303–1307.

Kritzer, J.P., C.R. Davies and B.D. Mapstone. 2001. Characterizing fish populations: effects of sample size and population structure on the precision of demographic parameter estimates. Can. J. Fish. Aquat. Sci. 58: 1557–1568.

Kume, G., Y. Kubo, T. Yoshimura, T. Kiriyama and A. Yamaguchi. 2009. Life history characteristics of the protogynous parrotfish *Calotomus japonicus* from northwest Kyushu, Japan. Ichthyol. Res. 57: 113–120.

Kuparinen, A. and J. Merila. 2007. Detecting and managing fisheries-induced evolution. **??** 22: 652–659.

Law, R. 2000. Fishing, selection, and phenotypic evolution. ICES. J. Mar. Sci. 57: 659–668.

Law, R. 2007. Fisheries-induced evolution: present status and future directions. Mar. Ecol. Prog. Ser. 335: 271–277.

Legendre, P. and L. Legendre. 2012. Numerical ecology, 3 edn. Elsevier, Amsterdam.

Lessa, R., C.R. Da Silva, J.F. Dias and F.M. Santana. 2016. Demography of the Agassiz's parrotfish *Sparisoma frondosum* (Agassiz, 1831) in north-eastern Brazil. J. Mar. Biol. Assoc. UK. 96: 1157–1166.

Lester, N.P., B.J. Shuter and P.A. Abrams. 2004. Interpreting the von Bertalanffy model of somatic growth in fishes: the cost of reproduction. Proc. R. Soc. Lond. B. Bio. 271: 1625–1631.

Levin, S.A. 2000. Multiple scales and the maintenance of biodiversity. Ecosystems 3: 498–506.

Lokrantz, J., M. Nystrom, M. Thyresson and C. Johansson. 2008. The non-linear relationship between body size and function in parrotfishes. Coral Reefs 27: 967–974.

Longenecker, K.E.N., Y.L. Chan, R.J. Toonen, D.B. Carlon, T.L. Hunt, A.M. Friedlander and E.E. Demartini. 2014. Archaeological evidence of validity of fish populations on unexploited reefs as proxy targets for modern populations. Conserv. Biol. 28: 1322–1330.

Longhurst, A. and D. Pauly. 1987. Ecology of tropical oceans. Academic Press, San Diego.

Lou, D.C. 1992. Validation of annual growth bands in the otolith of tropical parrotfishes (*Scarus schlegeli* Bleeker). J. Fish. Biol. 41: 775–790.

Lou, D.C., B.D. Mapstone, G.R. Russ, C.R. Davies and G.A. Begg. 2005. Using otolith weight-age relationships to predict age-based metrics of coral reef fish populations at different spatial scales. Fish. Res. 71: 279–294.

Moltschaniwshyj, N. and M. Cappo. 2009. Alternatives to sectioned otoliths: the use of other structures and chemical techniques to estimate age and growth for marine vertebrates and invertebrates. pp. 133–173. *In*: B.S. Green, B.D. Mapstone, G. Carlos and G.A. Begg (eds.). Tropical Fish Otoliths: Information for Assessment, Management and Ecology, Springer, Netherlands.

Mumby, P.J. 2006. The impact of exploiting grazers (Scaridae) on the dynamics of Caribbean coral reefs. Ecol. Appl. 16: 747–769.

Mumby, P.J. 2014. Stratifying herbivore fisheries by habitat to avoid ecosystem overfishing of coral reefs. Fish. Fish. 17: 266–278.

Mumby, P.J., C.P. Dahlgren, A.R. Harborne, C.V. Kappel, F. Micheli, D.R. Brumbaugh, K.E. Holmes, J.M. Mendes, K. Broad and J.N. Sanchirico. 2006. Fishing, trophic cascades, and the process of grazing on coral reefs. Science. 311: 98–101.

Munday, P.L., J.W. White and R.R. Warner. 2006. A social basis for the development of primary males in a sex-changing fish. Proc. R. Soc. Lond. B. Bio. 273: 2845–2851.

Munday, P.L., G.P. Jones, M.S. Pratchett and A.J. Williams. 2008. Climate change and the future for coral reef fishes. Fish. Fish. 9: 261–285.

Muñoz, R.C. and R.R. Warner. 2003. Alternative contexts of sex change with social control in the bucktooth parrotfish, *Sparisoma radians*. Environ. Biol. Fish. 68: 307–319.

Muñoz, R.C., B.J. Zgliczynski, B.Z. Teer and L.J. Laughlin. 2014. Spawning aggregation behavior and reproductive ecology of the giant bumphead parrotfish, *Bolbometopon muricatum*, in a remote marine reserve. Peer J 2, e681.

Munro, J.L. 1983. Caribbean coral reef fishery resources. ICLARM, Manila.

Paddack, M.J., S. Sponaugle and R.K. Cowen. 2009. Small-scale demographic variation in the stoplight parrotfish *Sparisoma viride*. J. Fish. Biol. 75: 2509–2526.

Pannella, G. 1974. Otolith growth patterns: an aid in age determination in temperate and tropical fishes. pp. 28–39. *In*: T.B. Bagenal (ed.). The Ageing of Fish. Unwin Brothers Unlimited, Surrey, England.

Pauly, D., V. Christensen, J. Dalsgaard, R. Froese and F. Torres Jr. 1998. Fishing down marine food webs. Science 279: 860–863.

Petersen, C.W. and R.R. Warner. 2002. The ecological context of reproductive behavior. pp. 103–118. *In*: P.F. Sale (ed.). Coral Reef Fishes: Dynamics and Diversity in a Complex Ecosystem. Academic Press, San Deigo, CA.

Randall, J.E. 1963. Notes on the systematics of parrotfishes (Scaridae), with emphasis on sexual dichromatism. Copeia 1963: 225–237.

Ratner, S. and R. Lande. 2001. Demographic and evolutionary responses to selective harvesting in populations with discrete generations. Ecology 82: 3093–3014. **??**

Reinboth, R. 1968. Protogynie bei Papageifischen (Scaridae). Z. Naturforsh. 23: 852–855.

Rhodes, K.L., M.H. Tupper and C.B. Wichilmel. 2008. Characterization and management of the commercial sector of the Pohnpei coral reef fishery, Micronesia. Coral Reefs 27: 443–454.

Ricklefs, R.E. 1987. Community diversity: relative roles of local and regional processes. Science 235: 167–171.

Robertson, D.R. and R.R.Warner. 1978. Sexual patterns in the labroid fishes of the western Caribbean. II: the parrotfishes (Scaridae). Smithsonian Contributions to Zoology 255: 1–26.

Robertson, D.R. and K.L. Cramer. 2014. Defining and dividing the greater Caribbean: insights from the biogeography of shorefishes. PLoS One 9, e102918.

Robertson, D.R., R. Reinboth and R.W. Bruce. 1982. Gonochorism, protogynous sex-change and spawning in three Sparisomatinine parrotfishes from the western Indian Ocean. Bull. Mar. Sci. 32: 868–879.

Robertson, D.R., J.L. Ackerman, J.H. Choat, J.L. Posada and J. Pitt. 2005. Ocean surgeonfish *Acanthurus bahianus*. I. The geography of demography. Mar. Ecol. Prog. Ser. 295: 229–244.

Robertson, D.R., F. Karg, R. Leao de Moura, B.C. Victor and G. Bernardi. 2006. Mechanisms of speciation and faunal enrichment in Atlantic parrotfishes. Mol. Phylogenet. Evol. 40: 795–807.

Roff, D.A. 1980. A motion for the retirement of the von Bertalanffy function. Can. J. Fish. Aquat. Sci. 37: 127–129.

Rowe, S. and J.A. Hutchings. 2003. Mating systems and the conservation of commercially exploited marine fish. Trends. Ecol. Evol. 18: 567–572.

Ruttenberg, B.I., S.L. Hamilton, S.M. Walsh, M.K. Donovan, A. Friedlander, E. DeMartini, E. Sala and S.A. Sandin. 2011. Predator-induced demographic shifts in coral reef fish assemblages. PLoS ONE 6, e21062.

Sadovy de Mitcheson, Y. and M. Liu. 2008. Functional hermaphroditism in teleosts. Fish. Fish. 9: 1–43.

Sale, P.F. 1980. The ecology of fishes on coral reefs. Oceanog. Mar. Biol. 18: 367–421.

Sale, P.F. 1998. Appropriate spatial scales for studies of reef-fish ecology. Aust. J. Ecol. 23: 202–208.

Scott, R.D. and J. Heikkonen. 2012. Estimating age at first maturity from change-points in growth rate. Mar. Ecol. Prog. Ser. 450: 147–157.

Smith, L.L., J.L. Fessler, M.E. Alfaro, J.T. Streelman and M.W. Westneat. 2008. Phylogenetic relationships and the evolution of regulatory gene sequences in the parrotfishes. Mol. Phylogenet. Evol. 49: 136–152.

Streelman, J.T., M. Alfaro, M.W. Westneat, D.R. Bellwood and S.A. Karl. 2002. Evolutionary history of the parrotfishes: biogeography, ecomorphology, and comparative diversity. Evolution 56: 961–971.

Taylor, B.M. 2014. Drivers of protogynous sex change differ across spatial scales. Proc. R. Soc. Lond. B. Bio. 281: 2013–2423.

Taylor, B.M. and J.H. Choat. 2014. Comparative demography of commercially important parrotfish species from Micronesia. J. Fish. Biol. 84: 383–402.

Taylor, B.M. and C. Pardee. 2017. Growth and maturation of the redlip parrotfish *Scarus rubroviolaceus*. J. Fish. Biol. 90: 2452–2461.

Taylor, B.M., P. Houk, G.R. Russ and J.H. Choat. 2014. Life histories predict vulnerability to overexploitation in parrotfishes. Coral Reefs 33: 869–878.

Taylor, B.M., S.J. Lindfield and J.H. Choat. 2015. Hierarchical and scale-dependent effects of fishing pressure and environment on the structure and size distribution of parrotfish communities. Ecography 38: 520–530.

Thorson, J.T., J.M. Cope and W.S. Patrick. 2014. Assessing the quality of life history information in publicly available databases. Ecol. Appl. 24: 217–226.

Trip, E.L., J.H. Choat, D.T. Wilson and D.R. Robertson. 2008. Inter-oceanic analysis of demographic variation in a widely distributed Indo-Pacific coral reef fish. Mar. Ecol. Prog. Ser. 373: 97–109.

Trip, E.D.L., P. Craig, A. Green and J.H. Choat. 2014a. Recruitment dynamics and first year growth of the coral reef surgeonfish *Ctenochaetus striatus*, with implications for acanthurid growth models. Coral Reefs 33: 879–889.

Trip, E.D.L., K.D. Clements, D. Raubenheimer and J.H. Choat. 2014b. Temperature-related variation in growth rate, size, maturation and life span in a marine herbivorous fish over a latitudinal gradient. J. Anim. Ecol. 83: 866–875.

Vallès, H. and H.A. Oxenford. 2014. Parrotfish size: a simple yet useful alternative indicator of fishing effects on Caribbean reefs. PLoS ONE 9, e86291.

Wakefield, C.B., A.J. Williams, S.J. Newman, M. Bunel, C.E. Dowling, C.A. Armstrong and T.J. Langlois. 2014. Rapid and reliable multivariate discrimination for two cryptic Eteline snappers using otolith morphometry. Fish. Res. 151: 100–106.

Warner, R.R. 1997. Evolutionary ecology: how to reconcile pelagic dispersal with local adaptation. Coral Reefs 16: S115–S120.

Warner, R.R. and I.F. Downs. 1977. Comparative life histories: growth vs. reproduction in normal males and sex-changing hermaphrodites of the striped parrotfish, *Scarus croicensis*. pp. 275–282. *In*: Third International Coral Reef Symposium, Miami. Rosential School of Marine and Atmospheric Science, Miami, Florida.

Warner, R.R. and D.R. Robertson. 1978. Sexual patterns in the labroid fishes of the western Caribbean. I: The wrasses (Labridae). Smithsonian Contributions to Zoology 254: 1–27.

Watson, D.L., E.S. Harvey, B.M. Fitzpatrick, T.J. Langlois and G. Shedrawi. 2010. Assessing reef fish assemblage structure: how do different stereo-video techniques compare? Mar. Biol. 157: 1237–1250.

Westneat, M.W. and M.E. Alfaro. 2005. Phylogenetic relationships and evolutionary history of the reef fish family Labridae. Mol. Phylogenet. Evol. 36: 370–390.

Wiens, J.A. 1989. Spatial scaling in ecology. Funct. Ecol. 3: 385–397.

Winemiller, K.O. and K.A. Rose. 1992. Patterns of life-history diversification in North American fishes: implications for population regulation. Can. J. Fish. Aquat. Sci. 49: 2196–2218.

CHAPTER

5

Vision and Colour Diversity in Parrotfishes

Ulrike E. Siebeck

School of Biomedical Sciences and Global Change Institute,
The University of Queensland, St. Lucia 4072 QLD, Australia
Email: u.siebeck@uq.edu.au

Introduction

Parrotfish (Labridae, Scarinae) live in relatively shallow waters around coral reefs and seagrass meadows, where the downwelling sunlight consists of a broad spectrum of wavelengths. Parrotfish make use of the full spectrum of light for their colours and patterns but have reduced visual sensitivity to the short wavelength part of the spectrum. In this chapter, I will first discuss what we know about the colours of parrotfish. This will be followed by a brief summary of key parameters of the underwater light environment, which sets the limits for vision and reduces the spectrum of colours available for communication. I will then go into what little we know about parrotfish visual system and visual behaviour.

Colours of Parrotfishes

What is Colour?

When light encounters an object, it is reflected, absorbed and/or transmitted depending on the properties of the object. While the colour of an object is determined by the spectrum of light it reflects, it is important to remember that colour and reflectance are not identical. Reflectance describes a physical property of an object while its spectral radiance is the product of this physical property as it interacts with the ambient light environment. The "actual" colour as perceived by the observer is a construct of the brain and depends on both, the observer's visual and perceptual system. The reflectance of an object can be quantified with spectro-radiometric methods (Marshall 2000b) while the perceived colour cannot easily be quantified and varies between different observers due to differences in their optics and spectral sensitivities. Despite perceptual differences between human observers, we have developed labels for different wavelength spectra so that we can talk about our perception of reflected light, i.e. colours. In this chapter, I will discuss both reflectance and colour, and it is important to note that whenever the term colour is used it refers to human perception only (e.g. blue refers to reflectance of wavelengths between 400-500 nm). If we want to understand how fish see colours, we have to take into account the properties of their visual system and ideally also the processing in the fish brain, an area of study that is still in its infancy. With information about spectral sensitivity of the visual system and the spectral properties of the illumination, we can model whether two objects are expected to be discriminable based on their reflectance spectra alone (Vorobyev and Osorio 1998). However, behavioural experiments are required to provide conclusive evidence (Kelber et al. 2003). When designing behavioural experiments to test visual abilities (e.g. colour vision, spatial resolution), it is important to keep in mind that the type of behavioural task (e.g. fish have separate processing channels for different types stimuli; e.g. Neumeyer et al. 1991, Siebeck et al. 2014), the illumination conditions (in dim light many animals have reduced colour vision; e.g. Neumeyer and Arnold 1989) as well as the ecology of the fish (e.g. Archerfish colour vision and acuity varies between the parts of the retina specialized for vision in air and water; Temple et al. 2010) influence the results.

Three parameters are commonly used to describe colours: brightness, hue and chroma (Endler 1990, Wyszecki and Stiles 2000). Brightness is defined as the total intensity of light reaching the eye from a colour patch at a given distance. Hue and chroma are associated with the physical properties of colour. Hue is the shade of colour, i.e. red, green, yellow etc. and is defined by the part of the reflectance spectrum that contains most photons. Chroma is a measure of the saturation of the colour and is a function of how rapidly the intensity changes with wavelength (Endler 1990). A reflectance spectrum with gradual changes will appear less saturated than a spectrum with steep slopes and large differences between different parts of the spectrum.

Colour patterns found on fish are created by two basic mechanisms that can occur in combination, or isolation of one another. Specialized colour cells, or chromatophores (xanthophores: yellow; melanophores: black; leucophores: white; erythrophores: orange/red, and very rarely also cyanophores: blue) in the dermis of fish either contain pigment ("pigment colours") or reflective/refractive structures ("structural colours"; iridophores: iridescent colours, most blues and UV; Cott 1940, Fox and Vevers 1960). Recently, two novel types of chromatophores have been described, erythro-iridophores, which contain both pigment and reflecting platelets (Goda et al. 2011), and chromatophores which contain fluorescent red pigment (Wucherer and Michiels 2012).

Blue and ultraviolet (UV) colours are generally of structural origin (for detailed review of blue colours see Bagnara et al. 2007). A true blue pigment colour has so far only been found in callionymid fish (Goda and Fujii 1995). Billiverdin, a bile pigment and product of haem catabolism has been found to give blood, bones collagen and scales a blue-green colour in various fishes (Jüttner et al. 2013), including terminal phase parrotfishes (Yamaguchi et al. 1977). Structural colours are created by interference phenomena, similar to the colours on butterfly wings (e.g. Ghiradella et al. 1972), or as a result of Tyndall (=Rayleigh) scattering. In the case of interference phenomena, stacks of crystals, usually guanine (or guanine mixed with hypoxanthine), with a high refractive index are interspersed with cell material of low refractive index and are thought to be responsible for the wavelength specific reflection of light. The distance between the layers determines which wavelengths are reflected, the smaller the distance, the shorter the reflected wavelengths (Land 1972, Jordan et al. 2014). In the case of scattering phenomena, the incident light is scattered by fine particles (e.g. guanine) smaller than the wavelengths of the light. Short wavelengths are reflected, while long wavelengths are absorbed by a melanin layer situated behind the scattering layer (Fox and Vevers 1960).

Parrotfish Colours

Similar to their bird namesakes, parrotfish exhibit a large diversity of colours and colour patterns. In comparison to this diversity in colour patterns, the morphology of these fish is relatively uniform so that colours are often used to discriminate species and life phases

Fig. 1. Colours of parrotfishes: (a) night colours are often different (darker, dark spots) from colour patterns displayed during the day; (b) iridescent corneas are found in many shallow water teleosts; (c-g) the colours and patterns of many parrotfish species change with maturity and sex (c) juveniles of many parrotfish species form mixed species schools displaying uniform or striped patterns; (d) female (initial phase) *Scarus frenatus* display different colours and patterns compared to terminal phase males (e, f); individual differences in patterns exist in terminal phase males. The differences are most notable on the head and face of the fish; (g) parrotfish larvae from three species (top to bottom: *Scarus iseri, Sparisoma atomarium* and *Sparisoma chrysopterum*); (h) some parrotfish species are more uniformly coloured. Instead of sexual dichromatism (different colour patterns), sexual dimorphisms (different body shapes) are found in such species. Picture credits: a-f, h U.E. Siebeck and g from (Baldwin 2013), reprinted with the permission of the author and publisher (license number 3431640620008).

(Choat 1969, Choat et al. 2012). The colours of few species have been described objectively, i.e. independently of the human visual system. The fifteen species (*Cetoscarus ocellatus* (formely *Ce. bicolor*), *Chlororus microrhinos, Ch. spilurus* (formerly *Ch. sordidus*), *Scarus chameleon, Sc. frenatus, Sc. ghobban, Sc. globiceps, Sc. longipinnis, Sc. niger, Sc. psittacus, Sc. rivulatus, Sc. spinus, Sc. taeniopterus, Sparisoma aurofrenatum* and *Sp. viride*) for which spectral measurements exist have colours belonging to 16 of the 21 colour categories described in (Marshall 2000b). Parrotfish and wrasses have complex (multiple spectral peaks) as well as simple colours (single spectral peak), which are often combined to form complex patterns. In many species, highly contrasting colours are displayed in close proximity while other species are more uniformly coloured (Fig. 1).

The spectral range of colours (ca. 350 nm – 750 nm; UV through to far red) displayed by parrotfish is larger than what the human eye can appreciate, due to our limited spectral sensitivity (400 nm – 700 nm; blue through to red). *Sparisoma aurofrenatum* and *Scarus chameleon* have some UV-blue reflective body areas (<400 nm) while 12 of the 15 parrotfish measured by Marshall (Marshall 2000b) displayed colours reflecting far red (>700 nm). Many parrotfish have UV-absorbing components ("natural sunscreens", Mycosporine-like Amino Acids) in their external mucus (Eckes et al. 2008), which prevent these species from displaying UV-reflective colours. So even if there are elements in the skin, which reflect UV, this short wavelength UV component is unlikely to be detectable due to the mucus layer absorbing UV (ca. 280 nm – 380 nm, depending on the MAA complement present). It is important to take the mucus layer into consideration when taking spectral measurements of the skin as the mucus is water soluble and readily rubs off, thus altering the reflectance properties.

Many parrotfish are sexually dichromatic and change colour as well as sex as they go through different life phases (i.e. juvenile, initial phase, and terminal phase), which has led to much taxonomic difficulty and confusion in the past (Choat 1969, Bellwood and Choat 1989). The larval stages of parrotfish are already pigmented and display colour patterns, however very few studies exist describing larval fish pigmentation and/or colouration (but see Leis and Carson-Ewart 2000, Baldwin 2013). Parrotfish larvae have typical patterns of several linear series of erythrophores and melanophores (Fig. 1g). The exact patterns vary with genus so that pattern differences may be used to evaluate phylogenetic relationships (Baldwin 2013). As juveniles, some parrotfish, such as *Ch. spilurus, Sc. psittacus, Sc. rivulatus* and *Sc. globiceps,* occur in mixed species schools and can have a relatively drab uniform colour or striped pattern (Fig. 1c), which they can rapidly change during behavioural interactions (physiological colour pattern change, PCP – see sections below on colour changes and function; Crook 1997a, b). As a consequence, these juveniles are very difficult to identify visually in the field. Many diagnostic characteristics used for the identification of adults are not present in juveniles, further adding to the problem of their correct identification (Bellwood and Choat 1989). In contrast, the more solitary juveniles of other species (e.g. *Sc. niger, Sc. frenatus* and *Sc. altipinnis)* have constant and distinctive colour patterns (Crook 1997a). For a detailed description of how to identify early post-settlement juvenile parrotfish based on their colour patterns, see Bellwood and Choat (1989).

As juveniles change into their first adult phase (initial phase) and subsequently into the second adult phase (terminal phase), their colour patterns can change dramatically, often becoming increasingly blue/green dominated. The blue-green colour seems to be due to biliverdin contained in the scales of the fish (Yamaguchi et al. 1977). Billiverdin has also been found in bones of fish (Jüttner et al. 2013) and may well be responsible for the green coloration of the teeth as well as pharyngeal teeth of many parrotfish. In the terminal phase, the colour patterns are often most distinctive, at least when observed at close distance by human observers (Randall et al. 1997). Interestingly, it is difficult to take pictures of parrotfish in their natural habitat which show their colourful patterns. The underwater light environment put together with the complimentary nature of the colours used in complex, high spatial frequency patterns, leads to blurring, or blending together of the colours with distance until the overall colour almost matches the underwater space light (Marshall 2000a).

Function of Colour

The function of the coloration of fish has attracted much interest over many years (e.g. Longley 1917, Lorenz 1962, Marshall 2000b, Siebeck 2004, Salzburger et al. 2006), and we are still far from an in-depth understanding, probably because specific functions might vary between different species, and, maybe more importantly, we have only recently begun to address this question from the perspective of the visual system of the intended receivers, rather than our own (Lythgoe 1979, Bennett et al. 1994, Endler and Basolo 1998). Despite the probably species-specific or even behavior-specific nature of colour patterns, several general principles have been proposed, including the possibility that certain colour patterns have no behavioural relevance at all (Marshall 2000a, b).

Camouflage is essential to the survival of fish. Whether or not an animal appears camouflaged to an observer depends on the visual system of the observer as well as on the ambient light environment and the background against which the fish is seen. It is therefore important to remember that animals which appear conspicuous to us may in fact be camouflaged to the intended receiver, such as potential predators, or may be camouflaged only when seen against certain backgrounds or even distances from the observer (Marshall 2000a). The masters of camouflage include the stonefish which resemble stones so closely that it is almost impossible to identify specific features such as the eye, even once the general shape of the fish has been recognized. This extreme type background matching only works for sedentary or slow moving fish and is not found in parrotfishes. The closest examples are species which occur in seagrass and macroalgal beds, such as *Calotomus spinidens* and *Leptoscarus vaigiensis* (Bonaldo et al. 2014, Lim et al. 2016), which have colour patterns which, at least to our eyes, match the background against which they are seen relatively well (Randall et al. 1997). Another type of perfect camouflage would occur if the reflectance spectrum of a fish in the water column matched the background absolutely when viewed from any direction (Muntz 1990). While this is never completely achieved, several attempts have been made in that direction. Many fish are darker on their dorsal surfaces than on their ventral ones (countershading) making them less conspicuous against both downwelling light when viewed from below and the dark background when viewed from above (Cott 1940). Most parrotfish are found close to the substrate and are probably rarely viewed from underneath, which may explain why there are not distinct examples of countershading.

Another way to reduce conspicuousness is to be compressed and have silvery surfaces that act as vertical mirrors reflecting the entire incident light, which is mostly seen in pelagic fish (Denton 1970). Vertical mirrors work because the underwater light field in the open water is almost symmetrical about the vertical axis and the reflected light is thus approximately identical to the light field behind the fish, which means that the contrast between the fish and its background is dramatically reduced (Denton 1970, 1971, Johnsen 2014). Camouflage can also be achieved with disruptive coloration, drawing attention to individual elements of the pattern while at the same time concealing the outline of the fish (Longley 1917, Cott 1940, Muntz 1990). The striped patterns of many juvenile parrotfish as well as the bold coloration of some adult parrotfish (e.g. *Sc. schlegeli* terminal phase) may be examples of attempts to break up the overall outline of the fish.

While camouflage is important to reduce the risk of predation, in many fish, bright colour patterns have been shown to convey information to either conspecifics or heterospecifics and are thus involved in **communication** (for review see Osorio and Vorobyev 2008). It is important to remember that colour patterns that appear conspicuous to us, may in fact be well camouflaged when seen by potential predators in the distance. This is due to the different visual system of fish compared to our own visual system, specifically, the different spectral sensitivities and the much lower visual resolution in fish, as well as the added effect of the underwater light environment (see below). Marshall (2000a) evaluated the use of bright colour patterns for simultaneous conspicuousness and camouflage at different spatial scales and found that the colours of labriform fish blend together between 1 m and 5 m distance depending on the properties of various fish visual systems. In some fish, the resulting additive colour closely resembles the background colour of open water, which means that these fish have the possibility to regulate their conspicuousness through behaviour. Their colours will be conspicuous over close distances, especially when seen against the reef background and are therefore ideally suited for communication, however over larger distances, they will be well camouflaged, especially when seen against the background of open water.

Colour communication is a largely understudied area in parrotfishes despite the rich diversity and complex nature of colour patterns (Choat 1969, Thresher 1984, Bellwood and Choat 1989). Most of what we know comes from observations in the wild, or is inferred from the fact that many parrotfish display sexual dichromatism. The existence of 'sneaker males' that display the female typical colour patterns (Thresher 1984) demonstrates that colour patterns are assumed by the fish to reliably communicate sexual identity. This only works if sneaker males are relatively rare, otherwise colour would lose its diagnostic feature for gender identification. In species without sexual dichromatism, body shapes may differ between the sexes and thus shape can act as a cue which may be used for visual discrimination of gender (e.g. *Scarus coeruleus* and *Sc. perrico*; Thresher 1984).

Parrotfish can spawn in pairs or in groups. Terminal phase males often defend territories and harems, and are more likely to be involved in pair spawning with an initial phase female than in group spawning with multiple initial phase males and females (Thresher 1984). It is possible that the individually different colour patterns play a role in mate selection as terminal phase males actively court certain females before spawning with them (Fig. 1e, f). There is some evidence that the colour patterns of territory forming terminal males intensify during spawning activities (Thresher 1984). A study looking at the influence of sexual selection and feeding morphology suggested that, in the territorial and strongly sexually dichromatic genera, *Scarus* and *Chlorurus*, diversification is shaped predominantly by sexual selection and thus colour patterns (Kazancioglu et al. 2009). This finding was not confirmed by a recent analysis of parrotfish evolutionary history, which showed that there is no single driver for diversification in parrotfish but that there are complex interactions between habitat, reproductive biology (colours, sexual selection) and ecology (Choat et al. 2012).

Most parrotfish change aspects of their coloration, either within a few seconds during specific behaviours (e.g. *Ch. spilurus* when joining other schooling juveniles, Crook 1997a, b), or at various stages throughout their life history (e.g. when they change sex). In general, colour changes can occur in response to the environment (e.g. a dark fish will pale if placed into a light coloured bucket, or a light fish will darken if exposed to UV due to melanin dispersion or "sun tanning"; primary physiological change), as part of behavioural interactions or communication (secondary physiological change), and during ontogenetic development (morphological colour change; Leclercq et al. 2010). Some types of colour changes are under the neural and/or endocrine control of the animals, e.g. changes during behavioural interactions, while others, such as ontogenetic colour changes, or sun tanning are not (for detailed reviews on fish colour changes see Leclercq et al. 2010, Nilsson Sköld et al. 2013).

Ontogenetic (Morphological) Colour Change

The larval stages of many species are already pigmented and display colour patterns (Baldwin 2013). These patterns change, often several times, as the fish mature and change sex (Fig. 1). Various hypotheses exist about why juveniles have different colours from adults, such as the 'intraspecific camouflage' hypothesis, where the idea is that juveniles gain access to adult territories as they are not recognized by conspecific adults (e.g. Fricke 1973, Booth 1990, Mahon 1994), and the 'adult habituation' hypothesis, which applies mainly to territorial fish. The latter hypothesis suggests that the juvenile colours attract the attention of the adults, which initially chase the juveniles but subsequently habituate to their presence so that they can establish themselves within adult territories (Thresher 1978). Most parrotfish are protogynous hermaphrodites and change colour shortly after changing from initial phase females into terminal phase males. Interestingly, some fishes start life as males (i.e., primary males) and have similar coloration as females, allowing them to gain access to females during group spawning events (Choat and Robertson 1975).

Behavioural Colour Change

Many fish change colour depending on their circumstances. Night colours are often different from day colours (Fig. 1a) and fast colour changes are commonly observed during communication, for example during courtship (Thresher 1984). In parrotfishes, the best studied behavioural colour change is that of *Ch. spilurus* (Crook 1997a, b, 1999a, b).

Chlorurus spilurus juveniles display three main physiological colour patterns (PCPs), striped, uniformly dark coloured and species-specific "bulls-eye" (white stripe across caudal peduncle with a central dark spot). Crook (1997a) found that the likelihood with which each pattern was displayed during field observations depended on fish size, schooling and feeding, and to a much lesser degree on the structural complexity of the background against which the fish were seen. Small fish swimming and feeding in schools were most often displaying the uniform or striped pattern, thus masking their species identity in similarly coloured heterospecific schools and reducing the likelihood of detection by predators, which often prefer rare prey (Almany et al. 2007). Larger individuals on the other hand tended to display the bullseye pattern. Crook hypothesized that the more conspicuous bullseye pattern may also be an adaptation to predator avoidance as the bullseye can be regarded as an 'eye-spot'. Such eye spots are found in a number of animals, including fish (e.g. *Pomacentrus amboinensis*; Gagliano 2008) and are thought to divert the attention of the predator away from the head of the fish or alternatively to intimidate potential predators (for review of eyespot functions see Stevens 2005). Put together, this indicates that parrotfish may be employing variable strategies of camouflage. Rather than attempting to blend into the background, which may be difficult for continuously moving fish feeding on the substrate, they either minimize their predation risk by blending into a school of other fish, or when alone by displaying the bullseye pattern.

Interactions with other species also influenced the PCP displayed as well as whether the fish stayed with a school or swam off alone, showing that these patterns may also have a role in communication in this species (Crook 1999a). Display of the bullseye in the school may indicate to the school that a particular fish is about to leave them. The extensive observations of *Ch. spilurus* by Crook have produced a wealth of correlative data, however experiments directly testing hypotheses about the function of colour and colour change are lacking. This is not an easy feat in parrotfish as they are difficult to maintain in aquaria, where conditions can be best controlled.

Mechanisms of Colour Change

Colour change in fish is relatively well understood, and can be achieved by dispersing or concentrating pigment granules within the chromatophores, or by changing the distance between the crystals in the multilayer stacks within the iridophores (Land 1972, Lythgoe and Shand 1989). Often, both mechanisms work together to produce the overall colour and colour change (Kasukawa et al. 1985, 1986, Fujii et al. 1989, Oshima et al. 1989). Iridophores can be motile (distance between stacks can be changed) and may be controlled by the sympathetic adrenergic system (Kasukawa et al. 1986) or by a combination of both, nervous and endocrine systems (Oshima et al. 1989). In species lacking motile iridophores, any colour change is due to dispersion and aggregation of pigment in the chromatophores only (Kasukawa and Oshima 1987).

The mechanism of colour change has not been directly investigated in parrotfish. The cream brown colours of the schooling juveniles suggest that pigment colours contained in chromatophores are responsible. It is unclear how the colour change between the different PCPs is controlled. There is evidence for hormonal control of the ontogenetic colour change between the initial and terminal phase of the stoplight parrotfish, *Sp. viride* (Cardwell and Liley 1991). The androgen 11-ketotestosterone (11-KT) was found in increased levels just before the sex change and again just before the colour change in *Sp. viride*, as the fish were changing into terminal phase males (Cardwell and Liley 1991). The authors further confirmed the hormonal control of colour change in these fish by initiating sex and colour change following administration of 11-KT to mature females.

Summary – Colours

In general, all fish have to solve the apparent trade-off between being conspicuous to intended receivers, such as potential mates, and at the same time being inconspicuous to potential predators. As a result, the colours we see today have evolved under the constraints of attempting to maximize both camouflage and conspicuousness, which, at first glance, appears difficult to achieve. It is however possible in cases where the visual systems of the intended observers and those of relevant predators or other "illegitimate receivers" (Alcock 2009) differ enough so that colours can be used for communication that are difficult to detect by potential predators while being conspicuous to intended receivers. The best-studied example of such a system is the communication with UV colours and patterns, which is often referred to as 'secret' or 'private communication' (Cummings et al. 2003, Siebeck 2004). Parrotfish may have developed another way to limit detection by predators, which usually detect their prey over larger distances. As juveniles they minimize detection by matching their colour patterns to form similarly coloured multi-species schools. As adults, their complimentary colours combined into complex patterns, attenuate in a way that the colours blur together and thus provide little contrast seen against the background (Marshall 2000a).

The Underwater Light Environment

The underwater light environment is highly variable and depends on the depth as well as the quality of the water (Jerlov 1976). Water essentially acts as a monochromator such that with increasing depth, the spectrum of available light is reduced until a narrow band of wavelengths around 475 nm is all that remains at several hundred metres depth, at least in clear waters (Jerlov 1976). In more turbid waters, or waters rich in organic material (DOM, dissolved organic material), the light spectrum is attenuated more rapidly compared to in clear waters. Both, the quantity and quality of the material suspended in the water column influences the light environment at any given depth. The light spectrum in coastal waters, for example, is generally long-wavelength shifted (perceived as greener by the human visual system), compared to the spectrum at the same depth in offshore waters, and light is more strongly attenuated with depth in coastal relative to offshore waters (Jerlov 1976). The reason for spectral shifts in different types of waters is due to the wavelength-specific nature of the attenuation process. As light comes in contact with water and other molecules, long wavelength light is most strongly affected by absorption while short wavelength light is mostly affected by scattering (Jerlov 1976). In the habitat of parrotfishes living on and around coral reefs, the spectrum of light includes ultraviolet wavelengths through to longer wavelengths (perceived as red by the human visual system).

Vision

An object is only visible if there is sufficient contrast between it and the background against which it is seen, and if sufficient numbers of photons reach the photoreceptor cells. The downwelling light spectrum as well as the light reflected off objects is filtered not only on its passage through water, but also once it reaches the eye. Light has to pass through the ocular media (cornea, lens and vitreous) on its way to the retina; where light energy is converted into action potentials, the language of the nervous system and the brain. Photoreceptors with different sensitivities and spectral properties in the retina are responsible for the detection and conversion of light. It is the comparison of the output of different photoreceptors tuned to different wavelengths which is analysed in the visual system and eventually leads to the perception of colour. Visual processing by the retina and brain of fishes is an area that is not well studied. The best-studied fish are zebrafishes as they serve as model organisms for a wide range of mostly neuroscience questions (Nikolaou et al. 2012).

Ocular Media

As light enters the eye, it has to pass through the cornea, lens and vitreous humour. Depending on the filter properties of these structures, the light spectrum and intensity is changed and/or reduced. Ultraviolet light is well known for its harmful properties, and many animals, including fish, protect their sensitive visual tissues in the retina by preventing UV from reaching these tissues (Douglas and McGuigan 1989, Siebeck and Marshall 2001, 2007, Losey et al. 2003). In most cases, the lens is the limiting filter of the eye (Siebeck and Marshall 2001) but UV-absorbing properties have also been found in the cornea (Siebeck and Marshall 2000, Siebeck et al. 2003) and vitreous of marine fish (Nelson et al. 2001). Mycosporine-like amino acids (MAAs) and/or carotenoids are responsible for the UV-absorbing properties of the ocular media (Dunlap et al. 1989, Siebeck et al. 2003). MAAs have also been found in the external mucus of many reef fishes, where they are thought to act as natural sunscreens (Zamzow and Losey 2002, Eckes et al. 2008).

The specific combination and quantities of the different MAAs found in fish, such as Palythine (320 nm), Asterina-330 (330 nm), Palythinol (332 nm) and Palythene (360 nm), determine how much UV light the ocular media absorb/transmit. Measuring ocular media transmission across different species of fish and/or ontogenetic series within a species of fish is a relatively quick way to establish the potential for UV vision (Siebeck and Marshall 2001, Losey et al. 2003). Transmission properties are generally reported as T50 values, i.e. the wavelength at which 50% of the maximal transmission is reached (see Douglas and McGuigan 1989, Siebeck and Marshall 2001). Based on the assessment of ocular media transmission properties, around 50% of all measured marine fish have a T50 value <400 nm and therefore possess the potential for UV vision (Douglas and McGuigan 1989, Siebeck and Marshall 2001, 2007).

All tested parrotfish species have ocular media which, taken together, block UV wavelengths (Table 1). The limiting filter is the lens with T50 values between 420 – 430 nm, while the cornea of parrotfish transmits some UV (T50 values between 380 – 394 nm; Siebeck and Marshall 2001). Parrotfishes and wrasses have the highest lens T50 values measured so far (Siebeck and Marshall 2000, 2001, 2007) and their lenses appear visibly yellow to human observers. A large range of possible lens T50 values has been found in marine fish.

Table 1. Transmission cut offs (T50 values, nm) for whole eyes, lenses and corneas

Species	Whole eye	Lens	Cornea	N	References
Calotomus carolinus	422	412	-	1+	1
Chlorurus perspicillatus	429	428	-	1+	1
Chlorurus microrhinos	427	426	389	1	2
Chlorurus spilurus TP	426-428	426-428	386-392	4+	1,2
Chlorurus spilurus IP	423	420	388	1	2
Scarus sp. Juvenile	423	421	388	1	2
Scarus iseri	424	423	394	2	2
Scarus dubius	429	422	-	1+	1
Scarus ghobban		424		1	2
Scarus schlegeli	427	426	389	1	2
Scarus psittacus TP	428	429	391	2	2
Scarus rivulatus	430-434	428	386	2	2
Scarus taeniopterus	431	429	386	1	2
Sparisoma aurofraenatum	420-433	420-431	384-390	4	2
Sparisoma viride IP	427	425	386	1	2
Sparisoma viride	426-432	425-432	380-391	4	2

Species: TP (terminal phase), IP (initial phase) and juvenile indicate cases for which the colour phase was noted; T50 values for **whole eyes, lenses** and **corneas** are given as ranges when more than one specimen was measured and the values differed; **N** indicates the number of specimens measured; + indicates cases for which the actual number of specimens is unknown – Losey et al. 2003 did not indicate how many specimens they measured; **References:** 1. Losey et al. 2003; 2 Siebeck and Marshall 2001.

At the other end of the range, the lowest lens T50 values (<310 nm) have been reported for damselfish species (Siebeck and Marshall 2007). While there is evidence for carotenoids in wrasse corneas (Siebeck and Marshall 2000), parrotfish cornea transmission curves do not show the three intermediate maxima and their corneas do not have the yellow coloration typical for carotenoids. Instead, parrotfish cornea transmission curves belong to class II, which have a single cut-off and a gradual change of transmission between maximal and minimal transmission (Siebeck and Marshall 2001).

The corneas of several parrotfish species are iridescent (Fig. 1b), a common feature of many shallow water teleosts (Lythgoe 1974, Siebeck et al. 2003). As in skin iridophores, iridescent colours in the cornea are produced by Rayleigh scattering (see section on structural colours above) and involve multilayer stacks of crystals interspersed with cellular material. Corneal iridophores can be formed by various different structures in the cornea, such as endoplasmic reticulum, connective tissue, protoplasm or collagen fibrils and the specific origin varies with species (Lythgoe 1975). *Scarus taeniopterus* was found to have the most commonly found type of corneal iridophore, in which the iridescent layer is found between Descemet's membrane and the stroma of the cornea (Lythgoe 1975). Four types of corneal iridophores described in fish are physiologically active, which means that light levels influence the appearance of the cornea as well as its transmission/reflectance properties (Shand 1988). In some species, the orientation of multilayer stacks is such that the spectrum and intensity of light is not altered. However, this is only true for light passing through the cornea at angles of incidence near to the normal. In contrast, light from oblique angles, such as direct sunlight from above is reflected away from the eye. Therefore iridescent corneas have been postulated to act as sunshades in these fish (Lythgoe 1974, 1975). It is also possible that the iridescent colours of the cornea act to camouflage the eye, which would otherwise stand out as a dark features on the colourful background of the parrotfish skin surrounding the eye (Lythgoe 1975).

In many species, the lens transmission properties change with age, due to the combination of the increase in the path length (caused by lens growth) and a variable amount of UV-absorbing compounds being deposited in the lens (Douglas and McGuigan 1989, Siebeck and Marshall 2007). The general pattern in parrotfish, but not other Labridae, is that the T50 values increase with lens size (Fig. 2). A threefold increase in lens diameter leads to a 10 nm shift towards longer wavelengths (Fig. 2), which is small compared to some other species (e.g. lemon damselfish *Pomacentrus moluccensis*, for which a four-fold increase in lens diameter leads to a 60 nm shift in lens T50). Interestingly, there are families that show a strong ontogenetic long-wavelength shift of T50 values, some families which do not show this shift (e.g. serranid species such as the coral trout, *Plectropomus leopardus*) and also families with lenses that transmit more UV with age as a result of decreasing T50 values (e.g. some holocentrid species; Siebeck and Marshall 2007, Fig. 2). It is not clear how the different patterns are achieved, mostly because the mechanisms for MAA acquisition, potential storage and transport to the eyes are currently unknown. Metazoans do not appear to be capable of producing MAAs *de novo* and must acquire MAAs via the food chain (Mason et al. 1998, Zamzow 2004). There is, however, some indication that, once acquired, fish can modify the structure of the MAA (Kandel 2012). There is some indication that MAAs can be stored for up to a week, but it is unclear where this would take place and how MAAs might be mobilised again when required (Zamzow et al. 2013). When fish are exposed to conditions lacking UV wavelengths, MAA levels in the epithelial mucus decrease (Zamzow 2004). It is currently unknown whether the MAAs in the ocular media are also affected by the environmental light conditions of the habitat of the fish.

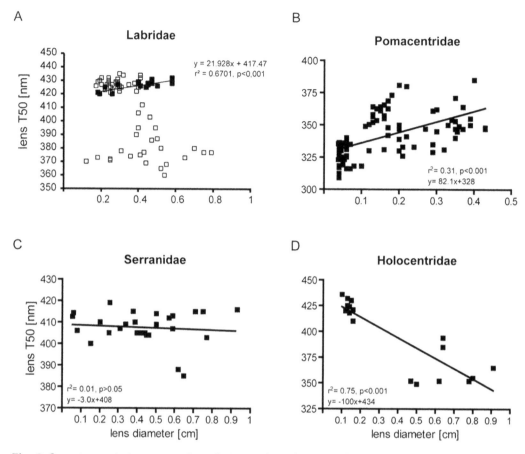

Fig. 2. Lens transmission properties relative to lens diameter for the (a) Labridae (parrotfishes: filled squares, wrasses: open squares), (b) Pomacentridae, (c) Serranidae, and (d) Holocentridae. Wavelengths, at which 50% of the maximal transmission is reached (T50 values) are given for all specimens measured. Overall, T50 values increase with increasing lens diameter. Different symbols indicate specimens of different species. Inset – regression equation and R^2 values.

Retina and Spectral Sensitivity

Teleost fish have a duplex retina, containing rods (dim vision) and cones (bright light and colour vision; Engström 1963a). Two different morphological cone types are found in fish, single and double cones. As the name suggests, double cones consist of two cones that are closely associated with each other. Unlike twin cones found in other teleosts, double cones have different spectral sensitivities (Engström 1963a, b). The discovery of close electric coupling between the two members of double cones has led to the assumption that the spectral sensitivity of the individual members is combined and thus that individual members cannot contribute separately to colour vision (Marchiafava 1985). This assumption has recently been challenged by a study showing that the double cones of the blackbar triggerfish, *Rhinecanthus aculeatus* do contribute to colour vision as separate channels (Pignatelli et al. 2010).

To date limited work has been published on eye morphology/retinal anatomy of parrotfishes. What little we know comes from a comparative study on settlement stage wrasses and parrotfishes by Lara (2001) who examined one *Scarus* sp. and four *Sparisoma* sp. larvae. Unlike wrasses, which have the unorganized cone arrangement typical for settlement stage larvae, settlement stage larvae of *Sparisoma* sp. and *Scarus* sp. have regular rows of narrow double cone pairs (Lara 2001). It is not known how this arrangement changes with age as the author did not examine juvenile or adult parrotfishes. Generally, the organisation of the photoreceptors in the retina increases until regular mosaics are formed (Collin and Pettigrew 1988a, b). Lara (2001) calculated visual acuity based on the distance between adjacent cone cells in the retinal preparations and found parrotfish larvae to have lower acuity than wrasse larvae (minimum separable angle across all settlement stage larvae, wrasses and parrotfishes, ranged between 34.6 and 86.6 min of arc), with wrasse larva acuity being at the lower end of the acuity of other published larval species. It is not clear how well those measurements reflect the behavioural acuity in these fish, as measurements are based on photoreceptors only, without taking the convergence at the level of the ganglion cells into account (ganglion cells usually integrate over more than one photoreceptor). Even when this convergence is taken into account as much as possible (convergence rates vary across the retina; Collin and Pettigrew 1989), behavioural acuity is often found to be different compared to calculations based on anatomical measurements (Champ et al. 2014). Many parrotfish have highly complex detailed patterns and it would be interesting to determine how well these fish can resolve these patterns and also, over which distances these patterns could be discriminated. Results would allow us to estimate if females can indeed discriminate between males on the basis of the individually different patterns and thus if the patterns contain clues (e.g. fitness) on the basis of which the females select their partners.

The spectral sensitivity of adult fish visual systems is generally well matched to the spectral properties of their habitat (Lythgoe 1979). A typical fish visual system is considered to include a cone type with spectral sensitivity matched to the wavelength of maximum downwelling light and a second cone type with spectral sensitivity off-set from the first (McFarland and Munz 1975, Loew and Lythgoe 1978, Lythgoe 1979, 1984). While the spectral sensitivities of parrotfish are currently unknown, some inferences can be made from the transmission properties of their ocular media. Parrotfish adults should be sensitive to blue through to red light and are unlikely to possess UV-sensitive cones. Larvae on the other hand have been hypothesized to have UV-sensitive cones (Lara 2001), however this was based purely on the organization of the photoreceptor mosaic in the retina, specifically the presence of corner cones. There are examples of fish species which have UV-sensitive cones as juveniles and which lose them before they fully mature (salmonids; e.g. Hawryshyn et al. 1989) so it is possible that larval parrotfish possess UV sensitivity while adults do not. Clearly, this needs to be confirmed with direct measurements of the cone spectral sensitivities of adult fishes.

Spectral sensitivity of photoreceptors can be measured using Microspectrophotometry (MSP; Bowmaker 1984). This technique is powerful as the absorbance of individual photoreceptors can be assessed. To date, only around 50 species of marine teleosts have been measured using this technique (Loew and Lythgoe 1978, Levine and MacNichol 1979, McFarland 1991, Shand 1993, Lythgoe et al. 1994, McFarland and Loew 1994, Britt et al. 2001, Losey et al. 2003).

Molecular genetics has been used to study the evolution of visual pigments and to identify the various opsins (pigment proteins) present in different photoreceptor types (e.g. Bowmaker 2008). In vertebrates, there are four spectrally distinct classes of cone visual pigments and a single rod opsin class. Depending on their specific tuning, these opsin classes produce visual pigments within a range of spectral sensitivities (SWS1 ~360 – 440 nm; SWS2 ~400 – 450 nm; RH1 ~480 – 510 nm; RH2 ~450 – 530 nm; M/LWS ~510

– 560 nm; Yokoyama 2008). Many teleost families have duplicated their opsin genes and as a result there may be several opsins within each class (Bowmaker 2008). While such a study has not been carried out for parrotfish, recent studies on damselfish and lampreys revealed five cone opsin genes (Rh2A/B, LWS, SWS1 and SWS2) and one rod opsin gene, Rh1 (Davies et al. 2007, Hofmann et al. 2012).

Visual pigment genetics and spectral sensitivities provide information about the potential an animal has for colour vision; however, behavioural experiments or experiments investigating visual processing are required to test this. If, for example, four spectral cone types are found, it is possible to conclude that the animal has the potential for tetrachromatic colour vision. The next step is then to test whether the signals from all cone types are compared, in which case tetrachromatic colour vision can be confirmed (Kelber et al. 2003). Behavioural experiments based on operant conditioning are frequently used to map out the spectral sensitivity of fish visual systems as well as test the potential for colour vision (e.g. Neumeyer 1984, Siebeck et al. 2008, Van-Eyk et al. 2011). Interestingly, the dimensionality of colour vision can be plastic and change depending on the brightness of the ambient illumination (e.g. Neumeyer and Arnold 1989). Also, the fish visual systems shows similarities with the human visual system in that separate processing channels exist for luminance and colour (Neumeyer et al. 1991, Siebeck et al. 2014), indicating that while differences exist in photoreceptor morphology and spectral types, the basic processes of vision may be conserved across vertebrates.

Overall, nothing is currently known about the ability of parrotfish to see and discriminate colours despite the multitude of colours and patterns the fish display. An increasing number of microspectrophotometry studies are being published; however so far they have not included parrotfish. Once spectral sensitivities are known, visual modelling can be used to predict which colours should be discriminable by a certain visual system (Osorio and Vorobyev 2005). The combination of visual modelling based on spectral sensitivity data with behavioural experiments is powerful as it allows the targeted testing of specific wavelength combinations, which is much more efficient than using behavioural experiments alone. It is important to remember that even if, for example, five spectral types of cones are present in a retina, it cannot be assumed that the animal is pentachromatic, unless behavioural experiments testing the underlying assumption (i.e. that the quantum catches of all receptors are compared) are carried out.

Polarization Vision

The sun radiates unpolarized light. Light becomes partially linearly polarized due to scattering in the atmosphere and water column, and also due to refraction and reflection at the water's surface (Horváth and Varjú 2004, Sabbah et al. 2005). The underwater polarization pattern is variable and depends on a range of variables, such as depth, quality of the water and, importantly, the position of the sun (Horváth and Varjú 2004, Sabbah et al. 2005, Waterman 2006). Many terrestrial and aquatic animals use polarized light or a solar compass for orientation and studies suggest that such visual cues may also be used by some reef fish (e.g. damselfish *Chromis viridis*; Hawryshyn et al. 2003, Mussi et al. 2005). The mechanism of polarization sensitivity in fish is unknown, but two hypotheses have been put forward: (i) internal reflection within double cones onto neighbouring ultraviolet-sensitive cones (Novales Flamarique et al. 1998) and (ii) axial dichroism of cone photoreceptors through tilting of the outer segment disc membranes (Roberts and Needham 2007). Interestingly, both mechanisms share two characteristic features: a regular, square cone photoreceptor mosaic and UV sensitivity. While it is currently unknown if parrotfish are sensitive to polarized light, it is unlikely if either of the above hypotheses are correct. Parrotfish are unlikely to have photoreceptors with maximal sensitivity in the UV due to their UV-blocking ocular media, and at least at the larval stage, parrotfish do not have a square photoreceptor mosaic (Lara 2001).

Summary – Vision

While detailed studies are lacking, parrotfishes appear to have a typical teleost retina with rods and cones. The only clue we have for the spectral sensitivity comes from ocular media studies which show that parrotfish have UV-blocking lenses. It is therefore likely that parrotfish are sensitive to wavelengths between 400 – 700 nm although photoreceptor sensitivities need to be determined via behavioural experiments and/or microspectrophotometric measurements before this can be confirmed. It appears that settlement stage parrotfish have lower visual acuities, and with that, can see less detail of the world around them compared to other teleost species. It is unknown how this changes throughout ontogenetic development and continued retinal growth.

Despite our very limited knowledge about the visual system and visual abilities of parrotfishes, the diversity of colours and colour patterns in combination with the changes in colour/pattern and sexual dimorphism clearly demonstrate that these features are important for parrotfishes. There is a wealth of future experiments waiting to be done on their visual system, the function of their colours and on the general visual behavior of parrotfishes. New methods for behavioural experiments in the field are necessary as parrotfishes are difficult to maintain in aquaria. This review has concentrated on visual cues in isolation of other sensory modalities, which of course is an unnatural scenario as a multitude of various sensory cues are available to any animal going about its business in its habitat. Ideally they should all be taken into account when evaluating the behaviour of an animal.

Acknowledgments

I would like to thank Justin Marshall for discussions about colour categories. The work was supported by the University of Queensland and the Australian Research Council (DP140100431).

References Cited

Alcock, J. 2009. Animal behavior: an evolutionary approach. 9th edition. Sinauer Associates, Sunderland, USA.

Almany, G.R., L.F. Peacock, C. Syms, M.I. McCormick and G.P. Jones. 2007. Predators target rare prey in coral reef fish assemblages. Oecologia 152: 751–761.

Bagnara, J.T., P.J. Fernandez and R. Fujii. 2007. On the blue coloration of vertebrates. Pigm. Cell Res. 20: 14–26.

Baldwin, C.C. 2013. The phylogenetic significance of colour patterns in marine teleost larvae. Zool. J. Linn. Soc. 168: 496–563.

Bellwood, D.R. and J.H. Choat. 1989. A description of the juvenile phase colour patterns of 24 parrotfish species (family Scaridae) from the Great Barrier Reef, Australia. Rec. Aust. Mus. 41: 1–41.

Bennett, A., I. Cuthill and K. Norris. 1994. Sexual selection and the mismeasure of color. Am. Nat. 144: 848–860.

Bonaldo, R.M., A.S. Hoey and D.R. Bellwood. 2014. The ecosystem roles of parrotfishes on tropical reefs. Oceanogr. Mar. Biol. Annu. Rev. 52: 81–132.

Booth, C.L. 1990. Evolutionary significance of ontogenetic colour change in animals. Biol. J. Linn. Soc. 40: 125–163.

Bowmaker, J.K. 1984. Microspectrophotometry of vertebrate photoreceptors: a brief review. Vision Res. 24: 1641–1650.

Bowmaker, J.K. 2008. Evolution of vertebrate visual pigments. Vision Res. 48: 2022–2041.

Britt, L.L., E.R. Loew and W.N. McFarland. 2001. Visual pigments in the early life stages of Pacific northwest marine fishes. J. Exp. Biol. 204: 2581–2587.

Cardwell, J.R. and N.R. Liley. 1991. Hormonal control of sex and color change in the stoplight parrotfish, *Sparisoma viride*. Gen. Comp. Endocrinol. 81: 7–20.

Champ, C.M., G. Wallis, M. Vorobyev, U.E. Siebeck and N.J. Marshall. 2014. Visual acuity in a species of coral reef fish: *Rhinecanthus aculeatus*. Brain Behav. Evol. 83: 31–42.

Choat, J.H. 1969. Studies on the biology of labroid fishes (Labridae and Scaridae) at Heron Island, Great Barrier Reef. PhD dissertation. University of Queensland, Brisbane, Australia.

Choat, J.H. and D.R. Robertson. 1975. Protogynous hermaphroditism in fishes of the family Scaridae. pp. 263–283. *In*: R. Reinboth (ed.). Intersexuality in the animal kingdom. Springer, Berlin.

Choat, J.H., O.S. Klanten, L. van Herwerden, D.R. Robertson and K.D. Clements. 2012. Patterns and processes in the evolutionary history of parrotfishes (Family Labridae). Biol. J. Linn. Soc. 107: 529–557.

Collin, S.P. and J.D. Pettigrew. 1988a. Retinal topography in reef teleosts. I. Some species with well-developed areae but poorly-developed streaks. Brain Behav. Evol. 31: 269–282.

Collin, S.P. and J.D. Pettigrew. 1988b. Retinal topography in reef teleosts. II. Some species with prominent horizontal streaks and high-density areae. Brain Behav. Evol. 31: 283–295.

Collin, S.P. and J.D. Pettigrew. 1989. Quantitative comparison of the limits on visual spatial resolution set by the ganglion cell layer in twelve species of reef teleosts. Brain Behav. Evol. 34: 184–192.

Cott, H.B. 1940. Adaptive Coloration in Animals. Methuen & Co. Ltd, London, UK.

Crook, A.C. 1997a. Colour patterns in a coral reef fish—Is background complexity important? J. Exp. Mar. Biol. Ecol. 217: 237–252.

Crook, A.C. 1997b. Determinants of the physiological colour patterns of juvenile parrotfish, *Chlorurus sordidus*. Anim. Behav. 53: 1251–1261.

Crook, A.C. 1999a. A quantitative analysis of the relationship between interspecific encounters, schooling behaviour and colouration in juvenile parrotfish (family Scaridae). Mar. Freshw. Behav. Physiol. 33: 1–19.

Crook, A.C. 1999b. Quantitative evidence for assortative schooling in a coral reef fish. Mar. Ecol. Prog. Ser. 176: 17–23.

Cummings, M.E., G.G. Rosenthal and M.J. Ryan. 2003. A private ultraviolet channel in visual communication. Proc. R. Soc. Lond. B Biol. Sci. 270: 897–904.

Davies, W.L., J.A. Cowing, L.S. Carvalho, I.C. Potter, A.E.O. Trezise, D.M. Hunt and S.P. Collin. 2007. Functional characterization, tuning, and regulation of visual pigment gene expression in an anadromous lamprey. FASEB J. 21: 2713–2724.

Denton, E.J. 1971. Reflectors in fishes. Sci. Am. 224: 64–72.

Denton, E.J. 1970. On the organization of reflecting surfaces in some marine animals. Philos. Trans. R. Soc. B Biol. Sci. 258: 285–313.

Douglas, R.H. and C.M. McGuigan. 1989. The spectral transmission of freshwater teleost ocular media—an interspecific comparison and a guide to potential ultraviolet sensitivity. Vision Res. 29: 871–879.

Dunlap, W.C., D.M. Williams, B.E. Chalker and A.T. Banaszak. 1989. Biochemical photoadaptation in vision: UV-absorbing pigments in fish eye tissues. Comparative Biochemistry and Physiology B Comparative Biochemistry 93: 601–607.

Eckes, M.J., U.E. Siebeck, S. Dove and A.S. Grutter. 2008. Ultraviolet sunscreens in reef fish mucus. Mar. Ecol. Prog. Ser. 353: 203–211.

Endler, J.A. 1990. On the measurement and classification of colour in studies of animal colour patterns. Biol. J. Linn. Soc. 41: 315–352.

Endler, J.A. and A.L. Basolo. 1998. Sensory ecology, receiver biases and sexual selection. Trends Ecol. Evol. 13: 415–420.

Engström, K. 1963a. Cone types and cone arrangements in teleost retinae. Acta Zool. (Stockh.) 44: 179–243.

Engström, K. 1963b. Structure, organization and ultrastructure of the visual cells in the teleost family Labridae. Acta Zool. (Stockh.) 44: 1–41.

Fox, H.M. and G. Vevers. 1960. The Nature of Animal Colours. Sidgwick & Jackson, London, UK.

Fricke, H.W. 1973. Individual partner recognition in fish: field studies on *Amphiprion bicinctus*. Naturwissenschaften 60: 204–206.

Fujii, R., H. Kasukawa, K. Miyaji and N. Oshima. 1989. Mechanisms of skin coloration and its changes in the blue-green damselfish, *Chromis viridis*: Physiology. Zool. Sci. 6: 477–486.

Gagliano, M. 2008. On the spot: the absence of predators reveals eyespot plasticity in a marine fish. Behav. Ecol. 19: 733–739.

Ghiradella, H., D. Aneshansley, T. Eisner, R.E. Silberglied and H.E. Hinton. 1972. Ultraviolet reflection of a male butterfly: interference color caused by thin-layer elaboration of wing scales. Science 178: 1214–1217.

Goda, M. and R. Fujii. 1995. Blue chromatophores in two species of callionymid fish. Zool. Sci. 12: 811–813.

Goda, M., M. Ohata, H. Ikoma, Y. Fujiyoshi, M. Sugimoto and R. Fujii. 2011. Integumental reddish-violet coloration owing to novel dichromatic chromatophores in the teleost fish, *Pseudochromis diadema*. Pigment Cell Melanoma Res. 24: 614–617.

Hawryshyn, C.W., M.G. Arnold, D.J. Chaisson and P.C. Martin. 1989. The ontogeny of ultraviolet photosensitivity in rainbow trout (*Salmo gairdneri*). Vis. Neurosci. 2: 247–254.

Hawryshyn, C.W., H.D. Moyer, W.T. Allison, T.J. Haimberger and W.N. McFarland. 2003. Multidimensional polarization sensitivity in damselfishes. J. Comp. Physiol. A 189: 213–220.

Hofmann, C.M., N.J. Marshall, K. Abdilleh, Z. Patel, U.E. Siebeck and K.L. Carleton. 2012. Opsin evolution in damselfish: convergence, reversal, and parallel evolution across tuning sites. J. Mol. Evol. 75: 79–91.

Horváth, G. and D. Varjú. 2004. Polarized light in animal vision: polarization patterns in nature. Springer, Berlin, Germany.

Jerlov, N.G. 1976. Marine optics. Elsevier, New York, USA.

Johnsen, S. 2014. Hide and seek in the open sea: pelagic camouflage and visual countermeasures. Annu. Rev. Mar. Sci. 6: 369–392.

Jordan, T.M., J.C. Partridge and N.W. Roberts. 2014. Disordered animal multilayer reflectors and the localization of light. J. Royal Soc. Interface 11: 20140948.

Jüttner, F., M. Stiesch and W. Ternes. 2013. Biliverdin: the blue-green pigment in the bones of the garfish (*Belone belone*) and eelpout (*Zoarces viviparus*). Eur. Food Res. Technol. 236: 943–953.

Kandel, F.L.M. 2012. Ultraviolet sunscreen on the coral reef: from coral to fish. PhD dissertation. University of Hawai'i at Manoa, Honolulu, Hawai'i.

Kasukawa, H. and N. Oshima. 1987. Divisionistic generation of skin hue and the change of shade in the scalycheek damselfish, *Pomacentrus lepidogenys*. Pigm. Cell Res. 1: 152–157.

Kasukawa, H., M. Sugimoto, N. Oshima and R. Fujii. 1985. Control of chromatophore movements in dermal chromatic units of blue damselfish—I. The melanophore. Comparative Biochemistry and Physiology – Part C. Comparative Pharmacology 81: 253–257.

Kasukawa, H., N. Oshima and R. Fujii. 1986. Control of chromatophore movements in dermal chromatic units of blue damselfish—II. The motile iridophore. Comparative Biochemistry and Physiology – Part C. Comparative Pharmacology 83: 1–7.

Kazancıoğlu, E., T.J. Near, R. Hanel and P.C. Wainwright. 2009. Influence of sexual selection and feeding functional morphology on diversification rate of parrotfishes (Scaridae). Proc. R. Soc. B Biol. Sci. 276: 3439–3446.

Kelber, A., M. Vorobyev and D. Osorio. 2003. Animal colour vision–behavioural tests and physiological concepts. Biol. Rev. 78: 81–118.

Land, M.F. 1972. The physics and biology of animal reflectors. Prog. Biophys. Mol. Biol. 24: 75–106.

Lara, M.R. 2001. Morphology of the eye and visual acuities in the settlement-intervals of some coral reef fishes (Labridae, Scaridae). Environ. Biol. Fishes 62: 365–378.

Leclercq, E., J.F. Taylor and H. Migaud. 2010. Morphological skin colour changes in teleosts. Fish Fish. 11: 159–193.

Leis, J.M. and B.M. Carson-Ewart. 2000. The larvae of Indo-Pacific coastal fishes: an identification guide to marine fish larvae. Brill, Leiden, The Netherlands.

Levine, J.S. and E.F. MacNichol. 1979. Visual pigments in teleost fishes: effects of habitat, microhabitat, and behavior on visual system evolution. Sens. Process. 3: 95–131.

Lim, I.E., S.K. Wilson, T.H. Holmes, M.M. Noble and C.J. Fulton. 2016. Specialization within a shifting habitat mosaic underpins the seasonal abundance of a tropical fish. Ecosphere 7: e01212.

Loew, E.R. and J.N. Lythgoe. 1978. The ecology of cone pigments in teleost fishes. Vision Res. 18: 715–722.

Longley, W.H. 1917. Studies upon the biological significance of animal coloration. II. A revised working hypothesis of mimicry. Am. Nat. 51: 257–285.

Lorenz, K. 1962. The function of colour in coral reef fishes. Proc. R. Inst. GB 39: 282–296.

Losey, G.S., W.N. McFarland, E.R. Loew, J.P. Zamzow, P.A. Nelson, N.J. Marshall and W.L. Montgomery. 2003. Visual biology of Hawaiian coral reef fishes. I. Ocular transmission and visual pigments. Copeia 203: 433–454.

Lythgoe, J.N. 1974. The ecology, function and phylogeny of iridescent multilayers in fish corneas. pp. 211–247. *In*: G.C. Evans, R. Bainbridge and O. Rackham (eds.). Light as an Ecological Factor: II. Blackwell Scientific Publications, Oxford, UK.

Lythgoe, J.N. 1975. The structure and function of iridescent corneas in teleost fishes. Proc. R. Soc. B Biol. Sci. 188: 437–457.

Lythgoe, J.N. 1979. Ecology of Vision. Clarendon Press, Oxford, UK.

Lythgoe, J.N. 1984. Visual pigments and environmental light. Vision Res. 24: 1539–1550.

Lythgoe, J.N. and J. Shand. 1989. The structural basis for iridescent colour changes in dermal and corneal iridophores in fish. J. Exp. Biol. 141: 313–325.

Lythgoe, J.N., W.R.A. Muntz, J.C. Partridge, J. Shand and D.M. Williams. 1994. The ecology of the visual pigments of snappers (Lutjanidae) on the Great Barrier Reef. J. Comp. Physiol. A 174: 461–467.

Mahon, J.L. 1994. Advantage of flexible juvenile coloration in two species of Labroides (Pisces: Labridae). Copeia 1994: 520–524.

Marchiafava, P.L. 1985. Cell coupling in double cones of the fish retina. Proc. R. Soc. B Biol. Sci. 226: 211–215.

Marshall, N.J. 2000a. Communication and camouflage with the same 'bright' colours in reef fishes. Philos. Trans. R. Soc. B Biol. Sci. 355: 1243–1248.

Marshall, N.J. 2000b. The visual ecology of reef fish colours. pp. 83–120 *In*: Y. Espmark, T. Amundsen and G. Rosenqvist (eds.). Animal Signals: Signalling and Signal Design in Animal Communication. Tapir Academic Press, Trondheim, Norway.

Mason, D.S., F. Schafer, J.M. Shick and W.C. Dunlap. 1998. Ultraviolet radiation-absorbing mycosporine-like amino acids (MAAs) are acquired from their diet by medaka fish (*Oryzias latipes*) but not by SKH-1 hairless mice. Comp. Biochem. Physiol. Part A Mol. Integr. Physiol. 120: 587–598.

McFarland, W.N. 1991. The visual world of coral reef fishes. pp. 16–38 *In*: P.F. Sale (ed.). The Ecology of Fishes on Coral Reefs. Academic Press, Inc, Sand Diego, USA.

McFarland, W.N. and E.R. Loew. 1994. Ultraviolet visual pigments in marine fishes of the family Pomacentridae. Vision Res. 34: 1393–1396.

McFarland, W.N. and F.W. Munz. 1975. Part III: The evolution of photopic visual pigments in fishes. Vision Res. 15: 1071–1080.

Muntz, W.R.A. 1990. Stimulus, environment and vision in fishes. pp. 491–511 *In*: R.H. Douglas and M.B.A. Djamgoz (eds.). The Visual System of Fish. Chapman and Hall Ltd, London, UK.

Mussi, M., T.J. Haimberger and C.W. Hawryshyn. 2005. Behavioural discrimination of polarized light in the damselfish *Chromis viridis* (family Pomacentridae). J. Exp. Biol. 208: 3037–3046.

Nelson, P.A., J.P. Zamzow and G.S. Losey. 2001. Ultraviolet blocking in the ocular humors of the teleost fish *Acanthocybium solandri* (Scombridae). Can. J. Zool. 79: 1714–1718.

Neumeyer, C. 1984. On spectral sensitivity in the goldfish: evidence for neural interactions between different "cone mechanisms". Vision Res. 24: 1223–1231.

Neumeyer, C. and K. Arnold. 1989. Tetrachromatic color vision in the goldfish becomes trichromatic under white adaptation light of moderate intensity. Vision Res. 29: 1719–1727.

Neumeyer, C., J.J. Wietsma and H. Spekreijse. 1991. Separate processing of "color" and "brightness" in goldfish. Vision Res. 31: 537–549.

Nikolaou, N., A.S. Lowe, A.S. Walker, F. Abbas, P.R. Hunter, I.D. Thompson and M.P. Meyer. 2012. Parametric functional maps of visual inputs to the tectum. Neuron 76: 317–324.

Nilsson Sköld, H., S. Aspengren and M. Wallin. 2013. Rapid color change in fish and amphibians– function, regulation, and emerging applications. Pigment Cell Melanoma Res. 26: 29–38.

Novales-Flamarique, I., C.W. Hawryshyn and F.I. Hárosi. 1998. Double-cone internal reflection as a basis for polarization detection in fish. J. Opt. Soc. Am. A 15: 349–358.

Oshima, N., H. Kasukawa and R. Fujii. 1989. Control of chromatophore movements in the blue-green damselfish, *Chromis viridis*. Comp. Biochem. Physiol. Part C Toxicol. Pharmcol. 93: 239–245.

Osorio, D. and M. Vorobyev. 2005. Photoreceptor sectral sensitivities in terrestrial animals: adaptations for luminance and colour vision. Proc. R. Soc. B Biol. Sci. 272: 1745–1752.

Osorio, D. and M. Vorobyev. 2008. A review of the evolution of animal colour vision and visual communication signals. Vision Res. 48: 2042–2051.

Pignatelli, V., C. Champ, N.J. Marshall and M. Vorobyev. 2010. Double cones are used for colour discrimination in the reef fish, *Rhinecanthus aculeatus*. Biol. Lett. 6: rsbl20091010.

Randall, J.E., G.R. Allen and R.C. Steene. 1997. Fishes of the Great Barrier Reef and Coral Sea. Crawford House Publishing Pty Ltd, Bathurst, Australia.

Roberts, N.W. and M.G. Needham. 2007. A mechanism of polarized light sensitivity in cone photoreceptors of the goldfish *Carassius auratus*. Biophys. J. 93: 3241–3248.

Sabbah, S., A. Lerner, C. Erlick and N. Shashar. 2005. Under water polarization vision—a physical examination. Recent Res. Dev. Exp. Theor. Biol 1: 123–176.

Salzburger, W., H. Niederstätter, A. Brandstätter, B. Berger, W. Parson, J. Snoeks and C. Sturmbauer. 2006. Colour-assortative mating among populations of *Tropheus moorii*, a cichlid fish from Lake Tanganyika, East Africa. Proc. R. Soc. B Biol. Sci. 273: 257–266.

Shand, J. 1988. Corneal iridescence in fishes: light-induced colour changes in relation to structure. J. Fish Biol. 32: 625–632.

Shand, J. 1993. Changes in the spectral absorption of cone visual pigments during the settlement of the goatfish *Upeneus tragula*: the loss of red sensitivity as a benthic existence begins. J. Comp. Physiol. A 173: 115–121.

Siebeck, U.E. 2004. Communication in coral reef fish: the role of ultraviolet colour patterns in damselfish territorial behaviour. Anim. Behav. 68: 273–282.

Siebeck, U.E. and N.J. Marshall. 2000. Transmission of ocular media in labrid fishes. Philos. Trans. R. Soc. B Biol. Sci. 355: 1257–1261.

Siebeck, U.E. and N.J. Marshall. 2001. Ocular media transmission of coral reef fish—can coral reef fish see ultraviolet light? Vision Res. 41: 133–149.

Siebeck, U.E. and N.J. Marshall. 2007. Potential ultraviolet vision in pre-settlement larvae and settled reef fish—A comparison across 23 families. Vision Res. 47: 2337–2352.

Siebeck, U.E., S.P. Collin, M. Ghoddusi and N.J. Marshall. 2003. Occlusable corneas in toadfishes: light transmission, movement and ultrastructure of pigment during light- and dark-adaptation. J. Exp. Biol. 206: 2177–2190.

Siebeck, U.E., G.M. Wallis and L. Litherland. 2008. Colour vision in coral reef fish. J. Exp. Biol. 211: 354–360.

Siebeck, U.E., G.M. Wallis, L. Litherland, O. Ganeshina and M. Vorobyev. 2014. Spectral and spatial selectivity of luminance vision in reef fish. Front. Neural Circuits 8: 118.

Stevens, M. 2005. The role of eyespots as anti-predator mechanisms, principally demonstrated in the Lepidoptera. Biol. Rev. 80: 573–588.

Temple, S., N.S. Hart, N.J. Marshall and S.P. Collin. 2010. A spitting image: specializations in archerfish eyes for vision at the interface between air and water. Proc. R. Soc. B Biol. Sci. 277: 2607–2615.

Thresher, R.E. 1978. Territoriality and aggression in the threespot damselfish (Pisces: Pomacentridae): an experimental study of causation. Z. Tierpsychol. 46: 401–434.

Thresher, R.E. 1984. Reproduction in Reef Fishes. TFH Publications Inc., Neptune City, USA.

Van-Eyk, S.M., U.E. Siebeck, C.M. Champ, J. Marshall and N.S. Hart. 2011. Behavioural evidence for colour vision in an elasmobranch. J. Exp. Biol. 214: 4186–4192.

Vorobyev, M. and D. Osorio. 1998. Receptor noise as a determinant of colour thresholds. Proc. R. Soc. B Biol. Sci. 265: 351–358.

Waterman, T.H. 2006. Reviving a neglected celestial underwater polarization compass for aquatic animals. Biol. Rev. 81: 111–115.

Wucherer, M.F. and N.K. Michiels. 2012. A fluorescent chromatophore changes the level of fluorescence in a reef fish. PLoS One 7: e37913.

Wyszecki, G. and W.S. Stiles. 2000. Color Science: Concepts and Methods, Quantitative Data and Formulae. 2nd edition. Wiley and Sons, New York, USA.

Yamaguchi, K., K. Kubo, K. Hashimoto and F. Matsuura. 1977. Linkages between chromophore and apoprotein in the biliverdin-protein of the scales of big blue parrotfish, *Scarus gibbus* Rüppell. Experientia 33: 583–584.

Yokoyama, S. 2008. Evolution of dim-light and color vision pigments. Annu. Rev. Genomics Hum. Genet. 9: 259–282.

Zamzow, J. 2004. Effects of diet, ultraviolet exposure, and gender on the ultraviolet absorbance of fish mucus and ocular structures. Mar. Biol. 144: 1057–1064.

Zamzow, J.P. and G.S. Losey. 2002. Ultraviolet radiation absorbance by coral reef fish mucus: photo-protection and visual communication. Environ. Biol. Fishes 63: 41–47.

Zamzow, J.P., U.E. Siebeck, M.J. Eckes and A.S. Grutter. 2013. Ultraviolet-B wavelengths regulate changes in UV absorption of cleaner fish *Labroides dimidiatus* mucus. PLoS One 8: e78527.

CHAPTER
6

Feeding in Parrotfishes: The Influence of Species, Body Size, and Temperature

Andrew S. Hoey

ARC Centre of Excellence for Coral Reef Studies, James Cook University,
Townsville, Queensland 4811, Australia
Email: Andrew.hoey1@jcu.edu.au

Introduction

Feeding, the process by which animals acquire nutrients, and hence energy for maintenance, growth and reproduction, is central to the health of individuals and populations. Indeed, variation in nutritional quality and quantity of diets are one of the most important determinants of animal fitness (Sterner and Elser 2002), and has led to several theories that relate the availability and/or quality of dietary resources to the foraging decisions of animals (e.g., Optimal Foraging Theory; Smith 1978). Although there has been debate as to the nutritional targets of parrotfishes, recent evidence suggests they are targeting protein-rich epilithic, endolithic and epiphytic microscopic phototrophs (Clements et al. 2017, Clements and Choat Chapter 3). The primary goal of feeding in parrotfishes must be to acquire sufficient nutrients from these dietary sources to fuel metabolic processes, and hence the spatial and temporal variation in the quality and quantity of these resources may shape the distribution and foraging patterns of parrotfishes (e.g., Russ et al. 2015). Despite sharing broadly similar diets, there is considerable inter- and intra-specific variation in the way parrotfishes feed, in particular the frequency of feeding and the extent to which they disturb benthic communities when feeding (e.g., Bellwood and Choat 1990, Bruggemann et al. 1994a, b).

In this chapter, I provide an overview of what is currently known of the feeding of parrotfishes. I acknowledge that acquisition of nutrition is central to feeding biology; however this is the focus of another chapter in this volume (Clements and Choat Chapter 3), and is beyond the scope of this chapter. Instead, in this chapter I focus on factors that influence the consequences of feeding, rather than its causes. I start by briefly describing differences in the morphology of the feeding apparatus of parrotfishes and relate these to differences in feeding rate and feeding impact among species. I then explore the influence of body size on the rate of feeding, and the disturbance of individual bites on the reef surface, both among and within species. Finally, I consider the potential influence of environmental conditions, namely temperature, on feeding in parrotfishes.

Inter-specific Variation in Feeding of Parrotfishes

Parrotfishes (Scarinae, Labridae) are a monophyletic group of approximately 100 species (Parenti and Randall 2011) that are distributed throughout the world's tropical oceans (Kulbicki et al. Chapter 10). It is their specialized feeding morphology, however, that has attracted much scientific interest and makes the parrotfish unique. Parrotfishes are characterised by coalesced oral teeth that form strong beak-like jaws, and a well-developed pharyngeal jaw apparatus (Gobalet 1989, Bellwood and Choat 1990). Importantly, the pharyngeal jaw apparatus of parrotfishes has several modifications from that of other members of the Labroidei that allow them to grind or mill, as opposed to crush, ingested calcareous material (Wainwright and Price Chapter 2). These modifications of the oral and pharyngeal jaws facilitate the unique feeding modes of parrotfishes and allow them to access dietary resources that are largely inaccessible to other herbivorous fishes (Bonaldo et al. 2014, Clements et al. 2017).

Despite the presence of fused beak-like oral jaws and modified pharyngeal jaws characterising the parrotfishes, it is the variation in these elements, coupled with differences in feeding behaviour, among species (Bellwood 1994) that has led to the differentiation of feeding modes among parrotfish species. Based largely on differences in the osteology and myology of the oral and pharyngeal jaws, parrotfishes are commonly classified into three main feeding modes, or functional groups: browsers, scrapers and excavators (Bellwood 1994, Green and Bellwood 2009), with these functional groups broadly aligning with major phylogenetic groupings (Streelman et al. 2002, Choat et al. 2012). Morphological and molecular phylogenies of parrotfish genera have identified two major clades: the scarinine (or reef) and sparisomatine (or seagrass) clades (Bellwood 1994, Streelman et al. 2002). The sparisomatine clade (i.e., *Calotomus*, *Cryptotomus*, *Leptoscarus*, *Nicholsia* and *Sparisoma*) is most abundant and speciose in the tropical Atlantic, with most species exhibiting few modifications of the feeding apparatus from the condition found in most other labrids (Bellwood 1994). The teeth on the oral jaws are discrete, arranged in oblique rows, with little or no evidence of external cementation, and the lower jaw and associated muscles exhibit the typical labrid form (Bellwood 1994). These features, together with observations of sparisomatine parrotfishes taking bites from the surfaces of marine macrophytes (seagrass and macroalgae), led to this group being classified as 'browsers' (but see Clements and Choat Chapter 3).

In contrast to the general sparisomatine condition, scarinine parrotfishes (*Bolbometopon*, *Chlorurus*, *Cetoscarus* and *Hipposcarus*, and *Scarus*) are most abundant and speciose in the Indo-Pacific, with most species feeding on hard calcareous surfaces colonised by algal turfs and associated trapped detrital material (i.e., epilithic algal matrix; EAM), and endolithic phototrophs. The teeth on the oral jaws form a mosaic of vertical and oblique rows, with varying levels of external cementation, the muscles that close the lower jaw are highly modified, and the pharyngeal jaws are more robust and the grinding surface and joint with the neurocranium are more elongate (for detailed descriptions see Bellwood 1994, Wainwright and Price Chapter 2). These modifications allow scarinine parrotfishes to remove pieces of the reef framework when feeding and grind these pieces of carbonate to sand (Gobalet Chapter 1, Wainwright and Price Chapter 2). Scarinine parrotfishes may be further divided based on the strength and mobility of the osteology and myology of the jaws. Members of the scarinine genera *Bolbometopon*, *Cetoscarus* and *Chlorurus* have robust jaws with simple and strong articulations among elements, and a relatively large and modified adductor mandibulae in which the A3 component inserts directly on the dentary (Bellwood and Choat 1990, Bellwood 1994). Collectively these features generate a forceful bite, and enable these species to 'excavate' the reef surface when feeding. In comparison, members of the scarinine genera *Hipposcarus* and *Scarus* have a gracile feeding apparatus with relatively mobile articulations in the oral jaw and smaller musculature in which the A1 and A2 components of the adductor mandibulae are fused (Bellwood 1994), and consequently these species tend to 'scrape' the reef surface when feeding. It should be noted that some sparisomatine parrotfishes exhibit several features characteristic of excavating (*Sparisoma amplum* and *Sparisoma viride*) and scraping (*Sparisoma aurofrenatum*) scarinine parrotfishes (Bruggemann et al. 1994a, Bernardi et al. 2000, Robertson et al. 2006).

How well do Functional Groups Reflect the Feeding Impact of Parrotfishes?

Reductionist approaches, such as the classification of parrotfishes by functional groups, although providing a useful starting point for exploring commonalities and describing broad patterns often obscure important interspecific differences. Differences among the three parrotfish functional groups have been related to the variation in the size of individual bites, the frequency of feeding, and hence the nature and intensity of the disturbance impact of these groups on the reef surface (e.g., Bellwood and Choat 1990, Bruggemann et al. 1994b, Hoey and Bellwood 2008). These generalities have often led to estimates of feeding impact of one species being applied across closely related taxa when estimating the impact of entire assemblages (e.g., Bellwood et al. 2003, Hoey and Bellwood 2008). But, how well do the functional groups relate to the feeding impact of component parrotfish species?

Another characteristic of many parrotfishes are the distinctive bite marks or 'feeding scars' they leave on the surfaces on which they feed. It is the presence of these feeding scars that has allowed the bite sizes of parrotfishes to be readily measured in the field, and the quantity of material they remove to be estimated. The area and depth of these feeding scars is, at least to some extent, related to functional group (Bellwood and Choat 1990). Feeding scars of scraping parrotfishes are characterised by two parallel feeding marks with the length being approximately 10 times as long as they are wide scars, and the depth of the scar generally < 1 mm (Hoey and Bellwood 2008). In contrast, feeding scars of excavating parrotfishes are approximately twice as long as they are wide, with four or more deep grooves running parallel to the major axis, and tend to be much deeper than those of scraping parrotfishes (Bellwood and Choat 1990, Hoey and Bellwood 2008). Parrotfishes with a browsing feeding mode generally do leave feeding scars on the reef substratum. Such descriptions may give the impression that all excavating and scraping parrotfishes leave bite marks when feeding; however for many species it is the exception rather than the rule.

Investigating the feeding of 24 parrotfish species on the Great Barrier Reef (GBR), Bellwood and Choat (1990) noted that while all seven excavating species (*Bolbometopon muricatum*, *Cetoscarus ocellatus*, *Chlorurus microrhinos*, *Chlorurus bleekeri*, *Chlorurus frontalis*, *Chlorurus japanensis* and *Chlorurus spilurus*) left bite marks when feeding, only three out of 17 scraping species left bite marks, and only occasionally. These qualitative observations are supported by quantitative assessment in Bonaire where adult individuals of the excavating *Sp. viride* were observed to leave bite marks almost twice as often as the scraping *Scarus vetula* (ca. 80 and 40 percent of bites, respectively; Bruggemann et al. 1994b). Similarly, recent observations of over 1,300 individual parrotfish bites from 20 species of parrotfish on Lizard Island, northern GBR revealed considerable variation in the frequency of bite scars among species. The frequency of scarring ranged from 28.8 to 82.4 percent of bites

(*Ch. spilurus* and *Ch. microrhinos*, respectively) for excavating species, and from less than three percent (*Scarus chameleon*, *Scarus globiceps*, *Scarus psittacus*, *Scarus schlegeli* and *Scarus spinus*) to 53.6 percent (*Scarus freantus*) for scraping taxa (Hoey and Fox unpublished data).

Together with the frequency of feeding scars, there are marked differences in the size of individual bites among parrotfish species. The amount of material removed from the reef substratum when feeding is a product of the fish's gape, and hence body length (discussed below), and the force applied to the substratum when biting. Given the forceful bite of excavating parrotfishes it is not surprising that they tend to scrape a larger area and remove a greater volume of reef substratum per bite than scraping parrotfishes (e.g., Bruggemann et al. 1994b, Bellwood 1995). For example, the area and volume of feeding scars of *Sc. vetula* were estimated to be ca. 30 and 86 percent smaller than those of *Sp. viride*, respectively (Bruggemann et al. 1994b). Greater variation in bite volume (and area) is evident among excavating taxa, ranging from 1.66 cm^3 (4.99 cm^2) in *B. muricatum* to 0.26 cm^3 (1.70 cm^2) in *Ch. microrhinos*, and 0.002 cm^3 (0.22 cm^2) in *Ch. spilurus* (Table 1). Similar variation in bite volume is evident among scraping parrotfish species, although data on bite area are relatively rare or not reported (Table 1).

The potential disturbance impact of an individual parrotfish is not only influenced by the size of the impact (or bite), but also the frequency at which those individual impacts are delivered. Feeding rate was one of the initial features used to differentiate parrotfishes into functional groups, with excavating taxa generally taking fewer bites than scraping taxa (Bellwood and Choat 1990). Indeed, reported feeding rates of excavating species are on average lower than those of scraping species, although there is considerable variation among species within each group (Table 1). For example, the feeding rates of excavating species range from 2.1 bites min^{-1} for *B. muricatum* to 24 bites min^{-1} for *Sp. viride*, and scraping species range from 7.1 to 66.0 for *Sc. chameleon* and *Sc. vetula*, respectively (Table 1). It is noteworthy that some of the species with the largest bite sizes have the lowest feeding rate, and conversely some of the species that rarely, if ever, produced a bite scar when feeding had the highest feeding rates. So what is the net effect of these differences in bite sizes and feeding rate on the disturbance impact of parrotfishes?

Table 1. Variation in feeding rate, bite area, and bite volume of parrotfishes among and within functional groups. Where two or more values were available a range is given

Species	Maximum length (cm)	Feeding rate (bites min^{-1})	Bite area (cm^2)	Bite volume (cm^3)
Excavators				
Bolbometopon muricatum	130	2.1[1]–6.08[2]	4.99[2]	1.66[2]
Cetoscarus bicolor	90	4.39[3]–5.88[4]		0.110[4]
Cetoscarus ocellatus	80	3.9[1]		
Chlorurus bleekeri	49	15.6[1]		
Chlorurus genazonatus	31	11.63[3]		
Chlorurus gibbus	70	6.38[4]–7.35[3]		0.114[4]
Chlorurus microrhinos	70	7.9[1]–9.6[1]	1.70[14]	0.26[14]
Chlorurus perspicillatus	60	12.89[5]–15.59[5]		0.05 - 0.07[5]
Chlorurus spilurus	40	13.8[1]–14.8[1]	0.22[14]	0.002[14]
Chlorurus sordidus	40	13.88[3]–15.3[4]		0.008[4]

(Contd.)

Chlorurus strongylocephalus	70	12.9[6]		
Sparisoma amplum	39	0.8[8]–7.0[9]		
Sparisoma viride	64	24.0[10]	1.03[15]	0.08[15]
Scrapers				
Scarus chameleon	31	7.1[1]–9.6[1]		
Scarus ferrugineus	41	11.88[4]–17.52[3]		0.009[4]
Scarus flavipectoralis	40	15.5[1]		
Scarus frenatus	47	10.72[4]–21.66[3]		0.011[4]
Scarus fuscopurpureus	38	14.99[3]		
Scarus ghobban	90	10.92[4]–32.5[16]		0.063[4]
Scarus globiceps	45	39.3[1]		
Scarus niger	40	11.3[6]–30.18[3]		0.002[4]
Scarus oviceps	35	12.9[1]		
Scarus persicus	50	21.4[3]		
Scarus psittacus	30	21.6[1]–25.02[3]		
Scarus rivulatus	40	17.3[1]–32.0[7]	0.01-0.65[7*]	
Scarus rubroviolaceus	70	10.1[1]–15.84[5]		0.05-0.09[5]
Scarus schlegeli	40	16.7[1]–20.1[1]		
Scarus spinus	30	45.0 (6.7)		
Scarus spp			0.104[2]	0.001[2]
Scarus iseri	35	40.4[11]		
Scarus trispinosus	35	9.8[12]–17.8[9]		
Scarus vetula	61	66.0[10]	0.36[10]	0.005[10]
Scarus zelindae	33	11.0[12]–19.42[12]		
Browsers				
Sparisoma axillare	37	1.0[8]–9.72[12]		
Sparisoma cretense	50	6.86[11]		
Sparisoma frondosum	34	3.0[8]–7.11[9]		
Sparisoma tuiupiranga	15	1.83[13]–5.38[13]		

Maximum lengths sourced from Froese and Pauly (2017).
Data sources: 1. Bellwood and Choat (1990), 2. Hoey and Bellwood (2008), 3. Hoey et al. (2016), 4. Alwany et al. (2009), 5. Ong and Holland (2010), 6. Lokrantz et al. (2008), 7. Bonaldo and Bellwood (2008), 8. Bonaldo et al. (2006), 9. Francini-Filho et al. (2008), 10. Bruggemann et al. (1994a), 11. Bonaldo et al. (2014), 12. Francini-Filho et al. (2010), 13. Ferreira et al. (1998), 14. Bellwood (1995), 15. Bruggemann et al. (1994b), Smith (2008). * indicates that the reported area and volume related to the algal material removed, as opposed to reef carbonates.

While differences in feeding rates may dampen the differences in bite sizes among species, there is still considerable variation in the potential area scraped and volume eroded among parrotfish taxa. For excavating parrotfishes, estimated annual per capita erosion rates vary from 23 kg per individual per yr for *Ch. spilurus*, 380 kg per individual per yr for *Ch. perspicillatus* (Ong and Holland 2010), 1,018 kg per individual per yr for *Ch. microrhinos* (Bellwood 1995), to over 5,500 kg per individual per yr for *B. muricatum* (Hoey and Bellwood 2008). This equates to a single *B. muricatum* excavating as much material as 240 *Ch. spilurus*, 14 *Ch. perspicillatus*, or 5.5 *Ch. microrhinos*. These patterns are even more pronounced when considering rates of coral predation. *Bolbometopon muricatum*, the largest parrotfish (up to 130 cm total length, TL), has been reported to take approximately half of its bites from live corals (Hoey and Bellwood 2008), while medium-bodied species (e.g., *Cetoscarus bicolor* and *Chlorurus gibbus* up to 80 cm TL) take up to 30 percent of bites and small-bodied species (e.g., *Chlorurus sordidus* up to 40 cm TL) take less than five percent of bites from live coral (e.g., Alwany et al. 2009, Hoey et al. 2016). Such estimates not only highlight the importance of *B. muricatum* in reef bioerosion and coral predation but also the overriding influence of body size.

Similar variation is evident among scraping parrotfishes, with estimated annual erosion rates varying from zero for relatively small species that rarely leave of feeding scars on the substratum (e.g., *Scarus globiceps*, *Scarus iseri*, *Scarus spinus*) to 380 kg per individual per yr for larger species such as *Sc. rubroviolaceus* (Ong and Holland 2010). Although few studies have directly compared the area of the reef surface that is scraped or cleared among individual parrotfish species, considerable variation may be expected given the differences in feeding rate and the proportion of bites that produce feeding scars.

Effect of Body Size on Functional Impact of Parrotfishes

Body size is an important and fundamental trait, influencing metabolic rates of individuals, predator-prey relationships, and hence population abundance and community structure across a range of taxa (e.g., Sterner and Elser 2002, Brown et al. 2004, Messmer et al. 2017, Dunic and Baum 2017). These relationships may be driven by allometry between metabolic rate and body size (e.g., Clarke and Johnston 1999, Brown et al. 2004), with smaller-bodied individuals and species typically having higher energy demands per unit of body mass. Further, allometric scaling of elements of the feeding apparatus, together with changes in feeding behaviour, may influence feeding mode and potential diet of organisms (e.g., Wainwright et al. 2004). Indeed, differences in body size may explain some of the interspecific variation in bite size, feeding rate, and functional impact of parrotfishes described above, and as such we may also expect to see similar intraspecific variation in feeding in parrotfishes.

As an organism grows, the rate of change in the elements of the feeding apparatus can have important implications for its feeding biology. An analysis of the feeding apparatus of 130 labrid species (including parrotfishes) from the GBR revealed that while some features of the feeding morphology scaled isometrically (e.g., gape of the oral jaws), others such as the mass of the adductor mandibulae and levator posterior (i.e., muscles that close the oral and pharyngeal jaws, respectively) showed strong positive allometry (Wainwright et al. 2004). Such changes suggest that the dimensions of individual bites will scale proportionally with fish length, while bite force will increase disproportionately to fish mass. Of the few studies that have investigated how the parrotfish body size relates to the proportion of bites that result in feeding scars, all have reported positive relationships (e.g., Bruggeman et al. 1994a, 1994b, Bonaldo and Bellwood 2008, Ong and Holland 2010). The nature of this relationship, however, appears to differ among species. For example, the proportion of bites that resulted in feeding scars increased from ca. 12 percent in small (5-14 cm fork length, FL) *Sc. vetula* and *Sp. viride* to ca. 30 and 75 percent in 25-34 cm FL *Sc. vetula* and *Sp. viride*, respectively (Bruggemann et al. 1996). Similarly, the proportion of bites that resulted in feeding scars increased with body size for *Ch. perspicillatus* and *Sc. rubroviolaceus*, but did not differ between species (Ong and Holland 2010). Comparisons of the feeding of six parrotfishes from Lizard Island, northern GBR emphasises this variation. Three of the six species show a marked increase in the proportion of bites leaving feeding scars above 20 cm total length (i.e., *Ch. microrhinos, Ch. spilurus, Sc. frenatus*), while others show a more gradual change (i.e., *Ce. ocellatus, Sc. frenatus*), and some (e.g., *Sc. schlegeli*) rarely produce feeding scars (Fig. 1). Irrespective of the rate of change, there was considerable interspecific variation in the proportion of bites resulting in scars for individual size classes (Fig. 1).

Given the scaling of elements of the feeding apparatus with body size (e.g., Wainwright et al. 2004) it is not surprising that the size of parrotfish feeding scars also scales with body size (e.g., Bruggemann et al. 1994a, b, Bonaldo and Bellwood 2008, Ong and Holland 2010). While these studies report the area of bite scars to be linearly related to the square of body length, and the volume of bite scars to be linearly related to body mass or the cube of body length, differences in methods used make comparisons among studies difficult. Examination of feeding scars for several parrotfish taxa from the northern GBR highlight differences in both the absolute bite size and the rate of change of bite area with body length among species (Fig. 2). The area of individual feeding scars of excavating taxa, and in particular large excavating taxa (Fig. 2b, c), are larger and tend to increase more rapidly with body length than smaller scraping taxa (Fig. 2e, f).

Together with changes in the potential disturbance impact of individual parrotfish bites, the feeding (or bite) rate of parrotfishes has been reported to vary with body size and/or life stage (e.g., Bruggemann et al. 1994a, 1994b, Ong and Holland 2010). While changes in feeding rate are undoubtedly linked to the nutritional requirements of individuals, it

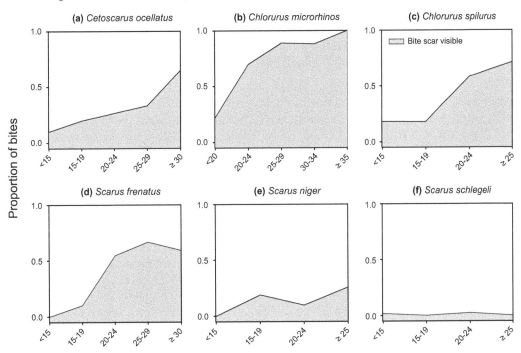

Fig. 1. Variation in proportion of parrotfish bites that produce visible feeding scars among species and size classes. (a) *Cetoscarus ocellatus* (n = 39), (b) *Chlorurus microrhinos* (n = 74), (c) *Chlorurus spilurus* (n = 125), (d) *Scarus frenatus* (n = 95), (e) *Scarus niger* (n = 132), (f) *Scarus schlegeli* (n = 226). Data are from direct observations of fishes feeding on reefs surrounding Lizard Island, northern Great Barrier Reef (Hoey and Fox unpublished).

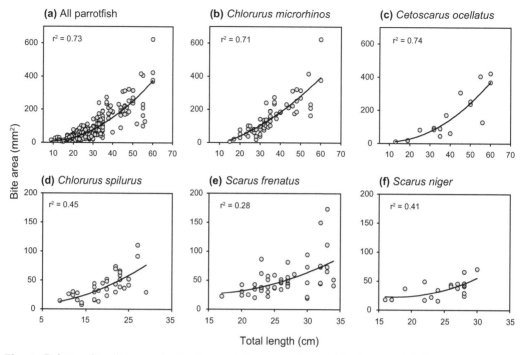

Fig. 2. Relationship between body size and the area of individual parrotfish bite scars. (a) all parrotfishes (n = 253), (b) *Chlorurus microrhinos* (n = 61) ,(c) *Cetoscarus ocellatus* (n = 16), (d) *Chlorurus spilurus* (n = 36), (e) *Scarus frenatus* (n = 51), (f) *Scarus niger* (n = 20). Data are from *in situ* measurements of fresh feeding scars on algal turf covered substratum on reefs surrounding Lizard Island, northern Great Barrier Reef (Hoey and Fox unpublished).

is beyond the scope of this chapter (see Clements and Choat Chapter 3 for a synthesis of the nutritional ecology of parrotfishes). Notwithstanding, any changes in the feeding rate of parrotfishes will directly influence the frequency of disturbance impacts on benthic communities.

Feeding rates of parrotfishes have generally been shown to be either independent of (e.g., Fox and Bellwood 2007), or negatively related to body size (e.g., Bruggemann et al. 1994c, Mumby 2006), with considerable variation within functional groups. For example, the bite rates of the excavating *Sp. viride* and *Ch. sordidus* have been reported to decrease with increasing body size (Bruggemann et al. 1994c, Lokrantz et al. 2008), while the bite rates of *Ch. stronglocephalus* and *Ch. perspicillatus* did not change with body size (Lokrantz et al. 2008, Ong and Holland 2010). Similarly, feeding rates of *Sc. rubroviolaceus* and *Sc. vetula* decreased with increasing body size (Bruggemann et al. 1996, Ong and Holland 2010), *Sc. niger* and *Sc. rivulatus* did not change (Fox and Bellwood 2007, Bonaldo and Bellwood 2008, Lokrantz et al. 2008), and *Sc. ferrugineus* initially increased, then decreased with increasing body size (Afeworki et al. 2013). Similarly, comparisons of feeding rates among 16 species of parrotfish from the northern GBR highlight the variability of the relationship between bite rate and body size (Fig. 3). Although bite rate was negatively related to body size for the majority of species, the bite rate of several species were independent of body size, and for one species, *Scarus flavipectoralis*, appeared to be positively related to body size (Fig. 3g).

The interspecific differences in the relationships between body size and bite rate are difficult to resolve, but may in part reflect differences in territoriality, nutritional requirements, feeding substrata or feeding habit among species and/or life stages. For example, decreased feeding rates of large terminal phase males of some species (e.g., *Sp. viride*: Bruggemann et al. 1994c) have been linked to increased time invested in territory defence. Differences in the quantity and quality of food, and/or the density of the reef substrata may also influence the nutritional yield per bite, and hence feeding rates. All of the feeding observations were conducted along the same stretch of reef, minimizing any large-scale differences in dietary resources. Further, species-specific relationships

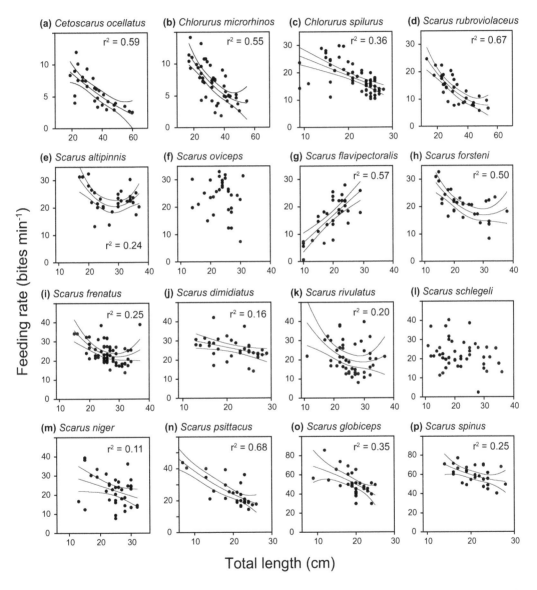

Fig. 3. Relationship between body size and feeding rate of 16 parrotfish species from Lizard Island, northern Great Barrier Reef. Data are from 3-min observations of focal individuals on an exposed reef crest (2-3 m depth), with a minimum of 30 observations per species. Lines represent best fit linear or first order polynomial regression, and upper and lower 95 percent confidence intervals. Lines are only included for significant relationships.

appear to be conserved across broad spatial scales (Fig. 4). Finally, feeding habit may have contributed to some of the variation in feeding rates. There were few commonalities among species for which feeding rate declined with increasing body size, ranging from large excavating species with low feeding rates (Fig. 3a, b) to small scraping species with

(a) *Chlorurus gibbus / strongylocephalus / microrhinos*

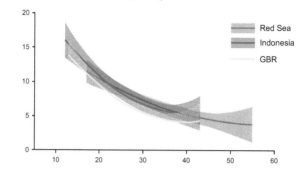

(b) *Chlorurus sordidus / spilurus*

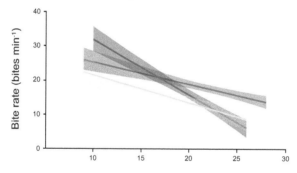

(c) *Scarus frenatus*

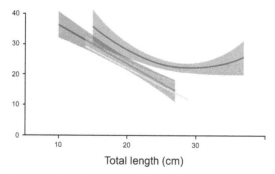

Fig. 4. Regional variation in the relationship between body size and feeding rate for three parrotfish species groups, (a) *Chlorurus gibbus/strongylocephalus/microrhinos*, (b) *Chlorurus sordidus/spilurus*, (c) *Scarus frenatus*. A minimum of 30 3-min focal individual observations were conducted for each species at each location. Central Red Sea: *Chlorurus gibbus* (n = 31), *Chlorurus sordidus* (n = 37), *Scarus frenatus* (n = 32); Sumatera, Indonesia: *Chlorurus strongylocephalus* (n = 32), *Chlorurus sordidus* (n = 31), *Scarus frenatus* (n = 32); Lizard Island, northern Great Barrier Reef: *Chlorurus microrhinos* (n = 46), *Chlorurus spilurus* (n = 59), *Scarus frenatus* (n = 59). Lines and shading represent best-fit linear or first order polynomial regression and 95 percent confidence intervals.

high feeding rates that rarely leave feeding scars (Fig. 3o, p). Interestingly, *Sc. flavipectoralis* was the only species in which feeding rate was positively related to body size, and also the only species that was regularly observed feeding on sandy substrates.

Clearly, body size is an important determinant of the disturbance impact of individual parrotfish, both among and within species. Despite considerable interspecific variation in the effect of fish size on feeding rates of parrotfishes, the large increases in bite size with fish size leads to dramatic increases in per capita rates of grazing and erosion by parrotfishes (Bruggemann et al. 1996, Bonaldo and Bellwood 2008, Lokrantz et al. 2008, Ong and Holland 2010). Previous assemblage-level assessments of the functional impacts of parrotfishes on coral reefs have generally not considered the influence of body size on the disturbance impacts of individual species (e.g., Hoey and Bellwood 2008, Alwany et al. 2009, see Mumby 2006 for exception). While this has largely been constrained by the lack of location- and species-specific data on feeding rates and bite sizes, the overriding influence of body size on grazing and erosion rates emphasizes the importance of incorporating the effects of body size into future assessments.

Effect of Temperature on Functional Impact of Parrotfishes

Environmental temperature, together with body size, is a key determinant of biological activities (e.g., Brown et al. 2004). Fishes are ectotherms and as such their rate of biochemical and cellular processes, metabolism and energy requirements are largely governed by environmental temperature (Pörtner and Farrell 2008). Therefore, any increase (or decrease) in environmental temperature will increase (or decrease) the energetic cost of maintenance, growth, activity and reproduction. A temperature-induced increase in metabolic demand may be compensated by increased quantity or quality of food, reduced energy expenditure, or a combination of both. For example, the serranid *Plectropomus leopardus* has been shown to reduce activity, increase food consumption and increase oxygen consumption under elevated temperatures in aquaria (Johansen et al. 2014, 2015, Clark et al. 2017). To my knowledge, no studies have examined the effect of temperature on the performance of parrotfish under controlled conditions. As such I will focus on those studies that have described changes in parrotfish feeding in response to natural changes in temperature.

Several studies have quantified the feeding of parrotfishes between seasons, with most reporting elevated feeding rates in the warmer months (Bellwood 1995, Afeworki et al. 2013). For example, feeding rates of *Ch. microrhinos* and *Ch. spilurus* were estimated to decrease by three and 24 percent, respectively, from summer to winter although water temperature data was not given (Bellwood 1995). Feeding data for *Sc. persicus* from the northern Arabian Sea indicate that a 6°C decrease in water temperature (from 29°C to 23°C) led to a 50-60 percent decrease in bite rate (Hoey et al. Chapter 12). Similarly, the feeding rates of *Sc. ghobban* (> 10 cm TL) on the temperate reefs of Kochi (33°N), southern Japan decline from ca. 15 bites min^{-1} in summer to less than five bites min^{-1} in late autumn and winter (Hoey et al. Chapter 12). Afeworki et al. (2013) also report a positive linear relationship between water temperature and feeding rate of initial phase (20-25 cm FL) *Sc. ferrugineus* in Eritrea, but a curve-linear relationship for terminal phase (30-35 cm FL) conspecifics with their feeding rate increasing up until ca. 32°C, after which it declined. The authors suggest this may indicate the thermal optima of 32°C for *Sc. ferrugineus*, although it is based on very few observations. In contrast to the typically positive relationships between parrotfish feeding and seasonal temperatures, Ong and Holland (2010) reported the feeding rates of *Sc. rubroviolaceus* and *Ch. perspicillatus* in Hawaii either don't change or increase by up to 14.6 percent from the warmer to the cooler months. It should be noted that these seasons only presented a 2°C change in temperature.

Broadly similar patterns between temperature and feeding rate are evident across latitudes and during upwelling events. A latitudinal comparison of the summer feeding rates of adult individuals (>15 cm TL) of three species of scraping parrotfishes (*Sc. ghobban, Sc. psittacus* and *Sc. schlegeli*) show declines of 39 to 71 percent between the northern and central GBR and Lord Howe Island, representing a temperature decrease of 4-5°C (Hoey et al. Chapter 12). Further, a seasonal upwelling that causes a relatively rapid 5°C decrease in water temperature on the Pacific coast of Panama has been shown to reduce the bite rate of *Sc. ghobban* by 50 percent, from 32.5 to 15.7 bites min^{-1} (Smith 2008). While the relative consistency of the response of feeding rate to temperature over both spatial (i.e., latitudinal) and temporal scales (i.e., seasonal, and upwelling versus non-upwelling) points toward the importance of temperature in shaping parrotfish feeding rates, it should be remembered that several other environmental factors typically covary with water temperature in natural settings.

Summary

This chapter has explored how the disturbance impacts (or effects) of parrotfish feeding on benthic reef communities vary among species, and with body size and temperature. In doing so, it has revealed the overwhelming effect of body size on the amount of material that is scraped or excavated from reef surfaces, and raises questions regarding the functional classification of this group. For example, per capita erosion rates of excavating parrotfishes range from 23.6 kg per individual per year for small *Chlorurus* spp. (e.g., *Ch. spilurus*), to 380 kg and 1015 kg per individual per year for medium and large *Chlorurus* spp. (e.g., *Ch. perspicillatus* and *Ch. microrhinos*, respectively), and over 5,500 kg per individual per year for *B. muricatum* (Bellwood 1995, Hoey and Bellwood 2008, Ong and Holland 2010). While several studies have acknowledged these differences among excavating species, differentiating between small-, medium- and large-bodied excavators, they have generally treated scraping parrotfishes as a largely homogenous group (e.g., Hoey and Bellwood 2008, Bellwood et al. 2012). As highlighted in the sections above there is considerable variation in the disturbances impacts among scraping parrotfishes, ranging from species such as *Sc. globiceps* and *Sc. spinus* that have rapid feeding rates (up to 80 bites min^{-1}) but rarely leave feeding scars on the substratum, to large species such as *Sc. rubroviolaceus* that have relatively slow feeding rates (10-15 bites min^{-1}) and remove amounts of material that are directly comparable to medium-bodied excavating parrotfishes (Ong and Holland 2010).

Further clouding the distinctions between functional groups are the ontogenetic changes in feeding rates and bites sizes within species. Most juvenile parrotfish, irrespective of functional group, feed in a way that is analogous to the feeding mode of scraping parrotfishes. Conversely, large-bodied individuals of some scraping species essentially function as excavators; their larger and heavier oral jaws and powerful musculature allowing them to penetrate deeper into the reef substratum when feeding. While there is growing acknowledgement of these ontogenetic differences in the functional classifications of parrotfishes (i.e., 'scrapers and small excavators', and 'excavators and large scrapers'; Green and Bellwood 2009, Heenan et al. 2016) the marked difference both among and within species question the utility of the widely used functional classifications. The distinction between excavating, scraping and browsing parrotfishes primarily relates to their feeding mode, and although they share some similarities in their ecology it is becoming increasingly clear that these feeding modes do not directly translate to their disturbance impact (or 'function') on reefs. There is a clear need to move beyond the current use of excavating, scraping and browsing parrotfishes when assessing the ecological roles of parrotfishes. Incorporating species- and size-specific metrics of feeding, together with the influence of environmental conditions (e.g., temperature) and nutrition will greatly expand our understanding of the causes and consequences of feeding among parrotfishes.

Acknowledgments

I thank A. Baird, D. Bellwood, M. Berumen, J. Burt, S. Campbell, N. Fadli, D. Feary, R. Fox, J. Hoey, J. Johansen, D. Nembhard, M. Pratchett, D. Rasher, and the staff of WCS Indonesia Marine Program and King Abdullah University of Science and Technology for logistical support and/or discussions on the ideas presented in this chapter, and the Australian Research Council for financial support.

References Cited

Afeworki, Y., J.J. Videler and J.H. Bruggemann. 2013. Seasonally changing habitat use patterns among roving herbivorous fishes in the southern Red Sea: the role of temperature and algal community structure. Coral Reefs 32: 475–485.

Alwany, M.A., E. Thaler and M. Stachowitsch. 2009. Parrotfish bioerosion on Egyptian red sea reefs. J. Exp. Mar. Biol. Ecol. 371: 170–176.

Bellwood, D.R. 1994. A phylogenetic study of the parrotfishes family Scaridae (Pisces: Labroidei), with a revision of the genera. Rec. Aust. Mus. 20: 1–86.

Bellwood, D.R. 1995. Direct estimate of bioerosion by two parrotfish species, *Chlorurus gibbus* and *C. sordidus*, on the Great Barrier Reef, Australia. Mar. Biol. 121: 419–429.

Bellwood, D.R. and J.H. Choat. 1990. A functional analysis of grazing in parrotfishes (family Scaridae): the ecological implications. Environ. Biol. Fishes 28: 189–214.

Bellwood, D.R., A.S. Hoey and J.H. Choat. 2003. Limited functional redundancy in high diversity systems: resilience and ecosystem function on coral reefs. Ecol. Lett. 6: 281–285.

Bellwood, D.R., A.S. Hoey and T.P. Hughes. 2012. Human activity selectively impacts the ecosystem roles of parrotfishes on coral reefs. Proc. R. Soc. B 271: 1621–1629.

Bernardi, G., D.R. Robertson, K.E. Clifton and E. Azzuro. 2000. Molecular systematics, zoogeography, and evolutionary ecology of the Atlantic parrotfish genus *Sparisoma*. Mol. Phylogen. Evol. 15: 292–300.

Bonaldo, R.M. and D.R. Bellwood. 2008. Size-dependent variation in the functional role of the parrotfish *Scarus rivulatus*. Mar. Ecol. Prog. Ser. 360: 237–244.

Bonaldo, R.M., J.P. Krajewski, C. Sazima and I. Sazima. 2006. Foraging activity and resource use by three parrotfish species at Fernando de Noronha Archipelago, tropical West Atlantic. Mar. Biol. 149: 423–433.

Bonaldo, R.M., A.S. Hoey and D.R. Bellwood. 2014. The ecosystem roles of parrotfishes on tropical reefs. Oceanogr. Mar. Biol. Annu. Rev. 52: 81–132.

Brown, J.H., J.F. Gillooly, A.P. Allen, V.M. Savage and G.B. West. 2004. Toward a metabolic theory of ecology. Ecology 85: 1771–1789.

Bruggemann, J.H., M.J.H. van Oppen and A.M. Breeman. 1994a. Foraging by the stoplight parrotfish *Sparisoma viride*. I. Food selection in different, socially determined habitats. Mar. Ecol. Prog. Ser. 106: 41–55.

Bruggemann, J.H., M.W.M. Kuyper and A.M. Breeman. 1994b. Comparative analysis of foraging and habitat use by the sympatric Caribbean parrotfish *Scarus vetula* and *Sparisoma viride*. Mar. Ecol. Prog. Ser. 112: 51–66.

Bruggemann, J.H., J. Begeman, E.M. Bosma, P. Verburg and A.M. Breeman. 1994c. Foraging by the stoplight parrotfish *Sparisoma viride*. II. Intake and assimilation of food, protein and energy. Mar. Ecol. Prog. Ser. 106: 57–71.

Bruggemann, J.H., A.M. van Kessel, J.M. van Rooij and A.M. Breeman. 1996. Bioerosion and sediment ingestion by the Caribbean parrotfish *Scarus vetula* and *Sparisoma viride*: implications of fish size, feeding mode and habitat use. Mar. Ecol. Prog. Ser. 134: 59–71.

Choat, J.H., L. Herwerden, D.R. Robertson and K.D. Clements. 2012. Patterns and processes in the evolutionary history of parrotfishes (Family Labridae). Biol. J. Linnean Soc. 107: 529–557.

Clark, T.D., V. Messmer, A.J. Tobin, A.S. Hoey and M.S. Pratchett. 2017. Rising temperatures may drive fishing-induced selection of low-performance phenotypes. Sci. Rep. 7: 40571.

Clarke, A. and N.M. Johnston. 1999. Scaling of metabolic rate with body mass and temperature in teleost fish. J. Animal Ecol. 68: 893–905.

Clements, K.D., D.P. German, J. Piché, A. Tribollet and J.H. Choat. 2017. Integrating ecological roles and trophic diversification on coral reefs: multiple lines of evidence identify parrotfishes as microphages. Biol. J. Linn. Soc. 120: 729–751.

Dunic, J.C. and J.K. Baum. 2017. Size structuring and allometric scaling relationships in coral reef fishes. J. Anim. Ecol. 86: 577–589.

Ferreira, C.E.L., A.C. Peret and R. Coutinho. 1998. Seasonal grazing rates and food processing by tropical herbivorous fishes. J. Fish Biol. 53: 222–235.

Fox, R.J. and D.R. Bellwood. 2007. Quantifying herbivory across a coral reef depth gradient. Mar. Ecol. Progr. Ser. 339: 49–59.

Francini-Filho, R.B., R.L. Moura, C.M. Ferreira and E.O.C. Coni. 2008. Live coral predation by parrotfishes (Perciformes: Scaridae) in the Abrolhos Bank, eastern Brazil, with comments on the classification of species into functional groups. Neotrop. Ichthyol. 6: 191–200.

Francini-Filho, R.B., C.M. Ferreira, E.O.C. Coni, R.L. Moura and L. Kaufman. 2010. Foraging activity of roving herbivorous reef fish (Acanthuridae and Scaridae) in eastern Brazil: influence of resource availability and interference competition. J. Mar. Biol. Assoc. UK 90: 481–492.

Froese, R. and D. Pauly (eds.). 2017. FishBase. www.fishbase.org, version (06/2017).

Gobalet, K.W. 1989. Morphology of the parrotfish pharyngeal jaw apparatus. Am. Zool. 29: 319–331.

Green, A.L. and D.R. Bellwood. 2009. Monitoring functional groups of herbivorous reef fishes as indicators of coral reef resilience—a practical guide for coral reef managers in the Asia Pacific region. IUCN working group on Climate Change and Coral Reefs. IUCN, Gland, Switzerland.

Heenan, A., A.S. Hoey, G.J. Williams and I.D. Williams. 2016. Natural bounds on herbivorous coral reef fishes. Proc. R. Soc. B 283: 20161716.

Hoey, A.S. and D.R. Bellwood. 2008. Cross-shelf variation in the role of parrotfishes on the Great Barrier Reef. Coral Reefs 27: 37–47.

Hoey, A.S., D.A. Feary, J.A. Burt, G. Vaughan, M.S. Pratchett and M.L. Berumen. 2016. Regional variation in the structure and function of parrotfishes on Arabian reefs. Mar. Poll. Bull. 105: 524–531.

Johansen, J.L., V. Messmer, D.J. Coker, A.S. Hoey and M.S. Pratchett. 2014. Increasing ocean temperatures reduce activity patterns of a large commercially important coral reef fish. Glob. Change Biol. 20: 1067–1074.

Johansen, J.L., M.S. Pratchett, V. Messmer, D.J. Coker, A.J. Tobin and A.S. Hoey. 2015. Large predatory coral trout species unlikely to meet increasing energetic demands in a warming ocean. Sci. Rep. 5: 13830.

Lokrantz, J., M. Nystrom, M. Thyresson and C. Johansson. 2008. The non-linear relationship between body size and function in parrotfishes. Coral Reefs 150: 1145–1152.

Messmer, V., M.S. Pratchett, A.S. Hoey, A.J. Tobin, D.J. Coker, S.J. Cooke and T.D. Clark. 2017. Global warming may disproportionately affect larger adults in a predatory coral reef fish. Glob. Change Biol. 23: 2230–2240.

Mumby, P.J. 2006. The impact of exploiting grazers (Scaridae) on the dynamics of Caribbean coral reefs. Ecol. Appl. 16: 747–769.

Ong, L. and K.N. Holland. 2010. Bioerosion of coral reefs by two Hawaiian parrotfishes: species, size, differences and fishery implications. Mar. Biol. 157: 1313–1323.

Parenti, P. and J.E. Randall. 2011. Checklist of the species of the families Labridae and Scaridae: an update. Smithiana Bull. 13: 29–44.

Pörtner, H.O. and A.P. Farrell. 2008. Physiology and climate change. Science 322: 690–692.

Robertson, D.R., F. Karg, R.L. Moura, B. Victor and G. Bernardi. 2006. Mechanisms of speciation and faunal enrichment in Atlantic parrotfishes. Mol. Phylogenet. Evol. 40: 795–807.

Russ, G.R., S.A. Questel, J.R. Rizzari and A.C. Alcala. 2015. The parrotfish–coral relationship: refuting the ubiquity of a prevailing paradigm. Mar. Biol. 162: 2029–2045.

Smith, J.M. 1978. Optimization theory in evolution. Annu. Rev. Ecol. Syst. 9: 31–56.

Smith, T.B. 2008. Temperature effects on herbivory for an Indo-Pacific parrotfish in Panamá: implications for coral–algal competition. Coral Reefs 27: 397–405.

Sterner, R.W. and J.J. Elser. 2002 Ecological stoichimetry: the biology of elements from molecules to the biosphere. Princeton University Press, Princeton, NJ.

Streelman, J.T., M. Alfaro, M.W. Westneat, D.R. Bellwood and S.A. Karl. 2002. Evolutionary history of the parrotfishes: biogeography, ecomorphology, and comparative diversity. Evolution 56: 961–971.

Wainwright, P.C., D.R. Bellwood, M.W. Westneat, J.R. Grubich and A.S. Hoey. 2004. A functional morphospace for the skull of labrid fishes: patterns of diversity in a complex biomechanical system. Biol. J. Linnean Soc. 82: 1–25.

Functional Variation among Parrotfishes: Are they Complementary or Redundant?

Deron E. Burkepile[1], Douglas B. Rasher[2], Thomas C. Adam[3], Andrew S. Hoey[4] and Mark E. Hay[5]

[1] Department of Ecology, Evolution, and Marine Biology, University of California, Santa Barbara
  Email: deron.burkepile@lifesci.ucsb.edu
[2] Bigelow Laboratory for Ocean Sciences, 60 Bigelow Drive, East Boothbay, ME, 04544, USA
  Email: drasher@bigelow.org
[3] Marine Science Institute, University of California, Santa Barbara, CA, 93106 USA
  Email: adam@lifesci.ucsb.edu
[4] Australian Research Council Centre of Excellence for Coral Reef Studies,
  James Cook University, Townsville, QLD 4811, Australia
  Email: Andrew.hoey1@jcu.edu.au
[5] School of Biology, and Aquatic Chemical Ecology Center, Georgia Institute of Technology,
  Atlanta, GA 30332, USA
  Email: mark.hay@biology.gatech.edu

Introduction

Herbivory is a critical process that underpins coral reef structure and function. High levels of herbivory favor corals by excluding upright macroalgae and cropping abundant filamentous algae (Steneck 1988, Hay 1991) that otherwise negatively impact settlement of coral larvae (Kuffner et al. 2006, Nugues and Szmant 2006, Birrell et al. 2008, Dixson et al. 2014) as well as growth and survivorship of adult corals (McCook et al. 2001, Jompa and McCook 2003, Rasher and Hay 2010, Vega Thurber et al. 2012). Thus, the removal of herbivores often results in a decline in coral abundance (e.g. Lewis 1986, Hughes et al. 2007, Burkepile and Hay 2008). As corals decline, species-rich and topographically complex coral reefs transition to flattened, species-poor reefs, often with compromised ecosystem function (Graham et al. 2006, Alvarez-Filip et al. 2009, Hughes et al. 2010). These transitions frequently result in feedbacks that negatively impact coral recruitment, growth and survivorship and keep reefs in a coral-depauperate, algal-dominated state (Mumby and Steneck 2008, Nyström et al. 2012, Adam et al. 2015a).

Parrotfishes in particular play significant roles in the herbivore guild by clearing macroalgae from reefs (Burkepile and Hay 2008), promoting coral settlement and growth (Mumby et al. 2007, Adam et al. 2011, Hoey et al. 2011), bioeroding carbonate and producing sediment (Bellwood 1995a, Bellwood 1995b, Mallela and Fox Chapter 8), and preying on corals (Rotjan and Lewis 2006, Burkepile 2012, Bonaldo and Rotjan Chapter 9). In some cases, parrotfishes may represent the critical levers that push reefs toward recovery (i.e.

to a coral-dominated state) following coral-destroying disturbances such as hurricanes or bleaching events, with their influence suggested to overwhelm that of other herbivores (Mumby 2006, Mumby and Hastings 2008).

Despite suggestions of the overwhelming importance of parrotfishes, several recent studies demonstrate the importance of herbivore diversity, with surgeonfishes and rabbitfishes playing roles that are complementary to most parrotfishes (Burkepile and Hay 2008, 2010, Fox and Bellwood 2008, 2013, Hoey and Bellwood 2009, Burkepile and Hay 2010, Fox and Bellwood 2013, Rasher et al. 2013, Brandl and Bellwood 2014). Even within the parrotfishes there are significant morphological and ecological differences (Bellwood 1994, Bruggemann et al. 1994a, 1996, Streelman et al. 2002) that likely modulate their impacts on benthic community structure and reef resilience (Bellwood et al. 2004). Indeed, impacts of parrotfish feeding vary with the stage of algal community development (Burkepile and Hay 2010, Hoey and Bellwood 2011), the relative abundance of alternative prey (Burkepile 2012), the habitat in which studies are conducted (Hoey and Bellwood 2008), and the abundance of other key herbivores in the system (Morrison 1988, Burkepile and Hay 2008). The functional roles of parrotfishes are crucial, yet complex, for maintaining coral reef structure and function.

Biological diversity has long been argued to be a source of ecological stability (e.g. Darwin 1859; MacArthur 1955), with increasing diversity providing a degree of insurance (or resilience) against natural and anthropogenic stressors (e.g. Holling 1973, Yachi and Loreau 1999, Folke et al. 2004). Thus, the diversity of species, especially the diversity of consumers, often has positive impacts on ecosystem function (Worm et al. 2006, Duffy et al. 2007, Stachowicz et al. 2007). On coral reefs, herbivore diversity, rather than just biomass or density, is critical for maintaining ecosystem function because herbivores differ in sensitivity to algal chemical, morphological and structural defenses (Hay et al. 1994, Schupp and Paul 1994, Burkepile and Hay 2008, Rasher et al. 2013). As herbivore species richness increases, the ability of any given algae to escape or deter all herbivores declines (Rasher et al. 2013). Herbivore diversity may be key for healthy coral reefs as disturbed reefs may be able to recover if the right mix of herbivores is present to remove algae and facilitate the recovery of coral populations (Burkepile and Hay 2008, Mumby and Steneck 2008, Cheal et al. 2010, Hughes et al. 2010, Rasher et al. 2013). Ultimately, the importance of herbivore diversity for reef structure and function will depend on how fishes' feeding patterns fall on the scale of being complementary (feeding on different foods and/or in different ways) to being redundant (considerable overlap in diet and feeding), although both patterns may be important for ecosystem function.

Complementary feeding by mixed species groups of herbivores can result in more efficient grazing in terrestrial, marine benthic, and pelagic systems (Burkepile 2013); on reefs, this produces greater suppression of macroalgae and thus facilitation of corals (Burkepile and Hay 2008, Rasher et al. 2013). However, redundancy, or overlap, in feeding patterns may also be important because when certain species decline due to overfishing, disease, or natural fluctuations in populations other species may be able to fill their vacated role in the herbivore guild (Rosenfeld 2002). This redundancy may impart some response diversity to these systems where different, but redundant, species respond differently to disturbances such as hurricanes or bleaching events thus maintaining intact ecological processes (e.g. Elmqvist et al. 2003, Pratchett et al. 2014).

The question of complementarity among herbivorous fishes most often focuses on what fishes eat and how they make their diet choices. Many of the original ideas of diet complementarity revolved around herbivore jaw morphology and how this may drive partitioning of the algae that different fishes can consume (Bellwood and Choat 1990,

Bellwood 1994, Wainwright et al. 2004, Bellwood et al. 2006). Focusing on how fishes eat is valuable, but incomplete, for understanding a species' diet and function. The external properties of herbivore feeding morphology are easy for humans to see, understand, and study. But digestive physiology, nutritional requirements, and resistance or susceptibility to chemical feeding deterrents are also important (Horn 1989, Schupp and Paul 1994, Clements et al. 2009, Clements and Choat Chapter 3) with multiple prey traits often interacting to determine the diet breadth of consumers (Duffy and Paul 1992, Hay et al. 1994, Cruz-Rivera and Hay 2003). In addition, there are many other aspects of the ecology of herbivorous fishes that will determine their impacts on benthic communities, including foraging range, preferred feeding substrata, responses to predators, and body size. Because the complementarity/redundancy of species exists across multiple different axes of a species' niche in addition to what they eat, examinations of functional relationships across multiple different niche axes will likely reveal complex patterns of complementarity. Here we focus on both the causes and outcomes of patterns of complementarity and redundancy of parrotfishes.

Mechanisms of Complementarity and Redundancy in Parrotfishes: More than a Mouthful

The idea that biodiversity can have positive impacts on ecosystem function is grounded in ideas of complementarity and redundancy of resource use (Loreau et al. 2001, Rosenfeld 2002, Stachowicz et al. 2008). Simply, if different species are complementary in their resource use (e.g. different in the nutrients they uptake or foods they eat), then this complementarity will result in positive relationships between species diversity and ecosystem processes such as primary production. The more complementary species are in their resource use the more likely there will be a strong positive link between biodiversity and ecosystem function where the loss of an individual species will compromise ecosystem function. Conversely, when species are more redundant (e.g. they use resources more similarly), the relationships between biodiversity and ecosystem function will be less straightforward and the loss of redundant species may have little effect on ecosystem function. Species redundancy is likely also important for ecosystem function because species capable of performing the same functions may compensate for each other over space and time as populations fluctuate (Duffy 2009).

However, these notions of complementarity and redundancy often only consider one or a limited subset of species' traits such as nutrient usage or diet breadth. When one considers a limited subset of traits, it is more likely that species will appear redundant. This ignores the fact that species occupy niches that exist in multiple dimensions, providing different axes for species to be complementary (Rosenfeld 2002). Recent empirical and theoretical work focusing on multiple traits that influence ecosystem function suggest that biodiversity supports multifunctionality of systems and that species may be more complementary of one another than previously believed (Duffy et al. 2003, Hector and Bagchi 2007, Gamfeldt et al. 2008). For coral reef herbivores, the notion of complementarity has mostly focused on the feeding morphology of different species, what these species eat, and their subsequent impacts on the benthic community (e.g. Mantyka and Bellwood 2007, Burkepile and Hay 2008, Hoey and Bellwood 2009). However, the feeding preferences of a species are only one of many metrics of complementarity. Many other aspects of the ecology of fishes such as body size, home range, preferred feeding substrata, and responses to predators will influence a species' impact on communities and their role in ecosystem function. Understanding the causes and consequences of variation in each of these traits is therefore fundamental for understanding the levels of complementarity and redundancy present within the herbivore guild.

Diet Selection

Diet selection is often the focus of whether parrotfishes are complementary or redundant to one another. Much of the comparative work on parrotfishes in terms of diet has focused on their differences in jaw morphology and musculature as a way to infer what different species are capable of feeding upon. This work led to classifying parrotfishes in three general functional groups: (1) scrapers that eat mostly filamentous turf algae and associated detritus, (2) browsers that eat mostly upright macroalgae, and (3) excavators that eat epilithic and endolithic organisms via the consumption of reef carbonates (e.g. Bellwood 1994, Streelman et al. 2002, Bonaldo et al. 2014; Fig. 1a,b). While these morphological traits help reveal the potential diet items that a species can eat, they do not tell you what individual species actually eat in practice or why they eat what they do.

Fig. 1. (A) *Sparisoma viride* from the Caribbean and (B) *Chlorurus bleekeri* from the Indo-Pacific are important excavators that remove significant amounts of calcium carbonate from the reef every year. (C) The morning catch of an illegal hookah divers' boat at the Silver Bank Sanctuary, Dominican Republic. Note the abundance of parrotfishes, especially terminal phase *Sparisoma viride*. Photo credits: Madelyn Roycroft (A), Cody Clements (B), and Jose Alejandro Alvarez (C).

An important aspect of knowing the differences in feeding patterns among herbivore species is understanding how macroalgal chemical and physical defenses affect herbivore species differentially (Hay 1997). Previous work with herbivorous fishes has shown that physical or chemical defenses often deter different groups of herbivores (Paul and Hay 1986, Hay et al. 1994, Schupp and Paul 1994), suggesting that macroalgal defenses are important for driving overall patterns of herbivory on reefs. However, a range of tolerances for macroalgal defenses is likely an important mechanism driving functional diversity in feeding habits of herbivorous fishes. For example, a recent study on reefs in Fiji revealed strong complementarity among different species of herbivores for particular species of macroalgae that were driven by both chemical and physical defenses (Rasher et al. 2013). In the study, herbivores exhibited differential susceptibility to chemicals extracted from different species of algae, with the two most chemically defended algae each consumed exclusively by a single, but different species of fish. This range of tolerances for a suite of different macroalgal defenses is likely an important mechanism driving functional diversity in feeding habits of herbivorous fishes.

While the role of algal defenses in determining complementarity in feeding preferences is known to be important, an underappreciated aspect of diet complementarity among parrotfishes is how they select their diets based on nutritional requirements. Both nitrogen and phosphorus appear to be limiting to herbivores (Mattson 1980, Cebrian et al. 2009), and stoichiometric theory dictates that herbivores should prioritize consumption of foods that most closely match their chemical composition (Sterner and Elser 2002). For example, phosphorous-rich herbivores, such as many bony and scaled fish (Lall 1991), should be particularly sensitive to the phosphorous content of food (Sterner and Elser 2002). In addition, the physiology of consumers also likely impacts their nutritional demands. For example, smaller animals or those with short gut passages lose a large fraction of ingested nutrients in their feces and are often nitrogen limited (Barboza et al. 2009). Thus, potential differences in body composition (and hence nutrient demand) and in digestive physiologies of herbivorous fishes suggest that the nitrogen and phosphorus content of macroalgae may be particularly important for mediating species-specific foraging decisions of fishes. Indeed, experiments comparing the consumption of nutrient enriched algae to controls have documented higher rates of grazing and biomass removal of enriched algae (Boyer et al. 2004, Burkepile and Hay 2009, Chan et al. 2012). Some authors have speculated that enrichment may even result in increased consumption of chemically defended algal species (Cruz-Rivera and Hay 2003; Sotka and Hay 2009; Chan et al. 2012). However the role of different macronutrients in driving differential foraging patterns of herbivorous fishes is virtually unknown and an area ripe for future investigation to help elucidate the mechanisms that potentially select for complementarity.

Spatial Patterns of Foraging

Most studies addressing complementarity in herbivorous fishes focus solely on the diet of fishes and ignore many other potential axes of complementarity such as home range size, habitat preference, and substrate preference. Understanding patterns of complementarity and redundancy among parrotfishes requires not only knowing what they eat, but also where they eat. Herbivorous fishes frequently show evidence of fine-scale partitioning in the microhabitats that they target while foraging (Hay 1985, Robertson and Gaines 1986, Bruggemann et al. 1994a, Brandl and Bellwood 2014). For example, Brandl and Bellwood (2014) examined the functional niches of 21 species of herbivorous fish on the Great Barrier Reef based on how fishes foraged in different microhabitats. Overall there was a high degree of complementarity among parrotfishes, surgeonfishes and rabbitfishes. Even when assessing only parrotfish species, Brandl and Bellwood (2014) suggested that there was little overlap in how species used microhabitats. These data suggest that while some species may have similar preferences for dietary resources, their impacts are likely exerted in different microhabitats on reefs.

A few studies have examined foraging preferences of select species of parrotfishes in great detail; this work has revealed that these species differ greatly in both the habitats they utilize for foraging, as well as the substrata and algal species/functional groups they target (Bruggemann et al. 1994a, 1996). In addition, there is evidence that different species vary greatly in their home range sizes, which will also influence the space they impact (Mumby and Wabnitz 2002). The range over which species forage can be a function of different habitat characteristics such as coral cover, with different species of parrotfish exhibiting different relationships (Nash et al. 2012). Finally, individual size also varies dramatically both within and between species, which will further influence the scales over which parrotfish forage with larger fishes typically foraging over larger spatial scales (Nash et al. 2013). It is clear that we need a more comprehensive understanding of the habitat and foraging preferences as well as the scales of movement of parrotfish species to develop a better idea of how complementary their habitat use is and how this relates to their diet and impact on the benthos.

Habitat Utilization and Life History

While many parrotfishes are strongly reef-associated throughout their ontogeny, some species utilize off-reef nursery habitats such as seagrasses, mangroves, or inshore reefs before migrating to coral reefs (Nagelkerken et al. 2000, 2002, Mumby et al. 2004, Machemer et al. 2012). Thus, species may be complementary in their need for different types of juvenile habitat that may be under differential threat from coastal development and global climate change. For example, the largest parrotfish in the Caribbean, *Scarus guacamaia*, appears to be dependent on the availability of mangrove habitat during its juvenile phase, and mangrove clearing may have led to its local extinction in portions of the Caribbean (Mumby et al. 2004). Similarly, in areas of the Indo-Pacific some parrotfishes appear to use inshore reefs as juveniles but then move out to the forereef as adults (e.g. Aswani and Hamilton 2004). This connectivity between different habitats should enhance the resilience of coral reefs by providing reefs with a source of herbivores following localized disturbances (Adam et al. 2011). Yet, we know relatively little about the dependence of particular herbivore species on specific nursery habitats and even less about the scale of movement that occurs among habitats. The lack of these data suggests that we need to know more about the basic life history of many parrotfishes to understand the scope of their complementarity. Theoretically, two species of parrotfishes could be redundant in their diets but have different life history patterns making them differentially susceptible to exploitation and/or habitat degradation. Patterns of complementarity in life-history traits could therefore lead to increased response diversity and hence resilience of coral reefs by ensuring that a single perturbation does not result in the loss of ecosystem function.

Responses to Predators

Predators can have large impacts on populations of reef fishes (e.g. White et al. 2010), with potential cascading indirect effects on the resources of their prey (e.g. trophic cascades). Given the large size range of parrotfishes, these species will be differentially susceptible to predation by larger fishes (e.g. Mumby et al. 2006). Given that different sized parrotfishes appear to play different roles in the herbivore guild (see section below), changing predation patterns via fishing or protection from fishing may result in dynamic changes to the complementary/redundancy feeding landscape on reefs. Yet, there has been little emphasis on integrating patterns of feeding and risk to predation in parrotfishes and this remains a fruitful area of research.

In addition to reducing prey abundance, predators also strongly impact prey traits (e.g. behavior, growth and reproduction), and recent research indicates that these non-consumptive effects (NCEs) frequently have far-reaching ecosystem level consequences (Dill et al. 2003, Preisser et al. 2005). NCEs arise due to trade-offs between the need to acquire food resources while avoiding being eaten. As a result of these trade-offs, prey often allocate less time to foraging or shift activity to safer but less profitable habitats when predation risk is high (Lima and Dill 1990). NCEs are expected to be important on coral reefs, where high structural complexity offers many refugia, potentially increasing the ability of prey to respond behaviorally to their predators.

Recent research on reefs in both the Caribbean and central Pacific suggests that herbivorous fishes may restrict their foraging ranges and behaviours when predators are abundant (Madin et al. 2010a, Catano et al. 2015, 2016, 2017, Rasher et al. 2017), and that these shifts can affect the spatial distribution of algae and potentially modifying coral-algal competition (Madin et al. 2010b). This work suggests the possibility that predators could create heterogeneous landscapes where hotspots of grazing activity potentially create habitat suitable for coral recruitment.

However, there are likely species-specific responses to predation risk. For example, recent work from the wider Caribbean suggests that parrotfishes and surgeonfishes may alter their behavior, and ultimately their diets, differently in responses to increasing predator abundance (Catano et al. 2014). The parrotfish *Sparisoma aurofrenatum* significantly altered its feeding behavior and showed increases in diet diversity under increasing risk of predation while the surgeonfish, *Acanthurus bahianus*, was relatively unaffected by increasing predator abundance. The difference in their responses may be due, in part, to foraging behavior as surgeonfishes forage more often in large schools making them more insensitive to predation. Thus, there may be a range of demographic and behavioral responses to predators from parrotfishes and other herbivores that yield other potential axes of complementarity.

Body Size

Many of the traits described above vary predictably with body size. Parrotfish species often show ontogenetic shifts in diet by shifting from a more carnivorous or omnivorous diet when small to a more herbivorous diet when larger (Bellwood 1988, Chen 2002). Indeed, for several parrotfish species, large individuals have been described to perform different functional roles than small individuals with consequences for processes such as sediment reworking, bioerosion, and potentially biogeochemical fluxes (Bonaldo and Bellwood 2008, Lokrantz et al. 2008, Plass-Johnson et al. 2013, Hoey Chapter 6). These studies suggest that individuals of a single species could be functionally complementary to each other depending on the size of those individuals. Thus, altering the size distribution of a species (e.g. via overfishing) could alter the functional impact of that species. This is because there are specific functions such as bioerosion, which are best achieved or can only be achieved by large fishes (Bellwood et al. 2003, 2012), and because similarly sized fishes are likely to have similar diet preferences (Burkepile and Hay 2011), forage over similar spatial scales (Nash et al. 2013), and respond similarly to predators (Rizzari et al. 2014, Catano et al. 2016). Of course, body size is also associated with life-history characteristics that influence susceptibility to a range of perturbations, including overfishing and habitat degradation (Bellwood et al. 2012, Nash et al. 2013, Edwards et al. 2014). Identifying the relationships between body size and the different ecological functions performed by parrotfishes will therefore enable better predictions of how different types of perturbations will influence the function of the herbivore guild.

Phylogenetic History

In addition to body size, many functional traits are likely to be determined in part by phylogenetic history (Clarke and Warwick 1999). For example, while relatively little is known about the digestive capabilities of parrotfishes (Clements et al. 2009), the activity of certain digestive enzymes has been linked to phylogeny in some freshwater herbivorous fishes (e.g. German et al. 2004). Further, the enzymes required to digest storage compounds from brown algae are rare in herbivorous fishes while those required to digest compounds from red and green algae are common (Montgomery and Gerking 1980). This conforms well to the observation that relatively few herbivorous fishes on coral reefs consume large amounts of brown algae (Choat et al. 2002) and suggests the possibility that the ability to produce particular enzymes could lead to phylogenetic constraints on diet in herbivorous fishes. Indeed, parrotfishes that tend to browse on brown macroalgae are almost exclusively from the seagrass associated, or sparisomatine, clade (i.e. *Calotomus, Cryptotomus, Leptoscarus, Nicholsina* and *Sparisoma*) suggesting a strong role of phylogenetic history in shaping patterns of complementarity among parrotfishes (Parenti and Randall 2011, Adam et al. 2015b).

Using Multiple Metrics to Assess Complementarity

On Caribbean coral reefs, recent research has revealed considerable complementarity among parrotfishes. Using detailed behavioral observations, Adam et al. (2015b) documented substantial functional diversity among nine species of parrotfishes in the Florida Keys. Not surprisingly, fishes in the genus *Scarus* targeted filamentous algal turf assemblages, crustose coralline algae, and endolithic algae and avoided macroalgae, while fishes in the genus *Sparisoma* preferentially targeted macroalgae. However, species with similar diets were dissimilar in other attributes, including the habitats they selected (e.g. spur and groove vs. coral pavement), the types of substrata they fed from (e.g. dead coral heads vs. coral rubble vs. sand), and the distances they moved while foraging (e.g. spatially intensive vs. spatially extensive foraging) (See Fig. 2). These observations suggest that species that appear to be functionally redundant when looking at diet alone exhibit high levels of complementarity when we consider multiple functional traits.

Adam et al. (2015b) also observed many aggressive interactions between closely related parrotfishes, suggesting that differences in habitat selection may have been driven by interspecific competition for food. These results suggest that phylogenetic constraints on diet may result in niche partitioning along other more flexible niche axes. Further, they indicate that a better understanding of competitive dynamics among parrotfishes and other herbivores is likely to lead to further insights about the levels of complementarity and redundancy within the herbivore guild.

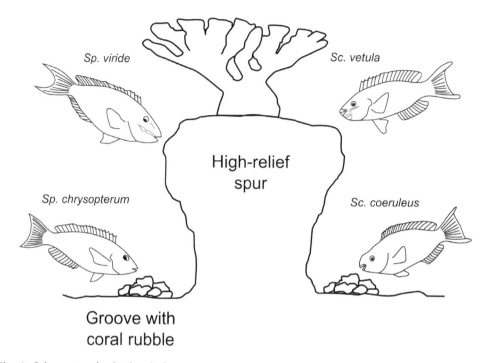

Fig. 2. Schematic of a high-relief spur and groove reef illustrating niche-partitioning among four species of Caribbean parrotfishes. The two species of *Scarus*, *Sc. vetula* and *Sc. coeruleus* both feed primarily on turf algae and associated detritus but are segregated in space, with *Sc. vetula* targeting sparse turfs growing on architecturally complex structures in high-relief areas and *Sc. coeruleus* feeding in lower relief areas on flat substrates with higher sediment loads. Likewise, species of *Sparisoma*, which feed to a greater degree on macroalgae, also feed in different areas of the reef and target different substrates. *Sp. viride* preferentially feeds on 3-dimensionally complex structures in high-relief areas while *Sp. chrysopterum* largely targets coral rubble and boulders in lower relief areas.

Impacts of Complementarity on Community Dynamics and Ecosystem Processes

As discussed in detail above, early observational studies often revealed distinctions or similarities among parrotfish species or genera with regard to their morphology, diet, and habitat utilization (e.g. Randall 1967, Lewis 1985, Bellwood and Choat 1990, Bruggemann et al. 1996, McAfee and Morgan 1996). The differences and similarities noted in these early studies were often so distinct that species could be assigned to discrete functional groups with the idea that different groups serve complementary ecological roles, and thus interactively shape overall patterns of benthic community structure, ecosystem processes, and resilience (Bellwood et al. 2004). Yet, the morphology and/or taxonomy that were sometimes used in early studies to infer functional role have proven inadequate for predicting functional overlap (Bellwood et al. 2006, Burkepile and Hay 2011), complicating our ability to predict complementarity effects in nature. Further, this functional group framework arose from comparing mostly observational studies that were unrelated in space or time and that could not detect the potential complexities or indirect effects of complementarity. More recently, scientists have employed manipulative experiments, or performed targeted sampling of diets and habitat use across gradients or in "natural experiments", to better determine herbivore function and detect complementarity effects in nature. In this section, we focus on these recent studies that can more unambiguously address the direct (e.g. effects on algal communities) and indirect (e.g. effects on corals via effects on algae) effects of complementarity among parrotfishes and between parrotfishes and other herbivores.

Effects of Parrotfish Complementarity

In the Caribbean, scientists have recently used manipulative experiments to reveal that complementary feeding between parrotfish significantly enhances reef resilience. Examining the individual vs. additive grazing impacts of the redband parrotfish *Sparisoma aurofrenatum* and princess parrotfish *Scarus taeniopterus* in single- vs. mixed-species enclosures, Burkepile and Hay (2008, 2010) used 8-10 month experiments to demonstrate that the two species cause unique and complementary changes to the algal community, with their relative importance depending on the successional state of the community. *Sc. taeniopterus* prevented macroalgal establishment and promoted the establishment of crustose coralline algae on new substrata but had little impact on established macroalgae occupying old substrata (Fig. 3). Conversely, *Sp. aurofrenatum* reduced the abundance of established upright macroalgae but fed less frequently on filamentous algal communities (Fig. 3). Because upright macroalgae and filamentous turfs have deleterious effects on corals (Birrell et al. 2008), only together did the two species exert a grazing effect that could potentially both prevent the spread of harmful algae and facilitate coral recruitment and growth. This complementarity effect appears critical to coral recovery following disturbance.

Complementarity among parrotfishes also dictates how rates of critical ecological processes differ across the Great Barrier Reef (GBR). By surveying a dramatic cross-shelf gradient of fish and benthic community structure, Hoey and Bellwood (2008) revealed that inner and outer shelf reefs are shaped by fundamentally different ecological processes and that such differences are due to cross-shelf variation in the relative abundances of parrotfish species with differing feeding modes (Fig. 4). Rates of substratum grazing and sediment reworking were highest and rates of bioerosion and coral predation lowest on inner shelf reefs, where small scraping species were abundant but excavators were rare. In contrast, rates of bioerosion and coral predation were highest on outer shelf reefs where large excavating species dominated. Hence, differences in the distribution of parrotfishes with complementary functions results in fundamentally different processes shaping each shelf system.

Other studies in the Caribbean and Indo-Pacific have shown differential distribution of parrotfishes within and across reefs (Russ 1984, 2003, Bruggemann et al. 1994a, Hoey et al. 2016) while others have shown strong spatial variation in herbivory across reef types (Hay 1981, 1985, Hoey and Bellwood 2010a, b). However, more work is needed to examine spatial differences in the complementarity of parrotfishes, as did Hoey and Bellwood (2008), across different types of reef systems to understand how turnover of parrotfish species across reef types may result in different rates of key ecological processes.

Effects of Complementarity between Parrotfishes and Other Herbivores

Perhaps equally striking are the ecological outcomes resulting from complementarity between parrotfishes and other herbivores. In the Caribbean, early observational studies showed differences between certain parrotfish and surgeonfish with regard to the macroalgae that they consume, with habitat-specific grazing rates explained by the

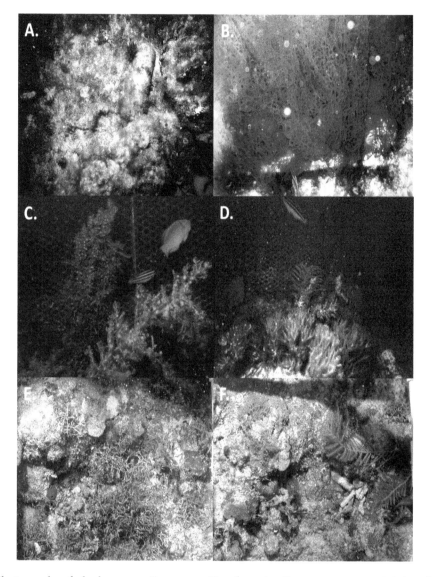

Fig. 3. Photographs of algal community composition from in situ mesocosms containing different combinations of parrotfish species. (A and B) Community dominated by filamentous algal turfs and *Kallymenia westii*, respectively, from mesocosms with only the parrotfish *Sparisoma aurofrenatum*. (C and D) Community dominated by *Sargassum* spp. and *Codium* spp., respectively, from mesocosms with only the parrotfish *Scarus taeniopterus*. (E) Community dominated by upright calcified algae such as *Halimeda* spp., *Amphiroa* spp. and *Jania* spp. from mesocosm with only the surgeonfish *Acanthurus bahianus*. (F) Community free of most upright macroalgae and dominated by small filamentous turf algae and crustose coralline algae from mesocosm with both parrotfishes *S. aurofrenatum* and *S. taeniopterus*. Photo credits: Deron Burkepile (B, E and F) and Mark Hay (A, C and D).

relative abundances of these two groups (Lewis 1985, Lewis and Wainwright 1985). Recently, Burkepile and Hay (2008) revealed this complementary browsing has indirect positive effects on corals, the foundation species of the ecosystem. Using a long-term manipulative experiment, they showed that when the parrotfish *Sp. aurofrenatum* and

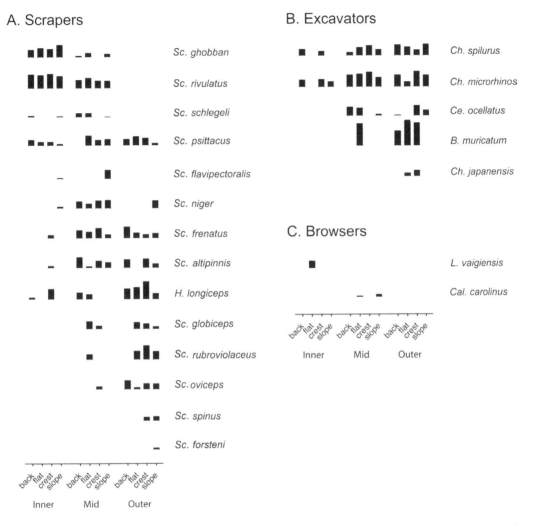

Fig. 4. Distribution of parrotfishes across the continental shelf in the northern Great Barrier Reef. Vertical bars represent the relative abundance of 21 parrotfish species across four habitats (reef slope, crest, flat and backreef) and three shelf positions (inner-, mid- and outer-shelf). Means are based on four replicate 10-min timed swims within each habitat, on two reefs within each shelf position. B = *Bolbometopon*, Ch = *Chlorurus*, Cal = *Calotomus*, Ce = *Cetoscarus*, H = *Hipposcarus*, L = *Leptoscarus*, Sc = *Scarus*

surgeonfish *A. bahianus* graze together, they reduce macroalgal abundance 75% more than either species alone. The result of this complementarity was 22% greater coral growth and 100% coral survival in the mixed species treatment. Removing either of the species (while still maintaining equal herbivore densities) resulted in dramatically different macroalgal communities with increased coral mortality (Fig. 3). Such findings suggest the benefits of mixed feeding documented in the enclosure study are present within more diverse herbivore assemblages found in nature.

In the Indo-Pacific, where herbivore diversity is highest, one might expect less complementarity and more redundancy relative to the Caribbean given the increased number of species. Yet, considerable research shows that complementary feeding between parrotfish and other herbivores also underpins reef resilience. For example, while scraping and excavating parrotfish play the critical role of preventing establishment of macroalgae on the GBR, the removal of mature macroalgae is carried out by a different suite of herbivores: surgeonfish in the genus *Naso,* rabbitfishes, and drummers (Fox and Bellwood 2008, Hoey and Bellwood 2010b, Hoey et al. 2013). In fact, removal of the leathery macroalga *Sargassum,* which dominate inner shelf reefs and suppresses coral health in the absence of herbivory (Hughes et al. 2007), is largely attributable to unicornfish in the genus *Naso* (Hoey and Bellwood 2009, 2010b). It also appears that the few parrotfishes that browse mature macroalgae in the Indo-Pacific play a complementary role to other browsers, generally targeting calcified macroalgae that surgeonfish and rabbitfish avoid (Schupp and Paul 1994, Mantyka and Bellwood 2007, Rasher et al. 2013). Parrotfish also differ from rabbitfish with regard to their microhabitat utilization (Fox and Bellwood 2013, Brandl and Bellwood 2014). Owing to their comparatively longer, narrower snouts and narrower heads, rabbitfish are capable of feeding in crevices that are inaccessible to parrotfish and surgeonfish. Consequently, their overlap in microhabitat use with the other groups when feeding is only 40-45%, compared with a 95% overlap that exists between some parrotfish and surgeonfish (Fox and Bellwood 2013). This multifunctional complementarity appears important to the resilience of Indo-Pacific reefs as some reefs with reduced herbivore diversity have undergone phase shifts to being dominated by macroalgae (Cheal et al. 2010, Rasher et al. 2013).

Complementary feeding among a variety of herbivores is also notable in Fiji, where it appears to have driven the recovery of reefs inside multiple no-take reserves. Shortly following a coral bleaching event that triggered system-wide coral mortality and a phase shift to macroalgae, several villages on the Coral Coast of Viti Levu established locally managed no-take marine reserves to aid ecosystem recovery and promote fisheries production in adjacent fishing grounds. Focusing on three well enforced reserves, Rasher et al. (2013) documented that reserve establishment had significantly enhanced herbivore diversity and biomass within reserves after 10 years, coincident with a near elimination of macroalgae and a return to coral dominance. In contrast, adjacent fishing areas had remained locked in a degraded macroalgal-dominated state. Using a series of experiments and surveys, they discovered the near elimination of macroalgae within reserves was likely due to complementary feeding among the four dominant browsers – the parrotfish *Chlorurus spilurus* (formerly *Ch. sordidus*), the unicornfishes *Naso unicornis* and *N. lituratus*, and the rabbitfish *Siganus argenteus*. Differential tolerances to macroalgal chemical and structural defenses drove this complementarity. Thus, a suite of macroalgal browsing species was required to eliminate the suite of variably defended macroalgae characterizing the degraded phase state. These four browser species, along with other herbivores, had remained rare or absent in fished areas but had increased in abundance significantly inside reserves, leading to macroalgal decline and coral recovery. Consistent with this interpretation, macroalgal abundance and diversity were negatively related to herbivore diversity across reefs. Fished reefs lacked the right abundance and mix of herbivores needed to initiate a phase shift reversal. Importantly, a different suite of parrotfishes and surgeonfishes grazed algal turfs and maintained the benthos in a cropped state suitable for coral recruitment and growth once the browsers cleared the macroalgae. Thus, this herbivore guild exhibited complementarity both within the browser functional group for removing abundant macroalgae and across the browser and grazer functional groups for either removing macroalgae once it became established or preventing its return once it became rare.

Urchins are another critical herbivore on many coral reefs worldwide and add important levels of redundancy to the herbivore guild. In the Caribbean, *Diadema antillarum* is the dominant reef-associated urchin, although its population levels have been extremely low across most of the Caribbean since a mass mortality event in the early 1980's (Lessios et al. 1984). However, before the mass mortality *D. antillarum* was one of the most important herbivores on Caribbean reefs, especially on reefs where parrotfishes were rare due to overfishing (Hay 1984). Both before and after the mass mortality, abundant *D. antillarum* often led to low macroalgal cover and increased coral recruitment and growth (e.g. Sammarco 1982, Carpenter 1986, Carpenter and Edmunds 2006, Idjadi et al. 2006), suggesting that these urchins perform a variety of functions unachievable by a single species of herbivorous fish.

The ability of *D. antillarum* to control a wide variety of algae is likely related to the fact that it is relatively undeterred by a suite of algal anti-herbivore defenses (Craft et al. 2013) and consequently has a very wide diet-breadth (Ogden 1976). Thus, *D. antillarum* appears able to consume a broad spectrum of algae that otherwise takes a suite of parrotfishes and surgeonfishes to consume. Further, urchins and fishes appear to forage on complementary spatial scales. Recent modeling work suggests that the high intensity, spatially-constrained foraging behavior of *D. antillarum* is more likely to facilitate coral settlement and coral dominance than the more spatially diffuse herbivory of large, roving fishes (Sandin and McNamara 2012). Thus, *D. antillarum* appears to have aspects of both redundancy and complementarity to parrotfishes and other herbivorous fishes, enabling *D. antillarum* to approximate the functional impact of many different species of fishes.

Similar patterns appear on other reefs around the world such as in the Indian Ocean where the urchin *Echinometra mathaei* kept reefs free of macroalgae even after most herbivorous fishes had been removed (McClanahan and Shafir 1990). However, coral reefs where urchins are the most abundant and important herbivore may be on a slippery slope to decline as overabundant urchins: (1) consume coral recruits and adult corals (Sammarco 1980, McClanahan and Shafir 1990), (2) slow reef growth and weaken the reef matrix due to bioerosion (Glynn et al. 1979), and (3) are vulnerable to mass mortalities and population crashes (Lessios et al. 1984, Uthicke et al. 2009). Thus, on many coral reefs, sea urchins are likely to be a poor substitute for a diverse and abundant fish community.

How Much Complementarity is There?

Ecosystem functions performed by one, or a few, species will be highly sensitive to changes in the abundance of those species (Walker 1992, Naeem 1998). Conversely, an ecosystem function with high levels of functional redundancy is expected to be relatively insensitive to the loss of a few individual species, as the losses of these species may be compensated by the actions of the remaining functionally similar species (Bellwood et al. 2004, Micheli et al. 2014). Thus, the extent of the complementarity and redundancy among parrotfishes and other herbivores will determine how robust different ecological processes are to disturbances to the fish fauna and, ultimately, how these disturbances affect the resilience of coral reefs.

Different Levels of Complementarity for Different Functions

Biogeographic history (e.g. Bernardi et al. 2000, Choat et al. 2012), and more recently human exploitation (Mumby 2006, Rasher et al. 2013, Taylor 2014), has shaped the considerable regional variation in the taxonomic and functional composition of parrotfishes. The most pronounced division in the parrotfish assemblage structure is between the Indo-Pacific and Atlantic Oceans, with assemblages in the Atlantic and Indo-Pacific being dominated by seagrass associated (i.e. *Sparisoma* clade) and reef associated (i.e. *Scarus* clade) taxa, respectively (Westneat and Alfaro 2005). Further, there is no overlap in species between the two regions, with 73 (out of 100) species and six (out of ten) genera being restricted to the Indo-Pacific (Parenti and Randall 2011, Bonaldo et al. 2014). Although these differences in taxonomic composition are reflected in the functional composition of parrotfishes in these regions, in particular the predominance of browsing taxa in the Atlantic, there are some striking similarities in the functional redundancy of parrotfishes among regions. Some species appear to be part of larger groups of functionally similar species (i.e. functionally redundant), while others appear to perform a unique and almost exclusive role on coral reefs (Bruggemann et al. 1996, Bellwood et al. 2003, 2012, Francini-Filho et al. 2008).

Browsing parrotfishes are almost exclusively from the seagrass associated, or sparisomatine, clade (i.e. *Calotomus*, *Cryptotomus*, *Leptoscarus*, *Nicholsina* and *Sparisoma*), and are most abundant and speciose in the tropical Atlantic, comprising 17 (14 species of *Sparisoma*, two species of *Nicholsina*, and *Cryptotomus roseus*) of 25 parrotfish species in this region (Parenti and Randall 2011). Dietary studies, feeding assays, and video observations show that these browsing parrotfishes are the dominant consumers of macroalgae and seagrass in the Atlantic, with considerable overlap in the diet of each species (e.g. Randall 1967, Lobel 1981, Lewis 1985, Burkepile and Hay 2011, Adam et al. 2015b). In contrast, browsing parrotfishes are relatively rare and less speciose on Indo-West Pacific reefs, comprising only six (five species of *Calotomus* and *Leptoscarus vaigiensis*) of the 75 parrotfish species in this region. Although relatively widespread across the Indo-Pacific and having a diet similar to their Atlantic counterparts (e.g. McClanahan et al. 1999), these species are typically rare and have a limited impact on the removal of macroalgae on most coral-dominated reefs (Hoey and Bellwood 2008, 2009, Michael et al. 2013). Instead, macroalgal browsing is dominated by unicornifshes (Acanthuridae: Nasinae), rabbitfishes (Siganidae), and drummers (Kyphosidae) on Indo-Pacific reefs (Robertson and Gaines 1986, Choat et al. 2002, Fox and Bellwood 2008, Hoey and Bellwood 2009, Vergés et al. 2012, Hoey et al. 2013, Michael et al. 2013; Loffler et al. 2015). Despite regional differences in the composition of browsing fishes and the relative contribution of parrotfishes to browsing, the diversity of species contributing to this process suggests there is some functional redundancy.

Scraping parrotfishes concentrate their feeding activities on the epilithic algal matrix (EAM), a conglomerate of filamentous algae, macroalgal propagules, detritus and associated infauna (Wilson et al. 2003), and are the most speciose functional group within the parrotfishes. Scraping parrotfishes are represented primarily by *Scarus* and *Hipposcarus* in the Indo-Pacific (46 of 75 species), and *Scarus* in the Atlantic (eight of 25 species), and supplemented by smaller individuals of excavating species in both regions. Further, *Sp. aurofrenatum* has also been recorded to function as a scraper on Caribbean reefs (McAfee and Morgan 1996, Bernardi et al. 2000). The relatively high species diversity within scraping parrotfishes, together with their preference for feeding on EAM covered substratum, suggest the loss of one or two species would not greatly impact the overall process. However, differences in feeding substrata and spatial distributions of scrapers may reduce the functional redundancy considerably (e.g. Brandl and Bellwood 2014, Adam et al. 2015b).

In marked contrast to browsing and scraping, bioerosion by parrotfishes is dominated by just one or two species on both Indo-Pacific and Atlantic reefs (Bellwood and Wainwright 2002). In the Indo-Pacific, parrotfishes were initially separated into scraping and excavating species based largely on morphological and evolutionary characters, with *Bolbometopon*, *Cetoscarus* and *Chlorurus* being identified as excavating genera (Bellwood

1994, Streelman et al. 2002). Based on this classification alone it could be assumed that there is considerable redundancy within this group. However, more recent studies have revealed that *Bolbometopon muricatum*, the largest of the parrotfish species (up to 130 cm in length), can remove over 5.5 tonnes of carbonate per year per individual (Bellwood et al. 2003, Hoey and Bellwood 2008). This is five-times greater than the largest species of *Chlorurus* (*Ch. microrhinos* 1.01 t yr^{-1}: Bellwood 1995b), over 14-times greater than that of medium sized *Chlorurus* (*Ch. perspicillatus* 0.38 t yr^{-1}: Ong and Holland 2010), and 220-times greater than that of small bodied species (*Ch. spilurus* 0.024 t yr^{-1}: Bellwood 1995b). Such variation in erosion rates means that whenever present, *B. muricatum* overwhelmingly dominates parrotfish bioerosion on Indo-Pacific reefs (Bellwood et al. 2003, 2012, Hoey and Bellwood 2008). Indeed, its feeding impact cannot be replaced by any other species.

Similarly, a single parrotfish species is largely responsible for external bioerosion across most of the tropical Atlantic. Early studies of parrotfish bioerosion on Caribbean reefs identified that *Sparisoma viride* was largely responsible for external bioerosion on reefs in the region (Scoffin et al. 1980, Bruggemann et al. 1994b, 1996). Although the volume of carbonate removed per individual (*Sp. viride* 0.14 t yr^{-1}: Bruggeman et al. 1996) is considerably lower than for large Indo-Pacific excavating species, *Sp. viride* in the Caribbean, and its sister taxon *Sp. amplum* on Brazilian reefs (Francini-Filho et al. 2008), may be the only excavating parrotfish species in the region. However, detailed feeding ecology of the two largest *Scarus* spp. in the Caribbean (*Sc. guacamaia* and *Sc. coelestinus*, both can grow to over 100 cm) is minimal due to their overexploitation and rarity across the region, but both species may be important bioeroders given their large body size. Yet, the limited redundancy in parrotfish bioerosion on tropical Indo-Pacific and Atlantic reefs indicates that these reef systems are extremely vulnerable to the loss of these single species with bioerosion rates declining precipitously when these species are exploited (Bellwood et al. 2003, Perry et al. 2014).

Levels of Complementarity in High vs. Low Diversity Systems

Herbivorous fish diversity is much higher in the Indo-Pacific than the Caribbean (Bellwood and Hughes 2001, Bellwood et al. 2004). Of the 100 species and ten genera of parrotfish from both regions, only 15 species in four genera inhabit the Caribbean (Roff and Mumby 2012). Likewise, of 88 species of surgeonfish, the Caribbean possesses only four species, all in a single genus, *Acanthurus*. Rabbitfish, of which there are 28 species in the Indo-Pacific, do not occur in the Caribbean (Roff and Mumby 2012). Therefore, one might predict niche partitioning (i.e. complementarity) among herbivores to be more pronounced in the Indo-Pacific simply as a consequence of greater herbivore diversity. Further, algal diversity also appears to be higher in the Indo-Pacific as compared to the Caribbean (Kerswell 2006), and a higher diversity of diet choices could facilitate finer partitioning of diet niches. Yet, to date there has been no formal comparison among regions. Extensive niche diversification has occurred within and among parrotfishes, surgeonfishes, and rabbitfishes on the GBR (Choat et al. 2002, Bellwood et al. 2004), and parrotfish in particular have evolved tremendous functional versatility in that system (Wainwright et al. 2004, Bellwood et al. 2006). Likely owing to the lower diversity of the region, parrotfish complementarity is also marked and important in the Caribbean (Burkepile and Hay 2011) and other low-diversity reef systems such as coastal South America (Bonaldo et al. 2006). Any difference in complementarity between regions is therefore likely to be a consequence of evolutionary history rather than fundamental differences in the process of herbivory.

Many Indo-Pacific reefs have generally resisted phase shifts following recent disturbance while many Caribbean reefs have not (Hughes 1994, Connell 1997, Bruno et al. 2009). A higher diversity of herbivores, and thus a higher potential for niche diversification and function redundancy, in the Indo-Pacific may explain, in part, the greater resilience of the region's reefs (Bellwood et al. 2004, Roff and Mumby 2012). However, redundancy–resilience relationships remain poorly resolved (Roff and Mumby 2012, Cheal et al. 2013). In fact, a consequence of the striking complementarity among Indo-Pacific herbivores appears to be low, not high, redundancy within functional groups; several important ecosystem processes (i.e. bioerosion, macroalgal removal) are shaped by the activities of only one or a few species (Bellwood et al. 2003, 2012, Hoey and Bellwood 2009). This is also likely true in the less diverse Caribbean (e.g. Bruggeman et al. 1996). Thus, in either region, the loss of a single or few species can result in dramatic changes to ecological processes that shape reef resilience. In general coastal ecosystems are defined by low functional redundancy, regardless of diversity (Micheli and Halpern 2005), but whether increases in complementarity actually leads to greater resilience requires additional investigation.

Effects of Fishing on Patterns of Complementarity/Redundancy

Fishing is arguably one of the oldest and most pressing anthropogenic effects on coral reefs (Jackson et al. 2001, Pandolfi et al. 2003), and one of the greatest drivers of fish assemblage structure in many reef systems (e.g. McClanahan 1994, Rasher et al. 2013, Edwards et al. 2014). Parrotfishes have long been targets of fishers across much of the Caribbean (Fig. 1c; Hawkins and Roberts 2004a, b), and are increasingly being targeted across the Indo-Pacific (Dalzell et al. 1996, Aswani and Hamilton 2004, Houk et al. 2012). Although there are regional differences in the predominant gear used (Hawkins and Roberts 2004a, Hicks and McClanahan 2012, Lindfield et al. 2014), fishers typically target the larger, and hence higher-valued, parrotfish species (Aswani and Hamilton 2004, Bellwood et al. 2012, Bejarano et al. 2013). For example, in the Solomon Islands fishers using spears target schools of resting *B. muricatum*, the world's largest parrotfish, with reports of individual fishers taking over 50 fishes in a single night (Aswani and Hamilton 2004).

Fishing pressure, together with the bias toward larger species and individuals, has resulted in marked declines in the biomass of parrotfishes in many regions (e.g. McClanahan 1994, Aswani and Hamilton 2004, Hawkins and Roberts 2004b, Floeter et al. 2006, Rhodes et al. 2008, Campbell et al. 2012), as well as changes in sex-ratios and life-history parameters (Hawkins and Roberts 2004b, Taylor 2014). On reefs with no or limited fishing for parrotfishes, biomass of the group can exceed 500 kg.ha^{-1} (Bruggemann et al. 1996, Bellwood et al. 2003, 2012, Edwards et al. 2014), and be an order of magnitude greater than on reefs with moderate-high levels of fishing (Hawkins and Roberts 2004a, Mumby et al. 2006, Bellwood et al. 2012).

Reductions in parrotfish biomass, in particular the loss of large-bodied species and individuals, appears to differentially affect ecosystem processes on reefs due to differences in redundancy for these processes. For example, fishing has a dramatic effect on parrotfish bioerosion on both Caribbean and Indo-Pacific reefs, with estimated rates of bioerosion declining exponentially in response to the loss of *B. muricatum* in the Indo-Pacific (Bellwood et al. 2012) and *Sp. viride* in the Caribbean (Bonaldo et al. 2014). This lack of redundancy for bioerosion by parrotfishes may result in a less stable reef matrix making reefs more vulnerable and less resilient to disturbances such as storms.

Scraping (or grazing) by parrotfishes appear to be more resilient to fishing, with overall rates of scraping showing no relation to fishing pressure across either Caribbean or

Indo-Pacific reefs (Bellwood et al. 2012, Bonaldo et al. 2014, Heenan et al. 2016). This lack of response may be related to the high redundancy within the group, and/or the apparent resilience of smaller-bodied, primarily scraping species, to fishing pressure. Although large-bodied species are the primary targets, small-bodied species are often captured and can be numerically abundant in catches, especially in areas where large-bodied species have been depleted (Rhodes et al. 2008). For example, the small-bodied scraping species (*Scarus iseri*, *Sc. taeniopterus* and *Sp. aurofrenatum*) either showed no change or increased abundance with increasing fishing pressure across the Caribbean (Hawkins and Roberts 2004a). Likewise, on Fijian reefs abundance of small scrapers such as *Scarus schlegeli* appeared more influenced by habitat availability and structure rather than fishing pressure (Wilson et al. 2008). Even when assessing patterns across 18 different Indo-Pacific reefs, the abundance of scraping parrotfishes showed no relationship to human population density (a proxy for fishing pressure), with the abundance of small-bodied parrotfishes being two-fold greater on heavily- versus lightly-fished reefs (Bellwood et al. 2012). Overall, these data suggest that different ecological processes are differentially susceptible to fishing pressure. However, a critical task will be to understand how much fishing pressure can be absorbed before these processes are significantly altered.

Conclusions and Future Directions

Clearly, there is a significant amount of complementarity and redundancy both within parrotfishes and between parrotfishes and other groups of herbivores. An important trend within the past few years shows the field moving away from simply trying to determine what algae and substrata different herbivores feed upon. More efforts have been devoted to trying to capture other metrics of complementarity such as home range and habitat use that inform the multifunctional relationships that exist among herbivorous fishes (e.g. Brandl and Bellwood 2014, Adam et al. 2015b). Continuing to explore these different aspects of complementarity among parrotfishes will continue to yield important information on the role of herbivore diversity on reefs. Not surprisingly, understanding the direct and indirect effects of complementarity on benthic dynamics is challenging as experiments manipulating the presence/absence of different herbivore species (e.g. Burkepile and Hay 2008) are logistically challenging and limiting when herbivore diversity is quite high such as in the Indo-Pacific. However, capitalizing on "experiments" from differential fishing pressure (e.g. Rasher et al. 2013) appears to be one way to address these questions in a more feasible manner. Finally, we would like to conclude with some suggestions for fruitful avenues of research that will help sort out the complex complementarity/redundancy relationships among parrotfishes.

An important step for transitioning from knowing that herbivores are complementary or redundant to understanding how this affects benthic dynamics will require moving beyond correlations between parrotfish/herbivore biomass and benthic community structure (e.g. Williams and Polunin 2001, Wismer et al. 2009, Burkepile et al. 2013). We need to be able to move to translating the multifunctionality of the herbivore assemblage into meaningful information of how it affects benthic reef dynamics. Given that we know different parrotfish species have different diets, feed in different places on a reef, and experience ontogenetic shifts in feeding impacts, the amount of herbivory occurring on a reef will not be a simple function of the total biomass of herbivorous fishes. A better understanding of herbivory can be achieved by combining data on bite rates, bite yield and area, and diet selection with abundances of different herbivore species. Such data would enable one to model processes such as scraping, browsing and bioerosion on a reef-wide scale from size-structured data collected by monitoring programs. Ultimately, these metrics would likely yield much more meaningful information about how the herbivore guild is affecting the benthos and yield more informative relationships with changes in the benthic community over time.

One important aspect for understanding complementarity among parrotfishes, especially the scraping parrotfishes, is a better understanding of the different communities that is commonly referred to as algal turfs or the EAM. Recent literature has pointed out that 'algal turfs' is a somewhat meaningless term ecologically (Connell et al. 2014) and needs to be better defined if we are to understand its place in the ecology of coral reefs. Algal turfs, or EAM, are likely to vary significantly in terms of algal species composition, canopy height, density of thalli, level of sedimentation, and importantly nutritional content (Connell et al. 2014, Clements and Choat Chapter 3). This extreme variation will likely attract some parrotfishes to particular types of algal turf communities. Thus, some of the redundancy that we may currently assume among these species as feeding on algal turfs may in fact be more complex complementarity as they are feeding on different types or components of algal turfs. Understanding these patterns is critical for understanding complementarity/redundancy effects on coral-friendly habitat as algal turf communities are often emphasized as being benign or beneficial for settling coral larvae (e.g. Birrell et al. 2008).

Along with the need to better understand what comprises algal turfs, or EAM, is a need to get a more complete picture of the diets of parrotfishes. Although we often make distinctions based on the surfaces from which species feed and/or the way in which they feed (e.g., scrapers, browsers, excavators), it is often less clear how diets may differ among species within these groups. For browsers, feeding assays with mature macroalgae (e.g. Lewis 1985, Mantyka and Bellwood 2007, Rasher et al. 2013) have allowed for a much better understanding of how different fishes can target different algal species. However, for scraping species that focus mostly on algal turfs, quantifying the algal species different parrotfishes consume is incredibly difficult given how parrotfishes finely grind their food while processing it in their pharyngeal mill (e.g. Choat et al. 2002). One interesting avenue of research would be to address these questions of diet complementarity via DNA barcoding (Valentini et al. 2009). DNA barcoding would allow species-level identification of algae from samples of gut contents and has already been used on a variety of different herbivores (e.g. Chelsky Budarf et al. 2011, García-Robledo et al. 2013).

Finally, integration of the multiple metrics of complementarity will allow for a better understanding of the complex nature of the relationships among parrotfishes. For example, predicting the impacts of different species of herbivores on benthic communities requires understanding not only what they eat but also how their feeding is distributed in space. Given the vast differences in phylogenetic history, body size, home range, and diet among species, there are many possible trait combinations. Understanding how all of these different lifestyles are important to the health of coral reefs is our ultimate challenge.

References Cited

Adam, T.C., D.E. Burkepile, B.I. Ruttenberg and M.J. Paddack. 2015a. Herbivory and the resilience of Caribbean coral reefs: knowledge gaps and implications for management. Mar. Ecol. Prog. Ser. 520: 1–20.

Adam, T.C., M.C. Kelley, B.I. Ruttenberg and D.E. Burkepile. 2015b. Resource partitioning along multiple niche axes drives functional diversity in parrotfishes on Caribbean coral reefs. Oecologia. 179: 1173–1185.

Adam, T.C., R.J. Schmitt, S.J. Holbrook, A.J. Brooks, P.J. Edmunds, R.C. Carpenter and G. Bernardi. 2011. Herbivory, connectivity, and ecosystem resilience: response of a coral reef to a large-scale perturbation. PLoS One 6: e23717.

Alvarez-Filip, L., N.K. Dulvy, J.A. Gill, I.M. Côté and A.R. Watkinson. 2009. Flattening of Caribbean coral reefs: region-wide declines in architectural complexity. Proc. R. Soc. B Biol. Sci. 276: 3019–3025.

Aswani, S. and R.J. Hamilton. 2004. Integrating indigenous ecological knowledge and customary sea tenure with marine and social science for conservation of bumphead parrotfish (*Bolbometopon muricatum*) in the Roviana Lagoon, Solomon Islands. Environ. Conserv. 31: 69–83.

Barboza, P.S., K.L. Parker and I.D. Hume. 2009. Integrative Wildlife Nutrition. Springer-Verlag, Heidelberg, Germany.

Bejarano, S., Y. Golbuu, T. Sapolu and P.J. Mumby. 2013. Ecological risk and the exploitation of herbivorous reef fish across Micronesia. Mar. Ecol. Prog. Ser. 482: 197–215.

Bellwood, D.R. 1988. Ontogenetic changes in the diet of early post-settlement *Scarus* species (Pisces, Scaridae). J. Fish Biol. 33: 213–219.

Bellwood, D.R. 1994. A phylogenetic study of the parrotfishes family Scaridae (Pisces: Labroidei), with a revision of genera. Rec. Aust. Mus. Suppl. 20: 1–86.

Bellwood, D.R. 1995a. Carbonate transport and within-reef patterns of bioerosion and sediment release by parrotfishes (family Scaridae) on the Great Barrier Reef. Mar. Ecol. Prog. Ser. 117: 127–136.

Bellwood, D.R. 1995b. Direct estimate of bioerosion by two parrotfish species, *Chlorurus gibbus* and *C. sordidus*, on the Great Barrier Reef, Australia. Mar. Biol. 121: 419–429.

Bellwood, D.R. and J.H. Choat. 1990. A functional analysis of grazing in parrotfishes (family Scaridae): the ecological implications. Environ. Biol. Fishes 28: 189–214.

Bellwood, D.R. and T.P. Hughes. 2001. Regional-scale assembly rules and biodiversity of coral reefs. Science 292: 1532–1535.

Bellwood, D.R. and P.C. Wainwright. 2002. The history and biogeography of fishes on coral reefs. pp. 5–32 *In*: P.F. Sale (ed.). Coral reef fishes: dynamics and diversity in a complex ecosystem. Academic Press, San Diego, USA.

Bellwood, D.R., A.S. Hoey and J.H. Choat. 2003. Limited functional redundancy in high diversity systems: resilience and ecosystem function on coral reefs. Ecol. Lett. 6: 281–285.

Bellwood, D.R., A.S. Hoey and T.P. Hughes. 2012. Human activity selectively impacts the ecosystem roles of parrotfishes on coral reefs. Proc. R. Soc. B Biol. Sci. 279: 1621–1629.

Bellwood, D.R., T.P. Hughes, C. Folke and M. Nyström. 2004. Confronting the coral reef crisis. Nature 429: 827–833.

Bellwood, D.R., P.C. Wainwright, C.J. Fulton and A.S. Hoey. 2006. Functional versatility supports coral reef biodiversity. Proc. R. Soc. B Biol. Sci. 273: 101–107.

Bernardi, G., D.R. Robertson, K.E. Clifton and E. Azzurro. 2000. Molecular systematics, zoogeography, and evolutionary ecology of the Atlantic parrotfish genus *Sparisoma*. Mol. Phylogenet. Evol. 15: 292–300.

Birrell, C.L., L.J. Mccook, B.L. Willis and G.A. Diaz-Pulido. 2008. Effects of benthic algae on the replenishment of corals and the implications for the resilience of coral reefs. Oceanogr. Mar. Biol. 46: 25–63.

Bonaldo, R.M. and D.R. Bellwood. 2008. Size-dependent variation in the functional role of the parrotfish *Scarus rivulatus* on the Great Barrier Reef, Australia. Mar. Ecol. Prog. Ser. 360: 237–244.

Bonaldo, R.M., A.S. Hoey and D.R. Bellwood. 2014. The ecosystem role of parrotfishes on tropical reefs. Oceanogr. Mar. Biol. 52: 81–132.

Bonaldo, R.M., J.P. Krajewski, C. Sazima and I. Sazima. 2006. Foraging activity and resource use by three parrotfish species at Fernando de Noronha Archipelago, tropical West Atlantic. Mar. Biol. 149: 423–433.

Boyer, K.E., P. Fong, A.R. Armitage and R.A. Cohen. 2004. Elevated nutrient content of tropical macroalgae increases rates of herbivory in coral, seagrass, and mangrove habitats. Coral Reefs 23: 530–538.

Brandl, S.J. and D.R. Bellwood. 2014. Individual-based analyses reveal limited functional overlap in a coral reef fish community. J. Anim. Ecol. 83: 661–670.

Bruggemann, J.H., M.W.M. Kuyper and A.M. Breeman. 1994a. Comparative analysis of foraging and habitat use by the sympatric Caribbean parrotfish *Scarus vetula* and *Sparisoma viride* (Scaridae). Mar. Ecol. Prog. Ser. 112: 51–66.

Bruggemann, J.H., J. Begeman, E.M. Bosma, P. Verburg and A.M. Breeman. 1994b. Foraging by the stoplight-parrotfish *Sparisoma viride*. II. Intake and assimilation of food, protein and energy. Mar. Ecol. Prog. Ser. 106: 57–71.

Bruggemann, J.H., A.M. vanKessel, J.M. vanRooij and A.M. Breeman. 1996. Bioerosion and sediment ingestion by the Caribbean parrotfish *Scarus vetula* and *Sparisoma viride*: Implications of fish size, feeding mode and habitat use. Mar. Ecol. Prog. Ser. 134: 59–71.

Bruno, J.F., H. Sweatman, W.F. Precht, E.R. Selig and V.G.W. Schutte. 2009. Assessing evidence of phase shifts from coral to macroalgal dominance on coral reefs. Ecology 90: 1478–1484.

Burkepile, D.E. 2012. Context-dependent corallivory by parrotfishes in a Caribbean reef ecosystem. Coral Reefs 31: 111–120.

Burkepile, D.E. 2013. Comparing aquatic and terrestrial grazing ecosystems: is the grass really greener? Oikos 122: 306–312.

Burkepile, D.E. and M.E. Hay. 2008. Herbivore species richness and feeding complementarity affect community structure and function on a coral reef. Proc. Natl. Acad. Sci. USA 105: 16201–16206.

Burkepile, D.E. and M.E. Hay. 2009. Nutrient versus herbivore control of macroalgal community development and coral growth on a Caribbean reef. Mar. Ecol. Prog. Ser. 389: 71–84.

Burkepile, D.E. and M.E. Hay. 2010. Impact of herbivore identity on algal succession and coral growth on a Caribbean Reef. PLoS One 5: e8963.

Burkepile, D.E. and M.E. Hay. 2011. Feeding complementarity versus redundancy among herbivorous fishes on a Caribbean reef. Coral Reefs 30: 351–362.

Burkepile, D.E., J.E. Allgeier, A.A. Shantz, C.E. Pritchard, N.P. Lemoine, L.H. Bhatti and C.A. Layman. 2013. Nutrient supply from fishes facilitates macroalgae and suppresses corals in a Caribbean coral reef ecosystem. Sci. Rep. 3: 1493.

Campbell, S.J., A.S. Hoey, J. Maynard, T. Kartawijaya, J. Cinner, N.A.J. Graham and A.H. Baird. 2012. Weak compliance undermines the success of no-take zones in a large government-controlled marine protected area. PLoS One 7: e50074.

Carpenter, R.C. 1986. Partitioning herbivory and its effects on coral reef algal communities. Ecol. Monogr. 56: 345–363.

Carpenter, R.C. and P.J. Edmunds. 2006. Local and regional scale recovery of *Diadema* promotes recruitment of scleractinian corals. Ecol. Lett. 9: 271–280.

Catano, L.B., A.A. Shantz and D.E. Burkepile. 2014. Predation, competition, and territorial damselfishes as drivers of herbivore foraging on Caribbean coral reefs. Mar. Ecol. Prog. Ser. 511: 193–207.

Catano, L.B., B.K. Gunn, M.C. Kelley and D.E. Burkepile. 2015. Predation risk, resource quality, and reef structural complexity shape territoriality in a coral reef herbivore. PLoS One 10: e0118764.

Catano, L.B., M.C. Rojas, R.J. Malossi, J.R. Peters, M.R. Heithaus, J.W. Fourqurean and D.E. Burkepile. 2016. Reefscapes of fear: predation risk and reef heterogeneity interact to shape herbivore foraging behavior. J. Anim. Ecol. 85: 146–156.

Catano, L.B., M.B. Barton, K.M. Boswell and D.E. Burkepile. 2017. Predator identity and time of day interact to shape risk-reward trade-off for herbivorous coral reef fishes. Oecologia 183: 763–773.

Cebrian, J., J.B. Shurin, E.T. Borer, B.J. Cardinale, J.T. Ngai, M.D. Smith and W.F. Fagan. 2009. Producer nutritional quality controls ecosystem trophic structure. PLoS One 4.

Chan, A., K. Lubarsky, K. Judy and P. Fong. 2012. Nutrient addition increases consumption rates of tropical algae with different initial palatabilities. Mar. Ecol. Prog. Ser. 465: 25–31.

Cheal, A.J., M.J. Emslie, M.A. MacNeil, I. Miller and H.P.A. Sweatman. 2013. Spatial variation in the functional characteristics of herbivorous fish communities and the resilience of coral reefs. Ecol. Appl. 23: 174–188.

Cheal, A.J., M.A. MacNeil, E. Cripps, M.J. Emslie, M. Jonker, B. Schaffelke and H.P.A. Sweatman. 2010. Coral-macroalgal phase shifts or reef resilience: links with diversity and functional roles of herbivorous fishes on the Great Barrier Reef. Coral Reefs 29: 1005–1015.

Chelsky Budarf, A., D.D. Burfeind, W.K.W. Loh and I.R. Tibbetts. 2011. Identification of seagrasses in the gut of a marine herbivorous fish using DNA barcoding and visual inspection techniques. J. Fish Biol. 79: 112–121.

Chen, L. 2002. Post-settlement diet shift of *Chlorurus sordidus* and *Scarus schlegeli* (Pisces: Scaridae). Zool. Stud. 41: 47–58.

Choat, J.H., K.D. Clements and W.D. Robbins. 2002. The trophic status of herbivorous fishes on coral reefs. Mar. Biol. 140: 613–623.

Choat, J.H., O.S. Klanten, L. van Herwerden, D.R. Robertson and K.D. Clements. 2012. Patterns and processes in the evolutionary history of parrotfishes (Family Labridae). Biol. J. Linn. Soc. 107: 529–557.

Clarke, K.R. and R.M. Warwick. 1999. The taxonomic distinctness measure of biodiversity: weighting of step lengths between hierarchical levels. Mar. Ecol. Prog. Ser. 184: 21–29.

Clements, K.D., D. Raubenheimer and J.H. Choat. 2009. Nutritional ecology of marine herbivorous fishes: ten years on. Funct. Ecol. 23: 79–92.

Connell, J.H. 1997. Disturbance and recovery of coral assemblages. Coral Reefs 16: S101–S113.

Connell, S.D., M.S. Foster and L. Airoldi. 2014. What are algal turfs? Towards a better description of turfs. Mar. Ecol. Prog. Ser. 495: 299–307.

Craft, J.D., V.J. Paul and E.E. Sotka. 2013. Biogeographic and phylogenetic effects on feeding resistance of generalist herbivores toward plant chemical defenses. Ecology 94: 18–24.

Cruz-Rivera, E. and M.E. Hay. 2003. Prey nutritional quality interacts with chemical defenses to affect consumer feeding and fitness. Ecol. Monogr. 73: 483–506.

Dalzell, P.J., T.J.H. Adams and N.V.C. Polunin. 1996. Coastal fisheries in the Pacific Islands. Oceanogr. Mar. Biol. 34: 395–531.

Darwin, C. 1859. On the Origin of Species. J. Murray, London.

Dill, L., M. Heithaus and C. Walters. 2003. Behaviorally mediated indirect interactions in marine communities and their conservation implications. Ecology 84: 1151–1157.

Dixson, D.L., D. Abrego and M.E. Hay. 2014. Chemically mediated behavior of recruiting corals and fishes: a tipping point that may limit reef recovery. Science 345: 892–897.

Duffy, J.E. 2009. Why biodiversity is important to the functioning of real-world ecosystems. Front. Ecol. Environ. 7: 437–444.

Duffy, J.E. and V.J. Paul. 1992. Prey nutritional quality and the effectiveness of chemical defenses against tropical reef fishes. Oecologia 90: 333–339.

Duffy, J.E., J.P. Richardson and E.A. Canuel. 2003. Grazer diversity effects on ecosystem functioning in seagrass beds. Ecol. Lett. 6: 637–645.

Duffy, J.E., B.J. Carinale, K.E. France, P.B. McIntyre, E. Thebault and M. Loreau. 2007. The functional role of biodiversity in ecosystems: incorporating trophic complexity. Ecol. Lett. 10: 522–538.

Edwards, C.B., A.M. Friedlander, A.G. Green, M.J. Hardt, E. Sala, H.P. Sweatman, I.D. Williams, B. Zgliczynski, S.A. Sandin and J.E. Smith. 2014. Global assessment of the status of coral reef herbivorous fishes: evidence for fishing effects. Proc. R. Soc. B Biol. Sci. 281: 20131835.

Elmqvist, T., C. Folke, M. Nyström, G. Peterson, J. Bengtsson, B. Walker and J. Norberg. 2003. Response diversity, ecosystem change, and resilience. Front. Ecol. Environ. 1: 488–494.

Floeter, S.R., B.S. Halpern and C.E.L. Ferreira. 2006. Effects of fishing and protection on Brazilian Reefs. Biol. Conserv. 128: 391–402.

Folke, C., S.R. Carpenter, B.H. Walker, M. Scheffer, T. Elmqvist, L. Gunderson and C.S. Holling. 2004. Regime shifts, resilience, and biodiversity in ecosystem management. Annu. Rev. Ecol. Evol. Syst. 35: 557–581.

Fox, R.J. and D.R. Bellwood. 2008. Remote video bioassays reveal the potential feeding impact of the rabbitfish *Siganus canaliculatus* (f: Siganidae) on an inner-shelf reef of the Great Barrier Reef. Coral Reefs 27: 605–615.

Fox, R.J. and D.R. Bellwood. 2013. Niche partitioning of feeding microhabitats produces a unique function for herbivorous rabbitfishes (Perciformes, Siganidae) on coral reefs. Coral Reefs 32: 13–23.

Francini-Filho, R.B., R.L. Moura, C.M. Ferreira and E.O.C. Coni. 2008. Live coral predation by parrotfishes (Perciformes: Scaridae) in the Abrolhos Bank, eastern Brazil, with comments on the classification of species into functional groups. Neotrop. Ichthyol. 6: 191–200.

Gamfeldt, L., H. Hillebrand and P.R. Jonsson. 2008. Multiple functions increase the importance of biodiversity for overall ecosystem functioning. Ecology 89: 1223–1231.

García-Robledo, C., D.L. Erickson, C.L. Staines, T.L. Erwin and W.J. Kress. 2013. Tropical plant–herbivore networks: reconstructing species interactions using DNA barcodes. PLoS One 8: e52967.

German, D.P., M.H. Horn and A. Gawlicka. 2004. Digestive enzyme activities in herbivorous and carnivorous prickleback fishes (Teleostei: Stichaeidae): ontogenetic, dietary, and phylogenetic effects. Physiol. Biochem. Zool. 77: 789–804.

Glynn, P.W., G.M. Wellington and C. Birkeland. 1979. Coral reef growth in the Galapagos: limitation by sea urchins. Science 203: 47–49.

Graham, N.A.J., S.K. Wilson, S. Jennings, N.V.C. Polunin, J.P. Bijoux and J. Robinson. 2006. Dynamic fragility of oceanic coral reef ecosystems. Proc. Natl. Acad. Sci. USA 103: 8425–8429.

Hawkins, J.P. and C.M. Roberts. 2004a. Effects of artisanal fishing on Caribbean coral reefs. Conserv. Biol. 18: 215–226.

Hawkins, J.P. and C.M. Roberts. 2004b. Effects of fishing on sex-changing Caribbean parrotfishes Biol. Conserv. 115: 213–226.

Hay, M.E. 1981. Spatial patterns of grazing intensity on a Caribbean barrier reef: herbivory and algal distribution. Aquat. Bot. 11: 97–109.

Hay, M.E. 1984. Patterns of fish and urchin grazing on Caribbean coral reefs: are previous results typical? Ecology 65: 446–454.

Hay, M.E. 1985. Spatial patterns of herbivore impact and their importance in maintaining algal species richness. Proc. 5th Int. Coral Reef Symp. 4: 29–34.

Hay, M.E. 1991. Fish-seaweed interactions on coral reefs: effects of herbivorous fishes and adaptations of their prey. pp. 96–119 *In*: P.F. Sale (ed.). The Ecology of Fishes on Coral Reefs. Academic Press, San Diego, USA.

Hay, M.E. 1997. The ecology and evolution of seaweed-herbivore interactions on coral reefs. Coral Reefs 16: S67–S76.

Hay, M.E., Q.E. Kappel and W. Fenical. 1994. Synergisms in plant defenses against herbivores – interactions of chemistry, calcification, and plant quality. Ecology 75: 1714–1726.

Hector, A. and R. Bagchi. 2007. Biodiversity and ecosystem multifunctionality. Nature 448: 188–192.

Heenan, A., A.S. Hoey, G.J. Williams and I.D. Williams. 2016. Natural bounds on herbivorous coral reef fishes. Proc. R. Soc. B 283: rspb20161716.

Hicks, C.C. and T.R. McClanahan. 2012. Assessing gear modifications needed to optimize yields in a heavily exploited, multi-species, seagrass and coral reef fishery. PLoS One 7: e36022.

Hoey, A.S. and D.R. Bellwood. 2008. Cross-shelf variation in the role of parrotfishes on the Great Barrier Reef. Coral Reefs 27: 37–47.

Hoey, A.S. and D.R. Bellwood. 2009. Limited functional redundancy in a high diversity system: single species dominates key ecological process on coral reefs. Ecosystems 12: 1316–1328.

Hoey, A.S. and D.R. Bellwood. 2010a. Among-habitat variation in herbivory on *Sargassum* spp. on a mid-shelf reef in the northern Great Barrier Reef. Mar. Biol. 157: 189–200.

Hoey, A.S. and D.R. Bellwood. 2010b. Cross-shelf variation in browsing intensity on the Great Barrier Reef. Coral Reefs 29: 499–508.

Hoey, A.S. and D.R. Bellwood. 2011. Suppression of herbivory by macroalgal density: a critical feedback on coral reefs? Ecol. Lett. 14: 267–273.

Hoey, A.S., M.S. Pratchett and C. Cvitanovic. 2011. High macroalgal cover and low coral recruitment undermines the potential resilience of the world's southernmost coral reef assemblages. PLoS One 6: e25824.

Hoey, A.S., S.J. Brandl and D.R. Bellwood. 2013. Diet and cross-shelf distribution of rabbitfishes (f. Siganidae) on the northern Great Barrier Reef: implications for ecosystem function. Coral Reefs 32: 973–984.

Hoey, A.S., D.A. Feary, J.A. Burt, G. Vaughan, M.S. Pratchett and M.L. Berumen. 2016. Regional variation in the structure and function of parrotfishes on Arabian reefs. Mar. Poll. Bull. 105: 524–531.

Holling, C.S. 1973. Resilience and stability of ecological systems. Annu. Rev. Ecol. Syst. 4: 1–23.

Horn, M.H. 1989. Biology of marine herbivorous fishes. Oceanogr. Mar. Biol. 27: 167–272.

Houk, P., K. Rhodes, J. Cuetos-Bueno, S. Lindfield, V. Fread and J.L. McIlwain. 2012. Commercial coral-reef fisheries across Micronesia: A need for improving management. Coral Reefs 31: 13–26.

Hughes, T. P. 1994. Catastrophes, phase shifts, and large-scale degradation of a Caribbean coral reef. Science 265: 1547–1551.

Hughes, T.P., M.J. Rodrigues, D.R. Bellwood, D. Ceccarelli, O. Hoegh-Guldberg, L. McCook, N. Moltschaniwskyj, M.S. Pratchett, R.S. Steneck and B.L. Willis. 2007. Phase shifts, herbivory, and the resilience of coral reefs to climate change. Curr. Biol. 17: 360–365.

Hughes, T.P., N.A.J. Graham, J.B.C. Jackson, P.J. Mumby and R.S. Steneck. 2010. Rising to the challenge of sustaining coral reef resilience. Trends Ecol. Evol. 25: 633–642.

Idjadi, J.A., S.C. Lee, J.F. Bruno, W.F. Precht, L. Allen-Requa and P.J. Edmunds. 2006. Rapid phase-shift reversal on a Jamaican coral reef. Coral Reefs 25: 209–211.

Jackson, J.B.C., M.X. Kirby, W.H. Berger, K.A. Bjorndal, L.W. Botsford, B.J. Bourque, R.H. Bradbury, R. Cooke, J. Erlandson, J.A. Estes, T.P. Hughes, S. Kidwell, C.B. Lange, H.S. Lenihan, J.M. Pandolfi, C.H. Peterson, R.S. Steneck, M.J. Tegner and R.R. Warner. 2001. Historical overfishing and the recent collapse of coastal ecosystems. Science 293: 629–637.

Jompa, J. and L.J. McCook. 2003. Contrasting effects of turf algae on corals: massive *Porites* spp. are unaffected by mixed-species turfs, but killed by the red alga *Anotrichium tenue*. Mar. Ecol. Prog. Ser. 258: 79–86.

Kerswell, A.P. 2006. Global biodiversity patterns of benthic marine algae. Ecology 87: 2479–2488.

Kuffner, I.B., L.J. Walters, M.A. Beccero, V.J. Paul, R. Ritson-Williams and K.S. Beach. 2006. Inhibition of coral recruitment by macroalgae and cyanobacteria. Mar. Ecol. Prog. Ser. 323: 107–117.

Lall, S.P. 1991. Digestibility, metabolism and excretion of dietary phosphorus in fish. *In*: C.B. Cowey and C.Y. Cho (eds.). Nutritional Strategies and Aquaculture Waste. University of Guelph, Ontario, Canada.

Lessios, H.A., D.R. Robertson and J.D. Cubit. 1984. Spread of *Diadema* mass mortality through the Caribbean. Science 226: 335–337.

Lewis, S.M. 1985. Herbivory on coral reefs: algal susceptibility to herbivorous fishes. Oecologia 65: 370–375.

Lewis, S.M. 1986. The role of herbivorous fishes in the organization of a Caribbean reef communitiy. Ecol. Monogr. 56: 183–200.

Lewis, S.M. and P.C. Wainwright. 1985. Herbivore abundance and grazing intensity on a Caribbean coral reef. J. Exp. Mar. Biol. Ecol. 87: 215–228.

Lima, S.L. and L.M. Dill. 1990. Behavioral decisions made under the risk of predation – a review and prospectus. Can. J. Zool. 68: 619–640.

Lindfield, S.J., J.L. McIlwain and E.S. Harvey. 2014. Depth refuge and the impacts of SCUBA spearfishing on coral reef fishes. PLoS One 9: e92628.

Lobel, P.S. 1981. Trophic biology of herbivorous reef fishes – alimentary pH and digestive capabilities. J. Fish Biol. 19: 365–397.

Loffler, Z., D.R. Bellwood and A.S. Hoey. 2015. Among-habitat algal selectivity by browsing herbivores on an inshore coral reef. Coral Reefs 34: 597–605.

Lokrantz, J., M. Nystrom, M. Thyresson and C. Johansson. 2008. The non-linear relationship between body size and function in parrotfishes. Coral Reefs 27: 967–974.

Loreau, M., S. Naeem, P. Inchausti, J. Bengtsson, J.P. Grime, A. Hector, D.U. Hooper, M.A. Huston, D. Raffaelli, B. Schmid, D. Tilman and D.A. Wardle. 2001. Biodiversity and ecosystem functioning: current knowledge and future challenges. Science 294: 804–808.

MacArthur, R. 1955. Fluctuations of animal populations, and a measure of community stability. Ecology 36: 533–536.

Machemer, E.G.P., J.F. Walter III, J.E. Serafy and D.W. Kerstetter. 2012. Importance of mangrove shorelines for rainbow parrotfish *Scarus guacamaia*: habitat suitability modeling in a subtropical bay. Aquat. Biol. 15: 87–98.

Madin, E.M.P., S.D. Gaines and R.R. Warner. 2010a. Field evidence for pervasive indirect effects of fishing on prey foraging behavior. Ecology 91: 3563–3571.

Madin, E.M.P., S.D. Gaines, J.S. Madin and R.R. Warner. 2010b. Fishing indirectly structures macroalgal assemblages by altering herbivore behavior. Am. Nat. 176: 785–801.

Mantyka, C.S. and D.R. Bellwood. 2007. Macroalgal grazing selectivity among herbivorous coral reef fishes. Mar. Ecol. Prog. Ser. 352: 177–185.

Mattson, W.J. 1980. Herbivory in relation to plant nitrogen content. Annu. Rev. Ecol. Syst. 11: 119–161.

McAfee, S.T. and S.G. Morgan. 1996. Resource use by five sympatric parrotfishes in the San Blas Archipelago, Panama. Mar. Biol. 125: 427–437.

McClanahan, T.R. 1994. Kenyan coral reef lagoon fish – effects of fishing, substrate complexity, and sea urchins. Coral Reefs 13: 231–241.

McClanahan, T.R. and S.H. Shafir. 1990. Causes and consequences of sea urchin abundance and diversity in Kenyan coral reef lagoons. Oecologia 83: 362–370.

McClanahan, T.R., V. Hendrick, M.J. Rodrigues and N.V.C. Polunin. 1999. Varying responses of herbivorous and invertebrate-feeding fishes to macroalgal reduction on a coral reef. Coral Reefs 18: 195–203.

McCook, L.J., J. Jompa and G. Diaz-Pulido. 2001. Competition between corals and algae on coral reefs: a review of evidence and mechanisms. Coral Reefs 19: 400–417.

Michael, P.J., G.A. Hyndes, M.A. Vanderklift and A. Vergés. 2013. Identity and behaviour of herbivorous fish influence large-scale spatial patterns of macroalgal herbivory in a coral reef. Mar. Ecol. Prog. Ser. 482: 227–240.

Micheli, F. and B.S. Halpern. 2005. Low functional redundancy in coastal marine assemblages. Ecol. Lett. 8: 391–400.

Micheli, F., P.J. Mumby, D.R. Brumbaugh, K. Broad, C.P. Dahlgren, A.R. Harborne, K.E. Holmes, C.V. Kappel, S.Y. Litvin and J.N. Sanchirico. 2014. High vulnerability of ecosystem function and services to diversity loss in Caribbean coral reefs. Biol. Conserv. 171: 186–194.

Montgomery, W.L. and S.D. Gerking. 1980. Marine macroalgae as foods for fishes: an evaluation of potential food quality. Environ. Biol. Fishes 5: 143–153.

Morrison, D. 1988. Comparing fish and urchin grazing in shallow and deeper coral reef algal communities. Ecology 69: 1367–1382.

Mumby, P.J. 2006. The impact of exploiting grazers (Scaridae) on the dynamics of Caribbean coral reefs. Ecol. Appl. 16: 747–769.

Mumby, P.J. and C.C.C. Wabnitz. 2002. Spatial patterns of aggression, territory size, and harem size in five sympatric Caribbean parrotfish species. Environ. Biol. Fishes 63: 265–279.

Mumby, P.J. and A. Hastings. 2008. The impact of ecosystem connectivity on coral reef resilience. J. Appl. Ecol. 45: 854–862.

Mumby, P.J. and R.S. Steneck. 2008. Coral reef management and conservation in light of rapidly evolving ecological paradigms. Trends Ecol. Evol. 23: 555–563.

Mumby, P.J., A.J. Edwards, J. Ernesto Arias-Gonzalez, K.C. Lindeman, P.G. Blackwell, A. Gall, M.I. Gorczynska, A.R. Harborne, C.L. Pescod, H. Renken, C.C.C. Wabnitz and G. Llewellyn. 2004. Mangroves enhance the biomass of coral reef fish communities in the Caribbean. Nature 427: 533–536.

Mumby, P.J., C.P. Dahlgren, A.R. Harborne, C.V. Kappel, F. Micheli, D.R. Brumbaugh, K.E. Holmes, J.M. Mendes, K. Broad, J.N. Sanchirico, K. Buch, S. Box, R.W. Stoffle and A.B. Gill. 2006. Fishing, trophic cascades, and the process of grazing on coral reefs. Science 311.

Mumby, P.J., A.R. Harborne, J. Williams, C.V. Kappel, D.R. Brumbaugh, F. Micheli, K.E. Holmes, C.P. Dahlgren, C.B. Paris and P.G. Blackwell. 2007. Trophic cascade facilitates coral recruitment in a marine reserve. Proc. Natl. Acad. Sci. USA 104: 8362–8367.

Naeem, S. 1998. Species redundancy and ecosystem reliability. Conserv. Biol. 12: 39–45.

Nagelkerken, I., G. van der Velde, M.W. Gorissen, G.J. Meijer, T. Van't Hof and C. den Hartog. 2000. Importance of mangroves, seagrass beds and the shallow coral reef as a nursery for important coral reef fishes, using a visual census technique. Estuar. Coast. Shelf Sci. 51: 31–44.

Nagelkerken, I., C.M. Roberts, G. Van Der Velde, M. Dorenbosch, M.C. Van Riel, E.C. De La Moriniere and P.H. Nienhuis. 2002. How important are mangroves and seagrass beds for coral-reef fish? The nursery hypothesis tested on an island scale. Mar. Ecol. Prog. Ser. 244: 299–305.

Nash, K.L., N.A.J. Graham, F.A. Januchowski-Hartley and D.R. Bellwood. 2012. Influence of habitat condition and competition on foraging behaviour of parrotfishes. Mar. Ecol. Prog. Ser. 457: 113–124.

Nash, K.L., N.A.J. Graham and D.R. Bellwood. 2013. Fish foraging patterns, vulnerability to fishing and implications for the management of ecosystem function across scales. Ecol. Appl. 23: 1632–1644.

Nugues, M.M. and A.M. Szmant. 2006. Coral settlement onto *Halimeda opuntia*: a fatal attraction to an ephemeral substrate? Coral Reefs 25: 585–591.

Nyström, M., A.V. Norström, T. Blenckner, M. la Torre-Castro, J.S. Eklöf, C. Folke, H. Österblom, R.S. Steneck, M. Thyresson and M. Troell. 2012. Confronting Feedbacks of Degraded Marine Ecosystems. Ecosystems 15: 695–710.

Ogden, J.C. 1976. Some aspects of herbivore-plant relationships on Caribbean reefs and seagrass beds. Aquat. Bot. 2: 103–116.

Ong, L. and K.N. Holland. 2010. Bioerosion of coral reefs by two Hawaiian parrotfishes: species, size differences and fishery implications. Mar. Biol. 157: 1313–1323.

Pandolfi, J.M., R.H. Bradbury, E. Sala, T.P. Hughes, K.A. Bjorndal, R.G. Cooke, D. McArdle, L. McClenachan, M.J.H. Newman, G. Paredes, R.R. Warner and J.B.C. Jackson. 2003. Global trajectories of the long-term decline of coral reef ecosystems. Science 301: 955–958.

Parenti, P. and J.E. Randall. 2011. Checklist of the species of the families Labridae and Scaridae: an update. Smithiana Bull. 13: 29–44.

Paul, V.J. and M.E. Hay. 1986. Seaweed susceptibility to herbivory: chemical and morphological correlates. Mar. Ecol. Prog. Ser. 33: 255–264.

Perry, C.T., G.N. Murphy, P.S. Kench, E.N. Edinger, S.G. Smithers, R.S. Steneck and P.J. Mumby. 2014. Changing dynamics of Caribbean reef carbonate budgets: emergence of reef bioeroders as critical controls on present and future reef growth potential. Proc. R. Soc. B Biol. Sci. 281.

Plass-Johnson, J.G., C.D. McQuaid and J.M. Hill. 2013. Stable isotope analysis indicates a lack of inter- and intra-specific dietary redundancy among ecologically important coral reef fishes. Coral Reefs 32: 429–440.

Pratchett, M.S., A.S. Hoey and S.K. Wilson 2014. Reef degradation and the loss of critical ecosystem goods and services provided by coral reef fishes. Curr. Opinion Environ. Sust. 7: 37–43.

Preisser, E.L., D.I. Bolnick and M.F. Benard. 2005. Scared to death? The effects of intimidation and consumption in predator-prey interactions. Ecology 86: 501–509.

Randall, J.E. 1967. Food habits of reef fishes of the West Indies. Stud. Trop. Oceanogr. 5: 665–847.

Rasher, D.B. and M.E. Hay. 2010. Chemically rich seaweeds poison corals when not controlled by herbivores. Proc. Natl. Acad. Sci. USA 107: 9683–9688.

Rasher, D.B., A.S. Hoey and M.E. Hay. 2013. Consumer diversity interacts with prey defenses to drive ecosystem function. Ecology 94: 1347–1358.

Rasher, D.B., A.S. Hoey and M.E. Hay. 2017. Cascading predator effects in a Fijian coral reef ecosystem. Sci. Rep. 7: 15684.

Rhodes, K.L., M.H. Tupper and C.B. Wichilmel. 2008. Characterization and management of the commercial sector of the Pohnpei coral reef fishery, Micronesia. Coral Reefs 27: 443–454.

Rizzari, J.R., A.J. Frisch, A.S. Hoey and M.I. McCormick. 2014. Not worth the risk: apex predators suppress herbivory on coral reefs. Oikos 123: 829–836.

Robertson, D.R. and S.D. Gaines. 1986. Interference competition structures habitat use in a local assemblage of coral reef surgeonfishes. Ecology 67: 1372–1383.

Roff, G. and P.J. Mumby. 2012. Global disparity in the resilience of coral reefs. Trends Ecol. Evol. 27: 404–413.

Rosenfeld, J.S. 2002. Functional redundancy in ecology and conservation. Oikos 98: 156–162.

Rotjan, R.D. and S.M. Lewis. 2006. Parrotfish abundance and selective corallivory on a Belizean coral reef. J. Exp. Mar. Biol. Ecol. 335: 292–301.

Russ, G.R. 1984. Distribution and abundance of herbivorous grazing fishes in the central Great Barrier Reef. II. Patterns of zonation of mid-shelf and outershelf reefs. Mar. Ecol. Prog. Ser. 20: 35–44.

Russ, G.R. 2003. Grazer biomass correlates more strongly with production than with biomass of algal turfs on a coral reef. Coral Reefs 22: 63–67.

Sammarco, P.W. 1980. *Diadema* and its relationship to coral spat mortality: grazing, competition, and biological disturbance. J. Exp. Mar. Biol. Ecol. 45: 245–272.

Sammarco, P.W. 1982. Echinoid grazing as a structuring force in coral communities: whole reef manipulations. J. Exp. Mar. Biol. Ecol. 61: 31–55.

Sandin, S.A. and D.E. McNamara. 2012. Spatial dynamics of benthic competition on coral reefs. Oecologia 168: 1079–1090.

Schupp, P.J. and V.J. Paul. 1994. Calcium carbonate and secondary metabolites in tropical seaweeds – variable effects on herbivorous fishes. Ecology 75: 1172–1185.

Scoffin, T.P., C.W. Stearn, D. Boucher, P. Frydl, C.M. Hawkins and I.G. Hunter. 1980. Calcium carbonate budget of a fringing reef on the west coast of Barbados. Part II. Erosion, sediments and internal structure. Bull. Mar. Sci. 30: 475–508.

Sotka, E.E. and M.E. Hay. 2009. Effects of herbivores, nutrient enrichment, and their interactions on macroalgal proliferation and coral growth. Coral Reefs 28: 555–568.

Stachowicz, J.J., J.F. Bruno and J.E. Duffy. 2007. Understanding the effects of marine biodiversity on communities and ecosystems. Annu. Rev. Ecol. Evol. Syst. 38: 739–766.

Stachowicz, J.J., R.J. Best, M.E.S. Bracken and M.H. Graham. 2008. Complementarity in marine biodiversity manipulations: Reconciling divergent evidence from field and mesocosm experiments. Proc. Natl. Acad. Sci. USA 105: 18842–18847.

Steneck, R.S. 1988. Herbivory on coral reefs: a synthesis. Proc. 6th Int. Coral Reef Symp. 1: 37–49.

Sterner, R.W. and J.J. Elser. 2002. Ecological Stoichiometry. Princeton University Press, Princeton, New Jersey, USA.

Streelman, J.T., M. Alfaro, M.W. Westneat, D.R. Bellwood and S.A. Karl. 2002. Evolutionary history of the parrotfishes: biogeography, ecomorphology, and comparative diversity. Evolution 56: 961–971.

Taylor, B.M. 2014. Drivers of protogynous sex change differ across spatial scales. Proc. R. Soc. B Biol. Sci. 281: 20132423.

Uthicke, S., B. Schaffelke and M. Byrne. 2009. A boom-bust phylum? Ecological and evolutionary consequences of density variations in echinoderms. Ecol. Monogr. 79: 3–24.

Valentini, A., F. Pompanon and P. Taberlet. 2009. DNA barcoding for ecologists. Trends Ecol. Evol. 24: 110–117.

Vega Thurber, R., D.E. Burkepile, A.M.S. Correa, A.R. Thurber, A.A. Shantz, R. Welsh, C. Pritchard and S. Rosales. 2012. Macroalgae decrease growth and alter microbial community structure of the reef-building coral, *Porites astreoides*. PLoS One 7: e44246.

Vergés, A., S. Bennett and D.R. Bellwood. 2012. Diversity among macroalgae-consuming fishes on coral reefs: a transcontinental comparison. PLoS One 7: e45543.

Wainwright, P.C., D.R. Bellwood, M.W. Westneat, J.R. Grubich and A.S. Hoey. 2004. A functional morphospace for the skull of labrid fishes: patterns of diversity in a complex biomechanical system. Biol. J. Linn. Soc. 82: 1–25.

Walker, B.H. 1992. Biodiversity and ecological redundancy. Conserv. Biol. 6: 18–23.

Westneat, M.W. and M.E. Alfaro. 2005. Phylogenetic relationships and evolutionary history of the reef fish family Labridae. Mol. Phylogenet. Evol. 36: 370–390.

White, J.W., J.F. Samhouri, A.C. Stier, C.L. Wormald, S.L. Hamilton and S.A. Sandin. 2010. Synthesizing mechanisms of density dependence in reef fishes: behavior, habitat configuration, and observational scale. Ecology 91: 1949–1961.

Williams, I.D. and N.V.C. Polunin. 2001. Large-scale associations between macroalgal cover and grazer biomass on mid-depth reefs in the Caribbean. Coral Reefs 19: 358–366.

Wilson, S.K., D.R. Bellwood, J.H. Choat and M.J. Furnas. 2003. Detritus in the epilithic algal matrix and its use by coral reef fishes. Oceanogr. Mar. Biol. 41: 279–310.

Wilson, S.K., R. Fisher, M.S. Pratchett, N.A.J. Graham, N.K. Dulvy, R.A. Turner, A. Cakacaka, N.V.C. Polunin and S.P. Rushton. 2008. Exploitation and habitat degradation as agents of change within coral reef fish communities. Global Change Biol. 14: 2796–2809.

Wismer, S., A.S. Hoey and D.R. Bellwood. 2009. Cross-shelf benthic community structure on the Great Barrier Reef: relationships between macroalgal cover and herbivore biomass. Mar. Ecol. Prog. Ser. 376: 45–54.

Worm, B., E.B. Barbier, N. Beaumont, J.E. Duffy, C. Folke, B.S. Halpern, J.B.C. Jackson, H.K. Lotze, F. Micheli, S.R. Palumbi, E. Sala, K.A. Selkoe, J.J. Stachowicz and R. Watson. 2006. Impacts of biodiversity loss on ocean ecosystem services. Science 314: 787–790.

Yachi, S. and M. Loreau. 1999. Biodiversity and ecosystem productivity in a fluctuating environment: The insurance hypothesis. Proc. Natl. Acad. Sci. USA 96: 1463–1468.

The Role of Parrotfishes in the Destruction and Construction of Coral Reefs

Jennie Mallela[1] and Rebecca J. Fox[2,3]

[1] Research School of Earth Sciences and Research School of Biology,
The Australian National University, Canberra, ACT 2600, Australia
Email: j.a.mallela93@members.leeds.ac.uk

[2] School of Life Sciences, University of Technology Sydney, Ultimo, NSW 2007, Australia

[3] Division of Ecology and Evolution, Research School of Biology, The Australian National
University Canberra, ACT 2600, Australia
Email: rebecca.fox-1@uts.edu.au

Introduction

"…there are living checks to the growth of coral reefs, and that the almost universal law of "consume and be consumed," holds good even with the polypifers forming those massive bulwarks, which are able to withstand the force of the open ocean." Darwin 1842

Coral reef bioerosion and sediment production are two of the most important functional roles of parrotfishes (Darwin 1842, Gygi 1975, Ogden 1977, Bellwood 1995a, b, Bonaldo et al. 2014). They are also among the most studied aspects of parrotfish biology and functional ecology, to the extent that news articles and even popular quiz culture (see http://www.comedy.co.uk/guide/tv/qi/episodes/8/13/) are replete with references to the contribution that parrotfish excrement makes to tropical sandy beaches. Much has been written about the mechanism of bioerosion by parrotfish in terms of the morphological innovations that enable them to perform this role (Frydl and Stearn 1978, Gobalet 1989, Bellwood and Choat 1990) and the implications for the trophic status of parrotfish of their nutritional ecology (Choat and Clements 1998, Choat et al. 2002, 2004, Clements et al. 2009). Both of these topics, as well as the impact of bioerosion on community succession (corallivory), are the subject of separate chapters within this volume (see Gobalet Chapter 1, Clements and Choat Chapter 3, and Bonaldo and Rotjan Chapter 9), and will therefore not be covered in detail here. Instead, this chapter details the bioerosion and sediment production functions of parrotfish within the wider context of coral reef geomorphology and coral reef carbonate budgets. The aim is to give the reader a geological perspective of the processes involved in reef framework development, focusing on carbonate production and destruction on reefs and the relative contribution of parrotfish to these functions. By setting the role of parrotfishes within this wider geological context, we have two goals: firstly, to emphasise that parrotfish, arguably more than any other group of fishes on coral reefs, are intrinsically tied to the framework and very substance of their habitat, putting them in the class of organisms known as ecosystem engineers (*sensu* Jones et al. 1994, 1997). Secondly, we hope to bring to the fore the wider issues of reef bioerosion and accretion, re-iterating the central, yet often neglected, importance of taking a carbonate building perspective (i.e. reef carbonate budget) to scientific and political debate over future impacts of global change on coral reef ecosystems.

Healthy coral reefs are composed of topographically complex, three dimensional carbonate structures which buffer shorelines from currents, waves and storms, replenish sand on adjacent coastlines and provide a home to a diverse array of organisms. The physical structure of all reefs represents the combined product of carbonate accretion, carbonate removal, sediment recycling, diagenesis and cementation (Goreau 1959; Stearn et al. 1977; Scoffin et al. 1980; Hubbard et al. 1990) (Fig. 1). Viewed at their most basic, all reefs are simply a balance between processes that produce calcium carbonate (primary production by corals and secondary production by calcareous encrusters, also known as sclerobionts, such as crustose bryozoans, coralline algae and calcareous worms), and those that remove it (destructive physical and chemical processes, and bioerosion by epilithic and endolithic organisms) (Chave et al. 1972, Stearn et al. 1977, Risk and MacGeachy 1978, Scoffin et al. 1980, Hutchings 1986, Kiene and Hutchings 1994, Chazottes 1995; Fig. 1). Consolidated reef material that is removed may then be converted to sediment and recycled back into the reef framework, deposited into neighbouring landmasses (e.g. beaches, reef islands) or transported off-reef and lost to the system (Hubbard et al. 1990). Constructive and destructive processes act at rates that determine the net rate of carbonate production within a particular reef system, which can be calculated via a budgetary approach, which simply sums the rates of carbonate addition and removal. Disturbances that bring about changes in the nature or rate at which these processes occur then impact on carbonate cycling and net carbonate production, bringing about associated changes in reef structure and framework performance.

Viewed from this perspective, the importance of processes such as bioerosion and sediment production are brought into sharp focus. Essentially, these are processes that fundamentally impact on the physical quantity and structure of the habitat around which reef ecosystems are built. It has been argued (Rose and Risk 1985, Edinger et al. 2000) that evaluations of reef health should be based on calculations of their carbonate budget status, thereby allowing for diagnosis of whether the framework is growing (a positive budget) or receding (a negative budget) through time and in danger of being denuded or 'drowned' (Neuman and Macintyre 1985, Blanchon and Shaw 1995). The biology of parrotfish is such that these organisms impact on both reef construction and destruction, giving them a special place in the determination of reef physical structures and reef carbonate budgets which will be explored in this chapter. Parrotfish, by virtue of their unique beak morphology, act as bioeroders of reef substratum as they feed. Their method of food processing through a secondary structure at the back of their jaw (known as a pharyngeal mill) and through an intestine lacking a stomach then results in the production and recycling of reef sediments. Parrotfish therefore contribute to processes on both sides of the carbonate budget ledger and are intricately tied to the overall performance of reefs (Fig. 1).

In this chapter, we explain the contribution of parrotfishes to reef destruction (via bioerosion) and reef construction (sediment production and reworking). We then bring the processes together within a "carbonate budget" framework, to illustrate how parrotfish bioerosion and sediment production roles sit within the overall battle of forces of construction and destruction that determine reef development. Then we explore how factors associated with global change are likely to impact on the carbonate budget with

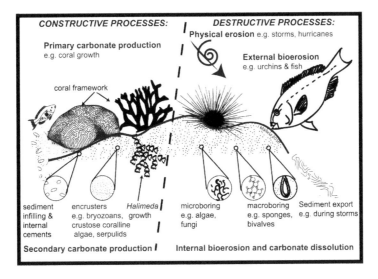

Processes	Influence on framework	Key references
CONSTRUCTIVE:		
Coral calcification	Scleractinian, hermatypic corals secrete $CaCO_3$ forming the primary building blocks of the reef framework. $CaCO_3$ production is dependent on extension rates and skeletal density and calcification is the product of both.	Stearn et al. 1977; Land 1979; Hubbard et al. 1990
Encrustation by calcareous sclerobionts	Encrusting organisms (*e.g.* bryozoans, crustose coralline algae, foraminifera, serpulids) recruit to the existing reefal substrate. They add $CaCO_3$ to the existing framework, rubble and unconsolidated sediments. This can strengthen, bind and consolidate reef framework.	Taylor 1990; Fagerstrom 1991; Rasser and Riegl 2002; Mallela 2013
Cementation	Wave energy forces seawater and sediment through the framework promoting submarine cementation by precipitating magnesium calcite and aragonite crystals. Infilling of interstitial cavities reduces porosity, immobilises interstitial sediment and strengthens the framework.	Bathurst 1966; see Scoffin 1992 for a review; Rasser and Riegl 2002; Manzello et al. 2008
Sediment retention	Allochthonous and autochthonous sediments can infill void space in the framework, small naturally occurring interstitial cavities in $CaCO_3$ skeletons, and galleries made by bioeroding organisms. This stabilises and consolidates reef framework, whilst the presence of interstitial sediments can also promote cementation of the primary reef framework.	Hubbard et al. 1990; Scoffin 1992; Rasser and Riegl 2002; Mallela and Perry 2007
DESTRUCTIVE:		
Grazing	The feeding activities of some grazing organisms (*e.g.* fish, echinoids) erode the outer surface of the reef and are capable of producing large quantities of carbonate sediment.	Gygi 1975; Bellwood and Choat 1990; Bak 1994; Bruggeman 1995; Conand et al. 1997
Microboring	The activities of microborers (*e.g.* cyanobacteria, chlorophytes and fungi) result in biochemical dissolution of carbonate substrates. This can result in dense, convoluted networks of small (*e.g.* diameter < 1mm) galleries that weaken the substrate and promote macroboring and grazing.	Golubic et al. 1975; Perkins and Tsentas 1976; Kobluk and Risk 1977; Hutchings 1986; Tribollet et al. 2002
Macroboring	Macroborers (*e.g.* bivalves, sponges, worms) physically and chemically erode carbonate substrates. Substrates become increasingly porous and boring produces significant quantities of reefal sediment.	Warme 1975; MacGeachy and Stearn 1976; Hutchings 1986; Boss and Liddell 1987
Physical erosion and sediment transport	High-energy events (*e.g.* hurricanes) can cause framework damage and generate large quantities of sediment and rubble. Unconsolidated substrates can then be transported from the reef by the local hydrodynamic regime.	Land 1979; Hubbard et al. 1990; Hubbard 1992; Bellwood 1995b
Chemical dissolution	$CaCO_3$ dissolution due to seawater chemistry. Dissolution of $CaCO_3$ is inversely related to grain size.	Golubic 1979; Scoffin 1992; Eyre et al. 2014

Fig. 1. Summary of reef building processes: carbonate production, carbonate erosion and sediment dynamics.

a view to understanding how the contribution of parrotfishes to reef construction and destruction might influence the dynamics of reef development in a changing world. Finally, we summarize the overall prognosis for reef development under future climate projections, the roles that bioerosion and sediment production by parrotfishes are likely to play under such conditions, and set out our suggestions for future research directions in this area.

Parrotfish as Carbonate Removers: Contribution to Reef Erosion

Forces of Reef Destruction: Physical, Chemical and Biological

Erosion occurs on all reefs as natural agents such as wind, water and the living biota abrade, scrape, dissolve and weather the reef. Carbonate erosion is therefore a critical process in shaping and controlling reef development. Erosion results in the loss of calcium carbonate from the reef framework, structural weakening, and the production of large volumes of carbonate sediment that is either re-incorporated back into, or exported from, the reef ecosystem (Frydl 1979, Hubbard et al. 1990, Bellwood 1995b). There are many processes that erode and recycle the physical structure of tropical reefs and these can be categorised as: physical, chemical and biological erosion.

Physical erosion on the reef can occur gradually and slowly over time (e.g. through the action of currents) or rapidly during high-energy events (e.g. storm induced waves) which can result in considerable framework damage and destruction (Rogers 1993, Scoffin 1993, Harmelin-Vivien 1994). During extreme high-energy events, chunks of reef framework can become dislodged generating sediment, rock and rubble. These unconsolidated carbonate substrates may tumble round the reef causing further destruction and may subsequently be exported from their reef of origin (Land 1979). Chemical erosion is the loss of reef carbonate as a result of dissolution by seawater chemistry (e.g. ocean acidification, Eyre et al. 2014) or can be mediated via biological activity of microboring organisms by the production of metabolic acid or excretion of enzymes (Tudhope and Risk 1985).

Biological erosion, or bioerosion (*sensu* Neumann 1966), occurs due to the activities of a diverse range of marine organisms from large parrotfish to microscopic fungi and bacteria which are all able to erode the reef by varying means (Frydl and Stearn 1978, May et al. 1982, Bellwood 1995a, b, Tribollet et al. 2002). Bioeroding organisms can be categorised as either epilithic (external eroders) or endolithic (internal eroders) (Fig. 1). Epilithic organisms such as grazing fish (e.g. parrotfish), urchins (e.g. *Diadema antillarium*), and molluscs (e.g. chitons) abrade, rasp and excavate the exterior of the reef during feeding activities (Steneck 1983). In contrast, endolithic organisms bore into and live within the reef framework (Golubic et al. 1975, MacGeachy and Stearn 1976, Warme 1977) and use a combination of physical and bio-chemical erosion to penetrate and then chemically corrode the substrate, effectively dissolving the carbonate from within (Golubic et al. 1975, 1981, Hutchings 1986). Endolithic bioeroding organisms are often classified as microborers (microscopic organisms: e.g. cyanobacteria, chlorophytes and fungi), and macroborers (e.g. sponges, bivalves and worms) and are capable of making a substantial contribution to rates of reef bioerosion. For example, Tribollet and Golubic (2005) calculated that the populations of endolithic microorganisms within dead corals on the northern Great Barrier Reef were capable of dissolving carbonate at the rate of more than 1 kg.m^{-2}.y^{-1}. In general, the relative contributions of epilithic and endolithic eroders to overall bioerosion rates varies according to reef location and state of health, with epilithic eroders tending to dominate the process on healthy, offshore reefs (Hutchings 1986) and parrotfish bioerosion dominating some Caribbean reefs, accounting for on average 76 percent of total bioerosion (Belize, Bonaire, Bahamas and Grand Cayman) (Perry et al. 2014). In contrast, on the Great Barrier Reef endoliths dominated bioeorsion on polluted or degraded reefs (Risk et al. 1995). On overfished reef settings in Jamaica, with low parrotfish and urchin densities, endoliths dominated carbonate bioerosion (Mallela and Perry 2007). Interestingly in Jamaica, microboring was the dominant bioerosive force in turbid reef settings accounting for 66 percent of total bioerosion with parrotfish only accounting for 3 percent. In contrast, macroborers dominated bioerosion on the clear water Jamaican fore-reef accounting for 58 percent of total bioerosion compared with 12 percent from parrotfish bioerosion (Mallela and Perry 2007).

Of the three forces of reef destruction (chemical, physical and biological), bioerosion typically plays a significant role in consolidated carbonate loss. In St Croix, US Virgins Islands, Hubbard et al. (1990) estimated that 60 percent of annual carbonate produced on the reef was reduced to sediment by bioerosion from urchins, fish and internal bioeroders. Over half of the sediment produced by bioerosion was then reincorporated back into the reef interior. The remainder moved into sand channels with the potential to be exported from the system during storms. The various processes of bioerosion are therefore an integral part of reef development influencing the rate and style of carbonate accumulation, substrate porosity, and the production, recycling and/or export of carbonate sediment in to, or out of, the reef matrix. Clearly, whilst bioerosion is initially a destructive process it also makes important contributions towards increasing reef framework complexity, consolidating the reef interior via sediment reincoporation and to replenishing reef sediments (Stearn et al. 1977, Scoffin et al. 1980, Hubbard et al. 1990, Harney and Fletcher 2003).

Parrotfish Bioerosion: How, What and Where are they Eroding?

Bioerosion by parrotfish occurs as a result of their unique jaw morphology and feeding behaviour. The primary food targets of parrotfish are algal, detrital, microbial and bacterial material contained within the epilithic algal matrix (EAM) covering reef surfaces and endoliths contained within consolidated reef carbonate matrix (Choat et al. 2002, 2004, Crossman et al. 2005, Clements et al. 2017). Parrotfish have fused teeth that form powerful beaks that are capable of breaking off, grating and crushing calcareous material. As a result of rasping their strong beaks across reef surfaces in order to feed, parrotfish erode and ingest a range of reef carbonate material including live coral conies, dead coral surfaces, reef rubble, sediment, crustose algal communities, encrusting foraminifera and bryozoans (Darwin 1845, Frydl and Stearn 1978, Steneck 1983, Alwany et al. 2009). A specially adapted pharyngeal jaw structure at the back of their mouth then crushes and grinds the reef carbonates, whilst their specialised digestive system extracts the food from the carbonate.

Different reef substrates are subject to varying levels of bioerosion and the composition of the underlying skeleton can be a key variable. Porous carbonates seem to be preferred and relationships between declining skeletal density and increased grazing activity (Tribollet et al. 2002) have been noted with parrotfish having a preference for regions of the colony with higher densities of macroborers (Rotjan and Lewis 2005) and/or those that are infested with microboring algae (Bruggemann et al. 1996). Substrates with high skeletal density and those covered with encrusting crustose coralline algae (CCA) are less frequently targeted and are subject to much lower rates of bioerosion by parrotfish (Bruggemann et al. 1996). Different substrates also elicit different biting modes. For example, when *Sparisoma viride* is feeding on coral it opens its mouth wide, reveals its teeth and bites at the substrate. Sometimes it holds on with its teeth and twists its head in order to break off chunks of coral. In contrast, when feeding on rubble encrusted with calcareous organisms (e.g. coralline algae and bryozoans) it peels off the calcareous layer and some of the underlying carbonate substrate (Frydl and Stearn 1978).

Two distinct biting modes (spot biting and focused biting, *sensu* Bruckner et al. 2000) result in the bioerosion of reef building corals and can be distinguished by the distinct lesions that they leave. Individual spot bites are commonly observed on Caribbean corals. Typically shallow and scattered, scars are caused by schools of parrotfish that take bites randomly from coral colonies. In contrast, focused bites are deeper, overlapping and target the same area repeatedly. As a result, focused bites are able to remove both tissue and skeleton from larger, radiating areas of coral colony. Erosion of live coral colonies by parrotfish tends to occur in relation to the relative abundance of preferred target coral species. In the Caribbean there are limited reports of parrotfish bioerosion inducing entire colony mortality, and those few cases are limited to delicate, palatable lagoonal corals such as *Porites divaracata* and *Porites porites furcata* and one dominant framework building reef species *Porites astreoides* in a particular habitat (the back reef; Rotjan and Lewis 2005, Mumby 2009). Other key framework building species that are commonly subject to parrotfish bioerosion include the *Orbicella* sp. (previously known as *Montastraea*), *Colpophyllia natans* and *Porites astreoides* (Bruckner and Bruckner 1998; Rotjan and Lewis 2006). On the Great Barrier Reef, massive *Porites* spp., a key primary reef builder, is a preferred target of species such as *Chlorurus microrhinos* (Bellwood 1985, Bonaldo and Bellwood 2011, Bonaldo et al. 2011). The Indo-Pacific's dominant parrotfish bioeroder, *Bolbometopon muricatum*, is known to target branching growth forms, including species from the genus *Acropora*, as well as *Porites cylindrica* (Bellwood 1985, Hoey and Bellwood 2008, Bonaldo and Bellwood 2011).

Quantifying Parrotfish Bioerosion: How Much and by Who?

Up to now we have referred to "parrotfish bioerosion" as a collective term. But the erosion impact of all parrotfish species is not equal. Specific differences in the muscle and jaw structure among species that give rise to functional distinctions and significant variations in maximum body size among species result in very different impacts on the reef substratum when individuals feed, with resulting differences in bioerosion capabilities. Two distinct functional groups of bioeroding parrotfish are now known to occur, with species being classified as "scrapers" or "excavators" (*sensu* Bellwood and Choat 1990) based on differences in their jaw morphology that give rise to variations in feeding behaviour. The functional classifications tend to follow taxonomic divisions, with scraping taxa belonging to the genera *Scarus* and *Hipposcarus* having a higher feeding rate, taking shallower bites from the substrate, and leaving relatively shallow feeding scars (bite marks in the carbonate framework) (Bellwood and Choat 1990, Bonaldo and Bellwood 2011, figure 5 in Bonaldo et al. 2014) (Fig. 2a). In contrast, the excavating taxa (*Bolbometopon, Cetoscarus, Chlorurus, Sparisoma amplum* and *Sp. viride*) prefer convex surfaces, and leave deeper scars which are often tightly clustered and may overlap (see figure 5 in Bonaldo et al. 2014) (Fig. 2b). Each group therefore has a very different impact on the reef and make different relative contributions to the removal of carbonate via bioerosion (Fig. 2a, b). 'Excavators' remove and ingest both the surface layer of the reef framework and the underlying substrate and therefore these tend to be the species that contribute significantly to bioerosion on reefs. There are some exceptions: for example, in the Caribbean *Scarus vetula* and *Sc. guacamaia* (both grazers) have been shown to erode reef substrate at substantial rate, simply because their large body size yields a heavier grazing scar (Gygi 1975, Bruggemann et al. 1996). But, in general, the bioerosion impact of excavating species far outweighs that of scraping parrotfish (Hoey and Bellwood 2008; Hoey et al. 2016a). Bruggemann et al. (1996) quantified erosion rates on a leeward section fringing reef of Bonaire for the two most common species of parrotfish present, the scraper *Sc. vetula* and the excavator *Sp. viride*. Although the relatively large body size of *Sc. vetula* meant that its scraping action was significant enough to make a contribution to reef bioerosion (rate of 2.42 kg m^2 y^1 measured in the shallow reef zone), the excavating bites of *Sp. viride* meant that similar-sized fish were ingesting about four times the amount of sediment per bite and *Sp. viride* was eroding reef substratum at a rate of 5.38 kg m^2 y^1 in that same reef zone. The example highlights two fundamental aspects of the process of parrotfish bioerosion as it relates to both reef carbonate budgets and reef conservation biology: size and species identity matter (see also Hoey Chapter 6). Several subsequent studies have calculated the bioerosion capabilities of individual species and Table 4 of Bonaldo et al. (2014) provides an excellent summary of the calculated range of contributions made to bioerosion rates on reefs by particular excavators and scrapers.

For any given reef, the total quantitative impact of parrotfish on reef destruction will be a function of species diversity, abundance and population size distribution. Parrotfish bioerosion rates vary along a range of spatial scales all the way from the local (within-reef abundances by habitat zone) to the regional-scale evolutionary and biogeographic processes that determine species' range distributions, highlighting both the importance of determining rates for particular locations and the danger of abstracting bioerosion rates between reefs and regions. At the local scale, quantitative rates of parrotfish bioerosion typically reach their maximum within reef crest habitats (Hoey and Bellwood 2008) as this is where parrotfish abundance and feeding impact is greatest (Fox and Bellwood 2007, Hoey and Bellwood 2008). Quantitative measures of parrotfish bioerosion were first made in the Caribbean in the 1970s as part of reef carbonate budget studies (Gygi 1975, Frydl and Stearn 1978, Scoffin et al. 1980) and revealed that the excavating species *Sp. viride* (Fig. 2c) dominates the overall contribution of parrotfish to bioerosion on reefs in this region.

In the Indo-Pacific, *B. muricatum* (Fig. 2d) dominates bioerosion by parrotfish when it is present on reefs (Bellwood et al. 2012). On the Great Barrier Reef, Australia, *B. muricatum* is capable of eroding reef crest habitats on the outer shelf reefs at a rate of 27.9 kg m^{-2} y^{-1} (Bellwood et al. 2003) and is responsible for 87.5 percent of total parrotfish bioerosion on outer shelf reefs (where it reaches its highest abundances; Hoey and Bellwood 2008). Overall it is responsible for 53.9 percent of the total cross-shelf erosion by parrotfish (Hoey and Bellwood 2008), an enormous contribution by a single species to such a critical physical process. The overall regional averages do distort the picture at the local level however, since *B. muricatum* is not present in all reef habitats and therefore does not contribute equally to the process of bioerosion across individual reef zones. On the Great Barrier Reef, other excavating species that contribute to the process of bioerosion are those within the genus *Chlorurus* (predominantly *Ch. microrhinos* and *Ch. spilurus*). *Ch. microrhinos* typically dominates bioerosion rates in those reef habitats and locations where *B. muricatum* is absent (Hoey and Bellwood 2008). In calculations of species-specific bioerosion rates on the Great Barrier Reef *Ch. microrhinos* (commonly observed lengths of 40-49 cm, maximum 70 cm) was found to account for 67.3 and 90.4 percent of parrotfish bioerosion on inner- and mid-shelf reefs respectively (Hoey and Bellwood 2008). Despite being just as abundant as *Ch. microrhinos* across the reef habitats and shelf locations studies, *Ch. spilurus* (with commonly

Fig. 2. (a) Example of parrotfish 'scraping' feeding scar within epilithic algal matrix showing the almost negligible erosional impact on the consolidated reef substratum. (b) Example of parrotfish "excavating" feeding scar on reef substratum showing how both the surface layer of the reef framework and the underlying substrate are removed. Note the significantly greater erosion impact of bites by "excavators" compared to "scrapers". (c) *Sparisoma viride,* the dominant bioeroding species of parrotfish in the Caribbean. (d) *Bolbometopon muricatum,* the dominant bioeroding species of parrotfish on reefs in the Indo-Pacific. Photo credits: (a, b) R.J. Fox, (c) Adona9 at the English language Wikipedia [GFDL (http://www.gnu.org/copyleft/fdl.html) or CC-BY-SA-3.0 (http://creativecommons.org/licenses/by-sa/3.0/)], via Wikimedia Commons, (d) Thomas Hubauer at Flickr (https://creativecommons.org/licenses/by/4.0) via Creative Commons.

observed sizes of 20-25 cm) was responsible for just 3.1% of cross-shelf parrotfish bioerosion (Hoey and Bellwood 2008), highlighting the importance of body size in determining the absolute quantitative impact that individual parrotfish species can have on rates of reef bioerosion.

Globally, parrotfish bioerosion is the result of a small subset of excavating species (Bonaldo et al. 2014) with a single species typically dominating specific reef regions. This highlights a worrying lack of functional redundancy within this key process on reefs (Bellwood et al. 2003, 2012). In addition, the fact that dominant parrotfish bioerorders are also those that attain the largest body size, unfortunately makes them the most attractive fishing targets and therefore the most vulnerable to overexploitation. The implications of this for reef bioerosion and carbonate budgets is discussed below.

Parrotfish as Carbonate Recyclers: Production and Reworking of Reef Sediments

Since parrotfish do not sequester carbonate, what goes in must come out. What goes in is either one or a mix of (i) consolidated reef carbonate (the result of the process of bioerosion described above) and (ii) loose sediments that are ingested alongside the algae and detrital material being targeted when parrotfish feed off the Epilithic Algal Matrix (EAM) (Purcell and Bellwood 2001, Crossman et al. 2001, Wilson et al. 2003). After passing through the parrotfish's unique pharyngeal jaw apparatus (see Gobalet Chapter 1), what comes out is either sediment that has been newly produced from bioeroded reef matrix, or sediment that is the result of reworking of existing reef sediments or a mixture of both (Bellwood 1996, Bruggemann et al. 1996, Frydl and Stearn 1978). The sight of a parrotfish evacuating its hindgut as a cloud of fine sediment is a familiar one on reefs (Fig. 3a), to the extent that the Hawaiian name for the female redlip parrotfish *Scarus rubroviolaceous* (uhu pālukaluka) translates as 'loose bowels'. The net result is the recycling of consolidated reef structure and the production and reworking of carbonate sediments (Frydl and Stearn 1978, Bellwood 1995b, Bruggemann et al. 1996, Bellwood et al. 2003, Alwany et al. 2009). The material excreted from the digestive tract is then either incorporated back into the reef matrix (via infilling of the framework) or exported out of the system by waves and currents. The processes of sediment ingestion and defecation can also result in the transport of sediments, usually as a net movement off reef, with parrotfish often defecating at specific sites that can be some distance away from feeding areas (Bellwood 1996). In this section we look in more detail at production, reworking and transport of sediment by parrotfishes: how they do it, how much of it they do in relation to other reef sediment producers, and what impact the sediment production has on reef processes.

Getting to the Crunch: Sediment Production and Reworking by Parrotfish

The process of sediment production and reworking by parrotfish involves the ingestion of consolidated and unconsolidated reef material via bioerosion as well as incidental ingestion of reef sediments trapped within the epilithic algal matrix (EAM), followed by the processing of this material through the pharyngeal jaws and sacculate digestive system and the subsequent excretion of the reworked sediment in their faeces (Fig. 3a). New sediment production by parrotfish comes from the breakdown of consolidated and unconsolidated reef substratum. It is therefore effectively the corollary of reef bioerosion and can also be described as carbonate recycling. Reworking of sediment, on the other hand, is merely sediment recycling and comes about through the incidental ingestion of existing reef sediments that are then altered by the process of being ground through the pharyngeal mill and digestive system. Frydl and Stearn (1978) calculated that, for the five Caribbean parrotfish species *Sc. vetula*, *Sc. iseri* (formerly *Sc. croicensis*), *Sc. taeniopterus*, *Sp. aurofrenatum* and *Sp. viride* an average 93 percent of their gut contents were composed of inorganic, sand sized, sediment. This highlights the unique capacity that parrotfish have for handling and processing reef carbonate and demonstrates the significance of their carbonate recycling and sediment reworking capabilities.

Sediment production varies between parrotfish species, most obviously in relation to the large variation in bioerosion rates between scraping versus excavating species (see

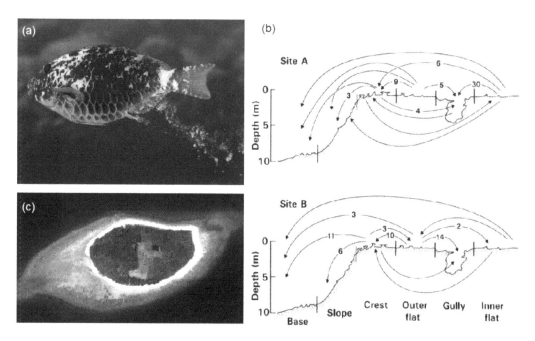

Fig. 3. The process of sediment production by parrotfish and its impacts on reef processes: (a) An individual of the species *Sparisoma frondosum* is photographed here in the act of depositing sediment back onto the reef. (Photo credit J.P. Krajewski). (b) The process of feeding and defecation may result in the transport of recycled or reworked material between reef habitats, as in the case of *Chlorurus microrhinos*, shown here to move between separate feeding and defecation sites on North Reef, Lizard Island, northern Great Barrier Reef (figure reproduced from Bellwood 1995b). Direction of arrows run between the reef zone where feeding ceased and the zone where defecation was observed. Numbers indicate the frequency of movements between two zones). (c) Sediment produced and reworked by parrotfish is the dominant constituent of many reef islands, such as Vakkaru in the Maldives shown here (Perry et al. 2015).

previous section). Even within the excavating functional guild, the relative contributions of newly bioeroded versus reworked material compared to the volume of sediment produced suggests that the relative importance of sediment reworking varies among species. In the Caribbean, estimates of the contribution of reworked sediment to total sediment production for *Sp. viride* range from 10-15 percent (Bruggemann et al. 1996) to around 30 percent (Gygi 1975, Scoffin et al. 1980). For Indo-Pacific excavating species, Bellwood (1996) showed that in *Ch. microrhinos* (previously *Ch. gibbus*) just 2.4 percent of the sediment produced comes from re-working, compared to 27.2 percent of the sediment produced by *Ch. spilurus* (previously *Ch. sordidus*) being reworked material. For the scraping species of parrotfishes, whose impact in terms of bioerosion is much lower, the relative contribution of sediment reworking is significantly greater. For example, in the Caribbean, the proportion of reworked material within sediments produced by *Sc. vetula* have been calculated to be in the region of 61-79 percent (Bruggemann et al. 1996). Hoey and Bellwood (2008) showed that, across the Great Barrier Reef, although rates of sediment reworking varied significantly among individual reef habitats, it was the scraping species belonging to the genus *Scarus* that were responsible for most of the sediment reworking. On reefs of the inner and mid-shelf, they were responsible for 91.5 and 60.2 percent respectively of reworked sediments. Much has been made in the previous section of the importance of excavating species in terms of their contribution to rates of bioerosion on reefs and certainly this also means they dominate the sediment production function in certain habitats (e.g. Hoey and Bellwood 2008, Perry et al. 2015). However, the high abundance and feeding rates of scraping species belonging to the genus *Scarus* mean that they have their own role to play. For example, Hoey and Bellwood (2008) observed that, along with high rates of sediment production in reef zones with high bioerosion, total sediment production was also high on the back reef zone of inner and mid-shelf reefs of the Great Barrier Reef (44.6 and 35.8 kg m^{-2} y^{-1} respectively) due to the high rates of sediment reworking by *Scarus* spp. in these particular habitats (e.g. 91 percent of total sediment production by parrotfish on inner shelf reefs was reworked sediment). The result suggests that scraping species belonging to the genus *Scarus* are also capable of making a significant contribution to sediment production within certain reef habitats.

But as with all biological systems, there are capacity constraints. Even for parrotfish, with their modified pharyngeal jaw, unique sacculate digestive system and no acidic stomach to be affected by the alkaline calcium carbonates in reef sediment (Choat 1991, Choat and Clements 1998), there are limits to the amount of crunch they are prepared to tolerate. Observations in the Caribbean note a reduction in parrotfish biomass, abundance and bioerosion at turbid water reef sites compared to adjacent clear water reef sites (Mallela et al. 2007, Mallela and Perry 2007) whilst research on the Great Barrier Reef suggests that 'scraping' parrotfishes are deterred from feeding on their preferred EAM-covered reef surfaces when high sediment loads are present within the algal matrix (Bellwood and Fulton 2008). When Bellwood and Fulton (2008) artificially reduced sediment loads within the EAM at sites along an outer reef flat on the Great Barrier Reef, the feeding rate (mean bites h^{-1}) of parrotfishes measured via remotely deployed video camera was more than double that recorded on neighbouring control plots containing natural sediment loads. Although the feeding rate increase for parrotfish was the smallest of the five dominant species of herbivorous fish reported by Bellwood and Fulton (2008), it still suggests that parrotfish are deterred by the prospect of ingesting high levels of carbonate sediments along with their target algal and detrital resources. The finding was supported by a similar experiment in which sediment was removed from reef base, crest and flat habitats at the same location on the Great Barrier Reef (Goatley and Bellwood 2012). Interestingly, parrotfish feeding rates increased most within the reef crest habitat, which had seen the smallest reduction in sediment loads (removal of just 150 g m^{-2} compared to the 7,800 g m^{-2} reduction in sediment load within reef flat experimental plots), showing just how responsive parrotfish can be to slight increases in sediment and how attractive areas of low sediment load are to them.

Bellwood and Fulton (2008) hypothesised that the preference for low sediment areas is likely to be based not on an inability to cope with high carbonate content, but rather that high sediment concentration necessarily means a lower concentration of organic material and therefore a lower quality food resource. Experimental results therefore suggest that there is a limit to the amount of sediment reworking that parrotfish are able to perform for reef systems. Feeding rates, and hence bioerosion, and their influence on benthic community succession and composition can be substantively influenced by moderate increases in sediment loads (Goatley and Bellwood 2013). This is bad news for a world in which reefs are being subjected to increases in terrigenous sediments as a result of greater terrestrial runoff and increases in marine sediments from resuspension of dredge spoils following seabed dredging.

Impacts of Parrotfish Sediment Production and Reworking on Reef Processes

Parrotfish also recycle a significant volume of carbonate material back into the marine environment. Recycled carbonates are typically incorporated back into the reef structure, transported off reef and/or on to neighbouring land masses. Carbonate recycling by parrotfish therefore impacts several reef geological processes including sediment transport, carbonate sediment reworking and the construction of reef islands.

Sediment Transport

Sediment resuspension and transport patterns on reefs are most significantly affected by hydrodynamic forces such as wind waves, swell waves, tidal- and wind-driven currents. In addition, some species of reef fish (e.g. parrotfish and surgeonfish), by virtue of their unique patterns of feeding and defecation, can also effect the transportation of sediment (e.g. Chartock 1983). Bellwood (1995b) examined patterns of feeding and sediment release by two excavating parrotfish species: *Ch. microrhinos* and *Ch. spilurus* (previously *Ch. gibbus* and *Ch. sordidus*, respectively). By comparing locations in which the fish took bites from the substratum with areas where they defecated, he showed that, for the smaller of the two excavators *Ch. spilurus*, defecation events in each reef zone took place in proportion to the number of bites taken in that zone. As a result, material was effectively being deposited back from where it had been removed. In the case of *Ch. microrhinos*, however, defecation events in deeper reef areas were disproportionately high, resulting in a net movement of material from the reef crest and flat off to the reef slope, base and gullies (Fig. 3b). In one third of cases, fish moved to a different reef zone (10-30 m away) in order to defecate. On the Great Barrier Reef, where the study was conducted, *Ch. microrhinos* is the second largest excavating parrotfish species, responsible for excavating an estimated 1.02 kg of material per individual per year (Bellwood 1995a, Hoey and Bellwood 2008). The impact of this net transport of material is likely to be considerable. For example, Bellwood (1995b) reports that, over a period of four days, a group of eight *Ch. microrhinos* defecated approximately 67 kg of material into two patches of reef (each approximately 6 m²). With preferred parrotfish defecation sites often being pits or indentations, the process of defecation and sediment transport therefore results in the in-filling of reef matrix.

Reef Framework Infilling by Reworked Carbonate

Reworking of the reef matrix can occur at different spatio-temporal rates with a large proportion of the resulting carbonate sediment being deposited in surficial and intra-framework sediments. A landmark study from St Croix, in the Caribbean examined seven cores taken from the reef framework and highlighted the importance of detrital material to long-term reef development and stability (Hubbard et al. 1990). The reef matrix was composed of a significant proportion of reef sediment. Over half (58 percent) of the carbonate sediment produced on the reef by bioerosion was then reincorporated back into the reef matrix with parrotfish accounting for 8 percent (0.02 kg m^{-2} y^{-1}) of this. The remaining 42 percent of sediment was estimated to fill in sand channels and/or be exported off reef. Subsequent work in Jamaica (Mallela 2004, Mallela and Perry 2007) estimated rates of net framework infilling by loose carbonate sediments to occur at rates of 0.2 and 0.1 kg m^{-2} y^{-1} at a clear and turbid water fore-reefs respectively. However, as parrotfish bioerosion was limited at both sites, accounting for only 12 and 3 percent of total bioerosion respectively, the majority of this sediment was created by other sources of bio-physical erosion (see

Table 2 for summaries). Of the loose carbonate sediment located within the reef framework, 27 percent originated from scleractinian, framework building corals (Mallela 2004). Other studies have found that the presence of interstitial sediment can trigger other secondary framework stabilising processes such as cementation of the reef framework (Scoffin, 1992), which consequently promotes secondary framework accretion. It also seems likely that these sediments may also trigger other diagenetic processes within framework crevices and consequently promote and enable subsequent framework accretion.

Alteration of Sediment Particle Size Distributions

The process of trituration of ingested sediment through the parrotfish pharyngeal mill has an impact on the size distribution of particles that make up the material. By the time that sediments have passed through the parrotfish intestinal system, they have undergone a decrease in average particle size. Comparing the particle size distributions of sediments in the intestines of *Ch. microrhinos* with those of sediments collected from known defecation sites of the species, Bellwood (1996) noted two things: firstly that the excreted sediment was dominated by the 125-250 µm and 250-500 µm size-classes, suggesting that *Ch. microrhinos* is a contributor of these particle sizes to reef sediments. Secondly, he noted a decrease in the fractions of fine sand (63-125 µm) and mud (<63 µm) at the defecation site compared to the sediments contained within the intestine, and suggested that these small particles were carried into the reef lagoon or off the reef. Parrotfish are known to be able to dissolve calcium carbonate in their gut (Smith and Paulson 1974) and so it seems likely that fine-grained carbonate sediment (e.g. micrite) loss could also occur via this process. By the same token, the particle size distribution of sediments within the gut of *Ch. spilurus* was similar to that of the sediments within the EAM, but with a higher proportion of fine sand and mud and a lower proportion of the largest particle size classes. Given that *Ch. spilurus* defecates within its feeding areas, the mismatch between the EAM and intestinal distributions can again be attributed to losses of fine sediments to the system during the process of food transit, carbonate dissolution and defecation. Hoey and Bellwood (2008) found a similar distribution of sediment particle sizes within the guts of scraping *Scarus* spp. to that observed for the excavating *Ch. spilurus*: high proportions of fine sand (63-125 µm) and mud (<63 µm) fractions and almost no sediment particles greater than 1000 µm. The process of bioerosion and sediment reworking by parrotfishes therefore represents a continual process by which sediment particle sizes are gradually decreased over time, until they reach the smallest fractions of fine sand and mud, at which point if chemical dissolution does not occur (Smith and Paulson 1974), the process of parrotfish excretion is likely to result in hydrological transport away from the reef and a net carbonate loss to the system (Bellwood 1996).

Reef Island Construction

Perhaps one of the most visible consequences of sediment production by reef organisms is the construction of reef islands. These fragile land masses are built entirely on a foundation of sediments deposited from the surrounding reef and remain reliant on the reef for inputs to sustain island growth and maintenance (Risk and Sluka 2000; Perry et al. 2011, 2013a, 2015). This effectively requires the ongoing conversion of consolidated reef matrix into sand-grade sediments to maintain a supply of material to the island. It was Darwin (1842) who first noted, when visiting Cocos-Keeling in the Indian Ocean, that fragments of coral and coral sand filled the hollows in the reef and adjacent lagoon areas. He also noted that large shoals of *Scarus* inhabiting the surf outside the reef and the lagoon fed on live corals.

Dissecting some of the parrotfish, and observing how their stomachs were distended by small coral fragments, he concluded that the fish were responsible, along with molluscs and worms, for grinding the coral material into the fine sediment present on the reef. Parrotfish have been shown to play a critical role in the supply of sediments to reef islands, such as the 1,192 coral islands that make up the Republic of Maldives in the Indian Ocean (Risk and Sluka 2000). In a recent study of Vakkaru Island (Maldives) and its surrounding reef, Perry et al. (2015) calculated that the outer reef flat was the dominant sediment-generating segment of reef (producing 75 percent of annual sediment totals despite only constituting 21 percent of reef area). Within this outer reef flat habitat, parrotfish (particularly excavators), were by far the dominant sediment producers, generating >85 percent of the total 5.7 kg $CaCO_3$ m^{-2} of sand-grade sediment. This sand-grade sediment, in turn, was found to be the dominant constituent of Vakkaru Island, demonstrating the role of parrotfish as critical engineering links in the process that allows consolidated reef framework to become reef island landform (Fig. 3c). For the approximately 350,000 inhabitants of the Maldives and the 600,000 tourists that visit the nation annually, parrotfish sediment production is both a very real and critical process.

Putting it all Together: The Impacts of Parrotfish on Reef Carbonate Budgets

Parrotfish act as critical ecosystem engineers of healthy reef systems and in this section we put together the impacts that their feeding and defecating activities have on reef carbonate budgets in terms of (a) reef building processes, (b) carbonate sediment dynamics and (c) carbonate loss. Table 1 summarises all these interactions and the effects that parrotfish have on organisms and processes that contribute to reef building and reef destruction.

Table 1. The role of parrotfishes in reef carbonate budgets

Reef building processes	*Parrotfish interactions*	*Reference*
Coral growth	Scraping and excavating activities result in carbonate loss. Tissue regeneration utilises coral energy resources in healing. This can result in reductions in growth rate, infection by disease and/or partial mortality. Exposed coral skeleton is available for colonisation by calcareous encrusting organisms and internal bioeroders.	Bruckner et al. 2000, Rotjan and Lewis 2005
Coralline algae and encruster growth	Grazing removes algae and sediment, creates new substrate, and promotes the development of crustose coralline algae and other encrusting, calcareous organisms (sclerobionts).	Birkeland 1977, Steneck 1988
Coral recruitment	Parrotfish feeding can facilitate coral recruitment by creating settlement space by removing algal turf and sediment and promoting the development of crustose coralline algae. Removal of algal turf and sediment also prevents competition, overgrowth and reduces post settlement mortality of recruits.	Birkeland 1977, Steneck 1988, Birrell et al. 2005, Box and Mumby 2007

(Contd.)

Primary framework builders	Excavating parrotfish prefer certain species of coral including massive and dome shaped primary framework builders (e.g. *Porites* in the Great Barrier Reef, *Montastrea* in Caribbean). This can result in reduced growth rates, increased incidence of disease and partial mortality and in extreme cases whole colony mortality.	Bythell et al. 1993, Bruckner et al. 2000, Rotjan and Lewis 2005
Carbonate sediment dynamics		
Sediment creation and recycling	Parrotfish recycle carbonate in the reef matrix, this infills natural gaps in the framework. Their faeces contains newly produced and reworked carbonate sediment. The modal particle size of excreted carbonate sediment is typically <250 µm (fine sand) and can be <63 µm (mud).	Frydl and Stearn 1978, Bellwood 1996, Mallela and Perry 2007
Sediment transport	Parrotfish promote net transport of sediment by removing sediment trapped in reef benthos, defecating on and off reef, and producing fine grained sediment. Fine grains are more prone to carbonate dissolution and hydrodynamic transport.	Frydl 1978, Bellwood 1996
Destructive processes		
External bioerosion	Excavating and scraping species physically remove calcium carbonate from *in situ* live and dead reef substrate.	Frydl 1978, Hutchings 1986
Internal bioerosion	Excavating activities provide space for colonisation by opportunistic endolithic borers. Erosion by parrotfish is further facilitated by the secondary porosity created by boring organisms. Some parrotfish species have a preference for macroborer infested substrates and this may play a role in limiting internal bioerosion.	Gygi 1975, Rotjan and Lewis 2005, Tribollet et al. 2002.
Sediment export	Parrotfish excrete carbonate sediment, a proportion will be lost as a result of hydrodynamic sorting and off-reef defecation.	Bellwood 1996, Hubbard et al. 1990
Carbonate dissolution	Parrotfish break down carbonate into sand and silt sized fractions which dissolve more rapidly.	Walter and Morse 1984, Eyre et al. 2014, Smith and Paulson 1974

Impacts on Reef Building Processes

Following the structure and terminology set out in Fig. 1, we first consider the role of parrotfishes in primary and secondary reef building processes. For reefs to persist, it is important that new reef-building organisms are able to recruit, to grow and contribute calcium carbonate to the framework. Parrotfish are able to facilitate the recruitment of corals and encrusting organisms via their feeding activities. As they remove algal turf and sediment, parrotfish provide fresh, biota-free substrates suitable for the settlement and development of coral recruits and calcareous encrusting communities (also called sclerobionts) (Steneck 1988; Birrell et al. 2005; Box and Mumby 2007). The removal of algae and sediment by parrotfish also limits competitive interactions, thereby promoting the post-settlement survival of slower growing calcareous organisms (Birkeland 1977, Mumby 2009, Mallela 2004, 2007, 2013).

Parrotfish bioerosion can also influence the growth rate (calcification), health and morphology of primary reef builders (corals). Many of the world's key reef building corals have massive or lobed morphologies (e.g. *Porites* and *Montastrea*) and bioeroding parrotfish have a preference for their convex (outwards curving) surfaces (Bellwood and Choat 1990; Bruckner et al. 2000). These massive and lobed species also tend to be slower growing than their branching or foliose counterparts. Whilst grazing scars tend to heal in one to two months, excavating scars are longer-lasting and may never heal (Bruckner et al. 2000). Scars leave a tissue-free zone vulnerable to disease and recruitment by other organisms (e.g. algal turf and internal bioeroders). If the colony grows up and around it, a depression prone to sediment infilling and/or smothering results. Coral tissue regrowth is also a costly energetic expenditure, and may come at the expense of coral growth (calcification). As a result, parrotfish corallivory can impact morphology, growth rates and survivorship in reef building colonies (Table 1; Bonaldo and Rotjan Chapter 9).

In sum, parrotfish bioerosion influences the recruitment, survival, diversity and health of primary reef building organisms (corals) in both positive and negative ways. Parrotfish can facilitate coral recruitment and in some regions a strong positive relationship has been observed between the grazing activity of parrotfish and the density of coral recruits (Mumby et al. 2007). However, their feeding activities can also damage and remove juvenile corals (Bak and Engel 1979, Box and Mumby 2007, Trapon et al. 2013) and, in extreme cases, can kill entire coral colonies (Bruckner and Bruckner 1998, Rotjan and Lewis 2005, Mumby 2009, Bonaldo et al. 2014).

Impacts on Carbonate Sediment Dynamics

Parrotfish also play a critical role in recycling carbonate in the reef matrix in the form of sediment. As highlighted during the earlier discussion of bioerosion, parrotfish feed on a wide array of reef carbonates which include the reef matrix, live and dead coral, rock, rubble and sand, they ingest carbonate. This passes through their digestive system becoming fine-grained sand and mud. A proportion of their faeces then re-enters the reef containing newly produced and reworked carbonate sediment which can be incorporated into the reef matrix (Frydl and Stearn 1978; Bellwood 1996; Mallela and Perry 2007). The role of carbonate sediment recycling, reincorporation and infilling within the reef matrix is often overlooked in many reefal carbonate budgets. Sediment infilling is however a vital reef process in terms of carbonate accumulation and facilitation of other secondary reef building processes (Scoffin 1992). In St Croix it was estimated that 58% of sediment formed during bioerosion was reincorporated back into the reef framework making a significant contribution to the carbonate budget (Hubbard et al. 1990).

Impacts on Carbonate Loss

Parrotfish are responsible for the erosion and loss of carbonates from the reef framework (Frydl and Stearn 1978; Scoffin et al. 1980). As they scrape and excavate the reef framework they remove sediment trapped in the reef benthos, remobilise and recycle it. This digested, fine grained, faecal material may be exported from the reef system by fish defecating off reef or as a result of hydrodynamic sorting. By breaking down reef carbonates and reducing grain size, parrotfish also facilitate the removal of sediments from the system by physical and chemical processes, since the finer grained carbonate sediment is more prone to hydrodynamic transport and carbonate dissolution (Bellwood 1996; Eyre et al. 2014).

Parrotfish bioerosion also impacts on rates of carbonate loss by influencing rates of endolithic (internal) bioerosion. When grazers and excavators rasp the surface of corals, their feeding action removes the surface layer of live coral tissue that protects the underlying skeleton. This facilitates the colonisation of the calcium carbonate skeleton by bioeroding endoliths. As the coral skeleton becomes infested by internal bioeroders, the skeleton becomes increasingly porous and more attractive to parrotfish as a grazing target and, hence, more vulnerable to internal and external bioerosion, resulting in a positive feedback loop that increases total bioerosion rates. This almost symbiotic relationship between the two classes of bioeroders was highlighted by Gygi (1975). Interestingly, subsequent studies have also observed a positive correlation between declining skeletal density and increased grazing activity (Tribollet et al. 2002), with parrotfish having a preference for regions of the colony with higher densities of endolithic borers (Rotjan and Lewis 2005).

The Relative Contribution of Parrotfish to Reef Carbonate Budgets

Much of the pioneering work on the role of parrotfish within a carbonate budget context was conducted in the Caribbean and Great Barrier Reef (Land 1979, Scoffin et al. 1980, Kiene 1988, Hubbard et al. 1990) and has provided us with a clear concept of the functional importance of parrotfish and other bioeroding organisms to the long-term health, development and accretion of our coral reefs. We know that healthy, accreting reefs are characterised by a positive carbonate budget value, signalling that carbonate accretion exceeds carbonate loss (Table 2). Interestingly, such reefs are also typically characterised by high parrotfish biomass (>500 kg ha^{-1}), with associated high rates of parrotfish bioerosion. However, on reefs that have been subjected to overfishing, it is typically the large, excavating and bioeroding species that are the first targets. The associated decline in parrotfish biomass and the removal of large individuals on such reefs has already resulted in significant reductions in external bioerosion that are showing up in carbonate budgets, to the point where parrotfish bioerosion is almost negligible on many overfished reefs. In a study of Jamaican reefs we see how external bioerosion by fish and urchins has become a secondary destructive force, attributed to overfishing and disease (Mallela and Perry 2007). Total parrotfish bioerosion was less than 0.02 kg CaCO$_3$ m^{-2} y^{-1}; in addition, this combined with very low abundance of bioeroding urchins (1.6 and 0.2 individuals m^{-2} at clear and turbid water sites respectively), caused internal bioerosion to become the dominant bio-destructive force.

By contrast, assessments of the Bahamas, Belize, Bonaire and Grand Cayman, all impacted to some degree by overfishing and declines in coral cover over the past three decades, show parrotfish bioerosion to still be the dominant bioeroding process (Perry et al. 2014). In the same study, Perry et al. (2014) found that overall bioerosion rates had declined by 75 percent compared to levels prior to the degradation of Caribbean reefs brought about by the combined impacts of overfishing, Hurricane Allen and the die-off of *Diadema*. Parrotfish bioerosion across the study locations had declined from an average rate of 4 kg CaCO$_3$ m^{-2} y^{-1} to 1.6 kg CaCO$_3$ m^{-2} y^{-1}, due to overfishing, but still made the largest contribution to overall reef bioerosion (80 percent of the 2 kg CaCO$_3$ m^{-2} y^{-1}). Such studies highlight that there has been a significant shift in not just the quantity, but the nature of bioerosion on Caribbean reefs which, in the 1970s and 80s, were dominated by *Diadema* urchin bioerosion (rates of 3-5 kg CaCO$_3$ m^{-2} y^{-1} constituting 80-90 percent of total bioerosion, Bak 1994).

Table 2. Summary of detailed carbonate budget studies highlighting the contribution of parrotfish (kg CaCO₃ m⁻² y⁻¹). Net CaCO₃ represents the sum of carbonate production and net carbonate sediment infilling (NIF) minus carbonate erosion rates (updated from Mallela 2004)

Net CaCO₃	Carbonate Accumulation				Carbonate Erosion						Site	Reference
	Coral	Encrusters	Cement	NIF	Parrotfish	Urchins	Macroboring	Microboring	Corallivores	Physical and biological erosion		
11.7	14.3						2.62				Gosong Cemara	Edinger et al. 2000
11.2	13.5						2.31				Pulau Kecil	Edinger et al. 2000
9.5 to -0.98	12.1 to 0.2	0.2 to 0.02*			2.8 to 0.95	≤0.02	0.07 to 0.005§	(0.29 to 0.55)ᵝ			Bonaire	Perry et al. 2012
9.2	11.2	0.6*			2.53	0.006	0.03				Wakatobi, Indonesia	Franco 2014
4.48	7.1	2.5*	1.8		0.02	5.3	1.4				Bellairs, Barbados	Scoffin et al. 1980
2.5	3.4						0.89				Lagun Marican	Edinger et al. 2000
1.94	1.8	0.05		0.01	0.01	0.0002	0.1	0.2			Turbid reef, Jamaica Turbid water	Mallela and Perry 2007 Mallela 2007
1.2	1.02	0.2		0.02	0.02	0.002	0.07	0.04			Clear water, Jamaica Clear water	Mallela and Perry 2007 Mallela 2007
1.1	3.1	2.1*								4.1	Discovery Bay, Jamaica	Land 1979

(Contd.)

Net CaCO₃	Carbonate Accumulation			Carbonate Erosion					Site	Reference
0.91	1.13	0.08	0.41	0.02	0.17	Macro + microb: 0.05			Cane Bay, St. Croix	Hubbard et al. 1990
0.89	0.71	0.51*						0.3	Kailua Bay, Hawaii	Harney and Fletcher 2003
0.57	3.1	0.72*	3.77	1.28	1.04	5.95	0.02		Uva Island, Fore Reef	Eakin 1996
0.45	2.95	0.60*	2.97	0.02	0.08	6.29	0.02		Uva Island, Back reef	Eakin 1996
0.31	0.42	1.71*	2.31	1.15	0.01	3.67	0		Uva Island, Reef flat	Eakin 1996
-0.56	0.01	1.67*	6.21	8.01	4.38	8.01	0		Uva Island, Reef Base	Eakin 1996
-0.7	1.67	0.003*		2.35	0.008	0.02[§]			Grenada, MPA	Franco, 2014
-0.8	4.3					5.09			Bondo	Edinger et al. 2000
-6.9	3.2					10.08			Pulau Panjang	Edinger et al. 2000

[§] Only sponge macroboring measured

* Only coralline algae measured

[β] Microboring not assessed at study sites, microborer rates taken from a study in the Bahamas (Vogel et al. 2000).

Overfished reefs, due to their close proximity to human habitation centres, are often simultaneously impacted by land-based pollution, with parrotfish abundance, biomass, feeding and bioerosion further reduced at turbid water, sediment impacted reef sites when compared to clear water sites (Mallela et al. 2007; Goatley and Bellwood 2012). These sediment-impacted and/or overfished reefs are characterised by a combination of low levels of parrotfish bioerosion, reduced grazing, elevated internal bioerosion and reduced framework accretion, which together can result in a negative net carbonate budget status (Table 2). The incidence of net negative carbonate budget status for individual reefs is increasing, particularly for reefs in the Caribbean. Perry et al. (2012) and Perry et al. (2013b) in recent calculations of carbonate budgets, showed that a quarter of their study reefs were in deficit (range -1.7 to -0.1 kg $CaCO_3$ m^{-2} y^{-1}) and a further quarter in low net positive states. Many reefs in the Caribbean are clearly entering a dangerous period of low reef growth potential. In the Indo-Pacific, some reefs have shifted to negative carbonate budget states after significant coral die-offs following warming events and associated coral bleaching (Eakin 1996, Edinger et al. 2000) (Table 2). Interestingly, the key driver in pushing the budget into the red was the subsequent high rates of bioerosion following environmental disturbance. The implications for reefs of increasing sea surface temperature and other global changes that have the potential to impact on net carbonate status are discussed below.

Forces of Construction and Destruction in a Changing World

We live in a changing world, and carbonate budget processes, and their interactions, reflect these changes. Growing coastal populations are placing increasing fishing pressure on our reefs, reducing fish biomass and impacting on functions such as bioerosion and sediment production. Increases in pollution levels, including greenhouse gases and catchment runoff, are reflected in deteriorating marine water chemistry, a reduction in fish biomass and declining reef accretion. Increasing concentrations of atmospheric greenhouse gases are changing the Earth's climate resulting in multiple, and often synergistic, effects: sea surface temperatures have increased approximately 0.5° C over the last 40 yrs (see Hoegh-Guldberg et al. 2014) and this trend is very likely to continue over the next century (Howes et al. 2015). Rising temperatures have already contributed to mean sea level rise (Rhein et al. 2013) while higher concentrations of carbon dioxide (CO_2) in the ocean is lowering the pH (ocean acidification). The average ocean surface pH has already dropped by about 0.1 units from a level of 8.179 in the pre-industrial era to 8.069 with a further decline of 0.3-0.5 pH units to 7.824 predicted by 2100 (Ciais et al. 2013). This section discusses how such changes are impacting reef carbonate budgets, focusing on the potential impacts of environmental drivers on parrotfish populations and on their ability to carry out functions of bioerosion and sediment production.

Implications of a Changing World for Carbonate Budgets: Receding Reefs?

It is clear that for the majority of reefs, where corals dominate framework growth and are responsible for the majority of carbonate production, anthropogenic-driven changes to the environment that impact on these baseline rates of calcification (including rate of growth, skeletal density and integrity) will have the most significant implications for the net carbonate budget. A comprehensive treatment of this topic falls outside the scope of this book and the current chapter, but the key drivers of global change and the mechanisms via which they are likely to impact on carbonate budgets are summarised for the reader in Table 3. What is immediately clear is that on both the construction and destruction sides of the equation, rising ocean temperatures, shifts in ocean chemistry (ocean acidification), overfishing and increasing levels of run-off and pollution are likely to have a negative impact on reefal carbonate budgets by compromising rates of accretion and enhancing rates of bioerosion and dissolution within the existing reef framework and sediments (Table 3).

There is high confidence among scientists that both higher ocean temperatures and ocean acidification will slow rates of carbonate production. The slow-down in carbonate production combined with the increase in carbonate loss means that we may well start to see some reef carbonate budgets start to head into the red, becoming eroding or 'receding reefs'. Wong et al. (2014) predicted with medium confidence that an atmospheric CO_2 concentration of 560 ppm will be the threshold at which global dissolution of reefs will occur. Perry et al. (2013b) calculate that, in the Caribbean, declines in live coral cover mean that contemporary accretion rates are below their long-term Holocene average for many reefs in the region. Across 19 sites in four countries, they calculated an average accretion rate of 1.36 mm y^{-1} (although with high variability from –1.17 to 11.93 mm y^{-1}). At the <5 m depth range, accretion rates were down to 0.6 mm y^{-1} from their long-term Holocene average of 3.6 mm y^{-1} and rates at the 5-10 m depth gradient were 2.1 mm y^{-1} compared to their long-term Holocene average of 3.8 mm y^{-1}. Whilst these particular Caribbean reefs were not yet showing signs of regional-scale loss (erosion), authors described them as being at an 'accretionary threshold'. These reduced rates of carbonate production combined with a loss of structural complexity (Alvarez-Filip et al. 2009, Perry et al. 2014) warn us that reefs are not well placed to handle further declines in accretion rates associated with increased warming and higher rates of dissolution from ocean acidification (Dove et al. 2013). In modelled simulations of carbonate budgets of Caribbean reefs in the future, Kennedy et al. (2013) demonstrate that, to maintain net positive carbonate budget status until 2080, these reefs will require both high rates of parrotfish grazing (to ensure net coral growth) and reductions in atmospheric greenhouse gas concentrations.

Of course, rates of accretion and bioerosion need to be set in the context of local sea-level histories. For example, Indo-Pacific reefs reached sea level approximately 6,000 yrs ago and have subsequently showed reduced rates of accretion compared to Caribbean reefs which have only just reached present-day sea level (Perry 2011). Slow accretion is therefore not necessarily a sign of poor health, but has to be viewed in the context of sea level change. Ocean thermal expansion and melting of glaciers has caused a rise in global mean sea level of 0.19 m (± 0.02) from 1901 to 2010 (Rhein et al. 2013). Over the next century sea level is predicted to rise at least 0.2-0.6 m. Although it has been suggested (Hamylton et al. 2014; Woodroffe and Webster 2014) that reefs will be able to keep pace with projected rates of sea level rise (and certainly there is evidence that in the past reefs have shown higher rates of accretion in response to rapidly rising sea levels (Smithers and Larcombe 2003, Browne et al. 2013), slower accretion rates associated with coral bleaching, disease and increased rates of carbonate dissolution (associated with ocean acidification) is likely to pose a challenge for some reefs to keep up with rising water levels). For others, rising sea-level may provide renewed opportunities for growth and further accretion.

Table 3. Summary of the potential impacts of global change on reef carbonate budgets. Budget impact indicates whether the alteration to particular reef processes are likely to have a positive (↑) or negative (↓) impact on net carbonate production. Impacts of each driver on a particular process are divided into the main categories illustrated in Fig. 1 of this chapter for ease of cross-reference

Driver	Implications for Constructive Processes	Budget impact (+/-)	Implications for Destructive Processes	Budget impact (+/-)
Ocean warming				
	Coral calcification: Increasing incidence and severity of bleaching events (Glynn 1993, Hoegh-Guldberg 1999), resulting in slowing of coral growth rates and higher coral mortality (De'ath et al. 2009, Mallela and Crabbe 2009, Eakin et al. 2010, Cantin and Lough 2014, Hoegh-Guldberg et al. 2014) and higher incidence of disease (Bruno et al. 2007). Effects are species-specific.	↓	**External bioerosion:** More dead surfaces available, but higher temps may interfere with organism recruitment and metabolism to slow rates of bioerosion	?(↓)
			Micro and macro-boring: More dead surfaces available leading to proliferation of endolithic communities (Fine and Loya 2002, Fine et al. 2004), but temperature may interfere with organism metabolism to slow rates of bioerosion (Fine et al. 2004, Tribollet et al. 2006).	?(↓)
			Physical erosion: Reductions in calcification, encrustation and cementation will result in increasing framework and skeletal porosity. A weakened reef framework will facilitate processes of physical erosion.	(↓)
Ocean acidification				
	Coral calcification: Lowers the rate at which carbonate is produced (Langdon and Atkinson 2005, Pandolfi et al. 2011, Crook et al. 2013),	↓	**External bioerosion:** Reductions in primary and secondary calcification likely to lead to weakened physical framework	?(↓)

(Contd.)

leading to reduced calcification (Kleypas et al. 2006, Kleypas and Yates 2009, Comeau et al. 2014a). Again, impacts are species-specific (Fabricius et al. 2011, Comeau et al. 2014b, Pörtner et al. 2014, Wong et al. 2014, Barkley et al. 2015).		that is more vulnerable to bioerosion (Scoffin 1992; Manzello et al. 2008). Anticipated increases in the rates of epilithic bioerosion (Kobluk and Risk 1977, Sammarco and Risk 1990). But, impact of OA on physiology, organism metabolism and population dynamics of external bioeroders yet to be determined. Overall impact therefore uncertain.	?(↓)
Encrustation: Crustose coralline algae (CCA) show decreases in calcification at species-specific thresholds of acidification (Kuffner et al. 2008, McCoy and Kamenos 2015, although see Ries et al. 2009).	→	**Micro and macro-boring**: anticipated increases in the rates of endolithic bioerosion (Kobluk and Risk 1977, Sammarco and Risk 1990, Fabricius et al. 2011, Wisshak et al. 2012, Crook et al. 2013, Barkley et al. 2015). But impact of OA on physiology, organism metabolism and population dynamics of borers yet to be fully determined. Overall impact uncertain.	(↓)
		Physical erosion: Reductions in calcification likely to lead to weakened physical framework that is more vulnerable to the physical destruction by storms (Scoffin 1992, 1993; Manzello et al. 2008).	(↓)
		Chemical dissolution: Acidification may result in dissolution of existing reef carbonate frameworks (Silbiger and Donahue 2014, Brinkman and Smith 2015, but see Shamberger et al. 2013 and Barkley et al. 2015).	(↓)

(Contd.)

Table 3: (Contd.)

Driver	Implications for Constructive Processes	Budget impact (+/-)	Implications for Destructive Processes	Budget impact (+/-)
Greater storm intensity and frequency				
	Coral calcification: Increased rainfall and flood events resulting in increased terrestrial runoff (e.g.sediment, nutrient, pesticides). Increased turbidity limits scleractinian coral contribution to reef accretion (Mallela and Perry 2007, Fabricus 2005)	→	**External bioerosion:** Sediment and nutrient run-off has been linked with reduced herbivore (including parrotfish) abundance and associated reductions in external bioerosion (Mallela et al. 2007, Mallela and Perry 2007). Higher sediment may also deter feeding by parrotfish (Bellwood and Fulton 2008, Goatley and Bellwood 2012).	←
	Encrustation: Increased catchment runoff limits encruster growth (Mallela and Perry 2007)	→	**Micro and macro-boring:** Increased catchment runoff favours endolithic erosion (Mallela and Perry 2007).	→
	Sediment retention	→	**Physical erosion:** Incidence of high-intensity storm activity likely to increase (Knutson et al. 2008, 2010), leading to higher levels of physical damage and erosion (Madin et al. 2008):	→
Pollution				
	Coral calcification: Pollution runoff (sedimentation, eutrophication and turbidity) limits coral growth, inhibits coral settlement and lowers skeletal density (Fabricius 2005 and references therein).	→	**External bioerosion:** Eutrophication pollution linked with reduced herbivore (including parrotfish) abundance and associated reductions in external bioerosion (Mallela et al. 2007, Mallela and Perry 2007).	←
	Encrustation: CCA abundance is negatively related to sedimentation (Kendrick 1991)	→	**Micro and macro-boring:** Increased nutrient availability and catchment runoff favours endolithic erosion (Rose and Risk 1985, Mallela and Perry 2007)	→

(Contd.)

Overfishing		
Coral calcification: Removal of algae by parrotfish limits competitive interactions, promoting settlement and survival of slower-growing calcareous organisms (Birkeland 1977, Steneck 1988, Box and Mumby 2007).	↓	
Sediment retention: Reduction in parrotfish biomass lowers amount of bioeroded material being reincorporated back into the reef matrix (Hubbard et al. 1990). However associated reduction in reworking of existing reef sediments to smaller particle sizes may slow the rate of sediment export via hydrological transport off-reef (Eyre et al. 2014).	?(↓)	
External bioerosion: Heavily-fished locations have lower rates of external bioerosion (Bellwood et al. 2012, Kennedy et al. 2013).	←	

Impacts on Parrotfish Populations and Functionality

Of all the anthropogenic stressors associated with global change, overfishing associated with population expansion most immediately impacts on parrotfish functionality. The impacts of increased fishing pressure on parrotfish and rates of bioerosion therefore have implications for reef accretion. Parrotfish are target species across much of their range (Jennings and Polunin 1995, Dalzell 1996, Hawkins et al. 2007, Passley et al. 2010) and their population structures are significantly impacted by human population density (Bellwood et al. 2012). In a survey of 18 reefs spanning the Indian and Pacific Oceans, Bellwood et al. (2012) found that the most heavily fished areas had lost almost all of their large parrotfishes, compared to lightly fished areas where 43-67 percent of individuals present were in the larger size categories of >25 cm, resulting in 50 percent more parrotfish biomass on these lightly fished reefs. For the heavily-fished locations, they calculated that bioerosion rates have fallen by between 20-50 percent compared with estimates of rates in the 1960s (see figure 5 in Bellwood et al. 2012). These declines are overwhelmingly driven at the individual species level, by the loss of the larger excavators such as *B. muricatum* and large *Chlorurus* that are disproportionately affected by fishing pressure by virtue of their large body size (Bellwood et al. 2003; Heenan et al. 2016). By contrast, scraping *Scarus* species showed no significant impact of increasing human population density (Bellwood et al. 2012). This is, of course, the group that has least impact in terms of parrotfish bioerosion; however, they still play a significant role in determining the net carbonate budget as their scraping action and high feeding rates moderate macro algal growth, promoting coral accretion.

Overfishing of Caribbean reefs in the decades leading up to the 1980s saw huge declines in parrotfish biomass with associated reductions in bioerosion rates, which might be thought of as good for reef carbonate budgets. However, the impact of declining parrotfish (and associated herbivory), alongside the demise of grazing urchins (*Diadema*) due to a pathogen, and destruction of reef framework by Hurricane Allen enabled the proliferation of fast-growing, macroalgae which, in turn, further impacted coral productivity, with severe implications for primary reef growth (Hughes 1994, Mallela and Perry 2007, Kennedy et al. 2013). The results of subsequent modelling simulations for the Caribbean over this period clearly show that the positive impact on the budget from reductions in parrotfish bioerosion (i.e. reduced destruction) are more than offset by the negative impacts of subsequent algal proliferation on coral growth and associated reductions in carbonate production (i.e. reduced construction) (Kennedy et al. 2013). The long-term impacts of harvesting of parrotfishes on reef accretion and reef carbonate budgets are clearly negative.

There is also clear evidence that the increases in terrestrial runoff onto reefs (e.g. sediment, nutrients, and pollutants such as pesticides) that are associated with increased storm activity and flood events have a negative impact on the carbonate budget. Kennedy et al. (2013) found that eutrophication associated with high levels of agricultural runoff and waste water on simulated Caribbean reef carbonate budgets had a significant impact on overall carbonate budget position, with simulated increases in nitrate concentration of 0.22 μmol l^{-1} resulting in the inability to achieve positive net carbonate production over the long term. In field-based carbonate budget calculations performed for healthy and runoff impacted reef sites in Jamaica, Mallela and Perry (2007) found that external bioerosion by parrotfish and urchins was reduced at runoff impacted sites compared to clear water sites due to a reduction in the biomass of these taxa. The impact of increased pollution and terrigenous sediment inputs on reef carbonate budgets are clearly negative.

By contrast, little is currently known about how parrotfish and their bioeroding and sediment production functions will adapt to, and impact upon, future reef building conditions under rising temperatures or changes in ocean water chemistry. Given that parrotfish show a preference for feeding off convex surfaces and less structurally complex areas of reef (Bellwood and Choat 1990, Fox and Bellwood 2013), will the decline in reef topographic complexity that accompanies coral community shifts in the wake of bleaching events (Alvarez-Filip et al. 2009, Bozec et al. 2014), and which will be facilitated by ocean acidification, impact on parrotfish feeding rates with increases in bioerosion? Will a future less stable, more porous reef matrix (as predicted under future ocean acidification scenarios) mean that less force is required for erosion, therefore facilitating parrotfish bioerosion (Kobluk and Risk 1977, Sammarco and Risk 1990)? These questions have not yet been asked, but the answers will have profound implications for the ability of parrotfish to continue to perform their functional role under future climate conditions. One potential future for parrotfishes sees their functional role as bioeroders enhanced, as calcareous substrates become increasingly porous, more prone to fracture and therefore easier to scrape and excavate.

However, global warming and ocean acidification could also have reverse effects on parrotfish functionality, mediated via potential (as yet undetermined) impacts on parrotfish population dynamics. Increased temperatures have the potential to impact on fish body size and reproduction (Howes et al. 2015) and there are a number of ways in which changing water chemistry could impact on parrotfish populations in terms of reproductive success (Hoey et al. 2016b), recruitment rates and survival (ability to detect predators) (Munday et al. 2009, 2014). We also currently know little about the potential impacts of rising temperatures and ocean acidification on the physiology of bioeroding organisms, including parrotfish. How will increasing water temperatures and acidification impact on organism metabolism and hence their capacity to bioerode? For example, ocean acidification has been associated with reduced respiratory capabilities in marine fish (Brauner and Baker 2009). Physiological and behavioural responses have so far only been tested in a limited number of model fish species and it is therefore not yet clear how parrotfish in particular will respond to ocean warming and acidification or how population structures and abundance estimates that underpin current carbonate budget bioerosion models may alter under future climatic conditions.

The potential variation that may lie around parameter estimates and forcing functions of existing modelling approaches to carbonate budget estimates represent a limitation to our current estimates of reef accretion and erosion in a changing world. Process-based, whole reef studies at sites naturally characterised by environmental conditions expected under future global change scenarios have provided some valuable insights into the processes of construction and destruction under such scenarios. For example, Barkley et al. (2015) found that, in Palau, coral communities along a natural pH gradient which mirrored predictions of anthropogenic-driven ocean acidification, displayed no difference in rates of coral calcification or skeletal density with increasing acidification but did exhibit an 11-fold increase in rates of coral bioerosion as pH decreased, primarily due to increasing macroboring. Similarly large increases have been reported from other reefs in naturally acidified waters (Fabricius et al. 2011, Crook et al. 2013) and tank experiments on bioeroding sponges (Wisshak et al. 2012), suggesting that the overall net direction for bioerosion in the future will be positive, with an associated loss of carbonate and negative impact on the carbonate budget (Table 3). Teasing apart the role that parrotfish will play in these overall changes in the carbonate budget will require further research into the specific impacts of temperature, water chemistry, pollution and sedimentation on their populations and physiology, in order to refine the assumptions currently implicit within reef carbonate budget models.

Summary and Future Research Directions

Which brings us to the future. In this chapter we have shown that parrotfish, by virtue of their biological link to processes of reef construction and destruction, play a critical role in determining the physical structuring and health of their habitat and, as such, constitute 'ecosystem engineers' (organisms that modify, create or destroy habitat and directly or indirectly modulate the availability of resources to other species, causing physical state changes in biotic or abiotic materials, *sensu* Jones et al. 1994, 1997). More than any other group of fishes in coral reef ecosystems, the feeding activities of parrotfish can substantially impact on the carbonate budget and determine whether a reef is in net accretionary, net erosional or a static state. By presenting parrotfish bioerosion and sediment production within the wider context of reef carbonate budgets, this chapter has aimed to extend the reader's appreciation for this tribe of fishes and to highlight the uniqueness of their functionality, beyond just the acceptance that parrotfish are good for the reef because they exert top-down control on the proliferation of algae.

The approach also highlights the synergistic nature of parrotfish feeding interactions with reef surfaces and other framework modifying biota. For example, the action of parrotfish rasping the reef substrate to feed on endoliths has the effect, not only of eroding existing structure, but of removing microboring organisms and exposing fresh surfaces to erosion, thereby removing the old and creating opportunities for new infestations and overgrowth. Seeing these important synergies and interactions confirms the need to study reef organisms not in isolation, but as part of overall processes, for example as part of processes of reef construction and destruction. Such an approach will be necessary if we are to understand how processes such as accretion and bioerosion may alter under future environmental conditions such as reduced fish biomass and diversity (e.g. from overfishing), declining water quality (e.g. increasing sediment and nutrient levels) and ocean acidification. Will reef destruction slow under such circumstances, with lower rates of external bioerosion offsetting the expected higher rates of carbonate dissolution and reduced rates of calcification? What will be the effects of elevated temperature and lower pH be on the metabolism of reef bioeroders (endolithic and epilithic) and how will this impact sediment production, recycling and carbonate dissolution? These are important questions that must be tackled within a process-based framework.

Setting parrotfish bioerosion within a carbonate budget context also serves to highlight their key functional role in reef development and how their interaction with other reef building and eroding organisms is vital in maintaining a healthy (accreting) reef budget. However, it is no longer enough to assert that reefs will recover if parrotfish populations recover. We have to consider how future environmental scenarios will impact on the interactive processes of carbonate production and carbonate erosion in order to safeguard the physical framework and topographic complexity of reef ecosystems in to the future.

Acknowledgments

We thank M. Risk, D. Bellwood and A. Hoey for advice and discussions over the years that have shaped the views expressed in this chapter. Special thanks to M. Risk for helpful comments on an earlier draft, and to M. Jennions and J-lab colleagues for all their support and encouragement of our endeavours.

References Cited

Alvarez-Filip, L., N.K. Dulvy, J.A. Gill, I.M. Côté and A.R. Watkinson. 2009. Flattening of Caribbean coral reefs: region-wide declines in architectural complexity. Proc. R. Soc. Lond. B Biol. Sci. 276: 3019–3025.

Alwany, M.A., E. Thaler and M. Stachowitsch M. 2009. Parrotfish bioerosion on Egyptian Red Sea reefs. J. Exp. Mar. Biol. Ecol. 371: 170–176.

Bak, R.P.M. 1994. Sea urchin bioerosion on coral reefs: place in the carbonate budget and relevant variables. Coral Reefs 13: 99–103.

Bak, R.P.M. and M.S. Engel. 1979. Distribution, abundance, survival of juvenile hermatypic corals (Scleractinia) and the importance of life history strategies in the parent coral community. Mar. Biol. 54: 341–352.

Barkley, H.C., A.L. Cohen, Y. Golbuu, V.R. Starczak, T.M. DeCarlo and K.E.F. Shamberger. 2015. Changes in coral reef communities across a natural gradient in seawater pH. Sci. Adv. 1: e1500328.

Bellwood, D.R. 1985. The functional morphology, systematic and behavioural ecology of parrotfishes (Family Scaridae). PhD thesis, James Cook University, Australia.

Bellwood, D.R. 1995a. Direct estimate of bioerosion by two parrotfish species, *Chlorurus gibbus* and *C. sordidus*, on the Great Barrier Reef, Australia. Mar. Biol. 121: 419–429.

Bellwood, D.R. 1995b. Carbonate transport and within-reef patterns of bioerosion and sediment release by parrotfishes (family Scaridae) on the Great Barrier Reef. Mar. Ecol. Prog. Ser. 117: 127–136.

Bellwood, D.R. 1996. Production and reworking of sediment by parrotfishes (family Scaridae) on the Great Barrier Reef, Australia. Mar. Biol. 125: 795–800.

Bellwood, D.R. and J.H. Choat. 1990. A functional analysis of grazing in parrotfishes (family Scaridae): the ecological implications. Environ. Biol. Fish. 28: 189–214.

Bellwood, D.R. and C.J. Fulton. 2008. Sediment-mediated suppression of herbivory on coral reefs: decreasing resilience to rising sea levels and climate change? Limnol. Oceanogr. 53: 2695–2701.

Bellwood, D.R., A.S. Hoey and J.H. Choat. 2003. Limited functional redundancy in high diversity systems: resilience and ecosystem function on coral reefs. Ecol. Lett. 6: 281–285.

Bellwood, D.R., A.S. Hoey and T.P. Hughes. 2012. Human activity selectively impacts the ecosystem roles of parrotfishes on coral reefs. Proc. Roy. Soc. B. 279: 1621–1629.

Birkeland, C. 1977. The importance of rate of biomass accumulation in early successional stages of benthic communities to the survival of coral recruits. Proc. 3rd Int. Coral Reef Symp. Miami: 15–21.

Birrell, C.L., L.J. McCook and B.L. Willis. 2005. Effects of algal turfs and sediment on coral settlement. Mar. Poll. Bull. 51: 408–414.

Blanchon, P. and J. Shaw. 1995. Reef drowning during the last deglaciation: evidence for catastrophic sea-level rise and ice-sheet collapse. Geology 23: 4–8.

Bonaldo, R.M. and D.R. Bellwood. 2011. Parrotfish predation on massive *Porites* on the Great Barrier Reef. Coral Reefs 30: 259–269.

Bonaldo, R.M., J.P. Krajewski and D.R. Bellwood. 2011. Relative impact of parrotfish grazing scars on massive *Porites* corals at Lizard Island, Great Barrier Reef. Mar. Ecol. Prog. Ser. 423: 223–233.

Bonaldo, R.M., A.S. Hoey and D.R. Bellwood. 2014. The ecosystem roles of parrotfishes on tropical reefs. Oceanogr. Mar. Biol. Ann. Rev. 52: 81–132.

Boss, S.K. and W.D. Liddell. 1987. Patterns of sediment composition of Jamaican fringing reef facies. Sedimentology 34: 77–87.

Box, S.J. and P.J. Mumby. 2007. Effect of macroalgal competition on growth and survival of juvenile Caribbean corals. Mar. Ecol. Prog. Ser. 342: 139–149.

Bozec, Y-M., L. Alvarez-Filip and P.J. Mumby. 2014. The dynamics of architectural complexity on coral reefs under climate change. Glob. Change Biol. 21: 223–235.

Brauner, C.J. and D.W. Baker. 2009. Patterns of acid-base regulation during exposure to hypercarbia in fishes. pp. 43–63 *In*: M.L.G. Wood (ed.). Cardio-Respiratory Control in Vertebrates, Berlin, Springer.

Brinkman, T.J. and A.M. Smith. 2015. Effect of climate change on crustose coralline algae at a temperate vent site, White Island, New Zealand. Mar. Freshw. Res. 66: 360–370.

Browne, N.K., S.G. Smithers and C.T. Perry. 2013. Carbonate and terrigenous sediment budgets for two inshore turbid reefs on the central Great Barrier Reef. Mar. Geol. 346: 101–123.

Bruckner, A.W. and R.J. Bruckner. 1998. Destruction of coral by Sparisoma viride. Coral Reefs 17: 350.

Bruckner, A.W., R.J. Bruckner and P. Sollins. 2000. Parrotfish predation on live coral: "spot biting" and "focused biting". Coral Reefs 19: 50.

Bruggemann, J.H. 1995. Parrotfish grazing on coral reefs: a trophic novelty. PhD Thesis. University of Groningen, Netherlands, p 213.

Bruggemann, J.H., A.M. van Kessel, J.M. van Rooij and A.M. Breeman. 1996. Bioerosion and sediment ingestion by the Caribbean parrotfish Scarus vetula and Sparisoma viride: implications of fish size, feeding mode and habitat use. Mar. Ecol. Prog. Ser. 134: 59–71.

Bruno, J.F., E.R. Selig, K.S. Casey, C.A. Page, B.L. Willis, C.D. Harvell, H. Sweatman and A.M. Melendy. 2007. Thermal stress and coral cover as drivers of coral disease outbreaks. PLoS Biol 5(6): e124.

Bythell, J.C., E.H. Gladfelter and M. Bythell. 1993. Chronic and catastrophic natural mortality of three common Caribbean corals. Coral Reefs 12: 143–152.

Cantin, N.E. and J.M. Lough. 2014. Surviving coral bleaching events: *Porites* growth anomalies on the Great Barrier Reef. PLoS One 9:e88720.

Chartock, M.A. 1983. The role of *Acanthurus guttatus* (Bloch and Schneider 1801) in cycling algal production to detritus. Biotropica 15: 117–121.

Chave, K., S.V. Smith and K.J. Roy. 1972. Carbonate production by coral reefs. Mar. Geol. 12: 123–140.

Chazottes, V., T. le Campion-Alsumard and M. Peyrot-Clausade. 1995. Bioerosion on coral reefs: interactions between macroborers, microborers and grazers (Moorea, French Polynesia). Palaeogeogr. Palaeoclimatol. Palaeoecol. 113: 189–198.

Choat, J.H. 1991. The biology of herbivorous fishes on coral reefs. pp. 120–155. *In*: P.F. Sale (ed.). Coral Reef Fishes, Dynamics and Diversity in a Complex Ecosystem. Academic Press, San Diego, USA.

Choat, J.H. and K.D. Clements. 1998. Vertebrate herbivores in marine and terrestrial environments: a nutritional ecology perspective. Ann. Rev. Ecol. Syst. 29: 375–403.

Choat, J.H., K.D. Clements and W.D Robbins. 2002. The trophic status of herbivorous fishes on coral reefs I: Dietary analyses. Mar. Biol. 140: 613–623.

Choat, J.H., W.D. Robbins and K.D. Clements. 2004. The trophic status of herbivorous fishes on coral reefs II: Food processing modes and trophodynamics. Mar. Biol. 145: 445–454.

Ciais, P., C. Sabine, G. Bala, L. Bopp, V. Brovkin, J. Canadell, A. Chhabra, R. DeFries, J. Galloway and M. Heimann.. 2013. Carbon and other biogeochemical cycles. pp. 465–570. *In*: T.F. Stocker, D. Qin, G-K. Plattner , M. Tignor, S.K. Allen, J. Boschung, A. Nauels, Y. Xia, V. Bex and P.M. Midgley. (eds.). Climate Change 2013: The Physical Science Basis. Contribution of Working Group I to the Fifth Assessment Report of the Intergovernmental Panel on Climate Change, Cambridge University Press, Cambridge, UK, New York, NY, USA.

Clements, K.D., D. Raubenheimer and J.H. Choat. 2009. Nutritional ecology of marine herbivorous fishes: ten years on. Func. Ecol. 23: 79–92.

Clements, K.D., D.P. German, J. Piché, A. Tribollet and J.H. Choat. 2017. Integrating ecological roles and trophic diversification on coral reefs: multiple lines of evidence identify parrotfishes as microphages. Biol. J. Linn. Soc. 120: 729–751.

Comeau, S., R.C. Carpenter, C.A. Lantz and P.J. Edmunds. 2014a. Ocean acidification accelerates dissolution of experimental coral reef communities. Biogeosci. Discuss. 11: 12323–12339.

Comeau, S., R.C. Carpenter, Y. Nojiri, H.M. Putnam, K. Sakai and P.J. Edmunds. 2014b. Pacific-wide contrast highlights resistance of reef calcifiers to ocean acidification. Proc. R. Soc. B: Biol. Sci. 281: 20141339.

Conand, C., P. Chabanet, P. Cuet and Y. Letourneur. 1997. The carbonate budget of a fringing reef in La Reunion Island (Indian Ocean): sea urchin and fish bioerosion and net calcification. Proc. 8th Int. Coral Reef Symp. 1: 953–958.

Crook, E.D., A.L. Cohen, M. Rebolledo-Vieyra, L. Hernandez and A. Paytan. 2013. Reduced calcification and lack of acclimatization by coral colonies growing in areas of persistent natural acidification. Proc. Natl. Acad. Sci. USA. 110: 11044–11049.

Crossman, D.J., J.H. Choat, K.D. Clements, T. Hardy and J. McConochie. 2001. Detritus as food for grazing fishes on coral reefs. Limnol. Oceanogr. 46: 1596–1605.

Crossman, D.J., J.H. Choat and K.D. Clements. 2005. Nutritional ecology of nominally herbivorous fishes on coral reefs. Mar. Ecol. Prog. Ser. 296: 129–142.

Dalzell, P. 1996. Catch rates, selectivity and yields of reef fishing. pp. 161–192. *In*: N.V.C. Polunin and C.M. Roberts (eds.). Reef Fisheries. Chapman and Hall, London, UK.

Darwin, C.R. 1842. On the Structure and Distribution of Coral Reefs. The Walter Scott Publishing Co. Ltd. London.

Darwin, C.R. 1845. Journal of Researches During the Voyage of the H.M.S. Beagle. T. Nelson and Sons, London.

De'ath, G., J.M. Lough, K.E. Fabricius. 2009. Declining coral calcification on the Great Barrier Reef. Science 323: 116–119.

Dove, S.G., D.I. Kline, O. Pantos, F.E. Angly, G.W. Tyson and O. Hoegh-Guldberg. 2013. Future reef decalcification under a business-as-usual CO_2 emission scenario. Proc. Natl. Acad. Sci. USA. 110: 15342–15347.

Eakin, C.M. 1996. Where have all the carbonates gone? A model comparison of calcium carbonate budgets before and after the 1982–1983 El Niño at Uva Island in the eastern Pacific. Coral Reefs 15: 109–119.

Eakin, C.M., J.A. Morgan, S.F. Heron, T.B. Smith, G. Liu, L. Alvarez-Filip et al. 2010. Caribbean corals in crisis: record thermal stress, bleaching, and mortality in 2005. PLoS One 5:e13969.

Edinger, E.N., G.V. Limmon, J. Jompa, W. Widjatmoko, J.M. Heikoop and M.J. Risk. 2000. Normal coral growth rates on dying reefs: are coral growth rather good indicators of reef health? Mar. Poll. Bull. 40: 404–425.

Eyre, B.D., A.J. Andersson and T. Cyronak. 2014. Benthic coral reef calcium carbonate dissolution in an acidifying ocean. Nature Clim. Change 4: 969–976.

Fabricius, K.E. 2005. Effects of terrestrial runoff on the ecology of corals and coral reefs: review and synthesis. Mar. Poll. Bull. 50: 125–146.

Fabricius, K.E., C. Langdon, S. Uthicke, C. Humphrey, S. Noonan, G. De'ath, R. Okazaki, N. Muehllehner, M.S. Glas and J.M. Lough. 2011. Losers and winners in coral reefs acclimatized to elevated carbon dioxide concentrations. Nature Climate Change 1: 165–169.

Fagerstrom, J.A. 1991. Reef-building guilds and a checklist for determining guild membership. Coral Reefs 10: 47–52.

Fine, M and Y. Loya. 2002. Endolothic algae: an alternative source of photoassimilates during coral bleaching. Proc. R. Soc. B. 269: 1205–1210.

Fine, M., L. Steindler and Y. Loya. 2004. Endolithic algae photoacclimate to increased irradience during coral bleaching. Mar. Freshw. Res. 55: 115–121.

Fox, R.J. and D.R. Bellwood. 2007. Quantifying herbivory across a coral reef depth gradient. Mar. Ecol. Prog. Ser. 339: 49–59.

Fox, R.J. and D.R. Bellwood. 2013. Niche partitioning of feeding microhabitats produces a unique function for herbivorous rabbitfishes (Perciformes, Siganidae) on coral reefs. Coral Reefs 32: 13–23.

Franco, C. 2014. Modelling the dynamics of $CaCO_3$ budgets in changing environments using a Bayesian Belief Network approach. PhD Thesis. Department of Biological Sciences, University of Essex, UK. p 349.

Frydl, P. 1979. The effect of parrotfish (Scaridae) on coral in Barbados (W.I). Internationale Revue der Gersmaten Hydobiologie 64: 737–748.

Frydl, P. and C.W. Stearn. 1978. Rate of bioerosion by parrotfish in Barbados Reef Environments. J Sediment. Petrol. 48: 1149–1158.

Glynn, P.W. 1993. Coral reef bleaching ecological perspectives. Coral Reefs 12: 1–17.

Goatley, C.H.R. and D.R. Bellwood. 2012. Sediment suppresses herbivory across a coral reef depth gradient. Biol. Lett. 8: 1016–1018.

Goatley, C.H.R. and D.R. Bellwood. 2013. Ecological consequences of sediment on high-energy coral reefs. PLoS One 8: e77737.

Gobalet, K.W. 1989. Morphology of the parrotfish pharyngeal jaw apparatus. Amer. Zool. 29: 319–331.

Golubic, S., R.D. Perkins and K.J. Lukas. 1975. Boring microorganisms and microborings in carbonate substrates. pp. 229–259. *In*: R.W. Frey (ed.). The Study of Trace Fossils: A Synthesis of Principles, Problems, and Procedures in Ichnology. Springer Verlag, New York.

Golubic, S. 1979. Carbonate dissolution. pp. 107–129. *In*: P.A. Trudinger and J. Swasine (eds.). Biogeochemical Recycling of Mineral Forming Elements. Elsevier Press, Oxford.

Golubic, S., I. Friedmann and J. Schneider. 1981. The lithobionic ecological niche, with special reference to microorganisms. Sediment. Geol. 51: 475–478.

Goreau, T.F. 1959. The ecology of Jamaican Coral Reefs I. Species composition, and zonation. Ecology 40: 67–90.

Gygi, R.A. 1975. *Sparisoma viride* (Bonnaterre), the stoplight parrotfish, a major sediment producer on coral reefs of Bermuda? Eclog Geol Helv 68: 327–359.

Hamylton, S.M., J.X. Leon, M.I. Saunders and C.D. Woodroffe. 2014. Simulating reef responses to sea-level rise at Lizard Island: a geospatial approach. Geomorphology 222: 151–161.

Harmelin-Vivien, M. 1994. The effects of storms and cyclones on coral reefs: a review. J. Coast. Res. 12: 211–231.

Harney, J.N. and C.H. Fletcher. 2003. A budget of carbonate framework and sediment production, Kailua Bay, Oahu, Hawaii. J. Sediment. Petrol. 73: 856–868.

Hawkins, J.P., C.M. Roberts, F.R. Gell and C. Dytham. 2007. Effects of trap fishing on reef fish communities. Aq. Conserv. Mar. Freshw. Ecosyst. 17: 111–132.

Heenan, A., A.S. Hoey, G.J. Williams and I.D. Williams 2016. Natural bounds on herbivorous coral reef fishes. Proc. R. Soc. B. 283: rspb20161716.

Hoegh-Guldberg, O. 1999. Climate change, coral bleaching and the future of the world's coral reefs. Mar. Freshwater Res. 50: 839–866.

Hoegh-Guldberg, O., R. Cai, E.S. Poloczanska, P.G. Brewer, S. Sundby, K. Hilmi, V.J. Fabry and S. Jung. 2014. The Ocean. pp. 1655–1731. *In*: V.R. Barros, C.B. Field, D.J. Dokken, M.D. Mastrandrea, K.J. Mach, T.E. Bilir, M. Chatterjee, K.L. Ebi, Y.O. Estraded, R.C. Genova, B. Girma, E.S. Kissel, A.N. Levy, S. MacCracken, P.R. Mastrandrea and L.L. White (eds.). Climate Change 2014: Impacts, Adaptation and Vulnerability. Part B: Regional Aspects. Contribution of Working Group II to the Fifth Assessment Report of the Intergovernmental Panel on Climate Change. Cambridge University Press, Cambridge, United Kingdom and New York, NY, USA.

Hoey, A.S. and D.R. Bellwood. 2008. Cross-shelf variation in the role of parrotfishes on the Great Barrier Reef. Coral Reefs. 27: 37–47.

Hoey, A.S., D.A. Feary, J.A. Burt, G. Vaughan, M.S. Pratchett and M.L. Berumen. 2016a. Regional variation in the structure and function of parrotfishes on Arabian reefs. Mar. Poll. Bull. 105: 524–531.

Hoey, A.S., E. Howells, J.L. Johansen, J-P.A. Hobbs, V. Messmer, D.M. McCowan, S.K. Wilson and M.S. Pratchett. 2016b. Recent advances in understanding the effects of climate change on coral reefs. Diversity 8: 12.

Howes, E.L., F. Joos, C.M. Eakin and J.-P. Gattuso. 2015. An updated synthesis of the observed and projected impacts of climate change on the chemical, physical and biological processes in the oceans. Front. Mar. Sci. 2: 36.

Hubbard, D.K. 1992. Hurricane induced sediment transport in open-shelf tropical systems – An example from St.Croix, U.S. Virgin Islands. J. Sediment. Petrol. 62: 946–960.

Hubbard, D.K., A.I. Miller and D. Scaturo. 1990. Production and cycling of calcium carbonate in a shelf-edge reef system (St. Croix, U.S. Virgin Islands): applications to the nature of reef systems in the fossil record. J. Sediment. Petrol. 60: 335–360.

Hughes, T.P. 1994. Catastrophes, phase shifts and large-scale degradation of a Caribbean coral reef. Science 265: 1547–1551.

Hutchings, P.A. 1986. Biological destruction of coral reefs. A review. Coral Reefs 4: 239–252.

Jennings, S. and N.V.C. Polunin. 1995. Comparative size and composition of yield from six Fijian reef fisheries. J. Fish Biol. 46: 28–46.

Jones, C.G., J.H. Lawton and M. Shachak. 1994. Organisms as ecosystem engineers. Oikos 69: 373–386.

Jones, C.G., J.H. Lawton and M. Shachak. 1997. Positive and negative effects of organisms as physical ecosystem engineers. Ecology 78: 1946–1957.

Kendrick, G.A. 1991. Recruitment of coralline crusts and filamentous turf algae in the Galapagos archipelago: effect of simulated scour, erosion and accretion. J. Exp. Mar. Biol. Ecol. 147: 47–63.

Kennedy, E.V., C.T. Perry, P.R. Halloran, R. Iglesias-Prieto, C.H.L. Schönberg, M. Wisshak, A.U. Form, J.P. Carricart-Ganivet, M. Fine, C.M. Eakin and P.J. Mumby. 2013. Avoiding coral reef functional collapse requires local and global action. Curr. Biol. 23: 912–918.

Kiene, W.E. 1988. A model of bioerosion on the Great Barrier Reef. Proc. 6th Int. Coral Reef Symp. Australia 3: 449–454.

Kiene, W.E. and P.A. Hutchings. 1994. Bioerosion experiments at Lizard Island, Great Barrier Reef. Coral Reefs 13: 91–98.

Kleypas, J.A., R.A. Feely, V.J. Fabry, C. Langdon, C.L. Sabine and L.L. Robbins. 2006. Impacts of ocean acidification on coral reefs and other marine calcifiers report: A guide for future research. Report of a workshop held 18–20 April 2005, St Petersburg, FL, sponsored by NSF, NOAA, and the US Geological Survey 1–88.

Kleypas, J.A. and K.K. Yates. 2009. Coral reefs and ocean acidification. Oceanography 22: 108–117.

Knutson, T.R., J.L. McBride, J. Chan, K. Emanuel, G. Holland, C. Landsea, I. Held, J.P. Kossin, A.K. Srivastava and M. Sugi. 2010. Tropical cyclones and climate change. Nat. Geosci 3: 157–163.

Knutson, T.R., J.J. Sirutis, S.T. Garner, G.A. Vecchi and I.M. Held. 2008. Simulated reduction in Atlantic hurricane frequency under twenty-first-century warming conditions. Nat. Geosci. 1: 359–364.

Kobluk, D.R. and M.J. Risk. 1977. Risk, rate and nature of infestation of a carbonate substratum by a boring alga. J. Exp. Mar. Biol. Ecol. 27: 107–115.

Kuffner, I.B., A.J. Andersson, P.L. Jokiel, K.S. Rodgers and F.T. Mackenzie. 2008. Decreased abundance of crustose coralline algae due to ocean acidification. Nature Geoscience 1: 114–117.

Land, L.S. 1979. The fate of reef derived sediment on the North Jamaican Island slope. Mar. Geol. 29: 55–71.

Langdon, C. and M.J. Atkinson. 2005. Effect of elevated pCO_2 on photosynthesis and calcification of corals and interactions with seasonal change in temperature/irradiance and nutrient enrichment. J. Geophys. Res. 110: C09S07.

MacGeachy, J.K. and C.W. Stearn. 1976. Boring by macro-organisms in the coral *Montastrea annularis* on Barbados reefs. Int. Rev. Der Gesmat. Hydrobiol. 61: 715–745.

Madin, J.S., M.J. O'Donnell and S.R. Connolly. 2008. Climate-mediated mechanical changes to post-disturbance coral assemblages. Biol. Lett. 4: 490–493.

Mallela, J. 2004. Coral reef communities and carbonate production in a fluvially-influenced embayment, Rio Bueno, Jamaica. Ph.D. Thesis, Manchester Metropolitan University, UK. p. 224.

Mallela, J. 2007. Coral reef encruster communities and carbonate production in cryptic and exposed coral reef habitats along a gradient of terrestrial disturbance. Coral Reefs 26: 775–785.

Mallela, J. 2013. Calcification by Reef-Building Sclerobionts. PLoS ONE 8:e60010.

Mallela, J. and M.J.C. Crabbe. 2009. Hurricanes and coral bleaching linked to changes in coral recruitment in Tobago. Mar. Env. Res. 68: 158–162.

Mallela, J. and C.T. Perry. 2007. Calcium carbonate budgets for two coral reefs affected by different terrestrial runoff regimes, Rio Bueno, Jamaica. Coral Reefs 26: 53–68.

Mallela, J., C.A. Roberts, C. Harrod and C.R. Goldspink. 2007. Distributional patterns and community structure of Caribbean coral reef fishes within a river-impacted bay. J. Fish Biol. 70: 523–537.

Manzello, D.P., J.A. Kleypas, D.A. Budd, C.M. Eakin, P.W. Glynn and C. Langdon. 2008. Poorly cemented coral reefs of the eastern tropical Pacific: Possible insights into reef development in a high-CO_2 world. Proc. Natl. Acad. Sci. USA. 105: 10450–10455.

May, J.A., I.G. Macintyre and R.D. Perkins. 1982. Distribution of microborers within planted substrates along a barrier reef transect, Carrie Bow Cay, Belize. pp. 93–107. *In*: K. Rutzler and I.G. Macintyre (eds.). The Atlantic Barrier Reef Ecosystem at Carrie Bow Cay, Belize, I. Structure and Communities. Smithsonian Institution Press, Washington.

McCoy, S.J. and N.A. Kamenos. 2015. Coralline algae (Rhodophyta) in a changing world: integrating ecological, physiological and geochemical responses to global change. J. Phycol. 51: 6–24.

Mumby, P.J. 2009. Herbivory versus corallivory: are parrotfish good or bad for Caribbean coral reefs? Coral Reefs 28: 683–690.

Mumby, P.J., A. Hastings and H.J. Edwards. 2007. Thresholds and the resilience of Caribbean coral reefs. Nature 450: 98–101.

Munday, P.L., D.L. Dixson, J. Donelson, G.P. Jones, M.S. Pratchett, G.V. Devitsina and K.B. Dowing. 2009. Ocean acidification impairs olfactory discrimination and homing ability of a marine fish. Proc. Nat. Acad. Sci. USA. 106: 1848–1852.

Munday, P.L., A.J. Cheal, D.L. Dixson, J.L. Rummer and K.E. Fabricius. 2014. Behavioural impairment in reef fishes caused by ocean acidification at CO_2 seeps. Nat. Clim. Change 4: 487–492.

Neumann, A.C. 1966. Observations on coastal erosion in Bermuda and measurements of the boring rate of the sponge *Cliona lampa*. Limnol. Oceanogr. 11: 92–108.

Neumann, A.C. and I.G. Macintyre. 1985. Reef response to sea level rise: keep-up, catch-up or give-up. Proc. 5th Int. Coral Reef Symp. Tahiti 3: 105–110.

Ogden, J.C. 1977. Carbonate-sediment production by parrot fish and sea urchins on Caribbean reefs. pp. 281–288. *In*: S.H. Frost, M.P. Weiss and J.B. Saunders (eds.). Reefs and related carbonates. American Association of Petroleum Geologists, Oklahoma.

Pandolfi, J.M., S.R. Connolly, D.J. Marshall and A.L. Cohen. 2011. Projecting coral reef futures under global warming and ocean acidification. Science 333: 418–422.

Passley, D., K. Aitken, G.-A. Perry. 2010. Characterisation of the Jamaican spearfishing sector. Proc. Gulf Caribb. Fish. Instit. 62: 235–240.

Perkins, R.D. and C.I. Tsentas. 1976. Microbial infestation of carbonate substrates planted on the St. Croix shelf, West Indies. Geol. Soc. Am. Bull. 87: 1615–1628.

Perry, C.T. 1999. Reef framework preservation in four contrasting modern reef environments, Discovery Bay, Jamaica. J. Coast. Res. 15: 796–812.

Perry, C. 2011. Carbonate budgets and reef framework accumulation. pp. 182–190. *In*: D. Hopley (ed.). Encyclopedia of Modern Coral Reefs: Structure, Form and Process. Springer, Dordrecht, Netherlands.

Perry, C.T., P.S. Kench, M.J. O'Leary, B.R. Reigl, S.G. Smithers and H. Yamano. 2011. Implications of reef ecosystem change for the stability and maintenance of coral reef islands? Glob. Change. Biol. 17: 3679–3696.

Perry, C.E., E.N. Edinger, P.S. Kench, G.N. Murphy, S.G. Smithers, R.S. Steneck and P.J. Mumby. 2012. Estimating rates of biologically driven coral reef framework production and erosion: a new census-based carbonate budget methodology and applications to the reefs of Bonaire. Coral Reefs 31: 853–868.

Perry, C.T., P.S. Kench, S.G. Smithers, H. Yamano, M.J. O'Leary and P. Gulliver. 2013a. Time scales and modes of reef lagoon infilling in the Maldives and controls on the onset of reef island formation. Geology 41: 1111–1114.

Perry, C.T., G.N. Murphy, P.S. Kench, S.G. Smithers, E.N. Edinger, R.S. Steneck and P.J. Mumby. 2013b. Caribbean-wide decline in carbonate production threatens coral reef growth. Nat. Comm. 4: 1402.

Perry, C.T., G.N. Murphy, P.S. Kench, E.N. Edinger, S.G. Smithers, R.S. Steneck and P.J. Mumby. 2014. Changing dynamics of Caribbean reef carbonate budgets: emergence of reef bioeroders as critical controls on present and future reef growth potential. Proc. R. Soc. B. 281: 20142018.

Perry, C.T., P.S. Kench, M.J. O'Leary, K.M. Morgan and F. Januchowski-Hartley. 2015. Linking reef ecology to island building: parrotfish identified as major producers of island-building sediment in the Maldives. Geology 43: 503–506.

Pörtner, H.-O., D. Karl, P.W. Boyd, W. Cheung, S.E. Lluch-Cota and Y. Norjiri. 2014. Ocean systems. pp. 411–484. *In*: C. Field, V. Barros, D. Dokken, K. Mach, M. Mastrandrea, T. Bilir, M. Chatterjee, K. Ebi, Y. Estrada and R. Genova (eds.). Climate Change 2014: Impacts, Adaptation and Vulnerability. Part A: Global and Sectoral Aspects. Contribution of Working Group II to the Fifth Assessment Report of the Intergovernmental Panel of Climate Change. Cambridge University Press, Cambridge, New York.

Purcell, S. W. and D.R. Bellwood. 2001. Spatial patterns of epilithic algal and detrital resources on a windward coral reef. Coral Reefs 20: 117–125.

Rasser M,W. and B. Riegl. 2002 Holocene coral reef rubble and its binding agents. Coral Reefs 21: 57–72.

Rhein, M., S.R. Rintoul, S. Aoki, E. Campos, D. Chambers, R.A. Feely, S. Gulev, G.C. Johnson, S.A. Josey, A. Kostianoy, C. Mauritzen, D. Roemmich, L.D. Talley and F. Wang. 2013. Observations: Ocean. *In*: T.F. Stocker, D. Qin, G.K. Platter, M. Tignor, S.K. Allen, A.J. Boschung, A. Nauels, Y. Xia, V. Bex and P.M. Midgley (eds.). Climate Change 2013: The Physical Science Basis. Contribution of Working Group I to the Fifth Assessment Report of the Intergovernmental Panel on Climate Change. Cambridge University Press, Cambridge.

Ries, J.B., A.L. Cohen and D.C. McCorkle. 2009. Marine calcifiers exhibit mixed responses to CO_2 induced ocean acidification. Geology 37: 1131–1134.

Risk, M.J. and J.K. MacGeachy. 1978. Aspects of bioerosion of modern Caribbean reefs. Revta. Biol. Trop. 26: 85–105.

Risk, M.J. and R. Sluka. 2000. The Maldives: a nation of atolls. pp. 325–352. *In*: T.R. McClanahan, C.R.C. Sheppard, D.O. Obura (eds.). Coral Reefs of the Indian Ocean: their ecology and conservation. Oxford University Press, New York.

Risk, M.J., P.W. Sammarco and E.N. Edinger. 1995. Bioerosion in *Acropora* across the continental shelf of the Great Barrier Reef. Coral Reefs 14: 79–86.

Rogers, C.S. 1993. Hurricanes and coral reefs: the intermediate disturbance hypothesis. Coral Reefs 12: 127–137.

Rose, C.S. and M.J. Risk. 1985. Increase in *Cliona delitirix* infestation of *Montastrea cavernosa* heads on an organically polluted portion of the Grand Cayman fringing reef. Mar. Ecol. 6: 345–363.

Rotjan, R.D. and S.M. Lewis. 2005. Selective predation by parrotfishes on the reef coral *Porites astreoides*. Mar. Ecol. Prog. Ser. 305: 193–201.

Rotjan, R.D. and S.M. Lewis. 2006. Parrotfish abundance and selective corallivory on a Belizean coral reef. J. Exp. Mar. Biol. Ecol. 335: 292–301.

Sammarco, P.W. and M.J. Risk. 1990. Large-scale patterns in internal bioerosion of *Porites*: cross continental shelf trends on the Great Barrier Reef. Mar. Ecol. Prog. Ser. 59: 145–156.

Scoffin, T.P. 1992. Taphonomy of coral reefs: a review. Coral Reefs 11: 57–77.

Scoffin, T.P. 1993. The geological effects of hurricanes on coral reefs and the interpretation of storm deposits. Coral Reefs 12: 203–221.

Scoffin, T.P., C.W. Stearn, D. Boucher, P. Frydl, C.M. Hawkins, I.G. Hunter and J.K. Macgreathy. 1980. Calcium carbonate budget of a fringing reef on the west coast of Barbados. Part II: Erosion, sediments, internal structure. Bull. Mar. Sci. 30: 475–508.

Shamberger, K.E.F., A.L. Cohen, Y. Golguu, D.C. McCorkle, S.J. Lentz and H.C. Barkley. 2013. Diverse coral communities in naturally acidified waters of a Western Pacific reef. Geophys. Res. Lett. 41.

Silbiger, N.J. and M.J. Donahue. 2014. Secondary calcification and dissolution respond differently to future ocean conditions. Biogeosci. Discuss. 11: 12799–12831.

Smith, R.J. and Paulson, A.C. 1974. Food transit times and gut pH in two Pacific parrotfish. Copeia. 3: 769–799.

Smithers, S. and P. Larcombe. 2003. Late Holocene initiation and growth of a nearshore turbid-zone coral reef: Paluma shoals central Great Barrier Reef, Australia. Coral Reefs 22: 499–505.

Stearn, C.W., T.P. Scoffin and W. Martindale. 1977. Calcium carbonate budget of a fringing reef on the west coast of Barbados. Part 1: Zonation and productivity. Bull. Mar. Sci. 27: 479–510.

Steneck, R.S. 1983. Escalating herbivory and resulting adaptive trends in calcareous algal crusts. Paleobiology 9: 44–61.

Steneck, R.S. 1988. Herbivory on coral reefs: a synthesis. Proc. 6th Int. Coral Reef Symp. Townsville, Australia. 1: 37–49.

Taylor, P.D. 1990. Encrusters. pp. 346–351. *In*: D.E.G. Briggs and P.R. Crowther (eds.). Palaeobiology. Blackwell Scientific Publications, Boston.

Trapon, M.L., M.S. Pratchett, A.S. Hoey and A.H. Baird 2013. Influence of fish grazing and sedimentation on the early post-settlement survival of the tabular coral Acropora cytherea. Coral Reefs 32: 1051–1059.

Tribollet, A. and S. Golubic. 2005. Cross-shelf differences in the pattern and pace of bioerosion of experimental carbonate substrates exposed for 3 years on the northern Great Barrier Reef, Australia. Coral Reefs 24: 422–434.

Tribollet, A., M.J. Atkinson and C. Langdon. 2006. Effects of elevated pCO_2 on epilithic and endolithic metabolism of reef carbonates. Glob. Change Biol. 12: 2200–2208.

Tribollet, A., G. Decherf, P.A. Hutchings and M. Peyrot-Clausade. 2002. Large scale spatial variability in bioerosion of experimental coral substrates on the Great Barrier Reef (Australia): importance of microborers. Coral Reefs 21: 424–432.

Tudhope, A.W and M.J. Risk. 1985. Rate of dissolution of carbonate sediments by microboring organisms, Davies Reef, Australia. J. Sediment. Petrol. 55: 440–447.

Vogel, K., M. Gektidis, S. Golubic, W.E. Kiene and G. Radtke. 2000. Experimental studies on microbial bioerosion at Lee Stocking Island, Bahamas and One Tree Island, Great Barrier Reef, Australia: implications for paleoecological reconstructions. Lethaia 33: 190–204.

Walter, L.M. and J.W. Morse. 1984. Reactive surface area of skeletal carbonates during dissolution: effect of grain size. J. Sediment. Res. 54: 1081–1090.

Warme, J.E. 1975. Borings as trace fossils and the processes of marine bioerosion. pp. 181–227. *In*: R.W. Frey (ed.). The Study of Trace Fossils: A Synthesis of Principles, Problems, and Procedures in Ichnology. Springer Verlag, New York.

Warme, J.E. 1977. Carbonate borers – Their role in reef ecology and preservation. pp. 261–279. *In*: S.H. Frost, M.P. Weiss and J.B. Saunders (eds.). Reefs and Related Environments – Ecology and Sedimentology. The American Association of Petroleum Geologists, Oklahoma.

Wilson, S.K., D.R. Bellwood, J.H. Choat and M.J. Furnas. 2003. Detritus in the epilithic algal matrix and its use by coral reef fishes. Oceanogr. Mar. Biol. Ann. Rev. 41: 279–309.

Wisshak, M., C.H.L. Schönberg, A. Form and A. Freiwald. 2012. Ocean acidification accelerates reef bioerosion. PLoS ONE 7(9): e45124.

Woodroffe, C.D. and J.M. Webster. 2014. Coral reefs and sea-level change. Mar. Geol. 352: 248–267.

Wong, P.P., I.J. Losada, J.P. Gattuso, J. Hinkel, A. Khattabi, K.L. McInnes, Y. Saito and A. Sallenger. 2014. Coastal systems and low-lying areas. pp. 361–409. *In*: C.B. Field, V.R. Barros, D.J. Dokken, K.J. Mach, M.D. Mastandrea, T.E. Bilir, M. Chatterjee, K.L. Ebi, Y.O. Estrada, R.C. Genova, B. Girma, E.S. Kissel, A.N. Levy, S. MacCracken, P.R. Mastrandrea and L.L. White (eds.). Climate Change 2014: Impacts, Adaptation and Vulnerability, Part A: Global and Sectoral Aspects. Contribution of Working Group II to the Fifth Assessment Report of the Intergovernmental Panel on Climate Change. Cambridge University Press, Cambridge, UK, New York.

CHAPTER
9

The Good, the Bad, and the Ugly: Parrotfishes as Coral Predators

Roberta M. Bonaldo[1] and Randi D. Rotjan[2]

[1] Grupo de História Natural de, Vertebrados Museu de Zoologia, Universidade Estadual de Campinas, Campinas, SP, Brazil, 13083-863
Email: robertabonaldo@gmail.com

[2] Department of Biology, Boston University, 5 Cummington Mall Road, Boston, MA 02215 USA
Email: rrotjan@bu.edu

Introduction

Corals are some of the most ubiquitous components of tropical reef benthic communities and are the foundation species of some of these ecosystems (Jones et al. 1994, Veron 2000). From tropical rocky shores with low coral cover in marginal locations to highly diverse coral reefs in the Indo-Pacific, corals are remarkably important ecosystem components, as they provide shelter (Beukers and Jones 1997) and food (Cole et al. 2008, Rotjan and Lewis 2008) to a wide variety of marine organisms. Although corals can occupy up to 80 percent of the benthic substratum on tropical reefs (Veron 2000), very few species feed directly on live corals. Indeed, coral predation (i.e., corallivory) is one of the most specialized feeding habits of marine species, with less than five percent of fish species (i.e., 128 species from 11 families) known to feed on live corals (Cole et al. 2008, Rotjan and Lewis 2008). Apart from the butterflyfishes (Chaetodontidae), of which half of the species feed predominantly on corals, only a small fraction of other reef fish species feed frequently on live coral colonies (Cole et al. 2008, Rotjan and Lewis 2008). There are also regional scale differences in the distribution of corallivorous fishes, with more abundant and diverse assemblages in the Indo-Pacific than the tropical Atlantic, and in low latitude compared to higher latitude or marginal locations (Rotjan and Lewis 2008, Hoey et al. 2011, 2016, Pratchett et al. 2013; but see Pratchett et al. 2014).

Parrotfishes (Scarinae, Labridae) are widely regarded as herbivorous fishes because of their frequent feeding on reef surfaces covered by benthic algae (Bruggemann et al. 1994b, Bonaldo et al. 2006, 2014, Alwany et al. 2009). The majority of parrotfishes feed predominantly on substrata covered by the epilithic algal matrix (the EAM *sensu* Wilson et al. 2003) and select this substratum over others available on the reef (McAfee and Morgan 1996, Lokrantz et al. 2008, Bonaldo et al. 2014). However, the robust, strong oral and pharyngeal jaws of the parrotfishes makes them one of the most versatile fish groups in tropical reefs, as they are able to feed on almost all reef surfaces, including live coral colonies (Bellwood and Choat 1990, Rotjan and Lewis 2008, Alwany et al. 2009). Indeed, the first evidence of vertebrate corallivory was provided by Darwin (1842), who recovered live corals from the stomachs of two *Scarus* species in the Indian Ocean.

Corallivory by parrotfishes has been observed in a number of species at a range of locations worldwide, with a diversity of coral taxa being fed upon. Coral predation, nevertheless, is not uniform among the parrotfishes. Most corallivorous parrotfishes are large bodied, typically excavating, species that bite deeper into the coral skeleton (e.g. Bruggemann et al. 1994b, Alwany et al. 2009, Bonaldo and Bellwood 2009; Fig. 1). Although corallivory by parrotfishes may be infrequent (Bonaldo et al. 2014), it has been shown to affect the growth (Frydl 1979) and morphology of coral colonies (Bonaldo et al. 2012, Welsh et al. 2015), the distribution of adult colonies of preferred prey species (Bonaldo and Bellwood 2011), and the survivorship of coral recruits (Trapon et al. 2013a). At the colony level, corallivory can negatively impact coral energy budgets, as colonies have limited energy stores that are allocated among growth, regeneration, and reproduction (Bak 1983, Harrison and Wallace 1990). Partial mortality of coral colonies by parrotfish grazing is a common and direct effect of the removal of live coral tissue and portions of the underlying skeleton (e.g. Cole et al. 2008, Rotjan and Lewis 2008, Bonaldo et al. 2011, Welsh et al.

Fig. 1. Some examples of parrotfishes feeding on live corals: the Spotted parrotfish, *Cetoscarus ocellatus* (A), and the Steephead parrotfish, *Chlorurus microrhinos* (B), from the Indo-Pacific, the Stoplight parrotfish, *Sparisoma viride* (C), from the Caribbean, and the Reef parrotfish, *Sparisoma amplum* (D), from Brazil. Photos: R.M. Bonaldo (A) and J.P. Krajewski (B-D).

2015). There are also records of total colony mortality as a result of intense parrotfish coral predation, but such events appear to be rare (Bruckner and Bruckner 2000, Bruckner et al. 2000). Together with partial colony mortality, energetically expensive reproduction is often compromised in favour of tissue regeneration processes in damaged colonies (Szmant-Froelich 1985, Rinkevich and Loya 1989, Van Veghel and Bak 1994).

In this chapter, we review parrotfish corallivory, with special focus to: (1) identify the main characteristics, drivers and consequences of coral predation by parrotfishes; (2) the likely implications of anthropogenic disturbances and reef degradation to parrotfishes, and to their roles as coral predators; and (3) the identification of current knowledge gaps in parrotfish corallivory and areas for future research.

Modes of Parrotfish Corallivory

Parrotfishes are among the most important corallivores within reef fish assemblages because of their unique feeding mode and hence impact on coral colonies. When feeding on corals, parrotfishes are able to remove large areas of live coral tissue and, in many cases, pieces of the underlying coral skeleton (Bruckner and Bruckner 2000, Hoey and Bellwood 2008, Bonaldo et al. 2011, Welsh et al. 2015). However, the impact caused by parrotfish corallivory is not uniform, as the feeding rate, bite area and depth, and thus the total amount of coral tissue removed, largely depends on the parrotfish species, life phase, and body size (Bellwood and Choat 1990, Lokrantz et al. 2008, Bonaldo et al. 2014, Hoey Chapter 6).

Corallivorous parrotfishes are usually classified into two main groups, considering their feeding behaviour, morphology, and impact on the feeding substratum: scrapers and excavators (Bellwood and Choat 1990, Bonaldo et al. 2014). Scrapers feed by scraping the surface and usually leave superficial marks on the substratum (Fig. 2a, c). In contrast, excavators possess stronger jaws and oral musculature, used for biting deeper into the substratum, and usually expose the underlying matrix when feeding on benthic surfaces (Bonaldo and Bellwood 2009, Bonaldo et al. 2014; Fig. 2b, d). Feeding marks of excavating species are approximately twice as long as they are wide, with four or more deep grooves running parallel to the major axis, while feeding marks of scraping parrotfishes usually consist in two parallel feeding marks that are often 10 times longer than they are wide (Bellwood and Choat 1990, Hoey and Bellwood 2008; Fig. 2a, b).

Most corallivorous parrotfishes are large excavators, such as *Bolbometopon*, *Cetoscarus* and *Chlorurus* in the Indo-Pacific (Bellwood et al. 2003, Hoey and Bellwood 2008) and Red Sea (Alwany et al. 2009), and *Sparisoma viride* and *Sp. amplum* in the Atlantic (Rotjan and Lewis 2006, Francini-Filho et al. 2008, Bonaldo et al. 2014). There are also records of large *Scarus* individuals feeding on corals, but the frequency of corallivory is generally lower than that of excavating species (e.g. Bruggemann et al. 1994b, McAfee and Morgan 1996, Francini-Filho et al. 2008). Indeed, body size seems an important pre-requisite for parrotfish corallivory, as even excavating species only start feed on live corals once individuals attain certain size; juveniles and small adults generally do not feed on live corals (Bruggemann et al. 1994b, 1996, Lokrantz et al. 2008).

The damage caused by parrotfish corallivory, as well as the size and healing time of the wounds on coral colonies, largely depends on parrotfish feeding mode. Bites from scraping parrotfishes usually heal in few weeks on 'healthy' coral colonies (e.g. Bruckner and Bruckner 2000, Bonaldo et al. 2012). On the other hand, the impact of bites from excavating species tend to be longer-lasting, not only because these marks are deeper, but also as excavating parrotfishes usually deliver multiple clustered bites, thus damaging larger

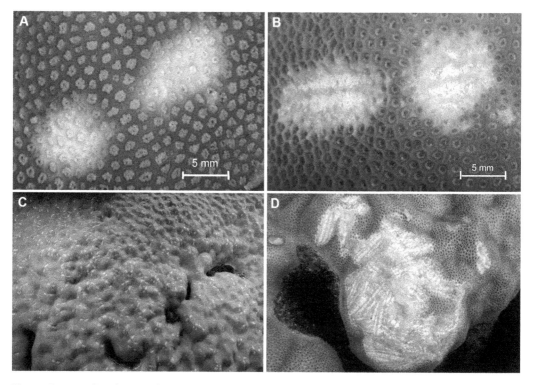

Fig. 2. Bite marks of parrotfishes on live coral colonies (massive *Porites* sp.): individual marks of scraping and excavating parrotfishes (A and B, respectively) and multiple marks of scraping and excavating parrotfishes (C and D, respectively). Photos: J.P. Krajewski.

areas on coral colonies ("focused biting" *sensu* Bruckner and Bruckner 2000). In some cases, the larger wounds caused by excavators may not recover at all, causing partial (Welsh et al. 2015) or total colony mortality (Bruckner and Bruckner 1998, 2000). On an inshore reef on the Great Barrier Reef, for example, individual bite marks of the steephead parrotfish, *Chlorurus microrhinos*, on massive *Porites* corals completely recovered within four months, whereas the healing probability of wounds composed of two or more clustered bites was below 50 percent (Welsh et al. 2015).

Only a small portion (less than one percent) of parrotfish bites is taken on live corals (McAfee and Morgan 1996, Francini-Filho et al. 2008, Bonaldo et al. 2014). Unlike the corallivorous butterflyfishes, there are no obligate corallivorous parrotfishes (i.e. species with more of 80 percent of their diet centered on live corals, *sensu* Cole et al. 2008) and parrotfishes that feed on live corals are, at most, facultative corallivores (Cole et al. 2008, Rotjan and Lewis 2008). However, a few species may take up to 50 percent of their bites on corals (Bellwood et al. 2003, Hoey and Bellwood 2008, Lokrantz et al. 2008, Alwany et al. 2009). The largest species in the group, the Bumphead parrotfish (*Bolbometopon muricatum*), feeds non-selectively, biting algal covered substrata and live corals in proportion to their availability (Hoey and Bellwood 2008). Indeed, *B. muricatum* is the only parrotfish that does not prefer algal covered surfaces to live corals for feeding (Bellwood et al. 2003, Hoey and Bellwood 2008). *B. muricatum* has, by far, the largest bite size among all parrotfishes with the bite volume of adult individuals being estimated to be 1.66 cm^3, corresponding to

3-10 fold the volume of other large excavators, such as *Ch. microrhinos* (0.56 cm^3; Bonaldo and Bellwood 2011), *Cetoscarus bicolor* and *Chlorurus gibbus* (0.11 and 0.11 cm^3, respectively, Alwany et al. 2009). Because of its large bite size and high proportion live corals in its diet, *B. muricatum* may dominate rates of coral predation and bioerosion, even in reefs where it is not the most abundant parrotfish species (Hoey and Bellwood 2008, Bellwood et al. 2012).

In addition to *B. muricatum*, other large excavting parrotfishes such as *Ce. bicolor* (to 80 cm TL) and *Ch. gibbus* (to 70 cm TL) in the Red Sea (Alwany et al. 2009), and *Chlorurus strongylocephalus* (to 70 cm TL) in Zanzibar (Lokrantz et al. 2008), may deliver 20 to 30 percent of their bites on live coral colonies (Table 1). However, the proportion of bites that is taken from live corals appears to vary among species and locations. On the Great Barrier Reef, *Cetoscarus ocellatus* (to 80 cm TL), the sister taxon of *Ce. bicolor,* and *Ch. microrhinos* (to 70 cm TL), sister taxon of both *Ch. gibbus* and *Ch. strongylocephalus*, have been reported to take fewer than 10 percent of their bites from live coral colonies (Bellwood and Choat 1990, Bonaldo and Bellwood 2008). In the Atlantic, the main excavating and corallivorous parrotfishes, *Sp. amplum* and *Sp. viride*, take fewer than five and two percent of their bites on live corals, respectively. Other large-bodied parrotfishes in the Atlantic have either been reported to take very few bites (less than one percent) on live corals (Francini-Filho et al. 2008, Bruggemann et al. 1994b), or not consume live coral at all (Bruggemann et al. 1994b, 1996, Rotjan and Lewis 2005; Table 1). These differences may be associated to local variations in parrotfish feeding preferences, temporal and local variations in the availability of preferred coral types (Bonaldo et al. 2012), or in the nutritional quality of alternate food types (Rotjan and Lewis 2005).

Global Perspectives on Parrotfish Corallivory

In addition to differences in the feeding modes associated to coral predation within the parrotfishes, the taxonomic composition of corallivorous parrotfishes varies among biogeographical regions. The most pronounced division in parrotfish assemblages is between the Indo-Pacific and the Atlantic, with the Indo-Pacific dominated by species from *Scarus* clade and the Atlantic by *Sparisoma* clade. Further, these regions do not share any parrotfish species and have limited overlap in genera (Parenti and Randall 2000, 2011, Streelman et al. 2002, Bonaldo et al. 2014). Excavating parrotfishes are largely restricted to the Indo-Pacific with the three main excavating genera (*Bolbometopon, Cetoscarus* and *Chlorurus*) and 95 percent of all excavating parrotfish species being exclusive to this region. Only two excavating species, *Sp. viride* and its sister taxon *Sp. amplum*, are present in the Atlantic (Bruggemann et al. 1994b, 1996, Francini-Filho et al. 2008).

Preferred coral groups may also vary between the Atlantic and the Indo-Pacific. In the Atlantic, parrotfishes feed preferentially on *Orbicella* (formerly *Montastrea* in this region) and *Porites* (Bruggemann et al. 1996, Rotjan and Lewis 2005, Francini-Filho et al. 2008), while in the Indo-Pacific parrotfishes feed mostly on *Porites* (Beissinger 1997, Bonaldo and Bellwood 2009, 2011), with exception of *B. muricatum*, which also targets *Acropora, Pocillopora* and *Montipora* colonies (Table 2) (Bellwood and Choat 1990, Bellwood et al. 2003, Hoey and Bellwood 2008). Recent evidence, however, suggests that corallivory may be context-dependent, whereby parrotfishes feed on the most preferred species when available, but will feed on other species (in order of preference) when needed (Rotjan and Lewis 2006, Burkepile 2012).

Table 1. Summary of dominant feeding substrata of parrotfishes on tropical reefs, presented as either the mean or range of bites (%) on each substratum type. Numbers in parenthesis are standard errors (extracted from Bonaldo et al. 2014)

Species*	Location	Proportion of bites (%)					
		EAM	Macroalgae	Live coral	CCA	Seagrass	Reference
Excavators							
Bolbometopon muricatum	Northern GBR, Australia	37.11	0	48.17	14.71	0	Hoey and Bellwood 2008
Chlorurus perspicillatus	Oahu, Hawaii	61	–	1	38	–	Ong and Holland 2010
Ch. sordidus	Zanzibar Island, Tanzania	91.9–100	0–2.7	0–6.3	0–13.3	–	Lokrantz et al. 2008
Ch. strongylocephalus	Zanzibar Island, Tanzania	80 – 100	0–10	0–26.7	0–10.5	–	Lokrantz et al. 2008
Sparisoma amplum	Abrolhos, Brazil	48.77	0.89 (0.9)	4.33 (1.4)	35.82 (3.8)	0.53	Bruckner and Bruckner 1998
	F Noronha, Brazil	60–65	7	0	0	0	Bonaldo et al. 2006
Sp. viride	Karpata, Bonaire	79–97	0	0–5.1	–	–	Bruggemann et al. 1994b
	San Blas, Panama	91	5	–	–	5	McAfee and Morgan 1996
Scrapers/browsers							
Scarus ferrugineus	Sheikh Said Island, Red Sea, Eritrea	92–97	–	2–8	0	0	Afeworki et al. 2011
Sc. iseri	San Blas, Panama	8.5	23	–	–	–	McAfee and Morgan 1996
Sc. niger	Zanzibar, Tanzania	92.3–100	0	0–9.1	0–7.7	–	Lokrantz et al. 2008
Sc. rivulatus	Orpheus Island, Australia	97.17	0	2.82	0	0	Bonaldo and Bellwood 2008
Sc. rubroviolaceus	Oahu, Hawaii	55	–	1	40	–	Ong and Holland 2010
Sc. trispinosus	Abrolhos, Brazil	50.99 (3.9)	4.32 (1.5)	1.3 (0.4)	34.74 (3.7)	7.15 (1.8)	Francini-Filho et al. 2010

(Contd.)

Sc. vetula	Karpata, Bonaire	67–99.8	0	0 – 0.2	–	–	Bruggemann et al. 1994b
Sc. zelindae	Abrolhos, Brazil	51.65 (3.8)	2.60 (1.4)	2.73 (1.2)	33.0 (3.7)	5.9 (1.5)	Francini-Filho et al. 2010
Sparisoma aurofrenatum	San Blas, Panama	26	23	–	–	23	McAfee and Morgan 1996
Sp. axillare	Abrolhos, Brazil	64.39 (3.3)	6.2 (1.7)	1.24 (0.5)	13.88 (1.9)	13.17 (2.64)	Francini-Filho et al. 2010
	F Noronha, Brazil	45–55	60–65	0	0	0	Bonaldo et al. 2006
Sp. chrysopterum	San Blas, Panama	11	11	–	–	43	McAfee and Morgan 1996
Sp. frondosum	Abrolhos, Brazil	53.54 (3.9)	5.7 (1.6)	1.98 (0.9)	16.98 (2.7)	18.79 (3.2)	Francini-Filho et al. 2010
	F Noronha, Brazil	49–55	33–36	0	0	0	Bonaldo et al. 2006
Sp. rubripinne	San Blas, Panama	17	14	–	–	23	McAfee and Morgan 1996

*Where necessary, species names used in original articles have been replaced by the currently valid name (Froese and Pauly 2014).

Table 2. Parrotfishes known to feed on live coral colonies and corals targeted by each species (adapted from Bonaldo et al. 2014)

Parrotfishes*	Location	Corals*	Source
Excavators			
Bolbometopon muricatum	Lizard Island, Australia	*Acropora cytherea, A. divaricata, A. hyacinthus, A. latistella, A. listeria, A. millepora, A. nasuta, A. secale, Anamastrea sp., Favites sp., Porites cylindrica*	Bellwood 1985
	Lizard Island, Australia	*Acropora* spp., massive *Porites* spp.	Bonaldo and Bellwood 2011
	Indo-Pacific	*Acropora* spp., *Montipora* spp., *Pocillopora verrucosa*	Bellwood et al. 2003, Hoey and Bellwood 2008
Cetoscarus bicolor	South Sinai, Red Sea, Egypt	*Porites* spp.	Alwany et al. 2009
Ce. ocellatus	GBR, Australia	*Montipora* sp., *Platygyra* spp., encrusting species	Bellwood 1985
Chlorurus gibbus	South Sinai, Red Sea, Egypt	Not specified	Alwany et al. 2009
Ch. microrhinos	GBR, Australia	Massive *Porites* spp.	Bonaldo et al. 2011, Bonaldo et al. 2012
Ch. perspicillatus	Oahu, Hawaii	Not specified	Ong and Holland 2010
Ch. spilurus	GBR, Australia	*Porites australiensis, Porites lobata, Porites lutea*	Bellwood 1985
Ch. strongylocephalus	Zanzibar, Tanzania	Not specified	Lokrantz et al. 2008
Sparisoma amplum	Abrolhos, Brazil	*Mussismilia braziliensis, Montastrea cavernosa, Siderastrea* spp.	Francini-Filho et al. 2008
Sp. viride	Karpata, Bonaire	Not specified	Bruggemann et al. 1994c, Bruggemann et al. 1994a
	Carrie Bow Cay, Belize	*Orbicella annularis, O. cavernosa, O. franksi, O. faveolata, Siderastrea siderastrea, Agaricia agaricites, Porites porites, Porites astroides, D. strigosa*	Rotjan and Lewis 2006
	Belize	*Porites porites*	Littler et al. 1989
	Carrie Bow Cay, Belize	*Porites astroides*	Rotjan and Lewis 2005

(Contd.)

Parrotfishes*	Location	Corals*	Source
Sp. viride (cont.)	Florida Keys	*Porites divaricata*	Miller and Hay 1998
	Rosario Islands, Colombia	*Orbicella annularis*	Sánchez et al. 2004
	Puerto Rico	*Colpophyllia natans*	Bruckner and Bruckner 1998
	St Croix	*Orbicella annularis*	Bruckner et al. 2000
Scrapers or browsers			
Scarus ferrugineus	South Sinai, Red Sea, Egypt	*Porites* spp.	Alwany et al. 2009
Sc. flavipectoralis	Lizard Island, Australia	Massive *Porites* spp.	Bonaldo and Bellwood 2011
Sc. frenatus	South Sinai, Red Sea, Egypt	Faviidae	Alwany et al. 2009, Bonaldo and Bellwood 2011
	Lizard Island, Australia	–	
	Heron Island, Australia	*Porites* spp.	Bellwood and Choat 1990
Sc. ghobban	South Sinai, Red Sea, Egypt	*Porites* spp.	
Sc. coelestinus	Caribbean	–	Randall 1967, 1974
Sc. guacamaia	Caribbean	–	Glynn 1997
Sc. niger	Lizard Island, Australia	Massive *Porites* spp.	Bonaldo and Bellwood 2011
	Zanzibar, Tanzania	–	Lokrantz et al. 2008
Sc. rivulatus	Lizard Island, Australia	*Porites* spp.	Bellwood 1985
Sc. rubroviolaceus	Oahu, Hawaii	–	Ong and Holland 2010
Sc. trispinosus	Abrolhos, Brazil	*Montastrea cavernosa, Mussismilia braziliensis, Mussismilia hartii, Favia gravida, Porites astreoides, Siderastrea* spp.	Francini-Filho et al. 2008
Sparisoma aurofrenatum	Florida Keys	*Porites divaricata*	Miller and Hay 1998

*Where necessary, species names used in original articles have been replaced by the currently valid name.

Targets of Corallivory

The nutritional importance of live corals as a food source for parrotfishes remains largely unknown. Most studies on parrotfish corallivory are focused either on patterns of parrotfish feeding behaviour and activity (e.g. McAfee and Morgan 1996, Bonaldo et al. 2006) or on the consequences of parrotfish feeding to the benthic community (e.g. Hoey and Bellwood 2008, Bonaldo et al. 2011). Studies that have attempted to distinguish the causes of corallivory from their consequences have not found a single explanation for why certain parrotfishes consume particular coral species, despite clear patterns of grazing (Rotjan and Dimond 2010).

The increased incidence of corallivory in large excavating parrotfishes (Bonaldo et al. 2014) suggests these species may use live corals more efficiently as a feeding substratum than smaller excavating and scraping species. The deeper bites of excavating species, especially larger individuals, may allow these species to access nutritional resources that are unavailable to species with shallower bites. In the Caribbean, for instance, the spotlight parrotfish *Sp. viride* preferentially feeds on *Orbicella annularis* colonies with more mature gonads, presumably because of their higher lipid content (Rotjan and Lewis 2009). Furthermore, on an inshore reef on the Great Barrier Reef, *Ch. micrhorhinos* inflicted more bites on massive *Porites* colonies during the predicted period of coral spawning than other times (Bonaldo et al. 2012).

Inadvertent and Facilitated Corallivory

Corallivory is the consumption of live corals, but most descriptions of corallivory in the literature are limited to consumption of adult coral tissue (e.g. Bruckner and Bruckner 2000, Bonaldo and Bellwood 2011). Yet, recently settled coral larvae and juvenile corals are also often consumed, as evidenced by numerous reports of intense grazing on coral spat and recruits (e.g. Penin et al. 2010, Korzen et al. 2011, Trapon et al. 2013a), juvenile corals (e.g. Lenihan et al. 2011), and small transplanted nubbins (e.g. Neudecker 1979, Littler et al. 1989, Grottoli-Everett and Wellington 1997, Miller and Hay 1998). While some of these reported grazing events may have been targeted efforts by parrotfishes, some may also have been inadvertent grazing by parrotfishes in targeting the EAM, or even by other herbivores with no reported corallivory component in their diet (Korzen et al. 2011). Many purported herbivores also likely and regularly consume a vast and poorly characterized (in terms of consumption rates) suite of faunal epibionts that cohabitate with preferred algal species (e.g., Kramer et al. 2013). Irrespective of whether it is targeted or inadvertent corallivory, extensive feeding by parrotfishes may influence the mortality and growth of early-life stages of corals, and hence the replenishment of coral populations.

Inadvertent corallivory may also occur on adult colonies, as there have been reports of parrotfishes feeding on macroborers (e.g., polychaetes, molluscs, or barnacles) within live corals (e.g. Frydl 1979, Rotjan and Lewis 2005). However, given the high reported rates of the same parrotfish species feeding on the same coral species in the absence of macroborers, it is likely that macroborer consumption may be incidental to corallivory, rather than vice-versa.

Impacts of Corallivory

The removal of live coral tissue is energetically costly for the coral itself and can cause total or partial mortality of coral colonies. Partial colony mortality is common in larger colonies and is a consequence of the damaged area failing to regenerate (e.g. Cole et al. 2008, Rotjan and Lewis 2008, Bonaldo et al. 2011, Welsh et al. 2015). Though impacts have been noted for single wounds, considerably greater impacts on colony health would be expected as a consequence of multiple bites, especially when inflicted in close proximity (Henry and Hart 2005, Welsh et al. 2015). Parrotfishes often inflict several bites in close proximity on the surface of corals, either as multiple bites in a single feeding bout or multiple feeding bouts (Bruckner and Brucker 2000, Welsh et al. 2015). The sublethal consequences of parrotfish feeding can be significant as energy is allocated away from growth and reproduction and toward regeneration of damaged tissues (Meesters et al. 1994, Henry and Hart 2005). Alternatively, maintaining colony growth may impair regeneration, which implies discrete trade-offs in energy allocation (Denis et al. 2013).

At the colony level, intensive corallivory can also negatively impact coral reproductive potential, as energy is diverted away from reproduction in favour of tissue regeneration processes (Szmant-Froelich 1985, Rinkevich and Loya 1989, Harrison and Wallace 1990, Van Veghel and Bak 1994). Van Veghel and Bak (1994) found that colonies of *Orbicella* inflicted with artificial lesions ~10 wk prior to spawning had fewer gonads per polyp, and fewer eggs per gonad in polyps surrounding the lesion, compared to polyps 20 cm away from the lesion area, and recently regenerated polyps had no reproductive activity.

The recovery of coral colonies to parrotfish corallivory appears to be related to their overall health, and energy availability. Generally, the neighbouring polyps bear the energetic costs of regeneration (Meesters et al. 1994, Oren et al. 1997). Thus, although larger colonies presumably have greater energy stores, the likelihood of regenerating damaged areas is approximately equal among differently sized colonies. Comparisons of intact versus healing tissues in *Montipora capitata* revealed that, in contrast to the gross morphology of wound repair, the two tissue types were histologically indistinguishable within two weeks of wound infliction (Work and Aeby 2010). Initial lesion size plays an important role in recovery potential, which is the likelihood that a coral will fully regenerate following tissue damage. For example, *Porites astreoides* in the Caribbean successfully repairs one cm² lesions approximately 50 percent of the time, whereas it does not completely re-grow tissue over five cm² lesions, either artificially-induced, or from parrotfish grazing (Bak and Steward-Van Es 1980, Rotjan and Lewis 2005). Similarly, other coral species recover more quickly or fully from smaller compared to larger lesions (Bak and Steward-Van Es 1980, Lester and Bak 1985, Oren et al. 1997, Croquer et al. 2002, Welsh et al. 2015).

Corals can completely recover from small feeding marks (e.g. Sánchez et al. 2004), and it is common to observe coral colonies in various stages of recovery from parrotfish corallivory (Fig. 3). However, recovery from corallivory is far from guaranteed (e.g. Rotjan and Lewis 2005, Bonaldo et al. 2011, Denis et al. 2011). Tissue loss with accompanying skeletal damage has lower regenerative success than tissue damage alone (Bak and Steward-Van Es 1980, Bak 1983, Croquer et al. 2002, Bonaldo et al. 2012). Studies mimicking the lesion scars of excavating parrotfish also found that lesion recovery was less than 20 percent (Denis et al. 2011) and is dependent on abiotic conditions, especially temperature (Edmunds and Lenihan 2010; reviewed by Rotjan and Lewis 2008). Because many corallivores repeatedly graze multiple sites on individual colonies (Rotjan and Dimond 2010), coral regeneration capabilities estimated from lesions inflicted at a single time point provide a best-case scenario for tissue regeneration. Many coral species have been observed to recover from repeated bouts of grazing with multiple lesions (e.g. Bruckner and Bruckner 2000, Welsh et al. 2015), but it remains unknown whether recovery potential is diminished over time or with increased grazing intensity.

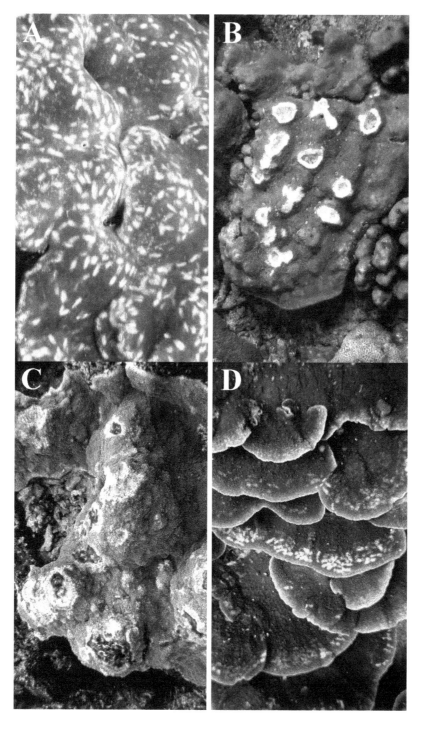

Fig. 3. Bite marks of parrotfishes on live coral colonies demonstrating repeated grazing where fresh scars are simultaneously present with older scars. Images depict various stages of tissue regeneration from fresh scars (A, B, C, D), to recovering tissue over older scars (A, D) to centrally-colonized by turf (B, C) or crustose coralline algae (B). Photos: R. Rotjan

Few studies have assessed the effects of parrotfish corallivory on coral populations. Overall, these studies indicate that, in areas of high parrotfish abundance, the distribution of coral colonies may be influenced by the intensive grazing by these fishes (e.g., Littler et al. 1989, Miller and Hay 1998, Rotjan and Lewis 2008, Rotjan & Dimond 2010, Bonaldo and Bellwood 2011). In the Caribbean, for example, parrotfish corallivory has been considered as one of the main determinants of the distribution of *P. astreoides* and *Porites porites furcata* (Little et al. 1989, Miller and Hay 1998). Likewise, on Lizard Island, a midshelf reef on the Great Barrier Reef, the abundance of massive *Porites* colonies has been negatively correlated with the abundance of parrotfishes in different reef zones (Bonaldo and Bellwood 2011). Indeed, in areas where parrotfish densities were high, massive *Porites* colonies were rare and small, and it was estimated that the surface area of *Porites* colonies was completely grazed every 270 d (Bonaldo and Bellwood 2011).

B. muricatum, the largest parrotfish, deserves special attention regarding parrotfish impacts to corals. Given its large bite size and high proportion of bites on live corals, *B. muricatum* has been shown to overwhelmingly dominate rates of coral predation and bioerosion in Indo-Pacific coral reefs, even in areas where it is not among the most abundant parrotfish species (Bellwood et al. 2003, 2012, Hoey and Bellwood 2008). While most corallivorous parrotfishes in the Indo-Pacific feed preferentially on *Porites* and other massive and submassive corals, *B. muricatum* feeds mostly on *Acropora* spp. (Table 2). Massive *Porites* and *Acropora* species have markedly different growth and regeneration rates (Veron 2000) and, as consequence, impacts of corallivory by *B. muricatum* and other parrotfishes on coral assemblages are likely to differ. However, to date no studies have assessed the dynamics of *B. muricatum* bite marks and little is known on the direct effects of corallivory by this species on the growth, survivorship, and reproduction of impacted coral colonies.

The effects of parrotfish grazing on recently settled and juvenile corals are complex. Studies have reported the density of juvenile corals to be positively (Hughes et al. 2007, Hoey et al. 2011, Penin et al. 2011) or negatively (Trapon et al. 2013a) related to the density of juvenile corals. Parrotfish grazing may facilitate the settlement of corals by clearing reef surfaces from benthic algae and sediments, but intense grazing by parrotfishes may contribute to the mortality of small corals through incidental predation. Therefore, at high parrotfish densities, the positive effects of parrotfish grazing in clearing the benthos may be countered by the increased mortality of coral recruits (Trapon et al. 2013b, c), while at low densities the lack of cleared substrata may inhibit settlement. Collectively this suggests that intermediate levels of parrotfish biomass may be optimal for the settlement, growth and survival of coral early life stages.

Corallivory on Changing Reefs

Overall, the effects of parrotfishes on corals are complex. Parrotfishes may enhance coral settlement by removing competitively dominant macroalgae and clearing space for settlement of coral larvae, but may simultaneously increase mortality of juveniles or new recruits via inadvertent or targeted corallivory. Parrotfishes may change or compromise the growth rates, morphology and survival of certain coral species, yet also may help maintain overall coral diversity by removing competitively dominant species or by creating space for new recruitment in a high-cover reef. Regardless of the exact nature of the interactions, parrotfishes undoubtedly influence coral demography on intact reefs.

On damaged reefs, which are increasingly the norm (e.g., Hughes et al. 2017), parrotfish-coral interactions are likely to become less predictable. Specifically, the availability and nutritional quality of preferred coral prey, as well as other prey items, are likely to change with reef condition (Rotjan and Lewis 2005, 2006, Burkepile 2012), and may in-turn influence movement and grazing patterns (Nash et al. 2016). Thus, the feeding ecology of a particular species may be shaped by local conditions rather than their optimal prey preferences.

There is an understandable and effective conservation strategy that advocates for increasing reef resilience by bolstering herbivorous populations (e.g. Mumby 2006, Mumby et al. 2006), but the unintended consequence may be, in some cases, a simultaneous increase in corallivory. In regions with low coral cover, low coral diversity, or low coral recruitment, each bout of corallivory has the potential to impact the overall coral community. In contrast, large parrotfishes have become increasingly rare in many regions as a result of overfishing (Bellwood et al. 2003, 2012, Heenan et al. 2016). In this sense, parrotfish assemblages in degraded reefs are usually dominated by small, fast-growing species, which rarely feed on coral colonies. Therefore, many ecological roles played by intact parrotfish assemblages, especially bioerosion and corallivory, are not fulfilled by the remaining species in overfished reefs (Bellwood et al. 2003, 2012). These modified coral reefs thus represent major challenges to our understanding of parrotfish corallivory, given the lower predictability of parrotfish impact patterns on corals and the potential risk of increased unanticipated changes and reductions in goods and services provided by coral reefs (cf. Bellwood et al. 2004).

In essence, much of the fundamental information needed to accurately predict the patterns and effects of coral predation is still lacking, but could be much improved with studies of nutritional ecology of parrotfish-coral interactions, chronic versus acute corallivory, and the potential for compensatory grazing. Furthermore, little is known on the specific feeding impact of individual corallivorous parrotfish species and the selectivity of corals as a foraging substratum. Once these fundamentals are known, we will be able to glean insight into the impact of corallivory on healthy reefs, as well as superimposed on a background of climate change and reef degradation.

Acknowledgments

We thank A.S. Hoey and two anonymous reviewers for valuable suggestions and J.P. Krajewski for the use of photographs.

References Cited

Afeworki, Y.A., J.H. Bruggemann and J.J. Videler. 2011. Limited flexibility in resource use in a coral reef grazer foraging on seasonally changing algal communities. Coral Reefs 30: 109–122.

Alwany, M.A., E. Thaler and M. Stachowitsch. 2009. Parrotfish bioerosion on Egyptian Red Sea reefs. J. Exp. Mar. Biol. Ecol. 371: 170–176.

Bak, R.P.M. 1983. Neoplasia, regeneration and growth in the reef-building coral *Acropora palmata*. Mar. Biol. 77: 221–227.

Bak, R.P.M. and Y. Steward-Van Es. 1980. Regeneration of superficial damage in the scleractinian corals *Agaricia agaricites f. purpurea* and *Porites astreoides*. Bull. Mar. Sci. 30: 883–887.

Beissinger, S.R. 1997. Integrating behavior and conservation biology: potentials and limitations. pp. 23–47. *In*: J.R. Clemmons and R. Buchholz (eds.). Behavioral Approaches to Conservation in the Wild. Cambridge University Press, Cambridge, UK.

Bellwood, D.R. 1985. The functional morphology, systematic and behavioural ecology of parrotfishes (Family Scaridae). PhD. Thesis, James Cook University, Australia.

Bellwood, D.R. and J.H. Choat. 1990. A functional analysis of grazing in parrotfishes (family Scaridae): the ecological implications. Environ. Biol. Fish. 28: 189–214.

Bellwood, D.R., A.S. Hoey and J.H. Choat. 2003. Limited functional redundancy in high diversity systems: resilience and ecosystem function on coral reefs. Ecol. Lett. 6: 281–285.

Bellwood, D.R., A.S. Hoey and T.P. Hughes. 2012. Human activity selectively impacts the ecosystem roles of parrotfishes on coral reefs. Proc. R. Soc. Lond. B. Bio. 279: 1621–1629.

Bellwood, D.R., T.P. Hughes, C. Folke and M. Nyström. 2004. Confronting the coral reef crisis: supporting biodiversity, functional groups and resilience. Nature 289: 827–833.

Beukers, J.S. and G.P. Jones. 1997. Habitat complexity modifies the impact of piscivores on a coral reef fish population. Oecologia 114: 50–59.

Bonaldo, R.M., J.P. Krajewski, C. Sazima and I. Sazima. 2006. Foraging activity and resource use by three parrotfish species at Fernando de Noronha Archipelago, tropical West Atlantic. Mar. Biol. 149: 423–433.

Bonaldo, R.M. and D.R. Bellwood. 2008. Size-dependent variation in the functional role of the parrotfish *Scarus rivulatus*. Mar. Ecol. Prog. Ser. 360: 237–244.

Bonaldo, R.M. and D.R. Bellwood. 2009. Dynamics of parrotfish grazing scars. Mar. Biol. 156: 771–777.

Bonaldo, R.M. and D.R. Bellwood. 2011. Parrotfish predation on massive *Porites* on the Great Barrier Reef. Coral Reefs 30: 259–269.

Bonaldo, R.M., J.P. Krajewski and D.R. Bellwood. 2011. Relative impact of parrotfish grazing scars on massive *Porites* on Lizard Island, Great Barrier Reef. Mar. Ecol. Progr. Ser. 423: 223–233.

Bonaldo, R.M., J.Q. Welsh and D.R. Bellwood. 2012. Spatial and temporal variation in coral predation by parrotfishes on the GBR: evidence from an inshore reef. Coral Reefs 31: 263–272.

Bonaldo, R.M., A.S. Hoey and D.R. Bellwood. 2014. The ecosystem roles of parrotfishes on tropical reefs. Oceanogr. Mar. Biol. 52: 81-132.

Bruckner, A.W. and R.J. Bruckner. 1998. Destruction of coral by *Sparisoma viride*. Coral Reefs 17: 350.

Bruckner, A.W. and R.J. Bruckner. 2000. Coral predation by *Sparisoma viride* and lack of relationship with coral disease. Proc. Ninth Int. Coral Reef Symp. Bali. 2: 23–27.

Bruckner, A.W., R.J. Bruckner and P. Sollins. 2000. Parrotfish predation on live coral: 'spot biting' and 'focused biting'. Coral Reefs 19: 50–50.

Bruggemann, J.H., M.J.H. van Oppen and A.M. Breeman. 1994a. Foraging by the stoplight parrotfish *Sparisoma viride* I. Food selection in different, socially determined habitats. Mar. Ecol. Prog. Ser. 106: 41–55.

Bruggemann, J.H., M.W.M. Kuyper and A.M. Breeman. 1994b. Comparative analysis of foraging and habitat use by the sympatric Caribbean parrotfish *Scarus vetula* and *Sparisoma viride*. Mar. Ecol. Prog. Ser. 112: 51–66.

Bruggemann, J.H., J. Begeman, E.M. Bosma, P. Verburg and A.M. Breeman. 1994c. Foraging by the stoplight parrotfish *Sparisoma viride* II. Intake and assimilation of food, protein and energy. Mar. Ecol. Prog. Ser. 106: 57–71.

Bruggemann, J.H., A.M. van Kessel, J.M. van Rooij and A.M. Breeman. 1996. Bioerosion and sediment ingestion by the Caribbean parrotfish *Scarus vetula* and *Sparisoma viride*: implications of fish size, feeding mode and habitat use. Mar. Ecol. Prog. Ser. 134: 59–71.

Burkepile, D.E. 2012. Context-dependent corallivory by parrotfishes in a Caribbean reef ecosystem. Coral Reefs 31: 111–120.

Cole, A.J., M.S. Pratchett and G.P. Jones. 2008. Diversity and functional importance of coral-feeding fishes on tropical coral reefs. Fish Fish. 9: 286–307.

Croquer, A., E. Villamizar and N. Noriega. 2002. Environmental factors affecting tissue regeneration of the reef-building coral *Montastrea annularis* (Faviidae) at Los Roques National Park, Venezuela. Rev. Biol. Trop. 50: 1055–1065.

Darwin, C.R. 1842. The structure and distribution of coral reefs. Smith, Elder & Co, London.

Denis, V., J. Debreuil, S. De Palmas, J. Richard, M.M. Guillaume and J.H. Bruggemann. 2011. Lesion regeneration capacities in populations of the massive coral *Porites lutea* at Réunion Island: environmental correlates. Mar. Ecol. Prog. Ser. 428: 105–111.

Denis, V., M.M. Guillaume, M. Goutx, S. de Palmas, J. Debreuil, A.C. Baker, R.K. Boonstra and J.H. Bruggemann. 2013. Fast growth may impair regeneration capacity in the branching coral *Acropora muricata*. PLoS One 8: e72618.

Edmunds, P.J. and H.S. Lenihan. 2010. Effect of sub-lethal damage to juvenile colonies of massive *Porites* spp. under contrasting regimes of temperature and water flow. Mar. Biol. 157: 887–897.

Francini-Filho, R.B., R.L. Moura, C.M. Ferreira and E.O.C. Coni. 2008. Live coral predation by parrotfishes (Perciformes: Scaridae) in the Abrolhos Bank, eastern Brazil, with comments on the classification of species into functional groups. Neotrop. Ichthyol. 6: 191–200.

Francini-Filho, R.B., C.M. Ferreira, E.O.C. Coni, R.L. Moura and L. Kaufman. 2010. Foraging activity of roving herbivorous reef fish (Acanthuridae and Scaridae) in eastern Brazil: influence of resource availability and interference competition. J. Mar. Biol. Assoc. UK 90: 481–492.

Froese, R. and D. Pauly. 2014. FishBase. Available at: http://www.fishbase.org.

Frydl, P. 1979. The effect of parrotfish (Scaridae) on coral in Barbados, W.I. Int. Revue Gesamt. Hydrobiol. 64: 737–748.

Glynn, P.W. 1997. Bioerosion and coral reef growth: a dynamic balance. pp. 68–95. *In*: C. Birkeland (ed.). Life and Death of Coral Reefs. Chapman and Hall, New York.

Grottoli-Everett, A.G. and G.M. Wellington. 1997. Fish predation on the scleractinian coral *Madracis mirabilis* controls its depth distribution in the Florida Keys, USA. Mar. Ecol. Prog. Ser. 160: 291–293.

Harrison, P.L. and C.C. Wallace. 1990. Reproduction, dispersal and recruitment of scleractinian corals. pp. 133–207. *In*: Z. Dubinsky (ed.). Ecosystems of the World: Coral Reefs. Elsevier Science Publishers, Amsterdan.

Heenan, A., A.S. Hoey, G.J. Williams and I.D. Williams. 2016. Natural bounds on herbivorous coral reef fishes. Proc. R. Soc. B. 283: rspb20161716.

Henry, L. and M. Hart. 2005. Regeneration from injury and resource allocation in sponges and corals – a review. Int. Rev. Hydrobiol. 90: 125–158.

Hoey, A.S. and D.R. Bellwood. 2008. Cross shelf variation in the role of parrotfishes on the Great Barrier Reef. Coral Reefs 12: 1316–1328.

Hoey, A.S., M.S. Pratchett and C. Cvitanovic. 2011. High macroalgal cover and low coral recruitment undermines the potential resilience of the world's southernmost coral reef assemblages. PLoS One e25824.

Hoey, A.S., D.A. Feary, J.A. Burt, G. Vaughan, M.S. Pratchett and M.L. Berumen. 2016. Regional variation in the structure and function of parrotfishes on Arabian reefs. Mar. Poll. Bull. 105: 524–531.

Hughes, T.P., M.J. Rodrigues, D.R. Bellwood, D. Ceccarelli, O. Hoegh-Guldberg, L. McCook, N. Moltschaniwskyj, M.S. Pratchett, R.S. Steneck and B. Willis. 2007. Phase shifts, herbivory, and the resilience of coral reefs to climate change. Curr. Biol. 17: 360–365.

Hughes, T.P., J.T. Kerry, M. Álvarez-Noriega, J.G. Álvarez-Romero, K.D. Anderson, A.H. Baird, A.H., R.C. Babcock, M. Beger, D.R. Bellwood, R. Berkelmans, T.C. Bridge, I.R. Butler, M. Byrne, N.E. Cantin, S. Comeau, S.R. Connolly, G.S. Cumming, S.J. Dalton, G. Diaz-Pulido, C.M. Eakin, W.F. Figueira, J.P. Gilmour, H.B. Harrison, S.F. Heron, A.S. Hoey, J.-P.A. Hobbs, M.O. Hoogenboom, E.V. Kennedy, C.-Y. Kuo, J.M. Lough, R.J. Lowe, G. Liu, M.T. McCulloch, H.A. Malcolm, M.J. McWilliam, J.M. Pandolfi, R.J. Pears, M.S. Pratchett, V. Schoepf, T. Simpson, W.J. Skirving, B. Sommer, G. Torda, D.R. Wachenfeld, B.L. Willis and S.K. Wilson. 2017. Global warming and recurrent mass bleaching of corals. Nature 543: 373–377.

Jones, C.G., J.H. Lawton and M. Shachak. 1994. Organisms as ecosystem engineers. Oikos 69:373–386.

Korzen, L., A. Israel and A. Abelson. 2011. Grazing effects of fish versus sea urchins on turf algae and coral recruits: possible implications for coral reef resilience and restoration. J. Mar. Biol. 2011: 960207.

Kramer, M.J., O. Bellwood and D.R. Bellwood. 2013. The trophic importance of algal turfs for coral reef fishes: the crustacean link. Coral Reefs 32: 575–583.

Lenihan, H.S., S.J. Holbrook, R.J. Schmitt and A.J. Brooks. 2011. Influence of corallivory, competition, and habitat structure on coral community shifts. Ecology 92: 1959–1971.

Lester, R.T. and R.P.M. Bak. 1985. Effects of environment on regeneration rate of tissue lesions in the reef coral *Montastrea annularis* (Scleractinia). Mar. Ecol. Prog. Ser. 24: 183–185.

Littler, M.M., P.R. Taylor and D.S. Littler. 1989. Complex interactions in the control of coral zonation on a Caribbean reef flat. Oecologia 80: 331–340.

Lokrantz, J., M. Nystrom, M. Thyresson and C. Johansson. 2008. The non-linear relationship between body size and function in parrotfishes. Coral Reefs 150: 1145–1152.

McAfee, S.T. and S.G. Morgan. 1996. Resource use by five sympatric parrotfishes in the San Blas Archipelago, Panama. Mar. Biol. 125: 427–437.

Meesters, E.H., M. Noordeloos and R.P.M. Bak. 1994. Damage and regeneration: links to growth in the reef-building coral *Montastrea annularis*. Mar. Ecol. Prog. Ser. 112: 119–128.

Miller, M.W. and M.E. Hay. 1998. Effects of fish predation and seaweed competition on the survival and growth of corals. Oecologia 113: 231–238.

Mumby, P.J. 2006. The impact of exploiting grazers (Scaridae) on the dynamics of Caribbean coral reefs. Ecol. Appl. 16: 747–769.

Mumby, P.J., C.P. Dahlgren, A.R. Harborne, C.V. Kappel, F. Micheli, D.R. Brumbaugh, K.E. Holmes, J.M. Mendes, K. Broad, J.N. Sanchirico, K. Buch, S. Box, R.W. Stoffle and A.B. Gill. 2006. Fishing, trophic cascades, and the process of grazing on coral reefs. Science 311: 98–101.

Nash, K.L., R.A. Abesamis, N.A. Graham, E.C. McClure and E. Moland. 2016. Drivers of herbivory on coral reefs: species, habitat and management effects. Mar. Ecol. Prog. Ser. 554: 129–140.

Neudecker, S. 1979. Effects of grazing and browsing fishes on the zonation of corals in Guam. Ecology 60: 666–672.

Ong, L. and K.N. Holland. 2010. Bioerosion of coral reefs by two Hawaiian parrotfishes: species, size, differences and fishery implications. Mar. Biol. 157: 1313–1323.

Oren, U., Benayahu Y. and Y. Loya. 1997. Effect of lesion size and shape on regeneration of the Red Sea coral *Favia favus*. Mar. Ecol. Prog. Ser. 146: 101–107.

Parenti, P. and J.E. Randall. 2000. An annotated checklist of the species of the labroid fish families Labridae and Scaridae. Ichthyol. Bull. 68: 1–97.

Parenti, P. and J.E. Randall. 2011. Checklist of the species of the families Labridae and Scaridae: an update. Smithiana Bulletin. 13: 29–44.

Penin, L., F. Michonneau, A.H. Baird, S.R. Connolly, M.S. Pratchett, M. Kayal and M. Adjeroud. 2010. Early post-settlement mortality and the structure of coral assemblages. Mar. Ecol. Prog. Ser. 408: 55–64.

Penin, L., F. Michonneau, A. Carroll and M. Adjeroud. 2011. Effects of predators and grazers exclusion on early post-settlement coral mortality. Hydrobiologia 663: 259–264.

Pratchett, M.S., A.S. Hoey, D.A. Feary, A.G. Bauman, J.A. Burt and B.M. Riegl. 2013. Functional composition of Chaetodon butterflyfishes at a peripheral and extreme coral reef location, the Persian Gulf. Mar. Pollut. Bull. 72: 333–341.

Pratchett, M.S., A.S. Hoey, C. Cvitanovic, J.P. Hobbs and C.J. Fulton. 2014. Abundance, diversity, and feeding behavior of coral reef butterflyfishes at Lord Howe Island. Ecol. Evol. 4: 3612–3625.

Randall, J.E. 1967. Food habits of reef fishes of the West Indies. Stud. Trop. Oceanogr. (Miami) 5: 665–847.

Randall, J.E. 1974. The effects of fishes on coral reefs. Proc. Second Int. Coral Reef Symp. Brisbane. 1: 159–166.

Rinkevich, B., Y. Loya. 1989. Reproduction in regenerating colonies of the coral *Stylophora pistilata*. pp. 257–265. *In*: E. Spanier et al. (eds.). Environmental Quality and Ecosystem Stability. Israel Society for Ecology and Environmental Quality Sciences, Jersulalem.

Rotjan, R.D. and S.M. Lewis. 2005. Selective predation by parrotfishes on the reef coral Porites astreroides. Mar. Ecol. Prog. Ser. 305: 193–201.

Rotjan, R.D. and S.M. Lewis. 2006. Parrotfish abundance and selective corallivory on a Belizean coral reef. J. Exp. Mar. Biol. Ecol. 335: 292–301.

Rotjan, R.D. and S.M. Lewis. 2008. The impact of coral predators on tropical reefs. Mar. Ecol. Prog. Ser. 367: 73–91.

Rotjan, R.D. and S.M. Lewis. 2009. Predators selectively graze reproductive structures in a clonal marine organism. Mar. Biol. 156: 569–577.

Rotjan, R.D. and J.L. Dimond. 2010. Discriminating causes from consequences of persistent parrotfish corallivory. J. Exp. Mar. Biol. Ecol. 390: 188–195.

Sánchez, J.A., M.F. Gil, L.H. Chasqui and E.M. Alvarado. 2004. Grazing dynamics on a Caribbean reef-building coral. Coral Reefs 23: 578–583.

Streelman, J.T., M. Alfaro, M.W. Westneat, D.R. Bellwood and S.A. Karl. 2002. Evolutionary history of the parrotfishes: biogeography, ecomorphology, and comparative diversity. Evolution 56: 961–971.

Szmant-Froelich, A.M. 1985. The effect of colony size on the reproductive ability of the Caribbean coral *Montastrea annularis*. Proc. Fifth Int. Coral Reef Symp. Tahiti. 4: 295–300.

Trapon, M.L., M.S. Pratchett and A.S. Hoey. 2013a. Spatial variation in abundance, size and orientation of juvenile corals related to the biomass of parrotfishes on the Great Barrier Reef, Australia. PLoS One 8: e57788.

Trapon, M.L., M.S. Pratchett, A.S. Hoey and A.H. Baird. 2013b. Influence of fish grazing and sedimentation on the early post-settlement survival of the tabular coral *Acropora cytherea*. Coral Reefs 32: 1051–1059.

Trapon, M.L., M.S. Pratchett, M. Adjeroud, A.S. Hoey and A.H. Baird. 2013c. Post-settlement growth and mortality rates of juvenile scleractinian coral in Moorea, French Polynesia versus Trunk Reef, Australia. Mar. Ecol. Prog. Ser. 488: 157–170.

Van Veghel, M.L.J. and R.P.M. Bak. 1994. Reproductive characteristics of the polymorphic Caribbean reef building coral *Montastrea annularis* III. Reproduction in damaged and regenerating colonies. Mar. Ecol. Prog. Ser. 109: 229–233.

Veron, J.E.N. 2000. Corals of the World. Townsville: Australian Institute of Marine Science.

Welsh, J.Q., R.M. Bonaldo and D.R. Bellwood. 2015. Clustered parrotfish feeding scar trigger partial coral mortality of massive *Porites* colonies on the inshore Great Barrier Reef. Coral Reefs 34: 81–86.

Wilson, S.K., D.R. Bellwood, J.H. Choat and M.J. Furnas. 2003. Detritus in the epilithic algal matrix and its use by coral reef fishes. Oceanogr. Mar. Biol. Annu. Rev. 41: 279–309.

Work, T.M. and G.S. Aeby. 2010. Wound repair in *Montipora capitata*. J. Invertebr. Pathol. 105: 116–119.

Geographic Variation in the Composition and Function of Parrotfishes

Michel Kulbicki[1], Alan M. Friedlander[2], David Mouillot[3] and Valeriano Parravicini[4]

[1] UMR "Entropie" IRD Labex Corail, Université de Perpignan, 66000 France
    Email: michel.kulbicki@ird.fr
[2] Pristine Seas, National Geographic Society, Washington DC
    Fisheries Ecology Research Lab, University of Hawaii, Honolulu, Hawaii 96822
    Email: friedlan@hawaii.edu
[3] Institut de Recherche pour le Développement, Université Montpellier,
    Montepellier 34095, France
    Email: david.mouillot@umontpellier.fr
[4] Ecole Pratique des Hautes Etudes, Université de Perpignan, 66000 France
    Email: valeriano.parravicini@ephe.sorbonne.fr

Introduction

At present one hundred species of parrotfishes are described worldwide (Parenti and Randall 2011, Eschmeyer 2015). Although originally recognized as an independent family (i.e. Scaridae), recent molecular evidence has shown that the parrotfishes are nested within the Labridae (Westneat and Alfaro 2005), a family that encompasses over 620 species. Although there is some debate regarding the taxonomic status of parrotfishes, here we consider them to be the Scarinae (a sub-family of the Labridae) with two major clades: Scarinine and Sparisomatine (Bonaldo et al. 2014). Parrotfishes are mainly tropical fishes, with rare representatives in sub-tropical and temperate regions. For example, in the eastern Atlantic *Sparisoma cretense* reaches latitudes of over 40°N, while in the western Atlantic several species (e.g. *Sparisoma axillare* and *Sp. frondosum*) reach latitudes of 30°S. Similarly in the Pacific, *Scarus ovifrons* has been reported from Tsushima island, southern Korea (34°N), while several species are reported from latitudes between 27 and 32°S (Easter, Rapa, Norfolk, Lord Howe, Kermadec, New South Wales, and South Africa). Most parrotfishes are found associated with reefs, either with or without coral, while juveniles of some species and a limited number of browsing species are found in macroalgae or seagrass adjacent to reefs (Randall 1983, Dorenbosch et al. 2006). Most parrotfishes live in shallow waters (0-20 m), with a few species occurring below 50 m. The maximum body size of parrotfishes varies from 13 cm total length (TL) for *Cryptotomus roseus* to 130 cm TL for *Bolbometopon muricatum*, with the vast majority (93 out of 100) of species being medium sized (i.e. between 20 and 80 cm TL).

Coral reefs worldwide are under extreme pressure from overfishing, pollution, disease, and climate change, leading to local and regional declines in the cover and composition of coral communities (Bruno and Selig 2007, Descombes et al. 2015, Hughes et al. 2017). The death of living corals provides opportunities for other sessile organisms, in particular algae, to colonize the dead coral skeletons. If unchecked this increased algal production can lead to the proliferation of large fleshy macroalgae that competes with corals for space (e.g. McCook et al. 2001, Birrell et al. 2005, Rasher et al. 2013). These declines in coral cover are invariably accompanied by declines in species that rely on corals, either directly or indirectly. In particular, reef fish assemblages tend to be more diverse when coral diversity and cover are high (Messmer et al. 2011, Holbrook et al. 2015, Richardson et al. 2017). Therefore, any organism that controls the settlement and growth of algae may be beneficial to corals, and consequently to biodiversity in general (Mumby et al. 2007; Hughes et al. 2007). Parrotfishes are widely regarded as a key group in preventing macroalgal overgrowth on coral reefs (reviewed by Bonaldo et al. 2014), through the removal of algal material and the clearing of space for the settlement of benthic organisms when feeding (e.g. Bonaldo & Bellwood 2009), the external bioerosion of hard substrata (Bellwood et al. 2003, 2012), and the suggested role of some species in the dispersal of algae (Vermeij et al. 2013, Tâmega et al. 2015). Conversely, parrotfishes have been identified as a source of mortality for recently settled corals (Trapon et al. 2013a), and some species feed directly on live coral colonies (Cole et al. 2008, Bonaldo and Rotjan Chapter 9). Therefore, parrotfishes have been the focus of much attention during the last decade as potentially important players in coral-algal dynamics and the resilience of reefs.

Despite an abundant literature on parrotfishes, there are still important gaps in the understanding of the assembly rules that shape parrotfish assemblages. There have been several studies that have described local and/or regional variation in the composition of parrotfish assemblages (e.g. Bellwood et al. 2003, 2012, Hoey and Bellwood 2008, Cheal et al. 2012, Hoey et al. 2016), and several comparisons of the potential role of herbivores, including parrotfishes, between the Atlantic and the Indo-Pacific realms (Bellwood et al. 2004, Schmidt 2013). However, a global comparative analysis of the life-histories, species richness, densities, and biomass of parrotfishes is lacking. In the present chapter, we investigate the organization and the functioning of parrotfish assemblages from the regional to the local scale, based on both existing literature and a recently assembled large and comprehensive database of field data. This investigation is organized along the following themes:

- The global distribution of parrotfishes: at present the major patterns in the distribution of parrotfishes are known (Bonaldo et al. 2014), but the assembly rules leading to these patterns have not been explored.
- From taxonomic to functional diversity: parrotfishes play a central role in several major ecological processes essential to the resilience of coral reefs. Although several major studies have investigated the ecological functions of parrotfishes, we still know little about the global geographical distribution of these functions
- From global to local: the rules that apply to global scale patterns and processes can be strongly influenced by local factors. In addition to species richness, abundance and biomass of parrotfishes are influenced by both large and local scale factors.

Global Geographic Distribution of Parrotfishes

Taxonomic Uncertainties

Despite their relatively large size, abundance, and importance to fisheries, many questions still remain on the taxonomy and distribution of parrotfishes. Most parrotfishes undergo ontogenetic changes in colour patterns, with juveniles, initial phase (primarily female), and terminal phase males having distinct colour patterns. This intra-specific variation in colour patterns was a source of considerable uncertainty as to the taxonomy of the group, with different life phases often being considered separate species. As a result, the global distribution of parrotfishes was uncertain until recently. Even today, new species are being described (e.g. *Sparisoma rocha, Sparisoma choati, Scarus hutchinsi*), which is very uncommon for large colourful species that inhabits shallow waters. In addition, a number of Indo-Pacific species that were initially thought to be a single species are now divided into two or more species based on genetic analyses (e.g. *Hipposcarus harid* and *Hipposcarus longiceps, Cetoscarus ocellatus* and *Cetoscarus bicolor, Chlorurus sordidus* and *Chlorurus spilurus*). This situation will likely continue due to advances in the genetic techniques, which can identify potential cryptic species within some taxa, such as *Scarus ghobban* (Bariche and Bernardi 2009, Choat et al. 2012) and *Scarus rubroviolaceus* (Fitzpatrick et al. 2011, Choat et al. 2012). Although the geographic distributions of most parrotfishes are known (Bonaldo et al. 2014), our understanding of the factors that shape these distributions is limited.

Global Species Richness

At present 100 species and ten genera (*Bolbometopon, Calotomus, Cetoscarus, Chlorurus, Cryptotomus, Hipposcarus, Leptoscarus, Nicholsina, Scarus* and *Sparisoma*) of parrotfishes are recognized (Table 1). Of these ten genera, only *Scarus* is found in all oceans. Two genera are restricted to the Atlantic (*Sparisoma* and *Cryptotomus*), one genus (*Nicholsina*) is found in both the Atlantic and Eastern Pacific, and five genera are unique to the Indo-Pacific (*Bolbometopon, Cetoscarus, Calotomus, Hipposcarus,* and *Chlorurus*). Over half of the known species are within the genus *Scarus* (52 spp.), while *Chlorurus* and *Sparisoma* have 18 and 15 species, respectively. All other genera consist of five or less species.

The majority of parrotfish species have a relatively broad geographic range, with few or no local endemics (Choat et al. 2012). The species with the most restricted range are *Scarus chinensis, Scarus gracilis, Sparisoma griseorubrum* and *Sp. rocha. Sc. chinensis* and *Sc. gracilis* are known only from type specimens (from south of Shangai, China) and consequently their distributions remain uncertain. *Sp. griseorubrum* is reported from Venezuela and probably occurs elsewhere in the Caribbean, but at present this species is poorly known, whereas *Sp. rocha* is restricted to Trindade Island and some seamounts in the Southwestern Atlantic (Pinheiro et al. 2015). Some species are limited to a specific region, such as the Hawaiian Islands, NW Indian Ocean (Red Sea and Persian Gulf), the eastern Atlantic, Marquesas (French Polynesia), Line Islands, and southern Japan. For example, three species are known only from the Hawaiian Islands, and five species from the Red Sea and Persian Gulf. Interestingly, there doesn't appear to be any commonalities of traits among these parrotfishes with restricted geographical ranges. Furthermore, geographical range is not related to maximum body size among parrotfishes, which is counter to the general trend in other reef fishes (Kulbicki et al. 2015).

Despite the broad geographic range of most parrotfish species, there are distinct regional differences in the composition of parrotfish assemblages, with the greatest difference being between the Atlantic and Indo-Pacific Oceans. Atlantic reefs are dominated by a combination of species from the genera *Scarus* and *Sparisoma*, with each genus accounting for ca. 50 percent of the parrotfish species at most locations; an exception is the eastern Atlantic where *Sparisoma* spp. dominate. In the Indian and Pacific Oceans, *Scarus* and *Chlorurus* are the most speciose genera, on average accounting for 59 percent and 24 percent of the parrotfish species richness, respectively.

Table 1. Total number of parrotfish species by genus and major biogeographical region (as defined by Kulbicki et al. 2013 and Spalding et al. 2007; see Fig. 2). % Reef fish: percentage represented by parrotfishes amongst all reef fishes of a region or realm

	Atlantic-East	Atlantic-West	Total Atlantic	East Pacific-EP	East Pacific-NEP	Total East Pacific	Indian-CIP	Indian-WI	Total Indian	Pacific-CIP	Pacific-CP	Pacific-NP	Pacific-SP	Total Pacific	All Oceans
Bolbometopon							1	1	1	1	1			1	1
Calotomus				1		1	2	3	3	3	4	1	1	4	5
Cetoscarus							1	1	1	1	1	1		2	2
Chlorurus							11	10	14	13	8		1	14	18
Cryptotomus		1	1												1
Hipposcarus							2	1	2	2	1			2	2
Leptoscarus							1	1	1	1	1		1	1	1
Nicholsina	1	2	3	1	1	1									3
Scarus	1	12	13	4	2	4	27	25	36	30	26	2	4	33	52
Sparisoma	4	13	15												15
ALL	6	28	32	6	3	6	45	42	58	51	42	4	7	57	100
% Reef fishes	1.5	3.2	2.6	1.7	0.8	1.1	1.9	2.3	2.1	1.4	1.3	0.6	0.9	1.4	1.5

Large-scale Factors and Parrotfish Species Richness

The number of parrotfish species in a given location is dependent on a number of large-scale factors. These factors can be split according to several major hypotheses: species richness may be determined by evolutionary history, dispersal and colonization capacity, habitat size, habitat diversity, temperature, and connectivity (see Parravicini et al. 2013 for detailed discussion). Analysis of the global distributions of parrotfishes show that species richness is influenced nearly equally by sea surface temperatures (SST), isolation, habitat size, and evolutionary history (Table 2a). This contrasts with that of total reef fish richness where habitat size (characterized by coral reef area and coastal length) and evolutionary history (characterized by biogeographical region) had significantly higher influence than other factors (Parravicini et al. 2013). In particular, SST and isolation had far less influence on total richness than they have for parrotfishes. In order to evaluate these differences with total richness, we tested the effects of the same factors on the relative richness of parrotfishes (as a percentage of total reef fish species richness). Habitat size had no influence, whereas evolutionary history, and SST were the most important factors in explaining relative parrotfish richness (Table 2b) The amount of deviance explained by these factors was far less than for parrotfish richness, with nearly no spatial component (0.4% for the SLM null model; tested by Moran's I index).

Table 2. Generalised Linear Model (GLM) and Spatial Linear Model (SLM) results assessing the relationship between selected predictor variables and (a) the absolute species richness and (b) relative species richness of parrotfishes. F and LR test the significance of each model compared to an only-intercept model (i.e. null hypothesis). Deviance explained in SLM is expressed as Nagelkerke R^2

Hypotheses	GLM				SLM			
	F	Deviance %	AIC	Moran's I	LR	Deviance %	AIC	Moran's I
A - Scarinae richness								
Null		0	2432	5.6***		55	2179	-7.1[ns]
Region	149.7***	49	2225	1.2***	19.7***	58	2164	-1.3[ns]
Area	223.5***	42	2265	3.5***	87.9***	67	2094	-1.1[ns]
Energy (SST)	89.4***	36	2294	4.7***	127.8***	70	2056	5.7[ns]
Isolation	98.8***	39	2282	0.5***	136.3***	71	2048	-2.3[ns]
Combined model	164.9***	77	1995	9.5***	242.2***	80	1951	0.1[ns]
B - Scarinae proportion								
Null		0	-2085	5.5***		0.4	-2099	6.4[ns]
Region	21.2***	12	-2121	7.8***	52.2***	19	-2147	1.2*
Area	4.9*	2	-2088	7.1***	12.0***	0.7	-2108	8.3[ns]
Energy (SST)	32.1***	17	-2140	7.3***	69.4***	23	-2164	1.8**
Isolation	13.5***	8	-2107	5.2***	25.8***	12	-2120	-9.9[ns]
Combined model	18.3***	27	-2169	9.6***	145.9***	40	-2230	8.0[ns]

*: significant at $p < 0.05$, **: significant at $p < 0.01$, ***: significant at $p < 0.001$, NS: not significant

Evolutionary history plays an essential role in the distribution of reef fish richness (Bellwood and Wainwright 2002, Pellissier et al. 2014, Leprieur et al. 2015) and parrotfishes are no exception (Streelman et al. 2002, Choat et al. 2012). In particular, Choat et al. (2012) showed that many sister species of *Scarus* and *Chlorurus* overlap in or near the Indo-Australian Archipelago (IAA), inducing a richness hotspot in that region. Our assessment of parrotfish richness supports this high diversity in the IAA (Fig. 1). However, relative parrotfish richness shows there are disproportionately more parrotfish species away from this centre of diversity, particularly the southern margins of reef fish distributions and in the central Pacific and western Indian Oceans (Fig. 2). The mechanisms for these patterns are currently unknown but likely to be related to SST, isolation and/or evolutionary history.

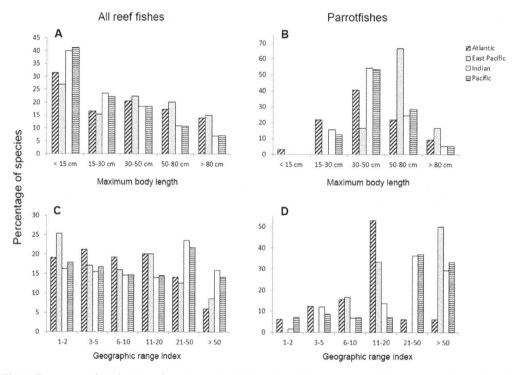

Fig. 1. Frequency distribution of maximum body length (a, b) and geographic range (c, d) for all reef fish species (a, c) and parrotfishes (b, d) according to realm (Atlantic, Eastern Pacific, Indian and Pacific oceans). The geographical range index is the number of checklists (out of 172, see Parravicini et al. 2014 and Kulbicki et al. 2013 for references) in which a species is recorded; low values indicate endemic species and high values indicate species with large geographical ranges.

The Functional Diversity of Parrotfishes and Its Global Distribution

Parrotfish Ecological Functions

Major Feeding Modes

Parrotfishes contribute to several ecological functions (Bonaldo et al. 2014, Burkepile et al. Chapter 7), and can be broadly classified into three major feeding categories: browsers,

Absolute Richness of Scarinae

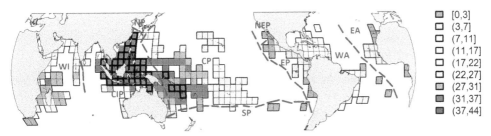

☐	[0,3]
☐	(3,7]
☐	(7,11]
☐	(11,17]
☐	(17,22]
☐	(22,27]
■	(27,31]
■	(31,37]
■	(37,44]

Relative Richness of Scarinae

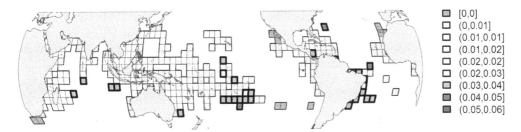

☐	[0,0]
☐	(0,0.01]
☐	(0.01,0.01]
☐	(0.01,0.02]
☐	(0.02,0.02]
☐	(0.02,0.03]
■	(0.03,0.04]
■	(0.04,0.05]
■	(0.05,0.06]

Fig. 2. Geographical distribution of parrotfish species richness and relative species richness (percent total reef fish species richness). The boundaries of the biogeographical regions are given on the top map (blue dotted lines). The regions are as follow (from west to east): WI: western Indian Ocean; CIP: central Indo-Pacific; CP: central Pacific; NP: north Pacific; SP: South Pacific; NEP: north eastern Pacific; EP: eastern Pacific; WA: western Atlantic; EA: eastern Atlantic. Each cell is 5°x5°. Only cells with parrotfish species are shown.

excavators and scrapers. Browsers feed on macroalgae or seagrass, and associated epiphytes. Browsing is considered the ancestral feeding mode of the parrotfishes (Streelman et al. 2002), and includes five genera (*Cryptotomus, Nicholsina, Calotomus, Leptoscarus* and *Sparisoma*). Regardless of whether browsers are feeding directly on macroalgae, seagrass, or the epiphytes growing on them, the net effect is removal of macroalgal or seagrass material. Browsing can reduce the standing biomass of these plants and therefore favour their competitors, such as corals and coralline algae. Scrapers and excavators typically take bites from reef surfaces covered with turf algal communities, although emerging research is showing that they are targeting protein-rich epilithic, endolithic and epiphytic microscopic phototrophs (Clements and Choat Chapter 3). Excavators typically bite deeper into the substratum than scraping parrotfishes, and both groups have been shown to remove macroalgal propagules and clear space for the settlement of other benthic organisms.

A number of parrotfish species are known to feed directly on corals (Bonaldo and Rotjan Chapter 9); however, corals are not a major dietary target of most parrotfishes (see reviews by Green and Bellwood 2009, Bonaldo et al. 2014). The only exception is the Indo-Pacific bumphead parrotfish, *B. muricatum,* which takes approximately half of its bites from live corals (Bellwood et al. 2003, 2012, Hoey and Bellwood 2008). Several other large species, including *Chlorurus gibbus, Chlorurus strongylocephalus, Chlorurus microrhinos, Ce. bicolor, Ce. ocellatus* and *Sparisoma viride* also feed on live corals, but this usually accounts for less than five percent of bites for these species (Bonaldo et al. 2014, Hoey et al. 2016).

Furthermore, incidental feeding by parrotfishes can influence the mortality of recently-settled corals (Trapon et al. 2013a), and the distribution of juvenile corals (Hoey et al. 2011, Trapon et al. 2013b, c).

Changes in Feeding Behavior with Size

Individual parrotfish species are typically classified into a single feeding mode; however, such groupings overlook the complexities of feeding in this group (Green and Bellwood, 2009, Molina-Urena 2009, Ong and Holland 2010, Bonaldo et al. 2014). In particular, many species change their feeding mode as they grow. For example, most excavating species function as scrapers when they are small, transitioning to an excavating feeding mode when their feeding apparatus is strong enough for this type of feeding (Bonaldo et al. 2014). Conversely, large individuals of some 'scraping' species may effectively function as excavators (e.g. *Sc. rubroviolaceus*, Ong and Holland 2010). Some studies also make the distinction between small- and large-bodied excavators, reflecting the volume of carbonates removed when feeding (e.g. Green and Bellwood 2009, Heenan et al. 2016). There are also several inconsistencies in the classification of some species. For instance, *Sparisoma* have been classified as browsers (except *Sparisoma amplum* and *Sp. viride*, Bonaldo et al. 2014), grazers (Cardoso et al. 2009), a combination of browsers and excavators (Molina-Urena 2009), or excavator-scrapers (Ferreira and Gonçalves 2006).

Global Geographic Distribution of Parrotfish Functional Diversity

An analysis of the spatial distribution of feeding modes of parrotfishes on a global scale using the classification proposed by Bonaldo et al. (2014) shows: (i) most browsers are found in the Atlantic where they are represented mainly by *Sparisoma*, whilst in the Indo-Pacific this feeding mode is represented by only two genera (*Calotomus* and *Leptoscarus*) and six species; (ii) excavators are nearly absent from the Atlantic, but represent nearly a third of the species in the Indo-Pacific; and (iii) scrapers are common everywhere, accounting for 40 percent of the species in the Atlantic and > 60 percent of species in the Indo-Pacific (Table 3). This suggests that the functional capacities of parrotfish species pools differ among biogeographical regions, and these differences may have implications for the dynamics of reefs in each region.

Local Variations in Parrotfish Assemblages

Local species assemblages are likely constrained by the species and life-history traits available in the regional species pool as predicted by local-regional diversity relationships (Loreau 2000). The characteristics of parrotfish assemblages (i.e. species composition, abundance, biomass, size distribution, and functional and phylogenetic structures) are influenced by the interaction between the species available at a regional level and a complex set of environmental factors (D'agata et al. 2014, Taylor et al. 2015, Heenan et al. 2016). The major factors known to influence the local distribution of reef fishes are: island type or sub-regions (Parravicini et al. 2013), reef type (e.g. Russ 1984a, Hoey and Bellwood 2008), exposure (e.g. Kulbicki et al. 2000b, Cheal et al. 2013), depth (e.g. Nagelkerken et al. 2001, Brokovich et al. 2008), coral cover (e.g. Chabanet et al. 1997, Pratchett et al. 2008), algal cover and biomass (e.g. Rossier and Kulbicki 2000, Hoey and Bellwood 2011), fishing or protection level (e.g. Friedlander et al. 2003, Williams et al. 2008). Isolating the effects of any of the factors on the functional and taxonomic composition, individual size, abundance, and biomass of parrotfish assemblages is difficult as they rarely operate in

Table 3. Number of parrotfish species per feeding mode (Bonaldo et al. 2014) and per size class (maximum body length) according to the biogeographical regions defined in Fig. 2

	Atlantic-EA	Atlantic-WA	Total Atlantic	East Pacific-EP	East Pacific-NEP	Total East Pacific	Indian-CIP	Indian-WI	Total Indian	Pacific-CIP	Pacific-CP	Pacific-NP	Pacific-SP	Total Pacific	ALL
Browsers	5	14	17	2	1	2	3	4	4	4	5	1	2	5	23
Excavators		2	2				14	13	17	16	11	1	1	17	23
Scrapers	2	12	13	4	2	4	29	24	36	32	27	1	4	34	54
< 15 cm		1	1												1
15-30 cm	2	5	7				6	7	9	7	6			7	16
30-50 cm	3	12	13	1	1	1	24	21	31	27	22	1	2	30	51
50-80 cm	1	6	7	4	2	4	13	10	14	15	12	2	4	16	25
> 80 cm	1	3	3	1		1	3	3	3	3	3		1	3	5
ALL	7	28	32	6	3	6	46	41	57	52	43	3	7	56	100

isolation. For example, fishing may have a profound influence on biomass, density and local distribution of large parrotfish species (e.g. Taylor et al. 2015, Heenan et al. 2016), masking any potential relationships with environmental factors. In particular, larger species may have abundances, biomasses and size distributions that differ markedly from natural baselines (e.g. *B. muricatum*, Bellwood et al. 2003, 2012, Kobayashi et al. 2011).

Most information on the local diversity, abundance, biomass, or size distribution of fishes does not focus specifically on parrotfishes. As a consequence, it is often difficult to have a general view of the local drivers. To overcome this, we combined information from 40 publications for the western Atlantic (Table 4), and our own analyses of a large database for the South Pacific (Fig. 3) that encompasses 120 sites across 99 islands or island groups.

Relationships between Local Assemblages and Regional Species Pools

As a general rule, species present in the regional pool are not necessarily observed in local assemblages. This may be because of unsuitable local habitat, extirpation by fishing, variations in recruitment and mortality patterns. The question is to identify if differences between regional and local species compositions are linked to identifiable factors and if these differences occur more frequently for some species traits than others.

Because species richness generally increases in an asymptotic manner with the sampled area (Connor and McCoy 1979), the analysis of the relationship between local and regional diversity requires a standardization of sampling effort. We therefore grouped transects at random within a site to reach a standard sampled area of 1750 m² for our Pacific dataset (this was not possible for the Atlantic studies). The maximum regional species richness was 36 species, yet the observed local diversity varied considerably within a region with a maximum of 22 species observed at a site. In other words not all parrotfish species will be represented on a given reef, with some species saturation at the highest regional diversity. Within their region of occurrence, large species were more frequently observed (58 percent)

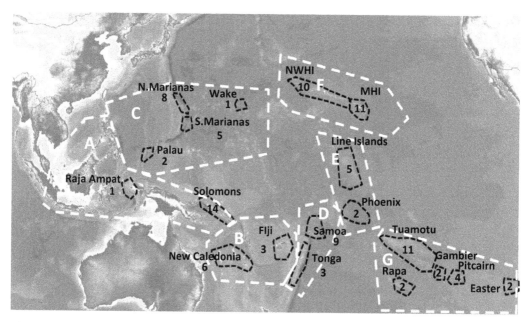

Fig. 3. Origin of the data used in the analyses on parrotfish diversity, density and biomass in the South Pacific (43 species, 150,000 parrotfish; 99 islands, 120 sites). NWHI: Northwestern Hawaiian Islands; MHI: Main Hawaiian Islands; N. Marianas: northern Marianas; S. Marianas: Southern Marianas. Numbers indicate the number of islands or zones per region. The sources of the data are indicated for each region by: F: Alan Friedlander; G: Alison Green; I: IRD; N: NOAA. Regions are: A: Coral Triangle (G); B: Melanesia (I); C: Micronesia (F, G, N); D: Samoa-Tonga (G, I, N); E: Line and Phoenix islands (F, N); F: Hawaiian archipelago (F, N); G: Polynesia (I, F).

than small ones (26 percent) and excavators were more commonly observed (58 percent) compared to browsers (29 percent). However, there is a great deal of variation within size classes and functional groups. For example, *Chlorurus japanensis* and *Ch. spilurus* despite being closely related species and sharing very similar geographical ranges (Choat et al. 2012) had very different frequencies of occurrence (22 and 83 percent, respectively). The largest difference in frequency of occurrence amongst closely related species was between *Calotomus spinidens* (4 percent) and *Calotomus carolinus* (69 percent). These differences may be related to differing habitat requirements and/or subtle variations in life-history strategies within Pacific parrotfishes. For example, *Ch. spilurus* occurs across a broad range of habitats on the Great Barrier Reef (GBR), while *Ch. japanensis* is largely restricted to the exposed reef crest of outer-shelf reefs (Hoey and Bellwood 2008). At present we are unable to identify which traits generate commonness or rarity amongst these species.

Relationship between Assemblage Parameters and Large-scale Environmental Factors

Species Frequency

The next question is to identify the factors influencing the occurrence of species or traits in local assemblages. In the Atlantic, three factors were examined: region (Caribbean versus Brazil), island type (small, large, and continental shelf), and level of protection from fishing (protected versus non-protected). Interestingly, there were no differences in the size structure of parrotfish assemblages across regions, protection from fishing, or island type. This is in contrast with studies that have shown large species tend to be more frequent on large islands (Luiz et al. 2013), and small species are usually more frequent in highly fished areas (Bellwood et al. 2012, Rasher et al. 2013, Kulbicki et al. 2015). In contrast, there were regional differences among feeding types. Browsers were more frequently observed in Brazil than the Caribbean, and conversely excavators were more frequently observed in the Caribbean than in Brazil. This shift may be related to changes in the dominant benthic substratum, from coral and carbonate structures in the Caribbean to rocky substrata in Brazil (e.g. Bonaldo et al. 2007). In the Pacific, the functional composition and size structure of local parrotfish assemblages did not differ from the proportions in the regional pools. Further, there was no detectable effect of regional (reef fish or coral) species richness, island size, isolation, SST or distance to the biodiversity center. The reasons for this lack of relationship are unknown at present, and require further investigation.

Abundance and Biomass

A comparison of the overall abundance (fish m^{-2}) and biomass (g m^{-2}) of parrotfishes between the western Atlantic and the Pacific indicates several major differences. Average abundance in the Atlantic was more than double that in the Pacific (0.21 vs 0.09 fish per m^2). In contrast, biomass was three times higher in the Pacific versus the western Atlantic (42 vs 14 g m^{-2}), meaning that the average size of individual parrotfish in the Pacific is more than six times greater than in the western Atlantic (450 vs 70 g). It is difficult to elucidate the origin of these differences. They could be linked to fishing pressure, which induces a reduction in average individual size. There are several lines of evidence that suggest fishing pressure is considerably higher and has operated over longer temporal scales in the Caribbean than in the Pacific (e.g. Jackson et al. 2001, Cinner et al. 2013). Alternatively, it could be linked to the life-history traits of parrotfishes in the two realms, with a higher proportion of large-bodied species in the Pacific (Fig. 1) and therefore lower abundances and larger individuals (as predicted by metabolic theory - Brown et al. 2004). Of course, the two factors could interact; for instance, the largest Atlantic parrotfish (*Scarus guacamaia*) has been extirpated over most of its range due to a combination of overfishing and the destruction of juvenile nursery habitat (i.e. mangroves) in many Caribbean locations.

The next question is to determine which factors influence the abundance and biomass of the various size and feeding groups of parrotfishes. In the Atlantic we found that the abundance of all parrotfish species: (i) decreased significantly from protected (0.39 fish m^{-2}) to unprotected (0.14 fish m^{-2}) areas; (ii) was higher in Brazil than in the Caribbean for a similar level of protection and island type; and (iii) was not affected by island size for a given level of protection from fishing (Table 4, Fig. 4).

In the Pacific, we were able to examine a wider spectrum of factors compared to the Atlantic (fishing level, region, island size, isolation and island type, distance to the biodiversity center, SST, fish and coral species pools; Table 5). Across the Pacific the abundance of parrotfishes was: (i) positively related to regional diversity, (ii) greater on high islands versus atolls, as previously noted by Pinca et al. (2011) and, (iii) positively related to SST (Table 5), which have also been reported for herbivores in Brazil (Floeter et al. 2005, but see Heenan et al. 2016). Therefore, Atlantic and Pacific realms show no convergence in the factors affecting the overall abundance of parrotfishes, although it should be acknowledged that the range of factors we examined differed between the two oceans. This lack of concordance may possibly be due to differences in regional reef fish

Table 4. Density (fish per m²) and biomass (g/m²) of parrotfish species for the Caribbean and Brazilian provinces. Biomass values are within parentheses. Each site was classified as according to the type of habitat (small or large island or continental shelf) and protection status (either protected or unprotected). "All species" is the average of the total densities or biomasses for parrotfish found in the literature. Total densities or biomasses will in general not be equal to the sum of the separate species as a number of studies have either analyzed a restricted number of species or group all parrotfish at the "family" level. Biomasses by Jackson et al. (2014) are indicated separately on the third line (biomass in g/m² and 95% confidence intervals)

	Caribbean Continent Fished	Caribbean Continent Protected	Brazil Continent Fished	Brazil Continent Protected	Caribbean Large Fished	Caribbean Large Protected	Caribbean Small Fished	Caribbean Small Protected	Brazil Small Protected
All species	0.08;(10)	0.299	0.296	0.28	0.188;(6.6)	1.167;(64.3)	0.097;(10.8)	0.152;(26.1)	0.349;(53.9)
Biomass (Jackson et al. 2014)	9.4 ± 3.0	50.3 ± 20.5			6.3 ± 1.4	18.3 ± 2.9	14.8 ± 3.0	16.9 ± 5.9	
Cryptotomus roseus	0.0002		0.0046				0.0096	0.015;(0.1)	
Nicholsina usta usta									
Sc. coelestinus	0.0005;(1.4)	0.0017					0.0002		
Sc. coeruleus	0.0004;(0.6)	0.0037					0.0013		
Sc. guacamaia	0.0002;(0.6)	0.0034			0.0001		0.0025	0.0001	
Sc. iseri	0.0404;(0.2)	0.1692			0.121;(2)	0.132;(20)	0.0451;(2)	0.0587;(0.5)	
Sc. taeniopterus	0.0062	0.0498	0.03		0.0164;(1.8)	0.191;(13.6)	0.0185;(0.6)	0.0216;(0.5)	
Sc. trispinosus			0.0001	0.0008					0.065;(88)
Sc. vetula	0.0016	0.0246				0.27	0.0109;(0.1)	0.0038;(1)	
Sc. zelindae			0.005	0.0031					0.0187;(1.8)
Sp. amplum			0.0065	0.0062					0.0096;(12.4)
Sp. atomarium	0.0038					0.001	0.0061	0.0066	
Sp. aurofrenatum	0.0285;(0.5)	0.0447			0.0355;(2.5)	0.2105;(7.6)	0.0253;(1.8)	0.044;(1.8)	
Sp. axillare	0.005;(0.2)	0.0031	0.1701	0.1817					0.1237;(19.9)
Sp. chrysopterum					0.0008	0.07	0.0021	0.0009;(0.1)	
Sp. frondosum			0.121	0.0228				0.0416	0.025;(21.5)
Sp. griseorubrum									
Sp. radians	0.0034		0.0315	0.0708	0.001	0.0001	0.0048	0.0093	0.039;(0.1)
Sp. rocha									0.001;(0.4)
Sp. rubripinne	0.0013;(0.9)	0.0288			0.018;(0.2)	0.33	0.0019;(0.3)	0.0016;(0.2)	
Sp. tuiupiranga			0.0835						

(Contd.)

Sp. viride	0.0108;(2.3)	0.0271			0.0648;(3.7)	0.37;(23.1)	0.0225;(2.1)	0.0196;(2.6)	
Browser	0.042;(1.8)	0.076	0.275	0.41	0.055;(2.7)	0.611;(7.6)	0.049;(2.3)	0.077;(2.4)	0.23;(26)
Excavators	0.01;(2.3)	0.027	0.006	0.006	0.064;(3.7)	0.37;(23.1)	0.022;(2.1)	0.019;(2.6)	0.009;(6.3)
Scappers	0.049;(3)	0.252	0.004	0.035	0.137;(3.8)	0.593;(33.6)	0.078;(2.9)	0.084;(2.2)	0.083;(22.5)
< 30 cm	0.036;(0.6)	0.044	0.07	0.119	0.036;(2.5)	0.211;(7.6)	0.045;(1.9)	0.074;(2)	0.039;(0.1)
31-50 cm	0.053;(1.4)	0.251	0.214	0.332	0.156;(4)	0.723;(33.6)	0.067;(3.1)	0.082;(1.5)	0.219;(32.1)
> 50 cm	0.013;(5)	0.06	0	0	0.064;(3.7)	0.64;(23.1)	0.037;(2.3)	0.023;(3.6)	0.065;(22)
References	1,3,7,8,9,17,23,24,25,26,31,33	8,10,13,18	35	4,5,7,12,23	6,24,25,27,28,30	29,39	8,15,19,20,25,27,34,38,40,42,43,44,45	19,32,37	2,11,14,21,22,36,35

1 – Bohnsack et al. 1999; 2 – Bonaldo et al. 2006; 3 – Brown-Saracino et al. 2007; 4 – Chaves and Monteiro-Neto 2009; 5 – Chaves et al. 2013; 6 – Chevalier and Cardena 2005; 7 – Cordeiro et al. 2016; 8 – Debrot et al. 2008; 9 – Dominici-Arosemena and Wolff 2005; 10 – Elise 2012; 11 – Ferreira and Gonçalves 2006; 12 – Floeter et al. 2007; 13 – Fonseca-Escalante 2003; 14 – Francini-Filho et al. 2010; 15 – Friedlander et al. 2013; 16 – Gonzalez-Sanson et al. 2009; 17 – Hernandez-Landa et al. 2015; 18 – Jaxion-Harm et al. 2012; 19 – Klomp and Kooistra 2003; 20 – Kopp et al. 2010; 21 – Krajewski and Floeter 2011; 22 – Longo et al. 2015; 23 – Mendonça-Neto et al. 2008; 24 – Molina-Urena 2009; 25 – Mumby 2006; 26 – Mumby and Wabnitz 2002; 27 – Nagelkerken et al. 2002; 28 – Nemeth and Appeldorn 2009; 29 – Pittman et al. 2008; 30 – Pittman et al. 2010; 31 – Nunez-Lara et al. 2003; 32 – Pattengill-Semmens and Gittings 2003; 33 – Peckol et al. 2003a; 34 – Peckol et al. 2003b; 35 – Perreira et al. 2014; 36 – Pinheiro et al. 2011; 37 – Pittman et al. 2008; 38 – Roff et al. 2011; 39 – Rojas et al. 2013; 40 – Rousseau 2010; 41 – Sedberry et al. 1999; 42 – Stockman 2010; 43 – Toller et al. 2010; 44 – van Rooij et al. 1996; 45 – Walter 2002

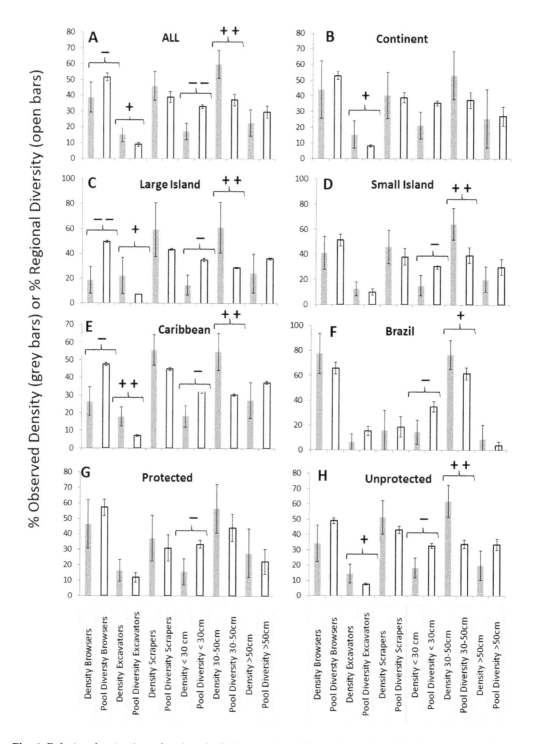

Fig. 4. Relative density (grey bars) and relative species richness (open bars) for the western Atlantic parrotfish. "–" or "+" difference significant at $p < 0.05$; "– –" or "++" difference significant at $p < 0.01$ from ANOVAs. B: browser; E: excavator; S: scraper; sizes are maximum adult size.

richness (Pacific: from 140 reef fish species at Easter Island to over 2,000 species in the Coral Triangle; western Atlantic: from 57 species at St Paul's Rock to 470 in Cuba).

Table 5. Relationships of local parrotfish assemblages in the Pacific with large scale environmental factors. The results are from GLMs. "–" or "+" difference significant at $p < 0.05$; "– –" or "++" difference significant at $p < 0.01$; "– – –" or "+++" difference significant at $p < 0.001$. "NA": not applicable. Empty cells imply a non-significant relationship. Region: see Fig. 3; Fishing level: MPA, Low, Medium, High; Island Type: Atoll versus non Atoll; SST: sea surface temperature; Distance BC: distance to the biodiversity center, taken as 0° × 120°E; Isolation: index built on the distance of the nearest reefs and their surface (Parravicini et al. 2013); Pool Fish Species Richness: number of reef fish species (all species) recorded in the checklist to which the site belongs; Relative Fish Species Pool Richness: proportion of species in the same class within the reef fish species pool (e.g. percent of browsers within the species pool when analyzing the percent of browsers in the local assemblages);
Pool Coral Species Richness: number of coral species for the region to which a site belongs

°: + for Island types indicates a higher value on high islands compared to atolls
*1: Coral Triangle = SW Pacific > all other regions; *2: Line Islands > Hawaii

	Fishing Level	Region	Island Size	Island Type°	SST	Distance BC	Isolation	Pool Fish Species Richness	Relative Pool Fish Species Richness	Pool Coral Species Richness
Density (fish m⁻²) – All				+	– –			+ + +	NA	
Biomass (g m⁻²) – All	– –	*1		+ + +	– –			+ + +	NA	
Average weight (g) – All	– – –	*2							NA	
Weight – Browser										
Weight – Excavator	– – –			– –						
Weight – Scrapper				– –						
Weight Species < 30 cm					–	+				+
Weight Species 30-50 cm	–									
Weight Species > 50 cm	– –									
Relative Density – Browser										
Relative Density – Excavator										
Relative Density – Scrapper				+ + +						
Relative Density Species < 30 cm	+ +			– – –						
Relative Density Species 30-50 cm	+									
Relative Density Species > 50 cm	– – –									
Relative Biomass – Browser										
Relative Biomass – Excavator	–			+ +						
Relative Biomass – Scrapper	+ +			– – –						
Relative Biomass Species < 30 cm	+ +									
Relative Biomass Species 30-50 cm								–	+	
Relative Biomass Species > 50 cm	– –									

The abundance of parrotfishes among feeding types and size classes varied among regions and was influenced by various factors. In the Atlantic we detected no variation in the abundance of the various feeding and size groups with level of protection from fishing (Fig. 4g, h). Browsers made up a higher proportion of the abundance in Brazil compared to the Caribbean, while excavators and scrapers dominated in the Caribbean (Fig. 4e, f). Large-bodied parrotfish species (> 50 cm total length, TL) were relatively more common on Caribbean reefs compared to Brazil, and may be related to the low diversity of these fishes in the Brazilian species pool (Fig. 4f). These patterns in abundance did not reflect changes in species richness (Fig. 4). For example, excavators generally comprised a greater proportion, and browsers a lower proportion of the total abundance than regional species richness. Similarly, the medium-bodied individuals (30-50 cm TL) comprised a greater proportion of total abundance compared to their species richness.

A similar analysis for the Pacific indicates that the density of feeding types and size classes were mainly affected by island type, and to a lesser extent, fishing pressure (Table 5). The abundance of scrapers was significantly greater on high islands compared to atolls, while small parrotfish species (<30 cm TL) had higher abundances on atolls than high islands. Fishing pressure also influenced the size structure of parrotfish assemblages with the relative abundance of small and medium parrotfishes (<50 cm TL) and larger parrotfishes (>50 cm TL) increasing and decreasing, respectively (Fig. 5). Comparable results have been reported for the South Pacific (Pinca et al. 2011) and Micronesia (Taylor et al. 2015).

Data from the Global Coral Reef Monitoring Network (GCRMN, Jackson et al. 2014) and from other published literature both show that biomass values for parrotfishes are much lower in the Caribbean, even in protected areas, compared to the Indo-Pacific (Table 5, Figs 6-8). Both datasets also show an effect of protection on overall Atlantic parrotfish biomass, however the effect is lower than for abundance, which is unusual since in most cases protection is more beneficial to biomass than density of reef fishes (Wantiez et al. 1997, Halpern 2003, Lester and Halpern 2008, Bellwood et al. 2012). In the western Atlantic, the highest estimates of parrotfish biomass were observed on the offshore Brazilian islands protected from fishing (Abrolhos, Noronha, Atoll Rocas, Trindade), reaching an average of 53.9 g/m². These biomass estimates are comparable to those for protected areas in the Pacific

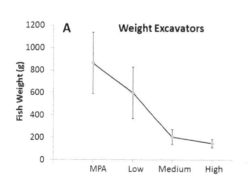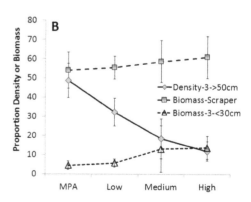

Fig. 5. Effect of fishing on Pacific reefs on (a) individual weight of excavating parrotfishes, (b) the relative density of large species (>50 cm TL) and relative biomass of scraping parrotfishes and small-bodied parrotfishes species (<30 cm TL). Error bars are 95% confidence intervals. Fishing levels are based upon persons per km² of reef (MPA: marine protected areas; Low: <5 ppkm²; Medium: 5-50 ppkm²; High: >50 ppkm²)

with similar regional species richness (Figs 7, 9). However, our analyses could not detect an effect of island type on overall parrotfish biomass. Examination of the proportion of the feeding type and size groups by biomass (Fig. 6) indicated that scrapers tended to make the largest contribution to biomass, followed by browsers, while smaller species (<30 cm TL) contributed less than the larger species. Comparing these proportions in biomass with the proportions in species richness from the regional pools (Fig. 6) indicates that browsers and small species (<30 cm TL) tend to contribute less to biomass than they contribute to regional diversity, the opposite was true for excavators and the largest species (>50 cm TL).

In the Pacific (Table 5, Fig. 5), the major factors influencing biomass were island type and fishing intensity. Not surprisingly, fishing had a higher impact on the biomass than the abundance of parrotfishes (Table 5). Comparable results have been reported for numerous

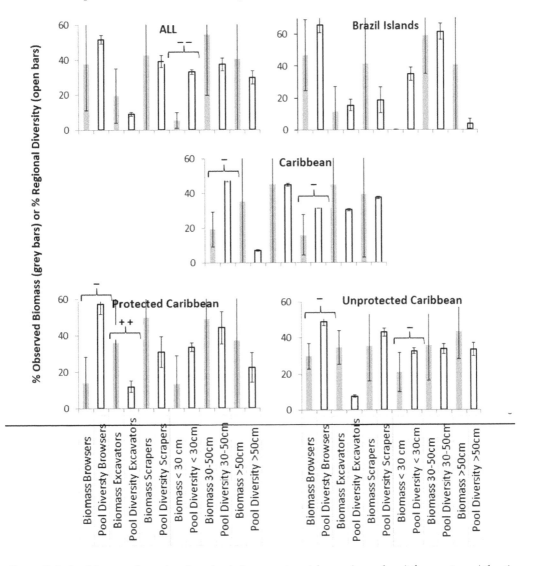

Fig. 6. Relative biomass (grey bars) and relative species richness (open bars) for western Atlantic parrotfish assemblages. "–" or "+" difference significant at $p < 0.05$; "– –" or "++" difference significant at $p < 0.01$ from ANOVAs. B: browser; E: excavator; S: scraper

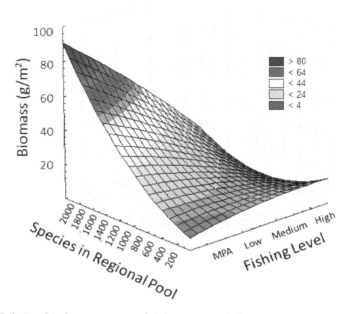

Fig. 7. Relationship between parrotfish biomass with fishing level and regional species richness in the Pacific (for data source see Fig. 3). Fishing level is based on human population density (MPA: marine protected area; Low: <5 persons per km²; Medium: 5-50 ppkm²; High: >50 ppkm²)

locations (e.g. Indo-Pacific Bellwood et al. 2003, 2012; New Caledonia: Carassou et al. 2013; Fiji: Rasher et al. 2013). In addition to fishing pressure, high islands supported higher relative biomass of excavators and lower biomass of scrapers than low islands or atolls. This contrasts with the findings of Heenan et al. (2016) who reported that the biomass of excavating, but not scraping, parrotfishes was greater on atolls than high islands in the central Pacific. Not surprisingly, the relative biomass of excavators (that tend to be large-bodied) and large-bodied (>50 cm TL) species decreased, and consequently the relative biomass of scrapers and small-bodied (<30 cm TL) species increased in areas under more intensive fishing pressure. The decline in biomass of the large species is due to both a decrease in their abundance and average body size. While fishing appears to have altered the functional composition of parrotfishes on Pacific reefs, no such shift was apparent for the western Atlantic reefs examined. The reason for these differential effects is not clear, but may relate to the differential relationships between body size and functional group between oceans.

Relationship between Assemblage Parameters and Local Scale Environmental Factors

At local scales, a striking difference between the Atlantic and the Indo-Pacific is the importance of mangroves and sea grass beds for parrotfishes. Several Atlantic parrotfish species use seagrass beds and, to a lesser extent, mangroves as nursery grounds or feeding areas (Randall 1983, Opitz 1996, Nagelkerken et al. 2000, Mumby et al. 2004). Although the reliance on these habitats varies among species and locations, some species are thought to be mangrove-dependent due to the exclusive use of these habitats by juveniles (e.g. *Sc.*

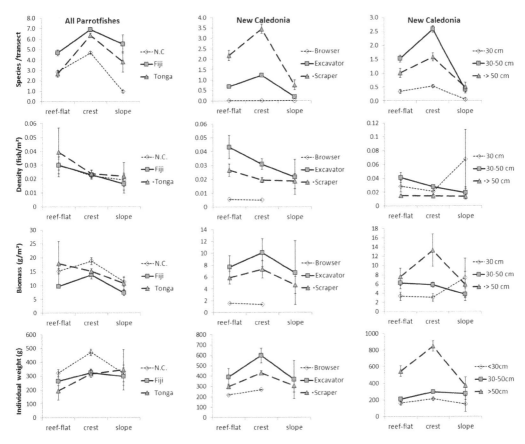

Fig. 8. Variations in parrotfish species richness (first row), density (second row), biomass (third row), and individual weight (fourth row) for all three regions (first column), according to feeding behaviour within New Caledonia (second column – Fiji and Tonga, not shown, show similar results) and according to species size in New Caledonia (third column – Fiji and Tonga, not shown, show similar results). Analysis of the database presented on Fig. 3 (2200 transects 50 m × 5 m). Error bars are 95% confidence intervals.

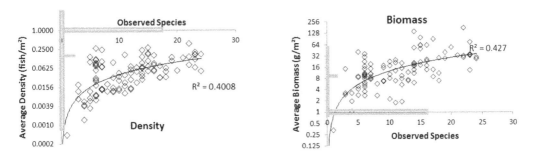

Fig. 9. Relationship between species richness and (a) density and (b) biomass of parrotfishes across the Pacific (based on data from 120 sites, see Fig. 3). For comparative purposes the grey bars on the x-axis show the species range in the Atlantic (1-16 species) and on the y-axis the density range (0.007-1.2 fish/m², mean 0.21) and the biomass range (0.8-61 g/m², mean 12) in the Atlantic.

guacamaia, Dorenbosch et al. 2006; Nagelkerken 2007). Interestingly, the species associated with seagrass beds and mangroves span all genera, feeding types, and body size (Molina-Urena 2009, Nagelkerken et al. 2000, 2002). Molina-Urena (2009) investigated the habitat use of 21 parrotfish species of the western Atlantic and found only three species were restricted to reef habitats (either rocky or coral reefs) and seven species were recorded from four or more of the nine habitats examined. In the Indo-Pacific there are reef-mangrove-seagrass configurations that closely resemble those found in the Atlantic, but present evidence suggests that only a few Indo-Pacific parrotfishes use mangroves or seagrass beds extensively either as juveniles or adults (Mellin et al. 2007, Nagelkerken 2007). These differences in the use of non-reef habitats is probably linked to evolutionary history and the temporal variability in the availability of these non-reef habitats among oceans. It should be noted that differences in tidal amplitude may have a major bearing on this relationship, with small tidal movements in the Caribbean (i.e. ~20 cm) meaning that these non-reefs habitats are always available to fishes, while the larger (up to 3 m) tidal amplitudes of the Indo-Pacific mean that these habitats are inaccessible for considerable periods each day.

Parrotfishes are found on all reef types, but their diversity, numerical abundance, biomass, and individual size varies according to reef type, with striking difference between the Atlantic and the Indo-Pacific. In the Atlantic, local differences in parrotfish assemblage structure have been related to wave exposure (e.g. Walter 2002, Peckol et al. 2003a,b, Pittman et al. 2010, Stockman 2010), but these differences are not consistent among locations. For instance, in the Abrolhos Archipelago, northeast Brazil parrotfish abundances are higher on exposed, as opposed to sheltered, reefs (Francini-Filho et al. 2010), whereas in the southern Brazil the pattern is reversed with parrotfish abundances greatest in sheltered sites (Floeter et al. 2007). These differences likely reflect inter-specific variation in habitat preferences. For instance, *Cr. roseus* was more abundant on exposed reefs in southern Brazil, whereas *Scarus trispinosus*, *Scarus zelindae*, *Sparisoma axillare* and *Sparisoma frondosum* were more abundant on exposed reefs at the Abrolhos Archipelago. Interestingly, *Sp. axillare* was most abundant on sheltered reefs in the south, but exposed reefs in the north (Floeter et al. 2007, Francini-Filho et al. 2010). This apparent shift is difficult to resolve but may be related to differences in reef structure, food availability or other environmental factors between the two regions. There are also inshore-offshore differences in the Atlantic, although the relationships appear to vary among locations. For example, Nemeth and Appeldorn (2009) found no difference in the biomass of parrotfishes between inner-shelf and mid-shelf reefs in Puerto Rico, whereas Francini-Filho et al. (2010) reported decreasing abundance from inshore to offshore reefs in Brazil. This variability in parrotfish assemblages at medium scales in the Atlantic could be due to differences in the availability of preferred habitats and depths (Hernandez-Landa et al. 2015), seascape configuration (Nagelkerken et al. 2000, Molina-Urena, 2009), as well as local habitat complexity and heterogeneity, coral and algae cover (e.g. Pittman et al. 2008, 2010).

In the Pacific the structure, diversity, abundance, biomass and individual weight of parrotfish assemblages are strongly dependent upon both among-reef (e.g. reef type, distance from coast) and within-reef (e.g. habitat heterogeneity, depth) variation. In contrast to the western Atlantic, most spatial patterns in parrotfish assemblages tend to be consistent across regions in the Pacific. One of the most prominent spatial patterns in parrotfish assemblages on Indo-Pacific reefs is the inshore-offshore gradient. On the GBR several studies have analyzed the cross-shelf distribution of parrotfishes, and have reported a general increase in the abundance (see Hoey and Bellwood 2008 for exception), biomass and species richness of parrotfishes from inshore to offshore reefs that is consistent among latitudes (Russ 1984a, b, Wismer et al. 2009, Hoey and Bellwood 2008, 2010, Cheal et al. 2012, 2013). Cheal et al. (2012) also found that parrotfish richness tended to decrease from north to south along the GBR, but this trend was not detectable for abundance. The cross-shelf differences are also evident on Ningaloo Reef, Western Australia (Johansson et al. 2013). These assemblage level patterns were driven by differences in individual species with some species occurring across the entire shelf (e.g. *Ch. spilurus*, *Scarus psittacus*), while others appeared closely linked to a particular shelf location (e.g. inner-shelf: *Sc. rivulatus*, mid-shelf: *Ch. microrhinos*, outer-shelf: *B. muricatum*, *Ch. japanensis*; Hoey and Bellwood 2008).

On a larger geographical scale, Pinca et al. (2011) found that parrotfish abundance was significantly lower and body size larger on outer barrier reefs compared to fringing reefs across 17 countries in the South Pacific, but found no relationship between reef type and biomass. In contrast, Heenan et al. (2016) reported that the biomass of excavating parrotfishes was higher on atolls compared to high islands, while the biomass of scraping parrotfishes was highest at intermediate complexities across 33 islands in the central and western Pacific. Comparisons of parrotfish assemblages across a subset of our Pacific sites which all have inshore-offshore gradients in reef types (New Caledonia, Fiji and Tonga) revealed that the total abundance, biomass and species richness of parrotfishes did not vary from inshore to offshore. However, the abundance and biomass of excavating parrotfishes increased significantly from inshore to offshore in all three regions (Fig. 10), in accordance with the findings from elsewhere (e.g. GBR: Hoey and Bellwood 2008, Ningaloo Reef: Johansson et al. 2013). Furthermore, the overall average weight of parrotfishes increased from inshore to offshore in all three regions (Fig. 10), and likely reflect the effects of fishing, since fishing pressure on parrotfishes is generally highest on coastal reefs. Similar cross-shelf increases in body weight have been documented for a range of reef fish species (Kulbicki et al. 2000a, Letourneur et al. 2000).

In the Pacific, several studies have considered within-reef distributions of parrotfishes. Bonaldo et al. (2014) noted that the within-reef distribution of parrotfishes is "remarkably stable", with the highest biomass in the shallowest areas, although the functional composition differed between the Indo-Pacific (including the Red Sea) and the western Atlantic. For example, Hoey and Bellwood (2008) found the biomass (but not abundance) of parrotfishes was greatest on the reef crest compared to reef slope, reef flat and back reef habitats within inner-, mid- and outer-shelf reefs of the northern GBR. Russ (1984b) also found species richness and abundance of parrotfishes to be highest on the reef crest of mid-shelf reefs in the central GBR. Numerous studies have also reported the abundance and biomass of parrotfishes to be greatest on the reef crest of inshore reefs of the GBR (e.g. Fox and Bellwood 2007, Loffler et al. 2015). An analysis of the previously mentioned database for New Caledonia, Fiji and Tonga showed that the species richness, biomass and individual weight of parrotfishes, but not abundance, was generally greatest on the reef crest, with these patterns generally holding across functional groups and body sizes (Fig. 8). This among habitat distribution has been related to the swimming abilities of parrotfishes (Bellwood et al. 2002) and the availability and productivity of dietary resources (Russ 2003). The fact that all parameters, except abundance, peaked on the crest could also suggest a mid-domain effect, with both deeper water and shallow water species being found together in the reef crest area, however the lower biomass, richness and body size on sheltered habitats of similar depth (e.g. back reef) suggest this is unlikely.

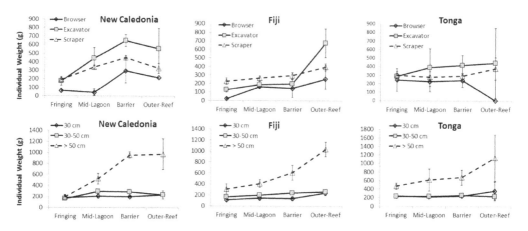

Fig. 10. Relationship between reef type, diet, species size and the average individual weight (g) of parrotfish for New Caledonia, Fiji and Tonga. Extracted from the database presented in Fig. 3. The error bars are 95% confidence intervals.

Conclusion

This chapter identifies important regional differences in the distribution of parrotfish species and functions. Superimposed on these regional patterns are local- and medium-scale variation in taxonomic and functional richness, abundance and biomass of parrotfishes that relate to environmental factors and fishing. Overall, it appears that evolutionary histories likely play a far more important role than previously thought. In particular, this chapter shows that not only is parrotfish species richness much lower in the Atlantic than the Pacific, but there are fundamental differences in the functional composition of parrotfishes between these oceans. Furthermore, the effects of fishing, island type and regional biogeography differ between the two realms. These results imply that the management of parrotfish assemblages should be different across realms and that the consequences of a reduction in parrotfish species richness, abundance or biomass on the ecological functioning of coral reefs may be very different between ocean realms.

There is still much to learn and investigate regarding the factors that shape the distribution and function of parrotfishes. In particular, we still know little about the functional differences among species. For example, there are nearly forty species of scraping parrotfishes in the Pacific, but despite differences in species size, school size and preferred habitats, we know little about the differences in ecological roles among these fishes. Similarly, we know very little about the intra-specific differences in life-history traits across their geographical range. Gust et al. (2002) showed that there can be important differences in growth rates within a species. Similarly, there are likely differences in behaviour, feeding rates and food preferences leading to geographical differences in the ecological roles of a given species. Parrotfishes tend to have wider ranges than the average reef fish species, but at present we still know very little about their pelagic larval duration, their recruitment success, or their preferred juvenile habitats. Similarly, we still know little about the reproduction of these fishes; specifically, conditions controlling spawning and the social organization of parrotfish groups. Habitat use, within or across reefs, is still poorly understood.

Lastly, new approaches are required to fully understand the ecology of this unique group. For instance, the integration of phylogeny and evolutionary history could be essential in understanding how these fishes may adapt to global changes (Descombes et al. 2015). Some of the approaches based on functional vulnerability (Parravicini et al. 2014) could also be of interest in defining conservation hotspots and coldspots (Cinner et al. 2016). However, one should be careful that predicting the adaptation of species to new environmental conditions based upon present geographical distribution and known ecological envelopes may lead to erroneous predictions (Parravicini et al. 2015). Parrotfishes are not alone in a reef fish assemblage, and it may be important to look at how parrotfishes interact within with other species across similar functional groups (e.g. mutualism), or other functional groups (e.g. cleaners, predators). Comparative analyses along geographical gradients and time series, as well as integrating spatial scales, from the local to the regional and global scales is essential if we are to better understand and manage this important group of species.

Acknowledgments

Special thanks go to Ivor Williams and Alison Green who provided access to the NOAA datasets and personal records, respectively, and for giving valuable comments that improved this chapter. The authors wish also to thank Roberta Bonaldo, M.C. José Luis Cabrera, Sergio Floeter, Cadu Ferreira, Ernesto Arias-Gonzales, Thiago Costa Mendes, and Simon Elise for their assistance on the biology and ecology of Atlantic parrotfish. We also wish to thank the three reviewers who gave constructive and positive comments, and last thank you to Andrew Hoey for being a very patient and understanding editor.

References Cited

Bariche, M. and G. Bernardi. 2009. Lack of a genetic bottleneck in a recent Lessepsian bioinvader, the blue-barred parrotfish, *Scarus ghobban*. Mol. Phylogen. Evol. 53: 592–595.

Bellwood, D.R. and P.C. Wainwright. 2002. The history and biogeography of fishes on coral reefs. pp. 5–32. *In*: P.F. Sale (ed.). Coral Reef Fishes: Dynamics and Diversity in a Complex Ecosystem. Elsevier, New York.

Bellwood, D.R., P.C. Wainwright, C.F. Fulton and A. Hoey. 2002. Assembly rules and functional groups at global biogeographical scales. Func. Ecol. 16: 557–562.

Bellwood, D.R., A.S. Hoey and J.H. Choat. 2003. Limited functional redundancy in high diversity systems: resilience and ecosystem function on coral reefs. Ecol. Lett. 6: 281–285.

Bellwood, D.R., T.P. Hughes, C. Folke and M. Nystrom. 2004. Confronting the coral reef crisis. Nature. 429: 827–833.

Bellwood, D.R., A.S. Hoey and T.P. Hughes. 2012. Human activity selectively impacts the ecosystem roles of parrotfishes on coral reefs. Proc. R. Soc. B. 279: 1621–1629

Birrell, C.L., L.J. McCook and B.L. Willis. 2005. Effects of algal turfs and sediment on coral settlement. Mar. Poll. Bull. 51: 408–414.

Bohnsack, J.A., D.B. McClellan, D.E. Harper, G.S. Davenport, G.J. Konoval, A.M. Eklund, J.P. Contillo, S.K. Bolden, P.C. Fischel, G.S. Sandorff and J.C. Javech. 1999. Baseline data for evaluating reef fish populations in the Florida Keys. NOAA Tech. Memo. NMFS-SEFSC, 427.

Bonaldo, R.M. and D.R. Bellwood. 2009. Dynamics of parrotfishes grazing scars. Mar. Biol. 156: 771–777.

Bonaldo, R.M., J.P. Krajewski, C. Sazima and I. Sazima. 2006. Foraging activity and resource use by three parrotfish species at Fernando de Noronha Archipelago, tropical West Atlantic. Mar. Biol. 149: 423–433.

Bonaldo, R.M., J.P. Krajewski, C. Sazima and I. Sazima. 2007. Dentition damage in parrotfishes feeding on hard surfaces at Fernando de Noronha Archipelago, southwest Atlantic Ocean. Mar. Ecol. Prog. Ser. 342: 249–254.

Bonaldo, R.M., A.S. Hoey and D.R. Bellwood. 2014. The ecosystem roles of parrotfishes on tropical reefs. Oceanogr. Mar. Biol. Annu. Rev. 52: 81–132.

Brokovich, E., S. Einbinder, N. Shashar, M. Kiflawi and S. Kark. 2008. Descending to the twilight-zone: changes in coral reef fish assemblages along a depth gradient down to 65 m. Mar. Ecol. Prog. Ser. 371: 253–262.

Brown, J.H., J.F. Gillooly, A.P. Allen, V.M. Savage and G.B. West. 2004. Toward a metabolic theory of ecology. Ecol. 85: 1771–1789.

Brown-Saracino, J., P. Peckol, H.A. Curran and M.L. Robbart, 2007. Spatial variation in sea urchins, fish predators, and bioerosion rates on coral reefs of Belize. Coral Reefs. 26: 71–78.

Bruno, J.F. and E.R. Selig 2007 Regional decline of coral cover in the Indo-Pacific: timing, extent, and subregional comparisons. PLoS One. 2: e711.

Carassou, L., M. Léopold, N. Guillemot, L. Wantiez and M. Kulbicki. 2013. Fishing and the resilience of coral-reefs: a global approach – application to the case of herbivorous fish in New Caledonia, southwest Pacific. PLoS One. 8: e60564.

Cardoso, S.C., M.C. Soares, H.A. Oxenford and I.M. Côté. 2009. Interspecific differences in foraging behavior and functional role of Caribbean parrotfish. Mar. Biodiv. Rec. 2: e148.

Chabanet, P., H. Ralambondrainy, M. Amanieu, G. Faure and R. Galzin, 1997. Relationships between coral reef substrata and fish. Coral Reefs. 16: 93–102.

Chaves, L.T.C. and C. Monteiro-Neto. 2009. Comparative analysis of rocky reef fish community structure in coastal islands of south-eastern Brazil. J. Mar. Biol. Assoc. U.K. 89: 609–619.

Chaves, L.T.C., P.H.C. Pereira and J.L.L. Feitosa. 2013. Coral reef fish association with macroalgal beds on a tropical reef system in North-eastern Brazil. Mar. Freshw. Res. 64: 1101–1111.

Cheal, A., M. Emslie, I. Miller and H. Sweatman. 2012. The distribution of herbivorous fishes on the Great Barrier Reef. Mar. Biol. 159: 1143–1154.

Cheal, A.J., M. Emslie, M.A. MacNeil, I. Miller and H. Sweatman. 2013. Spatial variation in the functional characteristics of herbivorous fish communities and the resilience of coral reefs. Ecol. Appl. 23: 174–188.

Chevalier, P.P. and A.L. Cárdenas. 2005. Variación espacial y temporal de las asociaciones de peces en arrecifes costeros de la costa oriental de la Bahía de Cochinos. I: Abundancia y Diversidad. Rev. Invest. Mar. 26: 45–57.

Choat, J., L. Herwerden, D.R. Robertson and K.D. Clements. 2012. Patterns and processes in the evolutionary history of parrotfishes (Family Labridae). Biol. J. Linnean Soc. 107: 529–557.

Cinner, J.E., N.A. Graham, C. Huchery and M.A. Macneil. 2013. Global effects of local human population density and distance to markets on the condition of coral reef fisheries. Conserv. Biol. 27: 453–458.

Cinner, J.E., C. Huchery, M.A. MacNeil, N.A.J. Graham, T.R. McClanahan, J. Maina, E. Maire, J.N. Kittinger, C.C. Hicks, C. Mora, E.H. Allison, S. D'Agata, A.S. Hoey, D.A. Feary, L. Crowder, I.D. Williams, M. Kulbicki, L. Vigliola, L. Wantiez, G. Edgar, R.D. Stuart-Smith, S.A. Sandin, A.L. Green, M.J. Hardt, M. Beger, A. Friedlander, S.J. Campbell, K.E. Holmes, S.K. Wilson, E. Brokovich, A.J. Brooks, J.J. Cruz-Motta, D.J. Booth, P. Chabanet, C. Gough, M. Tupper, S.C.A. Ferse, U.R. Sumaila and D. Mouillot. 2016. Bright spots among the world's coral reefs. Nature, 535: 416–419.

Cole, A.J., M.S. Pratchett and G.P. Jones. 2008. Diversity and functional importance of coral-feeding fishes on tropical coral reefs. Fish Fish. 9: 286–307.

Connor, E.F. and E.D. McCoy. 1979. The statistics and biology of the species-area relationship. Am. Nat. 113: 791–833.

Cordeiro, C.A.M.M., T.C. Mendes, A.R. Harborne and C.E.L. Ferreira. 2016. Spatial distribution of nominally herbivorous fishes across environmental gradients on Brazilian rocky reefs. J. Fish Biol. 89: 939–958.

D'agata, S., D. Mouillot, M. Kulbicki, S. Andréfouët, D.R. Bellwood, J.E. Cinner, P.F. Cowman, M. Kronen, S. Pinca and L. Vigliola. 2014. Human-mediated loss of phylogenetic and functional diversity in coral reef fishes. Curr. Biol. 24: 555–560.

Debrot, D., J.H. Choat, J.M. Posada and D.R. Robertson, 2008. High densities of the large bodied parrotfishes (Scaridae) at two Venezuelan offshore reefs: comparison among four localities in the Caribbean. Proc. 60th Gulf Caribbean Fish. Instit. Punta Cana, Dominican Republic, 335–338.

Descombes, P., M.S. Wisz, F. Leprieur, V. Parravicini, C. Heine, S.M. Olsen, D. Swingedouw, M. Kulbicki, D. Mouillot and L. Pellissier, 2015. Forecasted coral reef decline in marine biodiversity hotspots under climate change. Glob. Change Biol. 21: 2479–2487.

Dominici-Arosemena, A. and M. Wolff. 2005. Reef fish community structure in Bocas del Toro (Caribbean, Panama): gradients in habitat complexity and exposure. Caribb. J. Sci. 41: 613–637.

Dorenbosch, M., M.G.G. Grol, I. Nagelkerken and G. Van der Velde. 2006. Seagrass beds and mangroves as potential nurseries for the threatened Indo-Pacific humphead wrasse, Cheilinus undulatus and Caribbean rainbow parrotfish, *Scarus guacamaia*. Biol. Conserv. 129 : 277–282.

Elise, S. 2012. Variations spatio-temporelles de la structure du peuplement de poissons sur le recif de Dos Mosquises – Parc National de l'archipel des Roques, Venezuela –. Diplome de l'Ecole Pratique des Hautes Etudes, Perpignan. 122 pages and annexes.

Eschmeyer, W.N. 2015. Catalog of fishes. http://www.calacademy.org/scientists/catalog-of-fishes. Accessed July 2015.

Ferreira, C.E.L. and J.E.A. Gonçalves, 2006. Community structure and diet of roving herbivorous reef fishes in the Abrolhos Archipelago, south-western Atlantic. J. Fish Biol. 69: 1533–1551.

Fitzpatrick, J.M., D.B. Carlon, C. Lippe and D.R. Robertson. 2011. The West Pacific diversity hotspot as a source or sink for new species? Population genetic insights from the Indo-Pacific parrotfish Scarus rubroviolaceus. Mol. Ecol. 20: 219–234.

Floeter, S.R., M.D. Behrens, C.E.L. Ferreira, M.J. Paddack and M.H. Horn. 2005. Geographical gradients of marine herbivorous fishes: patterns and processes. Mar. Biol. 147: 1435–1447.

Floeter, S.R., W. Krohling, J.L. Gasparini, C.E. Ferreira and I.R. Zalmon, 2007. Reef fish community structure on coastal islands of the southeastern Brazil: the influence of exposure and benthic cover. Environ. Biol. Fish. 78: 147–160.

Fonseca-Escalante, A.C. 2003. A rapid assessment at Cahuita National Park, Costa Rica, 1999 (Part 1: stony corals and algae). Evaluación rápida del Parque Nacional Cahuita, Costa Rica, 1999 (Part 1: corales rocosos y algas). Atoll Res. Bull. 496: 248–257.

Fox, R.J. and D.R. Bellwood. 2007. Quantifying herbivory across a coral reef depth gradient. Mar. Ecol. Prog. Ser. 339: 49–59.

Francini-Filho, R.B., C.M. Ferreira, E.O.C. Coni, R.L. De Moura and L. Kaufman. 2010. Foraging activity of roving herbivorous reef fish (Acanthuridae and Scaridae) in eastern Brazil: influence of resource availability and interference competition. J. Mar. Biol. Assoc. U.K. 90: 481–492.

Friedlander, A.M., E.K. Brown, P.L. Jokiel, W.R. Smith and K.S. Rodgers. 2003. Effects of habitat, wave exposure, and marine protected area status on coral reef fish assemblages in the Hawaiian archipelago. Coral Reefs. 22: 291–305.

Friedlander, A.M., C.F.G. Jeffrey, S.D. Hile, S.J. Pittman, M.E. Monaco and C. Caldow (eds.). 2013. Coral reef ecosystems of St. John, U.S. Virgin Islands: Spatial and temporal patterns in fish and benthic communities (2001–2009). NOAA Technical Memorandum 152. Silver Spring, MD.

González-Sansón, G., C. Aguilar, I. Hernández, Y. Cabrera and A. Curry. 2009. The influence of habitat and fishing on reef fish assemblages in Cuba. Gulf Caribb. Res. 21: 13–21.

Green, A.L. and D.R. Bellwood. 2009. Monitoring functional groups of herbivorous reef fishes as indicators of coral reef resilience. A practical guide for coral reef managers in the Asia Pacific region. IUCN, Gland, Switzerland.

Gust, N., J. Choat and J. Ackerman. 2002. Demographic plasticity in tropical reef fishes. Mar. Biol. 140: 1039–1051.

Halpern, B.S. 2003. The impact of marine reserves: do reserves work and does reserve size matter?. Ecol. Appl. 13: 117–137.

Heenan, A., A.S. Hoey, G.J. Williams and I.D. Williams. 2016. Natural bounds on herbivorous coral reef fishes. Proc. R. Soc. B. 283: rspb.2016.1716.

Hernández-Landa, R.C., G. Acosta-González, E. Núñez-Lara and J.E Arias-González. 2015. Spatial distribution of surgeonfish and parrotfish in the north sector of the Mesoamerican Barrier Reef System. Mar. Ecol. 36: 432–446.

Hoey, A.S. and D.R. Bellwood. 2008. Cross-shelf variation in the role of parrotfishes on the Great Barrier Reef. Coral Reefs 27: 37–47.

Hoey, A.S. and D.R. Bellwood. 2010. Cross-shelf variation in browsing intensity on the Great Barrier Reef. Coral Reefs 29: 499–508.

Hoey, A.S. and D.R. Bellwood. 2011. Suppression of herbivory by macroalgal density: a critical feedback on coral reefs? Ecol. Lett. 14: 267–273.

Hoey, A.S., M.S. Pratchett and C. Cvitanovic. 2011. High macroalgal cover and low coral recruitment undermines the potential resilience of the world's southernmost coral reef assemblages. PLoS One 6: e25824.

Hoey, A.S., D.A. Feary, J.A. Burt, G. Vaughan, M.S. Pratchett and M.L. Berumen. 2016. Regional variation in the structure and function of parrotfishes on Arabian reefs. Mar. Poll. Bull. 105: 524–531.

Holbrook, S.J., R.J. Schmitt, V. Messmer, A.J. Brooks, M. Srinivasan, P.L. Munday and G.P. Jones. 2015. Reef fishes in biodiversity hotspots are at greatest risk from loss of coral species. PLoS One 10: e0124054.

Hughes, T.P., M.J. Rodrigues, D.R. Bellwood, D. Ceccarelli, O. Hoegh-Guldberg, L. McCook, N. Moltchaniwskyj, M.S. Pratchett, R.S. Steneck and B. Willis. 2007. Phase shifts, herbivory, and the resilience of coral reefs to climate change. Curr. Biol. 17: 360–365.

Hughes, T.P., J.T. Kerry, M. Álvarez-Noriega, J.G. Álvarez-Romero, K.D. Anderson, A.H. Baird, R.C. Babcock, M. Beger, D.R. Bellwood, R. Berkelmans, T.C. Bridge, I.R. Butler, M. Byrne, N.E. Cantin, S. Comeau, S.R. Connolly, G.S. Cumming, S.J. Dalton, G. Diaz-Pulido, C.M. Eakin, W.F. Figueira, J.P. Gilmour, H.B. Harrison, S.F. Heron, A.S. Hoey, J.-P.A. Hobbs, M.O. Hoogenboom, E.V. Kennedy, C.-Y. Kuo, J.M. Lough, R.J. Lowe, G. Liu, M.T. McCulloch, H.A. Malcolm, M.J. McWilliam, J.M. Pandolfi, R.J. Pears, M.S. Pratchett, V. Schoepf, T. Simpson, W.J. Skirving, B. Sommer, G. Torda, D.R. Wachenfeld, B.L. Willis and S.K. Wilson. 2017. Global warming and recurrent mass bleaching of corals. Nature 543: 373–377.

Jackson, J.B.C., M.K. Donovan, K.L. Cramer and V.V. Lam. (eds.). 2014. Status and Trends of Caribbean Coral Reefs: 1970–2012. Global Coral Reef Monitoring Network, IUCN, Gland, Switzerland.

Jackson, J.B., M.X. Kirby, W.H. Berger, K.A. Bjorndal, L.W. Botsford, B.J. Bourque, R.H. Bradbury, R. Cooke, J. Erlandson, J.A. Estes and T.P. Hughes. 2001. Historical overfishing and the recent collapse of coastal ecosystems. Science. 293: 629–637.

Jaxion-Harm, J., J. Saunders and M.R. Speight. 2012. Distribution of fish in seagrass, mangroves and coral reefs: life-stage dependent habitat use in Honduras. Int. J. Tropical Biol. Conserv. 60: 683–698.

Johansson, C.L., I.A. van de Leemput, M. Depczynski, A.S. Hoey and D.R. Bellwood. 2013. Key herbivores reveal limited functional redundancy on inshore coral reefs. Coral Reefs 32: 963–972.

Klomp, K.D. and D.J. Kooistra. 2003. A post-hurricane, rapid assessment of reefs in the windward Netherlands Antilles (stony corals, algae and fishes). Atoll Res. Bull. 496: 404–437.

Kobayashi, D.R., A. Friedlander, C. Grimes, R. Nichols and B. Zgliczynski. 2011. Bumphead Parrotfish (*Bolbometopon muricatum*) status review. U.S. Dep. Commer., NOAA Tech. Memo., NOAA-TM-NMFS-PIFSC-26.

Kopp, D., Y. Bouchon-Navaro, M. Louis, P. Legendre and C. Bouchon. 2010. Spatial and temporal variation in a Caribbean herbivorous fish assemblage. J. Coastal Res. 28: 63–72.

Krajewski, J.P. and S.R. Floeter. 2011. Reef fish community structure of the Fernando de Noronha Archipelago (Equatorial Western Atlantic): the influence of exposure and benthic composition. Environ. Biol. Fish. 92: 25–40.

Kulbicki, M., P. Labrosse and Y. Letourneur. 2000a. Stock assessment of commercial fishes in the northern New Caledonian lagoon - 2 - lagoon bottom and near reef fishes. Aq. Living Res. 13: 77–90.

Kulbicki, M., R. Galzin, M. Harmelin-Vivien, G. Mou Tham and S. Andréfouët. 2000b. Les communautés de poissons lagonaires dans les atolls des Tuamotu, principaux résultats du programme TYPATOLL (1995–1996). Nouméa, IRD, Doc.Sci. Tech. II3: 26–125.

Kulbicki, M., V. Parravicini, D.R. Bellwood, E. Arias-Gonza`lez, P. Chabanet, S.R. Floeter, A. Friedlander, J. McPherson, R.E. Myers, L. Vigliola and D. Mouillot. 2013. Global biogeography of reef fishes: a hierarchical quantitative delineation of regions. PLoS One 8 : e81847.

Kulbicki, M., D. Mouillot and V. Parravicini. 2015. Patterns and processes linked to body size. pp. 104–115. *In:* C. Mora (ed.). Ecology of Fishes on Coral Reefs. Oxford Univ. Press, Oxford.

Leprieur, F., S. Colosio, P. Descombes, V. Parravicini, M. Kulbicki, P.F. Cowman, D.R. Bellwood, D. Mouillot and L. Pellissier. 2015. Historical and contemporary determinants of global phylogenetic structure in tropical reef fish faunas. Ecography 38: 1–11.

Lester, S.E. and B.S. Halpern. 2008. Biological responses in marine no-take reserves versus partially protected areas. Mar. Ecol. Prog. Ser. 367: 49–56.

Letourneur, Y., M. Kulbicki and P. Labrosse. 2000. Fish stock assessment of the northern New Caledonian lagoons: 1–Structure and stocks of coral reef fish communities. Aq. Living Res. 13: 65–76.

Loffler, Z., D.R. Bellwood and A.S. Hoey. 2015. Among-habitat algal selectivity by browsing herbivores on an inshore coral reef. Coral Reefs 34: 597–605.

Longo, G.O., R.A. Morais, C.D.L. Martins, T.C. Mendes, A.W. Aued, D.V. Cândido, J.C. de Oliveira, J.C. Nunes, L. Fontoura, M.N. Sissini and M.M. Teschima. 2015. Between-habitat variation of benthic cover, reef fish assemblage and feeding pressure on the benthos at the only atoll in South Atlantic: Rocas Atoll, NE Brazil. PLoS One. 10: e0127176.

Loreau, M. 2000. Biodiversity and ecosystem functioning: recent theoretical advances. Oikos. 91: 3–17.

Luiz, O.J., A.P. Allena, D.R. Robertson, S.R. Floeter, M. Kulbicki, L. Vigliola, R. Becheler and J.S. Madin. 2013. Adult and larval traits as determinants of geographic range size among tropical reef fishes. Proc. Natl. Acad. Sci. U.S.A. 110: 16498–16502.

McCook, L., J. Jompa and G. Diaz-Pulido. 2001. Competition between corals and algae on coral reefs: a review of evidence and mechanisms. Coral Reefs. 19: 400–417.

Mellin, C., M. Kulbicki and D. Ponton. 2007. Seasonal and ontogenetic patterns of habitat use in coral reef fish juveniles. Estuar. Coast. Shelf Sci. 75: 481–491.

Mendonça-Neto, J.P.D., C. Monteiro-Neto and L.E. Moraes. 2008. Reef fish community structure on three islands of Itaipu, Southeast Brazil. Neotrop. Ichthyol. 6: 267–274.

Messmer, V., G.P. Jones, P.L. Munday, S.J. Holbrook, R.J. Schmitt and A.J. Brooks. 2011. Habitat biodiversity as a determinant of fish community structure on coral reefs. Ecology. 92: 2285–2298.

Molina-Ureña, H. 2009. Towards an ecosystem approach for non-target reef fishes: habitat uses and population dynamics of South Florida parrotfishes (Perciformes: Scaridae). Ph.D. dissertation. University of Miami, Florida (USA). 323 pp.

Mumby, P.J. 2006. The impact of exploiting grazers (Scaridae) on the dynamics of Caribbean coral reefs. Ecol. Appl. 16: 747–769.

Mumby, P.J. and C.C. Wabnitz. 2002. Spatial patterns of aggression, territory size, and harem size in five sympatric Caribbean parrotfish species. Environ. Biol.Fish. 63: 265–279.

Mumby, P.J., A.J. Edwards, J.E. Arias-González, K.C. Lindeman, P.G. Blackwell, A. Gall, M.I. Gorczynska, A.R. Harborne, C.L. Pescod, H. Renken, C.C.C. Wabnitz and G. Llewellyn. 2004. Mangroves enhance the biomass of coral reef fish communities in the Caribbean. Nature 427: 533–536.

Mumby, P.J., A. Hastings and H.J. Edwards. 2007. Thresholds and the resilience of Caribbean coral reefs. Nature. 450: 98–101.

Nagelkerken, I. 2007. Are non-estuarine mangroves connected to coral reefs through fish migration? Bull. Mar. Sci. 80: 595–607.

Nagelkerken, I., M. Dorenbosch, W.C.E.P. Verberk, E. Cocheret de la Moriniere and G. van der Velde. 2000. Importance of shallow-water biotopes of a Caribbean bay for juvenile coral reef fishes:

patterns in biotope association, community structure and spatial distribution. Mar. Ecol. Prog. Ser. 202: 175–192.

Nagelkerken, I., G. Van Der Velde and E.C. de la Morinière. 2001. Fish feeding guilds along a gradient of bay biotopes and coral reef depth zones. Aq. Ecol. 35: 73–86.

Nagelkerken, I., C.M. Roberts, G. Van Der Velde, M. Dorenbosch, M.C. Van Riel, E.C. De La Moriniere and P. H. Nienhuis. 2002. How important are mangroves and seagrass beds for coral-reef fish? The nursery hypothesis tested on an island scale. Mar. Ecol. Prog. Ser. 244: 299–305.

Nemeth, M. and R. Appeldoorn. 2009. The distribution of herbivorous coral reef fishes within fore-reef habitats: the role of depth, light and rugosity. Caribb. J. Sci. 45 : 247–253.

Nunez-Lara, E., C. Gonzalez-Salas, M.A. Ruiz-Zarate, R. Hernandez-Landa and J.E. Arias-Gonzalez. 2003. Condition of coral reef ecosystems in central-southern Quintana Roo (Part 2: Reef fish communities). Atoll Res. Bull. 496: 338–359.

Ong, L. and K.N. Holland. 2010. Bioerosion of coral reefs by two Hawaiian parrotfishes: species, size differences and fishery implications. Mar. Biol. 157: 1313–1323.

Opitz, S. 1996. Trophic interactions in Caribbean coral reefs. ICLARM. Manilla, Philippines.

Parenti, P. and J.E. Randall. 2011. Checklist of the species of the families Labridae and Scaridae: an update. Smithiana Bull. 13: 29–44.

Parravicini. V., M. Kulbicki, D.R. Bellwood, A.M. Friedlander, E. Arias-Gonzales, P. Chabanet, S.R. Floeter, L. Vigliola, S. D'Agata, R. Myers and D. Mouillot. 2013. Global patterns and predictors of tropical reef fish species richness. Ecography 36: 1–9.

Parravicini, V., S. Villéger, T.R. McClanahan, J.E. Arias-González, D.R. Bellwood, J. Belmaker, P. Chabanet, S.R. Floeter, A.M. Friedlander, F. Guilhaumon, L. Vigliola, M. Kulbicki and D. Mouillot. 2014. The vulnerability framework indicates alternative global protection priorities for coral reef fishes. Ecol. Lett. 17: 1101–1110.

Parravicini, V., E. Azzurro, M. Kulbicki and J. Belmaker. 2015. Niche shift can impair the ability to predict invasion risk in the marine realm: an illustration using Mediterranean fish invaders. Ecol. Lett. 18: 246–253.

Pattengill-Semmens, C.V. and S.R. Gittings. 2003. A rapid assessment of the Flower Garden Banks National Marine Sanctuary (stony corals, algae and fishes). Atoll Res. Bull. 496: 500–511.

Peckol, P.M., A. Curran, E. Floyd, M. Robbart, B.J. Greenstein and K. Buckman. 2003a. Assessment of selected reef sites in northern and south-Central Belize, including recovery from bleaching and hurricane disturbances (stony corals, algae and fish). Atoll Res. Bull. 496: 146–171.

Peckol, P.M., A. Curran, B.J. Greenstein, E.Y. Floyd and M.L. Robbart. 2003b. Assessment of coral reefs off San Salvador Island, Bahamas (stony corals, algae and fish populations). Atoll Res. Bull. 496: 124–145.

Pellissier, L., F. Leprieur, V. Parravicini, P. Cowman, M. Kulbicki, G. Litsios, S. Olesen, M. Wisz, D.R. Bellwood and D. Mouillot. 2014. Quaternary coral reef refugia preserved fish diversity. Science. 344: 1016–1019.

Pereira, P.H.C., R.L. Moraes, M.V.B. dos Santos, D.L. Lippi, J.L.L. Feitosa and M. Pedrosa. 2014. The influence of multiple factors upon reef fish abundance and species richness in a tropical coral complex. Ichthyol. Res. 61: 375–384.

Pinca, S., M. Kronen, F. Magron, B. McArdle, L. Vigliola, M. Kulbicki and S. Andréfouët. 2011. Relative importance of habitat and fishing in influencing reef fish communities across seventeen Pacific islands countries and territories. Fish. Res. 13: 361–379.

Pinheiro, H.T., C.E.L. Ferreira, J.C. Joyeux, R.G. Santos and P.A. Horta. 2011. Reef fish structure and distribution in a south-western Atlantic Ocean tropical island. J. Fish Biol. 79: 1984–2006.

Pinheiro, H.T., E. Mazzei, R.L. Moura, G.M. Amado-Filho, A. Carvalho-Filho, A.C. Braga, P.A. Costa, B.P. Ferreira, C.E.L. Ferreira, S.R. Floeter and R.B. Francini-Filho, 2015. Fish biodiversity of the Vitória-Trindade Seamount Chain, southwestern Atlantic: an updated database. PLoS One, 10: e0118180.

Pittman, S.J., S.D. Hile, C.F.G. Jeffrey, C. Caldow, M.S. Kendall, M.E. Monaco and Z. Hillis-Starr. 2008. Fish assemblages and benthic habitats of Buck Island Reef National Monument (St. Croix, U.S. Virgin Islands) and the surrounding seascape: A characterization of spatial and temporal patterns. NOAA Technical Memorandum NOS NCCOS 71. Silver Spring, MD.

Pittman, S.J., S.D. Hile, C.F.G. Jeffrey, R. Clark, K. Woody, B.D. Herlach, C. Caldow, M.E. Monaco and R. Appeldoorn. 2010. Coral reef ecosystems of Reserva Natural La Parguera (Puerto Rico): Spatial and temporal patterns in fish and benthic communities (2001–2007). NOAA Technical Memorandum NOS NCCOS 107. Silver Spring, MD.

Pratchett, M.S., P.L. Munday, S.K. Wilson, N.A.J. Graham, J.E. Cinner and D.R. Bellwood. 2008. Effects of climate-induced coral bleaching on coral-reef fishes – ecological and economic consequences. Ocean. Mar. Biol. Annu. Rev. 46: 251–296.

Randall, J.E. 1983. Caribbean reef fishes. T.F.H. Publications, Neptune City, N.J. (USA). 368 pp.

Rasher, D.B., A.S. Hoey and M.E. Hay. 2013. Consumer diversity interacts with prey defenses to drive ecosystem function. Ecology 94: 1347–1358.

Richardson, L.E., N.A. Graham, M.S. Pratchett and A.S. Hoey. 2017. Structural complexity mediates functional structure of reef fish assemblages among coral habitats. *Environ. Biol. Fish.* 100: 193–207.

Roff, G., M.H. Ledlie, J.C. Ortiz and P. J. Mumby. 2011. Spatial patterns of parrotfish corallivory in the Caribbean: the importance of coral taxa, density and size. PLoS One. 6: e29133.

Rojas, C.D., R. Claro Madruga, P.P. Chevalier Monteagudo, S. Perera Valderrama and H. Caballero Aragón. 2013. Estructura de las asociaciones de peces en los arrecifes coralinos del parque nacional Guanahacaribes, Cuba. Rev. Mar. Cost. 3: 153–169.

Rossier, O. and M. Kulbicki, 2000. A comparison of fish assemblages from two types of algae beds and coral reefs in the South-West lagoon of New Caledonia. Cybium 24: 3–26.

Rousseau, Y. 2010. Structure des peuplements ichthyologiques des récifs coralliens de La Martinique en relation avec la qualité de l'habitat. PhD Thesis, University of Perpignan, Perpignan, France.

Russ, G.R. 1984a. Distribution and abundance of herbivorous grazing fishes in the central Great Barrier Reef. I. Levels of variability across the entire continental shelf. Mar. Ecol. Prog. Ser. 20: 23–34.

Russ, G.R. 1984b. Distribution and abundance of herbivorous grazing fishes in the central Great Barrier Reef. II: Patterns of zonation of mid-shelf and outer-shelf reefs. Mar. Ecol. Prog. Ser. 20: 35–44.

Russ, G.R. 2003. Grazer biomass correlates more strongly with production than with biomass of algal turfs on a coral reef. Coral Reefs 22: 63–67.

Schmidt, C. 2013. As threats to corals grow, hints of resilience emerge. Science 339: 1517–1519.

Sedberry, G.R., H.J. Carter and P.A. Barrick 1999. A comparison of fish communities between protected and unprotected areas of the Belize reef ecosystem: implications for conservation and management. Proc. Gulf Caribb. Fish. Inst. 45: 95–127.

Spalding, M.D., H.E. Fox, G.R. Allen, N. Davidson, Z.A. Ferdana, M.A.X. Finlayson, B.S. Halpern, M.A. Jorge, A.L. Lombana, S.A. Lourie and K.D. Martin. 2007. Marine ecoregions of the world: a bioregionalization of coastal and shelf areas. BioScience. 57: 573–583.

Stockman, R. 2010. Comparison of reef fish populations at two sites off the southwest shore of the Commonwealth of Dominica. pp. 1–13. *In*: T. Lacher and W. Heyman (eds.). Study Abroad Dominica 2010. Dominica.

Streelman, J.T., M. Alfaro, M.W. Westneat, D.R. Bellwood and S.A. Karl. 2002. Evolutionary history of the parrotfishes: biogeography, ecomorphology, and comparative diversity. Evol. 56: 961–971.

Tâmega, F.T.S., M.A.O. Figueiredo, C.E.L. Ferreira and R.M. Bonaldo, 2015. Seaweed survival after consumption by the greenbeak parrotfish, *Scarus trispinosus*. Coral Reefs 35: 329–334.

Taylor, B.M., S.J. Lindfield and J.H. Choat. 2015. Hierarchical and scale-dependent effects of fishing pressure and environment on the structure and size distribution of parrotfish communities. Ecography. 38: 520–530.

Toller, W., A.O. Debrot, M.J.A. Vermeij and P.C. Hoetjes. 2010. Reef Fishes of Saba Bank, Netherlands Antilles: Assemblage Structure across a Gradient of Habitat Types. PLoS One 5: e9207.

Trapon, M.L., M.S. Pratchett, A.S. Hoey and A.H. Baird. 2013a. Influence of fish grazing and sedimentation on the early post-settlement survival of the tabular coral *Acropora cytherea*. Coral Reefs 32: 1051–1059.

Trapon, M.L., M.S. Pratchett, M. Adjeroud, A.S. Hoey and A.H. Baird. 2013b. Post-settlement growth and mortality rates of juvenile scleractinian coral in Moorea, French Polynesia versus Trunk Reef, Australia. Mar. Ecol. Prog. Ser. 488: 157–170.

Trapon, M.L., M.S. Pratchett and A.S. Hoey. 2013c. Spatial variation in abundance, size and orientation of juvenile corals related to the biomass of parrotfishes on the Great Barrier Reef, Australia. PLoS ONE 8: e57788.

van Rooij, J.M., E. de Jong, F. Vaandrager and J.J. Videler. 1996. Resource and habitat sharing by the stoplight parrotfish, Sparisoma viride, a Caribbean reef herbivore. Environ. Biol. Fish. 47: 81–91.

Vermeij, M.J.A., R.A. van der Heijden, J.G. Olthuis, K.L. Marhaver, J.E. Smith and P.M. Visser. 2013. Survival and dispersal of turf algae and macroalgae consumed by herbivorous coral reef fishes. Oecologia 171: 417–425.

Walter, R.P. 2002. Fish assemblages associated with coral patch reef communities at San Salvador, Bahamas. M.S. Thesis. State University of New York, Brockport, New York.

Wantiez, L., P. Thollot and M. Kulbicki. 1997. Effects of marine reserves on coral reef fish communities from five islands in New Caledonia. Coral Reefs 16: 215–224.

Westneat, M.W. and M.E. Alfaro. 2005. Phylogenetic relationships and evolutionary history of the reef fish family Labridae. Mol. Phylogen. Evol. 36: 370–390.

Williams, I.D., W.J. Walsh, R.E. Schroeder, A.M. Friedlander, B.L. Richards and K.A. Stamoulis. 2008. Assessing the importance of fishing impacts on Hawaiian coral reef fish assemblages along regional-scale human population gradients. Environ. Conserv. 35: 261–272.

Wismer, S., A.S. Hoey and D.R. Bellwood. 2009. Cross-shelf benthic community structure on the Great Barrier Reef: relationships between macroalgal cover and herbivore biomass. Mar. Ecol. Prog. Ser. 376: 45–54.

CHAPTER
11

Phenological Aspects of Parrotfish Ecology on Coral Reefs

Yohannes Afeworki[1] and Henrich Bruggemann[2]

[1] Department of Mathematics and Statistics, South Dakota State University,
P.O. Box 2225, Brookings, SD 57007, USA
Email: Yohannes.Tecleab@sdstate.edu
[2] UMR ENTROPIE Université de la Réunion-CNRS-IRD, CS 92003, 97744 Saint Denis,
La Réunion, France; Laboratoire d'Excellence CORAIL, www.labex-corail.fr
Email: henrich.bruggemann@univ-reunion.fr

Introduction

Spatial and temporal variation in resources exerts a strong influence on population processes in coral reef fishes. Increases in herbivorous fish populations have been recorded in response to increases in the cover of turf algae following large-scale coral mortality (Lindahl et al. 2001, Adam et al. 2011, Gilmour et al. 2013). Similarly, opportunistic feeding on coral spawn has been linked with improved condition and higher reproductive output in reef fishes (McCormick 2003). Spatial variation in resources is likewise linked to condition and population size in parrotfishes (Tootell and Steele 2015).

Many coral reefs are located on the continental shelves and are influenced by seasonal changes in rainfall, wind, irradiance, upwelling and temperature (McClanahan 1988, Diaz-Pulido and Garzon-Ferreira 2002, Done et al. 2007, Abesamis and Russ 2010, Fulton et al. 2014). In comparison to temperate and polar regions, the amplitude of seasonality in environmental conditions is generally smaller on coral reefs. Despite the smaller magnitude of the cycle in the tropics, algal communities display similar seasonality in productivity and species composition across a wide range of latitudes (Kain 1989, Diaz-Pulido and Garzon-Ferreira 2002, Ateweberhan et al. 2006a, Ferrari et al. 2012). For high latitude environments, ecological data are typically collected and interpreted in a seasonal context. Likewise, the pioneering works on coral reef algae have studied and found significant seasonal effects on productivity and grazing (Carpenter 1986, Klumpp and Mckinnon 1989, Polunin and Klumpp 1992). Frequently, however, environmental conditions in coral reefs have been assumed to be relatively stable. This has led to less emphasis on seasonal aspects of the ecology of coral reefs.

This chapter focuses on the seasonal changes in benthic algal communities on coral reefs and their effects on the ecology of parrotfishes. The consequences of seasonal changes in food resources can be expected to propagate through the food web but are likely to be more pronounced in the lower trophic levels, such as grazers (Clarke 1988). The effects of predictable temporal changes in food supply on foraging behaviour of parrotfishes and aspects of their life-history, such as growth and reproduction, have received limited attention. We begin by summarizing seasonal trends in coral reef benthic algae from different geographic regions. Then we discuss how parrotfishes handle the seasonal changes in food resources, and how changes in food abundance and intake affect the allocation of resources to reproduction and growth. To conclude, we present two case studies for which there is sufficient year-round sampling of resource abundance, feeding, growth and reproduction. One is the case of *Scarus iseri* in the Caribbean (Clifton 1995) and the second is that of *Scarus ferrugineus* in the southern Red Sea (Afeworki 2014).

Seasonality in Benthic Algae

After briefly outlining the primary dietary targets of parrotfishes, we will discuss the temporal dynamics of these sources of nutrition. Readers interested in greater detail of the feeding behaviour and nutritional ecology of parrotfishes can refer to Chapter 3 of this volume, as well as several reviews on the subject (Steneck 1988, Horn 1989, Choat 1991, Choat and Clements 1998, Wilson et al. 2003, Clements et al. 2009, Bonaldo et al. 2014). In short, based on jaw morphology, associated musculature and strength and the frequency of bites, three main grazing modes or functional groups have been defined in parrotfishes: browsing, excavating and scraping (Green and Bellwood 2009). Species belonging to seagrass or *Sparisoma* clade (i.e., *Calotomus, Cryptotomus, Leptoscarus, Nicholsia* and *Sparisoma)* are mainly browsers targeting macroalgae and/or seagrasses (Green and Bellwood 2009, Bonaldo et al. 2014). Members of the reef or *Scarus* clade, that includes the specious and pantropical genus *Scarus* and the genera *Bolbometopon, Chlorurus, Cetoscarus* and *Hipposcarus*, are mainly specialized in exploiting low-biomass algal turfs and associated trapped detrital material, referred to as the epilithic algal matrix (EAM), and endolithic organisms (Clements and Choat Chapter 3). Most parrotfishes belonging to these genera scrape the EAM from the carbonate matrix with little impact on the underlying substrate and are referred to as scrapers (Bellwood and Choat 1990, Bonaldo et al. 2011). However, large individuals and members of the Indo-Pacific genera *Bolbometopon, Chlorurus* and *Cetoscarus* are capable of denuding the calcareous substrate and ingesting substrate-bound components of the EAM, i.e. crustose coralline algae and boring endolithic algae (Bellwood and Choat 1990, Bellwood 1994, Bruggemann et al. 1994a). The latter are called excavating and some western Atlantic species such as *Sparisoma viride* belong to this category (Bruggemann et al. 1996). While most EAM-feeding parrotfishes take less than one percent of daily bites on living corals (Bruggemann et al. 1994b, Afeworki et al. 2011), corals represent an important dietary component (30-50 percent) of the largest excavating parrotfishes such as *Cetoscarus bicolor* and *Bolbometopon muricatum* (Bellwood et al. 2003, Hoey and Bellwood 2008, Alwany et al. 2009, Hoey et al. 2016). So, while marked differences in feeding behaviour and substrate choice exist among these functional groups, the majority of parrotfishes feed primarily off substrata covered by EAM.

Epilithic Algal Matrix

The EAM is a complex assemblage of short, turf-forming filamentous algae (< one cm high), macroalgal propagules, microalgae, cyanobacteria, and detritus that covers the surface of dead corals and consolidated reef matrix (Wanders 1977, Hatcher and Larkum 1983, Scott and Russ 1987, Wilson et al. 2003). The biomass and nutritional value of the detrital component may be comparable to or even higher than that of the turf algae (Crossman et al. 2001, Wilson et al. 2003). The remaining component of the EAM is the meiofauna such as harpacticoid copepods which can be abundant in sediment laden turfs (Kramer et al. 2012, 2013). Crustose coralline algae and endolithic algae are ubiquitous substrate-bound components of the EAM.

The productivity, cover and biomass of EAM across any given reef are known to vary seasonally and are typically higher during the warmer months (Carpenter 1981, Ateweberhan et al. 2006a). Field measurements of the productivity of the algal components of EAM generally show highest values in summer when both photosynthetically active radiation (PAR) and water temperature are highest (Adey and Steneck 1985, Carpenter 1985, 1986, Klumpp and Mckinnon 1989, Fig. 1). In fact, the degree of seasonality in productivity of the algal turfs on coral reefs is comparable to that of turf communities on temperate reefs (Fig. 2), although in locations where fish and urchin communities have not been depleted, the high grazing rates of reef herbivores maintain a low standing crop of benthic algae on reefs in all seasons (Ebeling and Hixon 1991, Russ 2003). Around 72 – 79 percent of the detritus within the EAM appears to be sourced *in situ* from the dissolved organic carbon (DOC) of the turf algae (Wilson et al. 2001, Wilson et al. 2003). Compared to other benthic primary producers, turf algae release more DOC which can amount to 10 percent of their daily production (Haas et al. 2010b, 2011). Thus, the detrital component of the EAM can be expected to have similar seasonal variation in production as the turf algae. Interestingly, bacteria (which comprise up to 25 percent of the detritus) also appear to attain higher biomass and production in summer compared to winter in the Great Barrier Reef (Wilson et al. 2003).

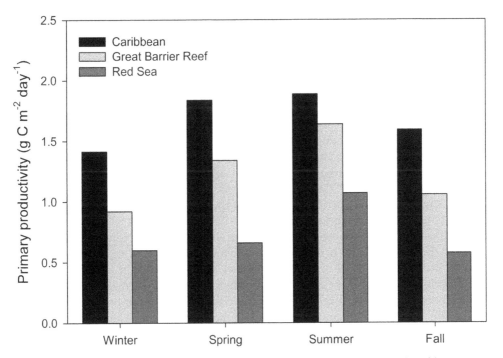

Fig. 1. Seasonal change in primary production of EAM in three major coral reef locations. Sources: Caribbean (van Rooij et al. 1998) ; Great Barrier Reef (Klumpp and Mckinnon 1992); Northern Red Sea (Rix et al. 2015).

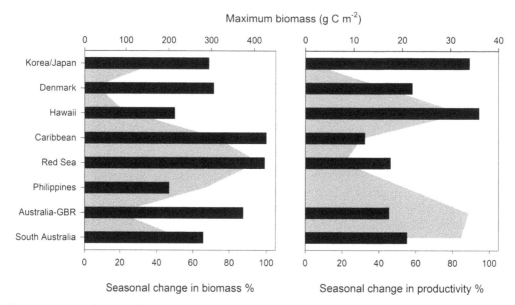

Fig. 2. Degree of seasonality expressed as percentage change in biomass (macroalgae, left-hand graph) or daily production (turf algae, right-hand graph). Grey area is the maximum estimated biomass at each site. Sources: Japan and Korea (Fuji et al. 1991, Kang et al. 2011); Denmark (Pedersen et al. 2005, Middelboe et al. 2006); Hawaii (de Wreede 1976, Smith et al. 2001); Caribbean (Carpenter 1985, Camacho and Hernández-Carmona 2012); Red Sea (Ateweberhan et al. 2006a, Rix et al. 2015); Philippines (Hurtado and Ragaza 1999); Great Barrier Reef (Klumpp and Mckinnon 1992, Schaffelke and Klumpp 1997); South Australia (Cheshire et al. 1996, Copertino et al. 2005).

On coral reefs the nutritional quality of the EAM has been estimated by the ratio of organic matter to sediment, with high sediment content deterring grazing by most herbivorous fishes, including parrotfishes (Purcell and Bellwood 2001, Bellwood and Fulton 2008, Goatley et al. 2010). Intense cropping of the EAM during the warmer seasons (Polunin and Klumpp 1992, Ateweberhan et al. 2006a, Afeworki et al. 2013b) makes this resource less prone to sediment loading and therefore of higher nutritional quality (Purcell 2000, Purcell and Bellwood 2001, Crossman et al. 2001). The EAM is, furthermore, probably of higher nutritional value in summer due to the increased nitrogen fixation by cyanobacteria (Wilkinson and Sammarco 1983, Wilkinson et al. 1985, Larkum et al. 1988, Williams and Carpenter 1997, Casareto et al. 2008, Rix et al. 2015). Taken together, these two factors suggest that the nutritional quality of the EAM is likely to be higher in summer than in the cool season (Afeworki et al. 2013b).

In addition to locally sourced detritus, the EAM receives suspended organic matter (SOM) from the water column originating from nearby macroalgae, seagrasses, mangroves and corals (Rivera-Monroy et al. 1998, Wilson et al. 2003, Heck et al. 2008, Haas et al. 2010a). The detrital input from coastal macrophytes to the EAM may be important and widespread since many coral reef systems comprise shallow zones where macrophytes dominate the benthic communities, at least seasonally (Reinthal and Macintyre 1994, Ateweberhan et al. 2005a, 2006b, McCook 1997, Unsworth et al. 2009, Heck et al. 2008, Wismer et al. 2009). In the southern Red Sea, 60-90 percent of the >1200 g m^{-2} dry mass of foliose and canopy-forming macroalgae is lost during a brief period of early spring (Ateweberhan et al. 2005a, 2005b, 2006b, 2006a). A similar phenomenon occurs in seagrass communities where a significant proportion (estimates range from 0-100 percent) of the daily production is exported as detritus to nearby ecosystems such as coral reefs (Heck et al. 2008, Naumann et al. 2013). Since production in these communities is seasonal, their contribution to the detritus will likewise be temporally variable. For example, in the Houtman-Abrolhos Islands, the organic content of the water column increases significantly following the massive shedding of macroalgal thalli at the end of summer (Crossland et al. 1984). Similarly, detritus on the fringing reefs of Reunion Island increases concurrently with benthic algal production during the hot season (Kolasinski et al. 2011).

The meiofauna associated with EAM has not been studied much, but recent estimates show that this component may be considerable, suggesting a significant role in coral reef trophodynamics (Logan et al. 2008, Kramer et al. 2012). Harpacticoid copepods appear to be the dominant group in the EAM and reef sediments (Logan et al. 2008) and may be nutritionally important to fishes (John et al. 1989). Parrotfishes ingest large numbers of the herpactocoid copepods and, in spite of a low contribution in terms of mass (Kramer et al. 2013), their diet contribution in terms of nutrients may be substantial considering that the meiofauna is nutritionally richer than algae or detritus. It has been suggested that the meiofauna of EAM on the GBR is temporally stable, however this is based on two studies that collected samples during two periods (November and March) and with limited spatial replication (Logan et al. 2008, Kramer et al. 2013). The lack of temporal variation may reflect similarity of the sampling periods chosen, rather than true temporal stability. Greater sampling over a range of water temperatures and environmental conditions, especially from the cool and hot seasons, and across a greater number of reefs is needed to ascertain if the meiofaunal component of the EAM is stable.

Crustose Coralline Algae

Crustose coralline algae are often associated with the EAM, occurring under or adjacent to turf algal assemblages, and are typically more abundant in areas of high wave energy (Klumpp and Mckinnon 1992, Wismer et al. 2009). Temporal dynamics of crustose corallines are generally the opposite of that of turf algae, with highest cover and biomass during the cooler parts of the year (Diaz-Pulido and Garzon-Ferreira 2002, Ateweberhan et al. 2006a, Afeworki et al. 2011, Ferrari et al. 2012). While turfs are tolerant to high irradiance and temperature, crustose corallines often abound where there is shade (Baynes 1999, Caragnano et al. 2009) or during the cool season (Naim 1993, Ateweberhan et al. 2006a). This is fundamentally associated with the difference in the parameters of the photosynthesis-irradiance relationship (P-I curves) between these two algal groups. Crustose corallines characteristically show low photosynthetic saturation irradiances and perform better at lower light levels (Marsh 1970), with photoinhibition setting in at low irradiance levels (52-205 $\mu E\ m^{-2}\ s^{-1}$) (Chisholm 2003). Turf algae are more tolerant and can achieve maximum production at higher light intensities (742-1184 $\mu E\ m^{-2}\ s^{-1}$) (Carpenter 1985). A case in point is the southern Red Sea where extreme temperature and irradiance lead to declines of crustose coralline in summer, while during the cooler seasons their cover and biomass rebound (Ateweberhan et al. 2006a). Seasonal shifts in dominance from crustose corallines to algal turfs at the onset of summer have also been reported on coral reefs in East Africa (McClanahan 1997). These shifts are probably the result of turfs overtopping the crustose corallines.

Macroalgae and Seagrasses

Macroalgae often dominate on shallow reef flats and crests of coastal reefs or in deeper fore reef zones (Reinthal and Macintyre 1994, Vuki and Price 1994, Ateweberhan et al.

2005a, 2006b, Hoey and Bellwood 2010, Rasher et al. 2013), while they are generally less conspicuous in more exposed reef settings (Marsh 1976, de Wreede 1976, McCook 1996, Hoey and Bellwood 2009). On degraded reefs, however, high macroalgal biomass often extends down the fore reef slopes (Hughes 1994, Hughes et al. 1999, Cheal et al. 2010).

The standing biomass of foliose and canopy-forming macroalgae generally shows a seasonal peak, the timing of which depends on latitude (de Wreede 1976). At low latitudes, such as the central Indo-Pacific, Caribbean, southern Red Sea and western Indian Ocean and some high latitude locations (e.g., Hawaii), macroalgae achieve peak abundance and development during the cooler months (de Wreede 1976, McClanahan 1988, Clifton 1997, Hurtado and Ragaza 1999, Ateweberhan et al. 2005a, 2005b). At higher latitudes, and on coral reefs of the GBR, peak development of foliose and canopy forming algae occurs during the warmest months (Crossland et al. 1984, Vuki and Price 1994, Schaffelke and Klumpp 1997). The degree of seasonality in biomass of fucoid macroalgae in coral reefs is similar to that of temperate subtidal habitats (Kain 1989, Fig 2). In addition, there seems to be no association between latitude and biomass of macroalgae in subtidal habitats (Konar et al. 2010, Fig. 2). Macroalgae of some coral reef habitats, such as the southern Red Sea and the upwelling regions of the Caribbean, are highly productive and highly seasonal exhibiting changes in biomass of near 100 percent.

Following growth and reproduction, macroalgal thalli degenerate, are increasingly covered by epiphytic algae and epifauna (Benayahu and Loya 1977, Martin-Smith 1993, Ateweberhan et al. 2006a, Lefevre and Bellwood 2010), detach and contribute to the detritus pool of the nearby reefs (Schaffelke and Klumpp 1997, Hurtado and Ragaza 1999, Ateweberhan et al. 2005a, 2006b). Like macroalgae, seagrass communities inhabit the shallow subtidal and intertidal zones of many coral reefs. Biomass, productivity and nutritional content of seagrass communities vary seasonally and are typically maximal in summer in most coral reef areas (Erftemeijer and Herman 1994, McKenzie 1994, Fourqurean et al. 2001, Unsworth et al. 2007b).

Both macroalgae and seagrass communities contribute to coral reef trophodynamics either by direct browsing of herbivores or through detrital pathways. Several parrotfish species reside in seagrass and macroalgal habitats; some are transient visitors moving from the reef to browse on the seagrass and macroalgae at high tide, while some may utilize these habitats as juveniles (McAfee and Morgan 1996, Valentine et al. 2007, Unsworth et al. 2007a, Heck et al. 2008, Paddack and Sponaugle 2008, Wilson et al. 2010, Afeworki et al. 2013a). Moreover detached seagrass blades and macroalgae thalli are transported to coral reefs where they can be directly consumed by herbivores fishes (Kilar and Norris 1988, Wernberg et al. 2006, Stimson 2013). Detritus on nearby reefs is therefore enriched by this seasonal flux of organic matter from macroalgae and seagrass communities (Marsh 1976, Hatcher 1983, 1984, Crossland et al. 1984, Ateweberhan et al. 2005a, 2006a). Scraping and excavating parrotfishes are expected to exploit this substantial, yet temporally variable, contribution of seagrass and macroalgae to the detritus in the EAM (discussed above).

Seasonality in Foraging

Food Selection

In nature, parrotfishes encounter a complex environment in which they must select among numerous dietary resources whose availability and nutritional composition are spatially and temporally variable (as described above). The pioneering works on herbivore foraging and diet preference typically focused on single nutrient maximization approaches and many herbivores were thought to select their diet to maximize their daily energy intake

(Belovsky 1984, 1986, Stephens and Krebs 1986). Recent advances are showing that rather than maximizing one aspect of their nutrition, herbivores aim to regulate and balance their nutrient budget including energy, minerals, proteins, etc. (Raubenheimer and Simpson 2003, Raubenheimer et al. 2009). In other words, animals seem to aim to meet their daily needs of each nutrient to achieve a balanced diet (Rubio et al. 2003). In this case an animal is likely to target the food items that provide nutrients that are deficient in the environment and/or needed in higher quantities to satisfy life functions (Pretorius et al. 2012).

In seasonally changing environments, the nutritional target of herbivores is known to change depending on resource availability and the condition of the animals (van Marken Lichtenbelt 1993, Pretorius et al. 2012, Shannon et al. 2013, Irwin et al. 2014). For example, the African elephant forages to maximize nitrogen intake in the wet season and energy in the dry season (Pretorius et al. 2012). Similarly, herbivorous lizards (van Marken Lichtenbelt 1993) and primates (Felton et al. 2009a, 2009b) show seasonal changes in their diet. It is notable that most information on nutritional ecology of herbivores is derived from terrestrial studies, with research into the nutritional ecology of marine herbivores lagging behind (Clements et al. 2009). From temperate marine environments there is evidence that herbivorous fishes vary their diet in response to changes in resources (Horn et al. 1986, Clements and Choat 1993, Caceres et al. 1994). For example, on Californian rocky shores, the stichaeid fishes *Cebidichthys violaceus* and *Xiphister coalita* show significant seasonal dietary shifts in response to changing availability of benthic algae (Horn et al. 1986). In winter both species maximize their energy intake as they primarily feed on the carbohydrate-rich perennial species *Iridaea flaccida*. During summer, protein-rich annual algae (*Microcladia coulteri*, *Porphyra perforata*, and *Smithora naiadum*) become available and both fishes target these to maximize their protein intake. Similar studies linking the feeding behaviour of herbivores to the temporal variation of algal resources within tropical marine environments such as coral reefs are essentially lacking (but see Afeworki et al. 2011).

Investigating whether coral reef grazers, and parrotfishes in particular, have similar responses to their terrestrial counterparts or whether they have evolved different mechanisms is of considerable interest. Evidence is accumulating that parrotfishes are capable of selecting nutritionally richer resources (Bruggemann et al. 1994b, Boyer et al. 2004, Goecker et al. 2005, Fong et al. 2006, Furman and Heck 2008). Choice experiments in the field and in the laboratory using *Thalassia testudinum* blades with different nitrogen content (collected from three sites in the Florida Keys) showed that the parrotfish *Sparisoma radians* preferentially fed on high nitrogen blades of *T. testudinum*, resulting in 27-97 percent tissue removal from the nitrogen-enriched plants compared to 3-10 percent removal from plants with lower nitrogen content (Goecker et al. 2005). Similarly, Burkepile and Hay (2009) conducted factorial experiments testing the effects of nutrient enrichment and herbivore exclusion on benthic communities on a fringing reef in the Florida Keys. Video recordings of *Scarus* spp. and *Sparisoma* spp. feeding revealed that these species grazed 3-10 times more from EAM on the nutrient-enriched cinder blocks than those growing on low-nutrient control blocks. Parrotfishes that occasionally feed on live corals also appear to selectively target nutritionally-rich components of the colony, such as areas that are infested with nitrogen rich macroborers (Rotjan and Lewis 2005) or polyps loaded with developing gonads (Rotjan and Lewis 2009).

Controlled laboratory experiments of food selection in rainbow trout, sea bream, and goldfish have demonstrated that the mechanism of food selection in fish is post-ingestive, i.e. external traits of the food are learned by association with its nutritional quality (Sánchez-Vázquez et al. 1998, 1999, Rubio et al. 2003, Vivas et al. 2006). Parrotfishes are diurnal feeders and vision is their primary cue for selecting and directing a bite at a substrate (Rice and

Westneat 2005). However, the decision to continue to bite on a particular substrate appears to be made after ingestion. Browsers like *Sp. radians*, but also scraping and excavating parrotfishes, appear to take 'exploratory' bites from the available substrates before focusing a large number of bites on preferred substrates (Bruggemann et al. 1994b, Goecker et al. 2005). Thus, the tendency for parrotfishes to direct longer forays (sensu Bellwood and Choat 1990) on preferred food items points to a feeding decision reached after the first bites have been ingested (Bruggemann et al. 1994a, 1994b, Afeworki et al. 2011). To investigate if visual cues are responsible for the preference of *Sp. radians* for nitrogen-enriched seagrass blades, Goecker et al. (2005) conducted choice experiments on two visually similar but nutritionally different foods (ground-seagrass and agar mixture moulded in the shape of sea grass blades). *Sp. radians* fed significantly more from the nitrogen rich seagrass-agar mixture than from the mixture with low nitrogen seagrass. This suggests that *Sp. radians* may be using olfaction or gustation to select nutrient-rich blades of the seagrass *Thallasia testudinum*. These reports strongly suggest that parrotfishes may have high sensitivity to nutrient concentrations in different food items and may target different combinations of the food resources to balance their daily nutritional requirements.

Variations in feeding preference by *Sc. ferrugineus* in response to seasonal changes in the community composition of the benthic algae have been reported for the southern Red Sea (Afeworki et al. 2011). Like EAM feeding parrotfishes from the tropical Atlantic (Bruggemann et al. 1994a, Bruggemann et al. 1994b), this species also feeds predominantly on turf algae growing on substrata with abundant endolithic algae, especially in summer (75 percent of bites). This food type has been shown to enable higher yields per bite and thus to be nutritionally more profitable than crustose corallines with or without algal turf overtopping it (Bruggemann et al. 1994c). The abundance of algal turfs growing on substrates infested with endolithic algae is low in the cool season and *Sc. ferrugineus* compensates by taking more bites from turfs growing on crustose corallines. In the Caribbean the sea urchin *Diadema antillarum* shows a similar response to the reduction in turf availability in winter by feeding on more corals and crustose corallines (Carpenter 1981). On the GBR, Lefevre and Bellwood (2011) studied intake from transplanted *Sargassum* bioassays and recorded increased feeding by *Scarus rivulatus* during the winter months. In contrast, there were virtually no bites taken from macroalgae for the remaining parts of the year. Increased macroalgal intake in winter by *Sc. rivulatus* suggests that there may be a seasonal change in diet composition possibly driven by declines in preferred EAM resources, or changes in the condition of the *Sargassum* (i.e., elevated abundance of epiphytic algae, microcrustaceans and the senescence of algal tissues) during winter. Ferrari et al. (2012) report a similar case in the Caribbean when in winter the macroalgae *Lobophora variegata* experiences more tissue loss to herbivory, although they didn't identify the species responsible for the increased consumption. As the cooler winter temperature is unlikely to have driven an increase in food intake, this rather appears to be a case of changing diet among the herbivores. The above cases suggest that species like *Sc. ferrugineus* are behaving like the terrestrial herbivores by widening their selection of food targets to include those of lower preference during the season when resources are limiting.

Seasonality in Feeding Rate

There is relatively more data on seasonal changes in feeding rate in parrotfishes. Those studies that have sampled the feeding rates of parrotfishes across seasons often report higher feeding rates in summer (Hatcher 1982, Carpenter 1986, Bellwood 1995, Afeworki et al. 2013b). Direct estimates of daily intake of ash free dry mass (AFDM) of EAM by parrotfishes peaks during the summer or spring in the western Atlantic (Ferreira et al.

1998) and in the southern Red Sea (Afeworki et al. 2013b). Estimates of EAM yield using experimental cages indicate that intake by parrotfishes and other reef grazers are highest in summer (Hatcher 1982, Afeworki et al. 2013a). Indirect evidence to support this is provided by Hansen et al. (1992) who measured higher EAM-derived detritus deposition in the winter into lagoon sediments at Davies Reef on the GBR. Although primary production of the EAM is higher in summer, the concomitant increase in herbivore grazing rate actually reduces the amount that eventually joins the detrital pool in the lagoon.

Parrotfish feeding rate is positively correlated to nutritional quality (Boyer et al. 2004, Goecker et al. 2005, Burkepile and Hay 2009) and productivity (Klumpp and Polunin 1990, Russ and McCook 1999, Russ 2003). These variables have been shown to be higher in summer and the observed increases in feeding rates during this season may be due to their effects combined with that of higher temperatures (see below). Other factors that may vary seasonally and affect parrotfish feeding rates include concentrations of inhibitory secondary metabolites and ash content (Steinberg 1989, Amade and Lemée 1998).

Besides nutritional content and potential yield per bite of food types, temperature is an important factor influencing feeding rate (Hatcher 1982, Smith 2008, Afeworki et al. 2013b). While feeding effort generally increases with temperature, limits to the metabolic scope of ectotherms may lead to declines in feeding rate under extremely high temperatures and this effect is more pronounced in large individuals (Thyrel et al. 1999, Pörtner and Knust 2007, Englund et al. 2011). There are indications that coral reef fish may be close to their critical thermal maximum and the upper limit for feeding and growth may be reached with rising global temperatures (Mora and Ospina 2001, Eme and Bennett 2009, Nilsson et al. 2009, Donelson et al. 2011). Afeworki et al. (2013b) report that such a thermal limit may already have been reached for large terminal phase males of *Sc. ferrugineus* in the southern Red Sea where summer temperatures are extremely high. Presently, summer sea surface temperatures (SST) regularly exceed 34 °C on fore reef slopes near Massawa (Ateweberhan et al. 2006a), possibly the result of significant warming in the Red Sea in the mid-1990s (Raitsos et al. 2011).

Seasonal Changes in Habitat use

Where there is seasonal variation in benthic algal communities among reef zones, roving herbivores may modify their habitat use. Paddack et al. (2006) reported seasonal changes in habitat use by *Scarus rubripinne, Sc. coelestinus* and *Kyphosus sectatrix* in the Florida Keys but the underlying causes are unknown. In the southern Red Sea, extreme summer temperatures may prevent parrotfishes and other grazing fishes from foraging on the shallow reef flats (Ateweberhan 2004, Ateweberhan et al. 2006a, Afeworki et al. 2013a). As a result, turf biomass reaches an annual maximum in these shallow zones (Ateweberhan et al. 2006a). In deeper reef zones, increased irradiance and temperature may facilitate higher productivity prompting reef grazers and parrotfishes to forage more in deeper zones during summer than in other seasons (Nemeth and Appeldoorn 2009, Afeworki et al. 2013a). The seasonally occurring dense canopies of macroalgae may also prevent some parrotfishes from foraging in these shallow reef zones (Hoey and Bellwood 2011, Afeworki et al. 2013a). The forgoing discussion is largely gleaned from studies that involved spatially and temporally repeated surveys. Studies focussed on unravelling the seasonal aspects of habitat use will furnish a better understanding of how herbivores cope with the seasonal changes in the distribution and abundance of their primary resources as well as the temporal dynamics of the trophic links between coral reefs and associated shallow water habitats.

Seasonality and Life-history

Distribution and abundance of species is determined by their population dynamics, which in turn is the sum of the processes of birth, death, emigration and immigration (Begon et al. 2009). In the case of fishes, births are defined as the number of larvae that settle and join the demersal population. This will be the cumulative result of the population size, individual fecundity, survival of the larvae in the pelagic environment (Doherty and Williams 1988) and post-settlement survival (Shulman 1987, Jones 1991, Almany and Webster 2006). Survival probability of fish at both the pelagic and the post-settlement phases appears to be strongly correlated with growth rate and body condition (Hoey and McCormick 2004, Beldade et al. 2012). Large females or those in good condition provision more yolk, which leads to larger sizes and better survivorship in larvae (McCormick 2003, 2006, Samhouri 2009, Beldade et al. 2012). For example, McCormick (2003) reported improved condition and reproductive output in a damselfish due to feeding on a seasonal pulse of coral spawn. Environmental conditions that influence the adult population may thus be equally important as the conditions during the pelagic and post-settlement phases in determining recruitment success. The very fact that the dietary targets of parrotfishes show seasonal variation (as discussed above) means that there will be seasonal influences on their body condition and life history traits such as growth and reproductive output.

Growth

In temperate environments where seasonality is pronounced, fish grow fastest during the spring when temperatures rise and resources abound (Casselman 1987, Jobling 1994). Even where the annual temperature range is small, seasonal changes in growth are recorded, related to temporal changes in resource abundance (Longhurst and Pauly 1987, Fishelson et al. 1987, Admassu and Casselman 2000, Johnson and Belk 2004, Afeworki et al. 2014). On coral reefs, spatial and temporal variation in resource levels are known to affect growth and body condition of reef fishes (Clifton 1995, Berumen et al. 2005, Afeworki et al. 2014). For example, planktivorous pomacentrids are able to utilize periodic pulses of coral propagules during mass coral spawning to store lipids (Pratchett et al. 2001). Hart and Russ (1996) demonstrated that on a reef impacted by crown of thorns starfish, *Acanthurus nigrofuscus* grew faster as a result of increased availability of EAM, their primary dietary resource. It is likely that growth in parrotfishes will follow a seasonal rhythm reflecting the productivity regime of the EAM discussed above.

Growth in parrotfishes has been primarily studied by age-length data where age (annuli) is read from sections of hard parts, primarily otoliths (Choat et al. 1996, 2003, Choat and Robertson 2002, Hamilton et al. 2008, also Taylor et al. Chapter 4). Beginning from the 1970s, research has shown that otoliths from tropical fish form annual alternating growth bands (Fowler 2009). The growth bands in otoliths are a result of differences in the proportions of the organic and aragonite components, with the opaque band containing relatively more organic matrix than the translucent/clear band (Watabe et al. 1982, Morales-Nin 2000, Hüssy et al. 2004). The opaque band is deposited during periods of higher food intake and growth (Hüssy and Mosegaard 2004, Høie and Folkvord 2006, Høie et al. 2008). Conversely, increased temperatures reduce otolith opacity by promoting deposition of the aragonite, especially when feeding is low (Hüssy et al. 2004). By looking at the timing of the opaque band formation in association with temperature histories of fish, the growth season and hence the season of increased feeding can be deduced (Fablet et al. 2011). If the seasons of feeding and optimal temperature coincide, the opaque zone will be deposited during this period. When the two do not concur in time, opaque bands tend to form during the fast growth period associated with increased food availability (Fablet et al. 2011). The latter may especially be true for coral reef fishes where the environmental temperature is generally above 20 °C, implying that lower temperatures may not be a constraint to growth. The following is an attempt to deduce the seasonality in growth and feeding in parrotfishes based on the timing of opaque zone formation. To understand the role of food availability in otolith banding, data from carnivorous, detritivorous/grazers, and planktivorous reef fishes from the GBR is compared.

On the GBR, the timing of the opaque band formation shows wide variation and appears to be associated with the trophic ecology of the species (Fig. 3). The season of opaque band formation in the carnivorous families Serranidae, Lutjanidae and Lethrinidae generally starts during Austral winter and ends in spring (Fig. 3), coinciding with a period of peak body condition. While some studies have shown that serranids feed at high rates in summer and have attributed this to the higher metabolic demands at elevated temperatures (Johansen et al. 2015, Messmer et al. 2017) or the increased availability of smaller-bodied newly settled fishes (Russell et al. 1977), the average size of prey ingested in winter is

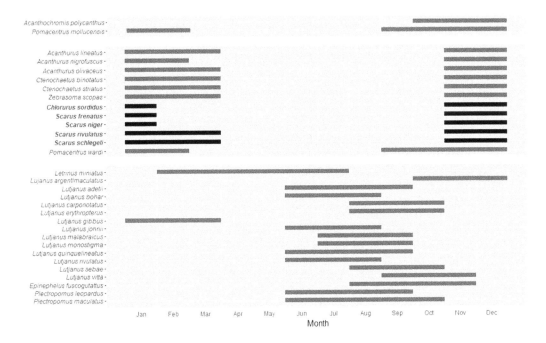

Fig. 3. Season of opaque band formation in otoliths of parrotfishes compared to other common coral reef fish families on the Great Barrier Reef, Australia. Species are arranged according to trophic grouping: Top panel: Planktivorous; Middle panel: Herbivorous/detritivorous; Bottom panel: Carnivorous. Parrotfish species are highlighted in bold and black bars. Sources: herbivorous fishes: (Fowler 1990, Fowler and Doherty 1992, Lou 1992a, 1992b, Choat et al. 1996, Hart and Russ 1996, Choat and Axe 1996, Newman et al. 1996, Cappo et al. 2000, Williams et al. 2005, Kingsford and Hughes 2005, Marriott and Mapstone 2006), carnivorous fishes: (Ferreira and Russ 1992, 1994, Fowler and Doherty 1992, Newman et al. 1996, Cappo et al. 2000, Williams et al. 2005, Marriott and Mapstone 2006, Pears et al. 2006); planktivorous fishes: (Fowler 1990, Fowler and Doherty 1992, Kingsford and Hughes 2005).

typically larger, and has been related to a better nutritional state during the cooler months (Beukers-Stewart and Jones 2004). Individuals of the grouper *Plectropomus leopardus* have been shown to feed more and be less active at higher temperatures in aquaria (Johansen et al. 2014, 2015) but in the field tend to feed more and have higher body fat deposits during the Austral winter (Ferreira 1993, St John 1995). Similarly, the snapper *Lutjanus carponotatus* achieves peak body condition in winter (Kritzer 2004). Validation of annulus formation in planktivorous damselfishes shows that the opaque zone is formed during summer (Fowler 1990, Fowler and Doherty 1992, Kingsford and Hughes 2005, Fowler 2009), coinciding with the highest abundance of zooplankton (Hamner et al. 1988). The opaque band in otoliths of parrotfishes and acanthurids is deposited during spring-summer, corresponding with the higher productivity of EAM. Temperature is unlikely to be the primary cause of the above patterns, since all of the above are resident coral reef fish that experience similar temperature regimes. Differences in feeding ecology or the timing of energetically expensive events (e.g., reproduction) may be important drivers of the difference in growth and otolith band formation. Further research in this area is clearly needed to understand the underlying causes of the lack of congruence in otolith band formation.

The evidence from otolith banding for a spring-summer growth optimum in parrotfishes is supported by studies that have used tag-recapture to analyse growth of individuals. For example, in the parrotfish *Sc. ferrugineus*, the season of fastest growth is spring - a season of abundant food and higher feeding rate but of lower temperature than summer (Afeworki et al. 2013b, 2014, Afeworki 2014). In this species, body condition and lipid content of the liver rise during the spring and reach peak values during the early summer (Afeworki 2014, Afeworki et al. 2014). Van Rooij et al. (1995a, 1995b) studied variation in body condition and growth in *Sp. viride* in Bonaire, the Caribbean, and found that all social categories of *Sp. viride* attained better body condition and all individuals (except largest territory holding males) showed peak growth during summer. This coincides with the season of high EAM productivity in the Caribbean (Adey and Steneck 1985, Carpenter 1986, Ferrari et al. 2012).

Reproduction

Seasonal change in benthic resources can influence the reproductive fitness of fishes. Research in other groups of fishes has shown that females that are in good physiological condition supply more nutrients to eggs (Kerrigan 1997, McCormick 2003, Donelson et al. 2008, Samhouri 2009, McDermott et al. 2011). The influence of this maternal input at the egg stage extends to the larvae, where larvae from such mothers have enhanced pelagic survival rates and contribute disproportionately to post-settlement populations (Hoey and McCormick 2004, Wright and Gibb 2005, Meekan et al. 2006). Female body condition also influences the quantity of eggs produced (Jorgensen et al. 2006, Rideout and Tomkiewicz 2011), either through changes in the frequency of spawning or on the batch fecundity (Kjesbu et al. 1991, McDermott et al. 2011). For example, females in poor condition are known to reduce the frequency of spawning within a season or even skip a whole reproductive season (Rideout et al. 2000).

Several observations suggest that these processes also occur in parrotfishes. Colin (1978) reported that, in *Sc. iseri*, the number of individuals participating in group spawning peaks in summer, suggesting that a higher proportion of females are in spawning condition during this season. The percentage of ripe females of *Sp. viride* within a Turks and Caicos population was shown to vary seasonally and was highest during the late winter and early spring season (Koltes 1993). Gonad analyses of several parrotfish species sampled off the

Jamaican coast by Munro et al. (1973) showed a decreasing proportion of reproductively active females from a winter peak. While ripe females occur year round on the GBR, their relative proportions have been shown to vary seasonally, peaking in winter for *Scarus schlegeli* and in spring for *Sc. rivulatus* (Lou 1992a). Recently, Afeworki (2014) recorded a lower proportion of ripe females in a southern Red Sea population of *Sc. ferrugineus* in summer than in winter, indicating that some females may skip spawning during the warmest months.

Similar observations have been made for other reef fishes. For example, a northern GBR population of *Letrinus miniatus* was shown to have a higher proportion of ripe females than a southern population (Williams et al. 2006). The proportion of reproductively active females of several species of damselfishes is lower during the wet season in Papua New Guinea (Srinivasan and Jones 2006). In Hawaii the surgeonfish *Zebrasoma flavescens* spawns daily but its batch fecundity is highest during its spring-summer peak spawning season (Bushnell et al. 2010). During winter, females of this surgeonfish skip spawning on some days and/or have low batch fecundity. The above studies provide evidence that parrotfishes and other reef fishes modulate their reproductive output, but it is unknown whether this is directly linked to food availability and body condition of the adult fish or whether it is simply a response to other environmental parameters, e.g. those prevailing in the pelagic environment.

Based on studies that involved sampling during most of the year, parrotfishes spawn year-round, but appear to have a distinct peak spawning period that varies between regions (Table 1). For example, the Caribbean parrotfish *Sp. viride* spawns daily throughout the year but with higher intensity during the cooler months (van Rooij et al. 1996). April-June is the peak recruitment season of parrotfishes and acanthurids in the Caribbean (Kopp et al. 2012), which corroborates the observations of a winter spawning maximum. In contrast, on the GBR the timing of peak recruitment of parrotfishes suggest they spawn during the warmer months (Russell et al. 1977). However, monthly samples of gonad weight in two species of parrotfishes from the GBR indicated a winter-spring peak (May-September) in spawning activity for *Sc. schlegeli* and a spring-summer peak (September – January) for *Sc. rivulatus* (Lou 1992a). Parrotfishes and many other reef taxa are reported to have a December-March (summer) peak in spawning at Kimbe Bay, Papua New Guinea (Claydon et al. 2014). Unlike the apparent species- and regional-specific spawning at low latitude sites, spawning in parrotfishes at higher latitudes appears to be more consistent and restricted to the warmer months. For example, in Japan the parrotfishes *Calotamos japonicus* and *Chlorurus spilurus* (previously *Ch. sordidus*) spawn from April-August (Yogo et al. 1980, Kume et al. 2010). *Sparisoma cretense* in the Mediterranean and the eastern Atlantic also spawn in summer (de Girolamo et al. 1999, Afonso et al. 2008). Local differences in time of peak spawning periods of coral reef fishes highlight the importance of environmental conditions experienced by adults. Year-round spawning at lower latitudes may be a bet-hedging strategy to spread the uncertainties in recruitment success. However, the existence of a seasonal peak in spawning effort points to the existence of a favourable season, during which reproductive output is maximized.

Early studies on the timing of reproduction in reef fishes suggested the importance of conditions that promote better larval survival during the pelagic phase (Johannes 1978, Doherty 1983, Doherty and Williams 1988). However, spawning seasonality in reef fish shows a wide variation and some of this variation is better explained by factors that influence the adult rather than the larval phase (Robertson 1991, Tyler and Stanton 1995, Goulet 1995, Abesamis et al. 2015). A case in point is Clifton's (1995) observation of asynchronous seasonal peaks in reproduction of two *Sc. iseri* populations in San Blas

Table 1. Records of monthly spawning in parrotfishes from different coral reef locations of the world. Peak spawning periods are indicated. Methods of data collection include 'spawning': direct underwater observation of spawning behaviour; 'recruitment': underwater recruit surveys; 'gonad maturation': gonadosomatic index and maturation of gonads assessed either histologically or macroscopically

Species	Location	Spawning time	Peak period*	Method	Source
Scarus iseri	Jamaica	Feb-Jun	Winter	Gonad maturation	Munro et al. (1973)
	Puerto Rico	Year round	Winter	Spawning	Colin and Clavijo (1988)
	Guadeloupe Island	unknown	Winter	Recruitment	Kopp et al. (2012)
Scarus taeniopterus	Puerto Rico	Year round	Insufficient data	Spawning	Colin and Clavijo (1988)
	Caribbean Mexico	Jun-Aug, January	Spring	Recruitment	Villegas-Sánchez et al. (2015)
Scarus vetula	Puerto Rico	Jan-Mar, Aug-Dec	Insufficient data	Spawning	Clavijo (1983)
Scarus spp.	Barbados	Year round	Summer	Recruitment	Valles et al. (2008)
Sparisoma aurofrenatum	Jamaica	Year round	Winter – spring	Gonad maturation	Munro et al. (1973)
	Puerto Rico	Year round	Insufficient data	Spawning	Colin and Clavijo (1988)
Sparisoma chrysopterum	Jamaica	Feb-Jun	Winter	Gonad maturation	Munro et al. (1973)
Sparisoma cretense	Greece	Jul-Sep	Summer	Gonad maturation	Petrakis and Papaconstantinou (1990)
Several parrotfish species	American Samoa	Jan-May	Winter	Gonad maturation	Page (1998)
Sparisoma viride	Jamaica	Feb-Jun	Winter	Gonad maturation	Munro et al. (1973)
	Turks and Caicos, BWI	Year round	Winter – spring	Gonad maturation	Koltes (1993)
	Florida Keys	Year round	Spring	Recruitment	Paddack and Sponaugle (2008)
	Bonaire	Year round	Winter	Spawning	van Rooij et al. (1996)
Sparisoma spp.	Barbados	Year round	Spring	Recruitment	Valles et al. (2008)
Calotomus japonicus	Japan	July-Oct	Summer	Gonad maturation	Kume et al. (2010)
Chlorurus bleekeri	Kimbe Bay, PNG	Year round	Summer	Spawning	Claydon et al. (2014)

(Contd.)

Species	Location			Method	Reference
Chlorurus spilurus	Japan	Spring-Summer	Summer	Spawning	Yogo et al. (1980)
	Philippines	Year round	Winter - spring	Recruitment	Abesamis and Russ (2010)
Scarus ferrugineus	Central Red Sea	unknown	Winter	Gonad maturation	Abdel-Aziz et al. (2012)
	Southern Red Sea	Year round	Winter	All three methods	Afeworki (2014)
Scarus niger	Philippines	Year round	Winter - spring	Recruitment	Abesamis and Russ (2010)
Scarus rivulatus	Great Barrier Reef	Year round	Spring-Summer	Gonad maturation	Lou (1992a)
Scarus quoyi	Kimbe Bay, PNG	Year round	Summer	Spawning	Claydon et al. (2014)
Scarus schlegeli	Great Barrier Reef	Year round	Winter - Spring	Gonad maturation	Lou (1992a)
Scarus spp.	Zanzibar	Year round	Winter, Oct	Gonad maturation	Nzioka (1979)
	Great Barrier Reef	Year round	Spring-Summer	Recruitment	Russell et al. (1977)

*For the studies that recorded recruitment, the peak spawning period was inferred by taking into account the pelagic larval duration of parrotfishes (29-47 d) (Lou 1993, Schultz and Cowen 1994, Sponaugle 2009, Ishihara and Tachihara 2011).

Archipelago, Panama. One population was from a sheltered reef which experiences increased EAM production and standing crop during the dry and stormy season. The second population was from a relatively exposed reef where the calm conditions during the Panamanian wet season led to higher EAM production and biomass. Feeding rate, gain in body mass, and gonad index of each *Sc. iseri* population closely tracked the site-specific patterns in EAM production. As a result, the peaks of reproduction in the two populations occurred in opposing seasons (Fig. 4). The close proximity of the populations—with only three km separation—minimizes the likelihood that the timing of peak reproduction is being determined by pelagic conditions.

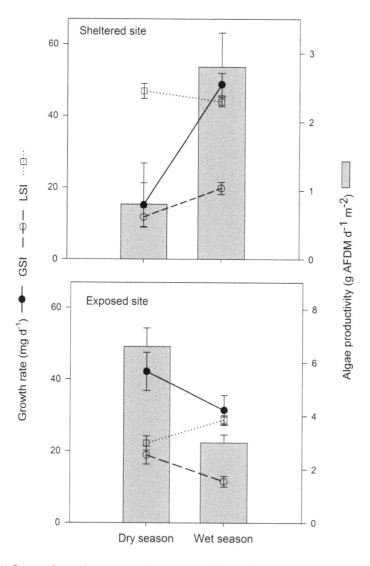

Fig. 4. Seasonal trends in growth, body condition (liver somatic index, LSI) and reproductive output (gonadosomatic index, GSI) in two nearby Caribbean populations of *Scarus iseri* compared to site-specific algal productivity (AFDM: ash-free dry mass of EAM) (source Clifton 1995).

Sc. ferrugineus in the southern Red Sea, provides further evidence of environmental conditions shaping the temporal pattern of reproductive effort via the resources available to the adults rather than their pelagic larvae. This region is strongly influenced by the Indian monsoon. As explained earlier, *Sc. ferrugineus* preferentially grazes on EAM growing on substrates with abundant endolithic algae rather than on EAM overtopping crustose corallines, as the former food type is nutritionally better and enables higher yields per bite (Afeworki et al. 2011). The seasonally changing availability of the high-yield food type combined with seasonal changes in grazing rate result in *Sc. ferrugineus* realising a 13-50 percent higher food intake in summer than in winter (Afeworki et al. 2013b). As a result, individuals of *Sc. ferrugineus* grow faster and gain more in liver and body weight in spring and summer than at other times of year (Afeworki et al. 2014, Fig. 5). Although reproduction takes place year-round, a peak in spawning activity occurs during the cool season (Afeworki 2014).

There is an important difference between the above case studies. In the two *Sc. iseri* populations the reproductive investment (GSI) closely tracks EAM production, suggesting that *Sc. iseri* behaves as an income breeder (Drent and Daan 1980). However further confirmation on this is needed as monthly data on gonad weight and spawning intensity are lacking (Clifton 1995). Moreover, liver weight and gonad weight have opposite trends (Fig. 4), suggesting that gonad maturation is being fuelled, at least partially, from energy reserves in the liver. In contrast, *Sc. ferrugineus* behaves like a capital breeder since the peak reproduction occurs during the season of low resource abundance and decreased food intake (Afeworki 2014, Afeworki et al. 2014). The evidence for the latter is based on monthly trends in gonad weight, oocyte maturation, egg stripping, daily observation of spawning

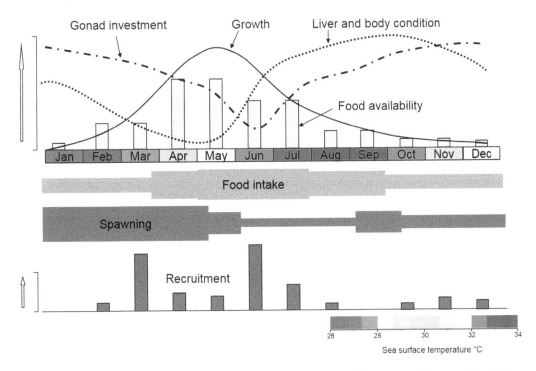

Fig. 5. Seasonal trends in availability of preferred algal resources and corresponding variation in food intake, growth, liver and body condition, spawning activity and recruitment in a *Scarus ferrugines* from the southern Red Sea (source Afeworki 2014).

behaviour, and monthly recruitment data (Afeworki 2014). This raises the question whether coral reef grazers that experience seasonal variation in resource abundance would behave as capital breeders like many ectothermic organisms (Bonnet et al. 1998, Varpe et al. 2009), or that reproductive cycles are decoupled from food availability. One likely factor that may select for a capital breeding strategy in parrotfishes is the possibility of a correlation between better survival rates for larvae and higher abundances of plankton, or calm sea conditions (Johannes 1978, Cushing 1990). For example, on the GBR, the period of peak plankton abundance appears to coincide with the time during which most species of reef fish would be undertaking the pelagic phase of their life cycle (Russell et al. 1977). Timing of spawning in many cases, however, does not seem to concur with calm conditions or peak pelagic productivity (Doherty 1983). Moreover, many species of reef fishes within a specific area can have different spawning peaks (Robertson 1991, Afeworki 2014, Abesamis et al. 2015). This may be because the larvae from different species have varying physical and biological requirements during the pelagic phase. However, evidence to support the hypothesis relating to differences in larval nutrition is currently lacking. Alternatively, the timing in reproduction may be due to changing environmental conditions affecting the reproducing adults (Robertson 1991). The frequently-observed prevalence of peak spawning during the season of lowest algal productivity suggests that an inevitable trade-off between time spent on reproduction versus feeding may play a role. This trade-off is expected to be prevalent in species that spawn daily, including many parrotfishes. By reducing time spent on reproduction in the summer when the availability of high-yield food sources peaks, and increasing time spent on foraging, parrotfishes may maximize their annual energy intake.

Like many other herbivores, parrotfishes spend over 90 percent of the day foraging (Montgomery et al. 1989, Bruggemann et al. 1994c, Bellwood 1995, van Rooij 1996). Spawning is a time-consuming activity taking up between 1-3 h day^{-1} in parrotfishes (van Rooij et al. 1996, Kuwamura et al. 2009, Afeworki 2014). Since feeding virtually ceases during spawning (Bonaldo et al. 2006, Afeworki et al. 2013b), there is clearly a conflict between the two activities and a strategy that minimizes spawning during the season of peak resource abundance may maximize energy intake and reproductive output. The trade-off between spawning and foraging and its influence on the timing of reproduction cannot currently be tested, since detailed data on feeding, body condition, gonad development and spawning is not yet available for most parrotfish species. However, the prevalence of a winter spawning peak in many parrotfish populations suggests that such a trade-off between feeding and spawning activities may well exist.

Conclusion

Given the major roles that parrotfishes play in the maintenance of the resilience capacity of coral reefs, a better understanding of how they cope with changes in resource abundance will improve modelling and scenarios for future reef states under various disturbance regimes. One approach to gain such understanding is to study the phenology of parrotfishes in relation to the seasonal variation of their food resources.

In this chapter we have shown that seasonal changes in algal and detrital resources can affect habitat use, feeding rate and food intake of parrotfishes. Most parrotfishes appear to have limited dietary flexibility, as detailed observations on several species, representative of scraping and excavating feeding modes and from the tropical Atlantic and the Indo-Pacific provinces, indicate a distinct preference for grazing on EAM growing on substrates infested with boring algae. The seasonal reduction of this preferred food type (generally during the cooler months) results in lowered food intake, with direct effects on growth rate and body condition. Local environmental conditions that determine resource availability for adults may influence the timing and intensity of reproductive activity, rather than regional pelagic conditions encountered by their larvae.

While seasonal changes in the community composition, productivity and standing crop of benthic algae on coral reefs have been detailed from numerous settings, relatively little attention has been dedicated to how such changes propagate through the food web. Given seawater temperature anomalies and severe cyclones are predicted to occur with increasing frequency, seasonal effects on resource abundance and feeding activity of consumers are likely to amplify. We advocate that there is ample potential for further research on the effects of seasonal shifts in resource abundance and distribution on the feeding ecology and life-history decisions of parrotfishes. Such studies will contribute to a better appreciation of their flexibility in the use of different algal resources and improve our understanding of annual variations in growth and condition and the timing of reproduction. More generally, a better understanding of the phenology of parrotfishes can help predict their responses to disturbances.

References Cited

Abdel-Aziz, E.H., F.A. Bawazeer, T.E. Ali and M. Al-Otaibi. 2012. Sexual patterns and protogynous sex reversal in the rusty parrotfish, *Scarus ferrugineus* (Scaridae): histological and physiological studies. Fish Physiol. Biochem. 38: 1211–1224.

Abesamis, R.A., C.R.L. Jadloc and G.R. Russ. 2015. Varying annual patterns of reproduction in four species of coral reef fish in a monsoonal environment. Mar. Biol. 162: 1993–2006.

Abesamis, R.A. and G.R. Russ. 2010. Patterns of recruitment of coral reef fishes in a monsoonal environment. Coral Reefs 29: 911–921.

Adam, T.C., R.J. Schmitt, S.J. Holbrook, A.J. Brooks, P.J. Edmunds, R.C. Carpenter and G. Bernardi. 2011. Herbivory, connectivity, and ecosystem resilience: response of a coral reef to a large-scale perturbation. PLoS One 6 (8): e23717.

Adey, W.H. and R.S. Steneck. 1985. Highly productive eastern Caribbean reefs: synergistic effects of biological, physical and geological factors. pp. 163–187. *In*: M.L. Reaka (ed.). The Ecology of Coral Reefs, Symposia Series for Undersea Research. National Oceanic and Atmospheric Administration, Rockville, Maryland, USA.

Admassu, D. and J.M. Casselman. 2000. Otolith age determination for adult tilapia, *Oreochromis niloticus* L. from Lake Awassa (Ethiopian Rift Valley) by interpreting biannuli and differentiating biannual recruitment. Hydrobiologia 418: 15–24.

Afeworki, Y. 2014. Population ecology of the rusty parrotfish *Scarus ferrugineus*, a dominant grazer on a seasonal coral reef. PhD Thesis, University of Groningen, The Netherlands.

Afeworki, Y., J.H. Bruggemann and J.J. Videler. 2011. Limited flexibility in resource use in a coral reef grazer foraging on seasonally changing algal communities. Coral Reefs 30: 109–122.

Afeworki, Y., J.J. Videler and J.H. Bruggemann. 2013a. Seasonally changing habitat use patterns among roving herbivorous fishes in the southern Red Sea: the role of temperature and algal community structure. Coral Reefs 32: 475–485.

Afeworki, Y., Z.A. Zekeria, J.J. Videler and J.H. Bruggemann. 2013b. Food intake by the parrotfish *Scarus ferrugineus* varies seasonally and is determined by temperature, size and territoriality. Mar. Ecol. Prog. Ser. 489: 213–224.

Afeworki, Y., J.J. Videler, Y.H. Berhane and J.H. Bruggemann. 2014. Seasonal and life-phase related differences in growth in *Scarus ferrugineus* on a southern Red Sea fringing reef. J. Fish Biol. 84: 1422–1438.

Afonso, P., T. Morato and R.S. Santos. 2008. Spatial patterns in reproductive traits of the temperate parrotfish *Sparisoma cretense*. Fish. Res. 90: 92–99.

Almany, G.R. and M.S. Webster. 2006. The predation gauntlet: early post-settlement mortality in reef fishes. Coral Reefs 25: 19–22.

Alwany, M.A., E. Thaler and M. Stachowitsch. 2009. Parrotfish bioerosion on Egyptian Red Sea reefs. J. Exp. Mar. Biol. Ecol. 371: 170–176.

Amade, P. and R. Lemée. 1998. Chemical defence of the mediterranean alga *Caulerpa taxifolia*: variations in caulerpenyne production. Aquat. Toxicol. 43: 287–300.

Ateweberhan, M. 2004. Seasonal dynamics of coral reef algae in the southern Red Sea: functional group and population ecology. PhD. thesis, University of Groningen, The Netherlands.

Ateweberhan, M., J.H. Bruggemann and A.M. Breeman. 2005a. Seasonal dynamics of *Sargassum ilicifolium* (Phaeophyta) on a shallow reef flat in the southern Red Sea (Eritrea). Mar. Ecol. Prog. Ser. 292: 159–171.

Ateweberhan, M., J.H. Bruggemann and A.M. Breeman. 2005b. Seasonal patterns of biomass, growth and reproduction in *Dictyota cervicornis* and *Stoechospermum polypodioides* (Dictyotales, Phaeophyta) on a shallow reef flat in the southern Red Sea (Eritrea). Bot. Mar. 48: 8–17.

Ateweberhan, M., J.H. Bruggemann and A.M. Breeman. 2006a. Effects of extreme seasonality on community structure and functional group dynamics of coral reef algae in the southern Red Sea (Eritrea). Coral Reefs 25: 391–406.

Ateweberhan, M., J.H. Bruggemann and A.M. Breeman. 2006b. Seasonal module dynamics of *Turbinaria triquetra* (Fucales, Phaeophyceae) in the southern Red Sea. J. Phycol. 42: 990–1001.

Baynes, T.W. 1999. Factors structuring a subtidal encrusting community in the southern Gulf of California. Bull. Mar. Sci. 64: 419–450.

Begon, M., M. Mortimer and D.J. Thompson. 2009. Population ecology: a unified study of animals and plants. John Wiley & Sons, New York, United States.

Beldade, R., S.J. Holbrook, R.J. Schmitt, S. Planes, D. Malone and G. Bernardi. 2012. Larger female fish contribute disproportionately more to self-replenishment. Proc. R. Soc. B-Biol. Sci. 279: 2116–2121.

Bellwood, D.R. 1994. A phylogenetic study of the parrotfish family Scaridae (Pisces: Labroidea), with a revision of genera. Rec. Aust. Mus. Suppl. 20: 1–86.

Bellwood, D.R. 1995. Direct estimate of bioerosion by two parrotfish species, *Chlorurus gibbus* and *C. Sordidus*, on the Great Barrier Reef, Australia. Mar. Biol. (Berlin, Ger.) 121: 419–429.

Bellwood, D.R. and J.H. Choat. 1990. A functional analysis of grazing in parrotfishes (Family Scaridae): the ecological implications. Environ. Biol. Fish. 28: 189–214.

Bellwood, D.R. and C.J. Fulton. 2008. Sediment-mediated suppression of herbivory on coral reefs: decreasing resilience to rising sea levels and climate change? Limnol. Oceanogr. 53: 2695–2701.

Bellwood, D.R., A.S. Hoey and J.H. Choat. 2003. Limited functional redundancy in high diversity systems: resilience and ecosystem function on coral reefs. Ecol. Lett. 6: 281–285.

Belovsky, G.E. 1984. Herbivore optimal foraging: a comparative test of three models. Am. Nat. 124: 97–115.

Belovsky, G.E. 1986. Optimal foraging and community structure: implications for a guild of generalist grassland herbivores. Oecologia 70: 35–52.

Benayahu, Y. and Y. Loya. 1977. Seasonal occurrence of benthic-algae communities and grazing regulation by sea urchins at the coral reefs of Eilat, Red Sea. Proceedings of the 3rd International Coral Reef Symposium, Miami, Florida.

Berumen, M.L., M.S. Pratchett and M.I. McCormick. 2005. Within-reef differences in diet and body condition of coral-feeding butterflyfishes (Chaetodontidae). Mar. Ecol. Prog. Ser. 287: 217–227.

Beukers-Stewart, B.D. and G.P. Jones. 2004. The influence of prey abundance on the feeding ecology of two piscivorous species of coral reef fish. J. Exp. Mar. Biol. Ecol. 299: 155–184.

Bonaldo, R.M., A.S. Hoey and D.R. Bellwood. 2014. The ecosystem roles of parrotfishes on tropical reefs. Oceanogr. Mar. Biol. 52: 81–132.

Bonaldo, R.M., J.P. Krajewski and D.R. Bellwood. 2011. Relative impact of parrotfish grazing scars on massive Porites corals at Lizard Island, Great Barrier Reef. Mar. Ecol. Prog. Ser. 423: 223–233.

Bonaldo, R.M., J.P. Krajewski, C. Sazima and I. Sazima. 2006. Foraging activity and resource use by three parrotfish species at Fernando de Noronha Archipelago, tropical West Atlantic. Mar. Biol. 149: 423–433.

Bonnet, X., B. Don and R. Shine. 1998. Capital versus income breeding: an ectothermic perspective. Oikos 83: 333–342.

Boyer, K., P. Fong, A. Armitage and R. Cohen. 2004. Elevated nutrient content of tropical macroalgae increases rates of herbivory in coral, seagrass, and mangrove habitats. Coral Reefs 23: 530–538.

Bruggemann, J.H., M.W.M. Kuyper and A.M. Breeman. 1994a. Comparative analysis of foraging and habitat use by the sympatric Caribbean parrotfish *Scarus vetula* and *Sparisoma viride* (Scaridae). Mar. Ecol. Prog. Ser. 112: 51–66.

Bruggemann, J.H., M.J.H. van Oppen and A.M. Breeman. 1994b. Foraging by the stoplight parrotfish *Sparisoma viride*. I. Food selection in different socially, determined habitats. Mar. Ecol. Prog. Ser. 106: 41–55.

Bruggemann, J.H., A.M. van Kessel, J.M. van Rooij and A.M. Breeman. 1996. Bioerosion and sediment ingestion by the Caribbean parrotfish *Scarus vetula* and *Sparisoma viride*: implications of fish size, feeding mode and habitat use. Mar. Ecol. Prog. Ser. 134: 59–71.

Bruggemann, J.H., J. Begeman, E.M. Bosma, P. Verburg and A.M. Breeman. 1994c. Foraging by the stoplight parrotfish *Sparisoma viride*. II. Intake and assimilation of food, protein and energy. Mar. Ecol. Prog. Ser. 106: 57–71.

Burkepile, D.E. and M.E. Hay. 2009. Nutrient versus herbivore control of macroalgal community development and coral growth on a Caribbean reef. Mar. Ecol. Prog. Ser. 389: 71–84.

Bushnell, M.E., J.T. Claisse and C.W. Laidley. 2010. Lunar and seasonal patterns in fecundity of an indeterminate, multiple-spawning surgeonfish, the yellow tang *Zebrasoma flavescens*. J. Fish Biol. 76: 1343–1361.

Caceres, C.W., L.S. Fuentes and F.P. Ojeda. 1994. Optimal feeding strategy of the temperate herbivorous fish *Aplodactylus punctatus* the effects of food availability on digestive and reproductive patterns. Oecologia 99: 118–123.

Camacho, O. and G. Hernández-Carmona. 2012. Phenology and alginates of two *Sargassum* species from the Caribbean coast of Colombia. Cienc. Mar. 38(2): 381–393.

Cappo, M., P. Eden, S.J. Newman and S. Robertson. 2000. A new approach to validation of periodicity and timing of opaque zone formation in the otoliths of eleven species of *Lutjanus* from the central Great Barrier Reef. Fish. Bull. 98: 474–488.

Caragnano, A., F. Colombo, G. Rodondi and D. Basso. 2009. 3-D distribution of nongeniculate corallinales: a case study from a reef crest of South Sinai (Red Sea, Egypt). Coral Reefs 28: 881–891.

Carpenter, R.C. 1981. Grazing by *Diadema antillarum* (Philippi) and its effects on the benthic algal community. J. Mar. Res. 39(4): 749–765.

Carpenter, R.C. 1985. Relationships between primary production and irradiance in coral reef algal communities. Limnol. Oceanogr. 30: 784–793.

Carpenter, R.C. 1986. Partitioning herbivory and its effects on coral reef algal communities. Ecol. Monogr. 56: 345–363.

Casareto, B.E., L. Charpy, M.J. Langlade, T. Suzuki, H. Ohba, M. Niraula and Y. Suzuki. 2008. Nitrogen fixation in coral reef environments. Proceedings of the 11th International Coral Reef Symposium, Ft. Lauderdale, Florida.

Casselman, J.M. 1987. Determination of age and growth. pp. 209–242. *In*: A.H. Weatherley and H.S. Gill (eds.). The Biology of Fish Growth, Academic Press London, London.

Cheal, A.J., M.A. MacNeil, E. Cripps, M.J. Emslie, M. Jonker, B. Schaffelke and H. Sweatman. 2010. Coral-macroalgal phase shifts or reef resilience: links with diversity and functional roles of herbivorous fishes on the Great Barrier Reef. Coral Reefs 29: 1005–1015.

Cheshire, A.C., G. Westphalen, A. Wenden, L.J. Scriven and B.C. Rowland. 1996. Photosynthesis and respiration of phaeophycean-dominated macroalgal communities in summer and winter. Aquat. Bot. 55: 159–170.

Chisholm, J.R.M. 2003. Primary productivity of reef-building crustose coralline algae. Limnol. Oceanogr. 48: 1376–1387.

Choat, J.H. 1991. The biology of herbivorous fishes on coral reefs. pp. 120–155. *In*: P.F. Sale (ed.). The Ecology of Fishes on Coral Reefs. Academic Press, San Diego.

Choat, J.H. and L. Axe. 1996. Growth and longevity in acanthurid fishes; an analysis of otolith increments. Mar. Ecol. Prog. Ser. 134: 15–26.

Choat, J.H. and K.D. Clements. 1998. Vertebrate herbivores in marine and terrestrial environments: a nutritional ecology perspective. Annu. Rev. Ecol. Syst. 29: 375–403.

Choat, J.H. and D.R. Robertson. 2002. Age-based studies. pp. 57–80. *In*: P.F. Sale (ed.). Coral Reef Fishes: Dynamics and Diversity in a Complex Ecosystem. Academic Press, San Diego, CA.

Choat, J.H., L. Axe and D. Lou. 1996. Growth and longevity in fishes of the family Scaridae. Mar. Ecol. Prog. Ser. 145: 33–41.

Choat, J.H., D.R. Robertson, J.L. Ackerman and J.M. Posada. 2003. An age-based demographic analysis of the Caribbean stoplight parrotfish *Sparisoma viride*. Mar. Ecol. Prog. Ser. 246: 265–277.

Clarke, A. 1988. Seasonality in the Antarctic marine-environment. Comp. Biochem. Physiol. B-Biochem. Mol. Biol. 90: 461–473.

Clavijo, I.E. 1983. Pair spawning and formation of a lek like mating system in the parrotfish *Scarus vetula*. Copeia 1983(1): 253–256.

Claydon, J.A.B., M.I. McCormick and G.P. Jones. 2014. Multispecies spawning sites for fishes on a low-latitude coral reef: spatial and temporal patterns. J. Fish Biol. 84: 1136–1163.

Clements, K.D. and J.H. Choat. 1993. Influence of season, ontogeny and tide on the diet of the temperate marine herbivorous fish *Odax pullus* (Odacidae). Mar Biol (Berl) 117: 213–220.

Clements, K.D., D. Raubenheimer and J.H. Choat. 2009. Nutritional ecology of marine herbivorous fishes: ten years on. Funct. Ecol. 23: 79–92.

Clifton, K.E. 1995. Asynchronous food availability on neighboring Caribbean coral reefs determines seasonal patterns of growth and reproduction for the herbivorous parrotfish *Scarus iserti*. Mar. Ecol. Prog. Ser. 116: 39–46.

Clifton, K.E. 1997. Mass spawning by green algae on coral reefs. Science 275: 1116–1118.

Colin, P.L. 1978. Daily and summer winter variation in mass spawning of striped parrotfish, *Scarus croicensis*. Fish. Bull. 76: 117–124.

Colin, P. L. and I. E. Clavijo. 1988. Spawning activity of fishes producing pelagic eggs on a shelf edge coral reef, southwestern Puerto Rico. Bull. Mar. Sci. 43: 249–279.

Copertino, M., S.D. Connell and A. Cheshire. 2005. The prevalence and production of turf-forming algae on a temperate subtidal coast. Phycologia 44: 241–248.

Crossland, C., B.G. Hatcher, M. Atkinson and S. Smith. 1984. Dissolved nutrients of a high-latitude coral reef, Houtman Abrolhos Islands, western Australia. Mar. Ecol. Prog. Ser. 14: 159–163.

Crossman, D.J., J.H. Choat, K.D. Clements, T. Hardy and J. McConochie. 2001. Detritus as food for grazing fishes on coral reefs. Limnol. Oceanogr. 46: 1596–1605.

Cushing, D.H. 1990. Plankton production and year-class strength in fish populations – an update of the match mismatch hypothesis. Adv. Mar. Biol. 26: 249–293.

de Girolamo, M., M. Scaggiante and M.B. Rasotto. 1999. Social organization and sexual pattern in the Mediterranean parrotfish *Sparisoma cretense* (Teleostei: Scaridae). Mar. Biol. 135: 353–360.

de Wreede, R.E. 1976. The phenology of three species of *Sargassum* (Sargassaceae, Phaeophyta) in Hawaii. Phycologia 15: 175–183.

Diaz-Pulido, G. and J. Garzon-Ferreira. 2002. Seasonality in algal assemblages on upwelling-influenced coral reefs in the Colombian Caribbean. Bot. Mar. 45: 284–292.

Doherty, P.J. and D.M. Williams. 1988. The replenishment of coral reef fish populations. Oceanogr. Mar. Biol. Annu. Rev. 26: 487–551.

Doherty, P. 1983. Diel, lunar and seasonal rhythms in the reproduction of two tropical damselfishes – *Pomacentrus flavicauda* and *Pomacentrus wardi*. Mar. Biol. 75: 215–224.

Done, T., E. Turak, M. Wakeford, L. DeVantier, A. McDonald and D. Fisk. 2007. Decadal changes in turbid-water coral communities at Pandora Reef: loss of resilience or too soon to tell? Coral Reefs 26: 789–805.

Donelson, J.M., M.I. McCormick and P.L. Munday. 2008. Parental condition affects early life-history of a coral reef fish. J. Exp. Mar. Biol. Ecol. 360: 109–116.

Donelson, J.M., P.L. Munday and M.I. McCormick. 2011. Acclimation to predicted ocean warming through developmental plasticity in a tropical reef fish. Global Change Biol. 17: 1712–1719.

Drent, R. and S. Daan. 1980. The prudent parent: energetic adjustments in avian breeding. Ardea 68: 225–252.

Ebeling, A.W. and M.A. Hixon. 1991. Tropical and temperate reef fishes: comparison of community structure. pp. 509–562. *In*: P.F. Sale (ed.). The Ecology of Fishes on Coral Reefs. Academic Press, San Diego.

Eme, J. and W.A. Bennett. 2009. Critical thermal tolerance polygons of tropical marine fishes from Sulawesi, Indonesia. J. Therm. Biol. 34: 220–225.

Englund, G., G. Ohlund, C.L. Hein and S. Diehl. 2011. Temperature dependence of the functional response. Ecol. Lett. 14: 914–921.

Erftemeijer, P.L. and P.M. Herman. 1994. Seasonal changes in environmental variables, biomass, production and nutrient contents in two contrasting tropical intertidal seagrass beds in South Sulawesi, Indonesia. Oecologia 99: 45–59.

Fablet, R., L. Pecquerie, H. de Pontual, H. HØie, R. Millner, H. Mosegaard and S.A.L.M. Kooijman. 2011. Shedding light on fish otolith biomineralization using a bioenergetic approach. PLoS One 6: e27055.

Felton, A.M., A. Felton, D.B. Lindenmayer and W.J. Foley. 2009a. Nutritional goals of wild primates. Funct. Ecol. 23: 70–78.

Felton, A.M., A. Felton, D. Raubenheimer, S.J. Simpson, W.J. Foley, J.T. Wood, I.R. Wallis and D.B. Lindenmayer. 2009b. Protein content of diets dictates the daily energy intake of a free-ranging primate. Behavioral Ecology 20: 685–690.

Ferrari, R., M. Gonzalez-Rivero, J. Ortiz and P. Mumby. 2012. Interaction of herbivory and seasonality on the dynamics of Caribbean macroalgae. Coral Reefs 31: 683–692.

Ferreira, B.P. 1993. Age, growth, reproduction and population biology of *Plectropomus* spp (Epinephelinae: Serranidae) on the Great Barrier Reef, Autralia. PhD Thesis, James Cook University, Australia.

Ferreira, B. and G.R. Russ. 1992. Age, growth and mortality of the inshore coral trout *Plectropomus maculatus* (Pisces: Serranidae) from the central Great Barrier Reef, Australia. Aust. J. Mar. Freshwater Res. 43: 1301–1312.

Ferreira, B. and G.R. Russ. 1994. Age validation and estimation of growth rate of the coral trout, *Plectropomus leopardus*, (Lacepede 1802) from Lizard Island, Northern Great Barrier Reef. Fish. Bull. 92: 46–57.

Ferreira, C.E.L., A.C. Peret and R. Coutinho. 1998. Seasonal grazing rates and food processing by tropical herbivorous fishes. J. Fish Biol. 53: 222–235.

Fishelson, L., L.W. Montgomery and A.H. Myrberg Jr. 1987. Biology of surgeonfish *Acanthurus nigrofuscus* with emphasis on changeover in diet and annual gonad cycles. Mar. Ecol. Prog. Ser. 39: 37–47.

Fong, P., T. Smith and M. Wartian. 2006. Epiphytic cyanobacteria maintain shifts to macroalgal dominance on coral reefs following ENSO disturbance. Ecology 87: 1162–1168.

Fourqurean, J.W., A. Willsie, C.D. Rose and L.M. Rutten. 2001. Spatial and temporal pattern in seagrass community composition and productivity in south Florida. Mar. Biol. 138: 341–354.

Fowler, A.J. 1990. Validation of annual growth increments in the otoliths of a small, tropical coral reef fish. Mar. Ecol. Prog. Ser. 64: 25–38.

Fowler, A.J. and P. Doherty. 1992. Validation of annual growth increments in the otoliths of two species of damselfish from the southern Great Barrier Reef. Aust. J. Mar. Freshwater Res. 43: 1057–1068.

Fowler, A. 2009. Age in years from otoliths of adult tropical fish. pp. 55–92. *In*: B.S. Green, B.D. Mapstone, G. Carlos and G.A. Begg (eds.). Tropical Fish Otoliths: Information for Assessment, Management and Ecology, Springer, New York.

Fuji, A., H. Watanabe, K. Ogura, T. Noda and S. Goshima. 1991. Abundance and productivity of microphytobenthos on a rocky shore in southern Hokkaido. Bull. Fac. Fish. Hokkaido Univ. 42: 136–146.

Fulton, C.J., M. Depczynski, T.H. Holmes, M.M. Noble, B. Radford, T. Wernberg and S.K. Wilson. 2014. Sea temperature shapes seasonal fluctuations in seaweed biomass within the Ningaloo coral reef ecosystem. Limnol. Oceanogr. 59(1): 156–166.

Furman, B.T. and K.L. Heck. 2008. Effects of nutrient enrichment and grazers on coral reefs: an experimental assessment. Mar. Ecol. Prog. Ser. 363: 89–101.

Gilmour, J.P., L.D. Smith, A.J. Heyward, A.H. Baird and M.S. Pratchett. 2013. Recovery of an isolated coral reef system following severe disturbance. Science 340: 69–71.

Goatley, C.H.R., D.R. Bellwood and O. Bellwood. 2010. Fishes on coral reefs: changing roles over the past 240 million years. Paleobiology 36: 415–427.

Goecker, M.E., K.L. Heck Jr and J.F. Valentine. 2005. Effects of nitrogen concentrations in turtlegrass *Thalassia testudinum* on consumption by the bucktooth parrotfish *Sparisoma radians*. Mar. Ecol. Prog. Ser. 286: 239–248.

Goulet, D. 1995. Temporal patterns of reproduction in the Red Sea damselfish *Amblyglyphidodon leucogaster*. Bull. Mar. Sci. 57: 582–595.

Green, A.L. and D.R. Bellwood. 2009. Monitoring functional groups of herbivorous reef fishes as indicators of coral reef resilience – A practical guide for coral reef managers in the Asia Pacific region. IUCN working group on Climate Change and Coral Reefs. IUCN, Gland, Switzerland.

Haas, A.F., C. Jantzen, M.S. Naumann, R. Iglesias-Prieto and C. Wild. 2010a. Organic matter release by the dominant primary producers in a Caribbean reef lagoon: implication for in situ O2 availability. Mar. Ecol. Prog. Ser. 409: 27–39.

Haas, A.F., M.S. Naumann, U. Struck, C. Mayr, M. el-Zibdah and C. Wild. 2010b. Organic matter release by coral reef associated benthic algae in the Northern Red Sea. J. Exp. Mar. Biol. Ecol. 389: 53–60.

Haas, A.F., C.E. Nelson, L. Wegley Kelly, C.A. Carlson, F. Rohwer, J.J. Leichter, A. Wyatt and J.E. Smith. 2011. Effects of coral reef benthic primary producers on dissolved organic carbon and microbial activity. PLoS One 6: e27973.

Hamilton, R., S. Adams and J.H. Choat. 2008. Sexual development and reproductive demography of the green humphead parrotfish (*Bolbometopon muricatum*) in the Solomon Islands. Coral Reefs 27: 153–163.

Hamner, W., M. Jones, J. Carleton, I. Hauri and D.M. Williams. 1988. Zooplankton, planktivorous fish, and water currents on a windward reef face: Great Barrier Reef, Australia. Bull. Mar. Sci. 42: 459–479.

Hansen, J.A., D.W. Klumpp, D.M. Alongi, P.K. Dayton and M.J. Riddle. 1992. Detrital pathways in a coral reef lagoon. Mar. Biol. 113: 363–372.

Hart, A.M. and G.R. Russ. 1996. Response of herbivorous fishes to crown-of-thorns starfish *Acanthaster planci* outbreaks. 3. Age, growth, mortality and maturity indices of *Acanthurus nigrofuscus*. Mar. Ecol. Prog. Ser. 136: 25–35.

Hatcher, B.G. 1982. The interaction between grazing organisms and the epilithic algal community of a coral reef: a quantitative assessment. Proceedings of the 4th International Coral Reef Symposium, Marine Science Center, University of the Philippines, Manila, Philippines.

Hatcher, B.G. 1983. Grazing in coral reef ecosystems. pp. 164–179. *In*: D.J. Barnes (ed.). Perspectives on Coral Reefs. Australian Institute of Marine Science, Townsville, Australia.

Hatcher, B.G. 1984. A maritime accident provides evidence for alternate stable states in benthic communities on coral reefs. Coral Reefs 3: 199–204.

Hatcher, B.G. and A.W.D. Larkum. 1983. An experimental analysis of factors controlling the standing crop of the epilithic algal community on a coral reef. J. Exp. Mar. Biol. Ecol. 69: 61–84.

Heck, K.L., T.J.B. Carruthers, C.M. Duarte, A.R. Hughes, G. Kendrick, R.J. Orth and S.W. Williams. 2008. Trophic transfers from seagrass meadows subsidize diverse marine and terrestrial consumers. Ecosystems 11: 1198–1210.

Hoey, A.S. and D.R. Bellwood. 2008. Cross-shelf variation in the role of parrotfishes on the Great Barrier Reef. Coral Reefs 27: 37–47.

Hoey, A.S. and D.R. Bellwood. 2009. Limited functional redundancy in a high diversity system: single species dominates key ecological process on coral reefs. Ecosystems 12: 1316–1328.

Hoey, A.S. and D.R. Bellwood. 2010. Cross-shelf variation in browsing intensity on the Great Barrier Reef. Coral Reefs 29: 499–508.

Hoey, A.S. and D.R. Bellwood. 2011. Suppression of herbivory by macroalgal density: a critical feedback on coral reefs? Ecol. Lett. 14: 267–273.

Hoey, A.S. and M.I. McCormick. 2004. Selective predation for low body condition at the larval-juvenile transition of a coral reef fish. Oecologia 139: 23–29.

Hoey, A.S., D.A. Feary, J.A. Burt, G. Vaughan, M.S. Pratchett and M.L. Berumen. 2016. Regional variation in the structure and function of parrotfishes on Arabian reefs. Mar. Poll. Bull. 105: 524–531.

Høie, H. and A. Folkvord. 2006. Estimating the timing of growth rings in Atlantic cod otoliths using stable oxygen isotopes. J. Fish Biol. 68: 826–837.

Høie, H., A. Folkvord, H. Mosegaard, L. Li, L. Clausen, B. Norberg and A. Geffen. 2008. Restricted fish feeding reduces cod otolith opacity. J. Appl. Ichthyol. 24: 138–143.

Horn, M.H. 1989. Biology of marine herbivorous fishes. Oceanogr. Mar. Biol. 27: 167–272.

Horn, M.H., M.A. Neighbors and S.N. Murray. 1986. Herbivore responses to a seasonally fluctuating food supply: growth potential of two temperate intertidal fishes based on the protein and energy assimilated from their macroalgal diets. J. Exp. Mar. Biol. Ecol. 103: 217–234.

Hughes, T.P. 1994. Catastrophes, phase shifts, and large-scale degradation of a Caribbean coral reef. Science 265: 1547–1551.

Hughes, T.P., A.M. Szmant, R. Steneck, R. Carpenter and S. Miller. 1999. Algal blooms on coral reefs: What are the causes? Limnol. Oceanogr. 44: 1583–1586.

Hurtado, A. and A. Ragaza. 1999. *Sargassum* studies in Currimao, Ilocos Norte, Northern Philippines I. Seasonal variations in the biomass of *Sargassum carpophyllum* J. Agardh, *Sargassum ilicifolium* (Turner) C. Agardh and *Sargassum siliquosum* J. Agardh (Phaeophyta, Sargassaceae). Bot. Mar. 42: 321–325.

Hüssy, K. and H. Mosegaard. 2004. Atlantic cod (*Gadus morhua*) growth and otolith accretion characteristics modelled in a bioenergetics context. Can. J. Fish. Aquat. Sci. 61: 1021–1031.

Hüssy, K., H. Mosegaard and F. Jessen. 2004. Effect of age and temperature on amino acid composition and the content of different protein types of juvenile Atlantic cod (*Gadus morhua*) otoliths. Can. J. Fish. Aquat. Sci. 61: 1012–1020.

Irwin, M.T., J.L. Raharison, D. Raubenheimer, C.A. Chapman and J.M. Rothman. 2014. Nutritional correlates of the "lean season": effects of seasonality and frugivory on the nutritional ecology of diademed sifakas. Am. J. Phys. Anthropol. 153: 78–91.

Ishihara, T. and K. Tachihara. 2011. Pelagic Larval Duration and Settlement Size of Apogonidae, Labridae, Scaridae, and Tripterygiidae Species in a Coral Lagoon of Okinawa Island, Southern Japan. Pac. Sci. 65: 87–93.

Jobling, M. 1994. Fish bioenergetics. Chapman and Hall, London.

Johannes, R.E. 1978. Reproductive strategies of coastal marine fishes in the tropics. Environ. Biol. Fish. 3: 65–84.

Johansen, J.L., V. Messmer, D.J. Coker, A.S. Hoey and M.S. Pratchett. 2014. Increasing ocean temperatures reduce activity patterns of a large commercially important coral reef fish. Global Change Biol. 20: 1067–1074.

Johansen, J.L., M.S. Pratchett, V. Messmer, D.J. Coker, A.J. Tobin and A.S. Hoey. 2015. Large predatory coral trout species unlikely to meet increasing energetic demands in a warming ocean. Scientific Reports 5: 13830.

John, J.S., G.P. Jones and P.F. Sale. 1989. Distribution and abundance of soft-sediment meiofauna and a predatory goby in a coral reef lagoon. Coral Reefs 8: 51–57.

Johnson, J. and M. Belk. 2004. Temperate Utah chub form valid otolith annuli in the absence of fluctuating water temperature. J. Fish Biol. 65: 293–298.

Jones, G.P. 1991. Postrecruitment processess in the ecology of coral reef populations: A multifactorial perspective. pp. 294–327. *In*: P.F. Sale (ed.). Ecology of Fishes on Coral Reefs. Academic Press, San Diego, CA, USA.

Jorgensen, C., B. Ernande, O. Fiksen and U. Dieckmann. 2006. The logic of skipped spawning in fish. Can. J. Fish. Aquat. Sci. 63: 200–211.

Kain, J. 1989. The seasons in the subtidal. British Phycological Journal 24: 203–215.

Kang, J., H. Choi and M. Kim. 2011. Macroalgal species composition and seasonal variation in biomass on Udo, Jeju Island, Korea. Algae 26: 333–342.

Kerrigan, B.A. 1997. Variability in larval development of the tropical reef fish *Pomacentrus amboinensis* (Pomacentridae): the parental legacy. Mar. Biol. 127: 395–402.

Kilar, J. and J. Norris. 1988. Composition, export, and import of drift vegetation on a tropical, plant-dominated, fringing-reef platform (Caribbean Panama). Coral Reefs 7: 93–103.

Kingsford, M. and J.M. Hughes. 2005. Patterns of growth, mortality, and size of the tropical damselfish *Acanthochromis polyacanthus* across the continental shelf of the Great Barrier Reef. Fish. Bull. 103: 561–573.

Kjesbu, O.S., J. Klungsoyr, H. Kryvi, P. Witthames and M. Walker. 1991. Fecundity, atresia, and egg size of captive Atlantic cod (*Gadus morhua*) in relation to proximate body composition. Can. J. Fish. Aquat. Sci. 48: 2333–2343.

Klumpp, D.W. and A.D. McKinnon. 1989. Temporal and spatial patterns in primary production of a coral-reef epilithic algal community. J. Exp. Mar. Biol. Ecol. 131: 1–22.

Klumpp, D.W. and N.V.C. Polunin. 1990. Algal production, grazers and habitat partitioning on a coral reef: positive correlation between grazing rate and food availability. pp. 372–388. *In*: M. Barnes and R.N. Gibson (eds.). Trophic Relationships in the Marine Environment. Aberdeen University Press, Aberdeen.

Klumpp, D.W. and A.D. McKinnon. 1992. Community structure, biomass and productivity of epilithic algal communities on the Great Barrier Reef: dynamics at different spatial scales. Mar. Ecol. Prog. Ser. 86: 77–89.

Kolasinski, J., K. Rogers, P. Cuet, B. Barry and P. Frouin. 2011. Sources of particulate organic matter at the ecosystem scale: a stable isotope and trace element study in a tropical coral reef. Mar. Ecol. Prog. Ser. 443: 77–93.

Koltes, K.H. 1993. Aspects of the reproductive biology and social structure of the stoplight parrotfish *Sparisoma viride*, at Grand Turk, Turks and Caicos Islands, Bwi. Bull. Mar. Sci. 52: 792–805.

Konar, B., K. Iken, J. Cruz-Motta, L. Benedetti-Cecchi, A. Knowlton, G. Pohle, P. Miloslavich, M. Edwards, T. Trott, E. Kimani, R. Riosmena-Rodriguez, M. Wong, S. Jenkins, A. Silva, I.S. Pinto and Y. Shirayama. 2010. Current patterns of macroalgal diversity and biomass in northern hemisphere rocky shores. PLoS One 5: e13195.

Kopp, D., Y. Bouchon-Navaro, M. Louis, P. Legendre and C. Bouchon. 2012. Spatial and temporal variation in a Caribbean herbivorous fish assemblage. J. Coast. Res. 28: 63–72.

Kramer, M.J., D.R. Bellwood and O. Bellwood. 2012. Cryptofauna of the epilithic algal matrix on an inshore coral reef, Great Barrier Reef. Coral Reefs 31: 1007–1015.

Kramer, M.J., O. Bellwood and D.R. Bellwood. 2013. The trophic importance of algal turfs for coral reef fishes: the crustacean link. Coral Reefs 32: 575–583.

Kritzer, J.P. 2004. Sex-specific growth and mortality, spawning season, and female maturation of the stripey bass (*Lutjanus carponotatus*) on the Great Barrier Reef. Fish. Bull. 102: 94–107.

Kume, G., Y. Kubo, T. Yoshimura, T. Kiriyama and A. Yamaguchi. 2010. Life history characteristics of the protogynous parrotfish *Calotomus japonicus* from northwest Kyushu, Japan. Ichthyol. Res. 57: 113–120.

Kuwamura, T., T. Sagawa and S. Suzuki. 2009. Interspecific variation in spawning time and male mating tactics of the parrotfishes on a fringing coral reef at Iriomote Island, Okinawa. Ichthyol. Res. 56: 354–362.

Larkum, A.W.D., I.R. Kennedy and W.J. Muller. 1988. Nitrogen fixation on a coral reef. Mar. Biol. 98: 143–155.

Lefevre, C.D. and D.R. Bellwood. 2010. Seasonality and dynamics in coral reef macroalgae: variation in condition and susceptibility to herbivory. Mar. Biol. 157: 955–965.

Lefevre, C.D. and D.R. Bellwood. 2011. Temporal variation in coral reef ecosystem processes: herbivory of macroalgae by fishes. Mar. Ecol. Prog. Ser. 422: 239–251.

Lindahl, U., M.C. Öhman and C.K. Schelten. 2001. The 1997/1998 mass mortality of corals: effects on fish communities on a Tanzanian coral reef. Mar. Pollut. Bull. 42: 127–131.

Logan, D., K. Townsend, K. Townsend and I. Tibbetts. 2008. Meiofauna sediment relations in leeward slope turf algae of Heron Island reef. Hydrobiologia 610: 269–276.

Longhurst, A.R. and D. Pauly. 1987. Ecology of Tropical Oceans. Academic Press INC., San Diego.

Lou, D.C. 1992a. Age specific patterns of growth and reproduction in tropical herbivorous fishes. PhD Thesis, James Cook University, Australia.

Lou, D.C. 1992b. Validation of annual growth bands in the otolith of tropical parrotfishes (*Scarus schlegeli* Bleeker). J. Fish Biol. 41: 775–790.

Lou, D.C. 1993. Growth in juvenile *Scarus rivulatus* and *Ctenochaetus binotatus*: a comparison of families Scaridae and Acanthuridae. J. Fish Biol. 42: 15–23.

Marriott, R.J. and B.D. Mapstone. 2006. Geographic influences on and the accuracy and precision of age estimates for the red bass, *Lutjanus bohar* (Forsskal 1775): a large tropical reef fish. Fish Res. 80: 322–328.

Marsh, J.A. 1976. Energetic role of algae in reef ecosystems. Micronesica 12: 13–21.

Marsh, J.A. 1970. Primary productivity of reef-building calcareous red algae. Ecology 51: 255–263.

Martin-Smith, K.M. 1993. Seasonal variation in tropical benthic *Sargassum* and associated motile epifauna. Proceedings of the 7th International Coral Reef Symposium, University of Guam Press, UOG Station, Guam.

McAfee, S.T. and S.G. Morgan. 1996. Resource use by five sympatric parrotfishes in the San Blas Archipelago, Panama. Mar. Biol. 125: 427–437.

McClanahan, T.R. 1988. Seasonality in east Africa's coastal waters. Mar. Ecol. Prog. Ser. 44: 191–199.

McClanahan, T.R. 1997. Primary succession of coral-reef algae: Differing patterns on fished versus unfished reefs. J. Exp. Mar. Biol. Ecol. 218: 77–102.

McCook, L.J. 1997. Effects of herbivory on zonation of *Sargassum* spp. within fringing reefs of the central Great Barrier Reef. Mar. Biol. 129: 713–722.

McCook, L. 1996. Effects of herbivores and water quality on *Sargassum* distribution on the central great barrier reef: cross-shelf transplants. Mar. Ecol. Ser. 139: 179–192.

McCormick, M.I. 2003. Consumption of coral propagules after mass spawning enhances larval quality of damselfish through maternal effects. Oecologia 136: 37–45.

McCormick, M.I. 2006. Mothers matter: crowding leads to stressed mothers and smaller offspring in marine fish. Ecology 87: 1104–1109.

McDermott, S.F., D.W. Cooper, J.L. Guthridge, I.B. Spies, M.F. Canino, P. Woods and N. Hillgruber. 2011. Effects of maternal growth on fecundity and egg quality of wild and captive Atka mackerel. Mar. Coast. Fish. 3: 324–335.

McKenzie, L.J. 1994. Seasonal changes in biomass and shoot characteristics of a *Zostera capricorni* Aschers. dominant meadow in Cairns Harbour, northern Queensland. Aust. J. Mar. Freshwater Res. 45: 1337–52.

Meekan, M., L. Vigliola, A. Hansen, P. Doherty, A. Halford and J. Carleton. 2006. Bigger is better: size-selective mortality throughout the life history of a fast-growing clupeid, *Spratelloides gracilis*. Mar. Ecol. Prog. Ser. 317: 237.

Messmer, V., M.S. Pratchett, A.S. Hoey, A.J. Tobin, D.J. Coker, S.J. Cooke and T.D. Clark. 2017. Global warming may disproportionately affect larger adults in a predatory coral reef fish. Global Change Biol. 23: 2230–2240.

Middelboe, A.L., K. Sand-Jensen and T. Binzer. 2006. Highly predictable photosynthetic production in natural macroalgal communities from incoming and absorbed light. Oecologia 150: 464–476.

Montgomery, W.L., A.A. Myrberg and L. Fishelson. 1989. Feeding ecology of surgeonfishes (Acanthuridae) in the northern Red Sea, with particular reference to *Acanthurus nigrofuscus* (Forsskal). J. Exp. Mar. Biol. Ecol. 132: 179–208.

Mora, C. and A.F. Ospina. 2001. Tolerance to high temperatures and potential impact of sea warming on reef fishes of Gorgona Island (tropical eastern Pacific). Mar. Biol. 139: 765–769.

Morales-Nin, B. 2000. Review of the growth regulation processes of otolith daily increment formation. Fish Res. 46: 53–67.

Munro, J., V. Gaut, R. Thompson and P. Reeson. 1973. Spawning seasons of Caribbean reef fishes. J. Fish Biol. 5: 69–84.

Naim, O. 1993. Seasonal responses of a fringing reef community to eutrophication (Reunion Island, Western Indian Ocean). Mar. Ecol. Prog. Ser. 99: 137–151.

Naumann, M.S., C. Jantzen, A.F. Haas, R. Iglesias-Prieto and C. Wild. 2013. Benthic primary production budget of a Caribbean reef lagoon (Puerto Morelos, Mexico). PLoS One 8: e82923.

Nemeth, M. and R. Appeldoorn. 2009. The distribution of herbivorous coral reef fishes within fore-reef habitats: the role of depth, light and rugosity. Caribb. J. Sci. 45: 247–253.

Newman, S.J., D.M. Williams and G.R. Russ. 1996. Age validation, growth and mortality rates of the tropical snappers (Pisces: Lutjanidae) *Lutjanus adetii* (Castelnau, 1873) and *L. quinquelineatus* (Bloch, 1790) from the central Great Barrier Reef, Australia. Mar. Freshwater Res. 47: 575–584.

Nilsson, G.E., N. Crawley, I.G. Lunde and P.L. Munday. 2009. Elevated temperature reduces the respiratory scope of coral reef fishes. Global Change Biol. 15: 1405–1412.

Nzioka, R.M. 1979. Observations on the spawning seasons of East African reef fishes. J. Fish Biol. 14: 329–342.

Paddack, M.J., R.K. Cowen and S. Sponaugle. 2006. Grazing pressure of herbivorous coral reef fishes on low coral-cover reefs. Coral Reefs 25: 461–472.

Paddack, M.J. and S. Sponaugle. 2008. Recruitment and habitat selection of newly settled *Sparisoma viride* to reefs with low coral cover. Mar. Ecol. Prog. Ser. 369: 205–212.

Page, M. 1998. The biology, community structure, growth and artisanal catch of parrotfishes of American Samoa. American Samoa Department of Marine and Wildlife Resources, Biological Report Series.

Pears, R.J., J.H. Choat, B.D. Mapstone and G.A. Begg. 2006. Demography of a large grouper, *Epinephelus fuscoguttatus*, from Australia's Great Barrier Reef: implications for fishery management. Mar. Ecol. Prog. Ser. 307: 259–272.

Pedersen, M.F., P.A. Stæhr, T. Wernberg and M.S. Thomsen. 2005. Biomass dynamics of exotic *Sargassum muticum* and native *Halidrys siliquosa* in Limfjorden, Denmark—implications of species replacements on turnover rates. Aquat. Bot. 83: 31–47.

Petrakis, G. and C. Papaconstantinou. 1990. Biology of *Sparisoma cretense* in the Dodecanese (Greece). J. Appl. Ichthyol. 6: 14–23.

Polunin, N.V.C. and D.W. Klumpp. 1992. Algal food supply and grazer demand in a very productive coral-reef zone. J. Exp. Mar. Biol. Ecol. 164: 1–15.

Pörtner, H.O. and R. Knust. 2007. Climate change affects marine fishes through the oxygen limitation of thermal tolerance. Science 315: 95–97.

Pratchett, M., N. Gust, G. Goby and S. Klanten. 2001. Consumption of coral propagules represents a significant trophic link between corals and reef fish. Coral Reefs 20: 13–17.

Pretorius, Y., J.D. Stigter, W.F. de Boer, S.E. van Wieren, C.B. de Jong, H.J. de Knegt, C.C. Grant, I. Heitkönig, N. Knox, E. Kohi, E. Mwakiwa, M.J.S. Peel, A.K. Skidmore, R. Slotow, C. van der Waal, F. van Langevelde and H.H.T. Prins. 2012. Diet selection of African elephant over time shows changing optimization currency. Oikos 121: 2110–2120.

Purcell, S.W. and D.R. Bellwood. 2001. Spatial patterns of epilithic algal and detrital resources on a windward coral reef. Coral Reefs 20: 117–125.

Purcell, S.W. 2000. Association of epilithic algae with sediment distribution on a windward reef in the northern Great Barrier Reef, Australia. Bull. Mar. Sci. 66: 199–214.

Raitsos, D.E., I. Hoteit, P.K. Prihartato, T. Chronis, G. Triantafyllou and Y. Abualnaja. 2011. Abrupt warming of the Red Sea. Geophys. Res. Lett. 38: L14601.

Rasher, D.B., A.S. Hoey and M.E. Hay. 2013. Consumer diversity interacts with prey defenses to drive ecosystem function. Ecology 94: 347–1358.

Raubenheimer, D. and S.J. Simpson. 2003. Nutrient balancing in grasshoppers: behavioural and physiological correlates of dietary breadth. J. Exp. Biol. 206: 1669–1681.

Raubenheimer, D., S.J. Simpson and D. Mayntz. 2009. Nutrition, ecology and nutritional ecology: toward an integrated framework. Funct. Ecol. 23: 4–16.

Reinthal, P.N. and I.G. Macintyre. 1994. Spatial and temporal variations in grazing pressure by herbivorous fishes: Tobacco Reef, Belize. Atoll Res. Bull. 425: 1–11.

Rice, A.N. and M.W. Westneat. 2005. Coordination of feeding, locomotor and visual systems in parrottishes (Teleostei: Labridae). J. Exp. Biol. 208: 3503–3518.

Rideout, R.M. and J. Tomkiewicz. 2011. Skipped spawning in fishes: more common than you might think. Mar. Coast. Fish. 3: 176–189.

Rideout, R., M. Burton and G. Rose. 2000. Observations on mass atresia and skipped spawning in northern Atlantic cod, from Smith Sound, Newfoundland. J. Fish Biol. 57: 1429–1440.

Rivera-Monroy, V.H., C.J. Madden, J.W. Day Jr., R.R. Twilley, F. Vera-Herrera and H. Alvarez-Guillén. 1998. Seasonal coupling of a tropical mangrove forest and an estuarine water column: enhancement of aquatic primary productivity. Hydrobiologia 379: 41–53.

Rix, L., V.N. Bednarz, U. Cardini, N. van Hoytema, F.A. Al-Horani, C. Wild and M.S. Naumann. 2015. Seasonality in dinitrogen fixation and primary productivity by coral reef framework substrates from the northern Red Sea. Mar. Ecol. Prog. Ser. 533: 79–92.

Robertson, D.R. 1991. The role of adult biology in the timing of spawning of tropical reef fishes. pp. 356–386. *In*: P.F. Sale (ed.). The Ecology of Fishes on Coral reefs. Academic Press, San Diego.

Rotjan, R.D. and S.M. Lewis. 2005. Selective predation by parrotfishes on the reef coral *Porites astreoides*. Mar. Ecol. Prog. Ser. 305: 193–201.

Rotjan, R. and S. Lewis. 2009. Predators selectively graze reproductive structures in a clonal marine organism. Mar. Biol. 156: 569–577.

Rubio, V.C., F.J. Sánchez-Vázquez and J.A. Madrid. 2003. Macronutrient selection through postingestive signals in sea bass fed on gelatine capsules. Physiol. Behav. 78: 795–803.

Russ, G.R. 2003. Grazer biomass correlates more strongly with production than with biomass of algal turfs on a coral reef. Coral Reefs 22: 63–67.

Russ, G.R. and L.J. McCook. 1999. Potential effects of a cyclone on benthic algal production and yield to grazers on coral reefs across the central Great Barrier Reef. J. Exp. Mar. Biol. Ecol. 235: 237–254.

Russell, B., G. Anderson and F. Talbot. 1977. Seasonality and recruitment of coral reef fishes. Aust. J. Mar. Fresh. Res. 28: 521–528.

Samhouri, J.F. 2009. Food supply influences offspring provisioning but not density-dependent fecundity in a marine fish. Ecology 90: 3478–3488.

Sánchez–Vázquez, F., T. Yamamoto, T. Akiyama, J. Madrid and M. Tabata. 1999. Macronutrient self-selection through demand-feeders in rainbow trout. Physiol. Behav. 66: 45–51.

Sánchez-Vázquez, F.J., T. Yamamoto, T. Akiyama, J.A. Madrid and M. Tabata. 1998. Selection of macronutrients by goldfish operating self-feeders. Physiol. Behav. 65: 211–218.

Schaffelke, B. and D.W. Klumpp. 1997. Biomass and productivity of tropical macroalgae on three nearshore fringing reefs in the central Great Barrier Reef, Australia. Bot. Mar. 40: 373–383.

Schultz, E. and R. Cowen. 1994. Recruitment of coral reef fishes to Bermuda: local retention or long distance transport. Mar. Ecol. Prog. Ser. 109: 15–28.

Scott, F.J. and G.R. Russ. 1987. Effects of grazing on species composition of the epilithic algal community on coral reefs of the central Great Barrier Reef. Mar. Ecol. Prog. Ser. 39: 293–304.

Shannon, G., R.L. Mackey and R. Slotow. 2013. Diet selection and seasonal dietary switch of a large sexually dimorphic herbivore. Acta Oecol. 46: 48–55.

Shulman, M. 1987. What controls tropical reef fish populations: recruitment or benthic mortality? An example in the Caribbean reef fish *Haemulon flavolineatum*. Mar. Ecol. Prog. Ser. 39: 233–242.

Smith, J., C. Smith and C. Hunter. 2001. An experimental analysis of the effects of herbivory and nutrient enrichment on benthic community dynamics on a Hawaiian reef. Coral Reefs 19: 332–342.

Smith, T.B. 2008. Temperature effects on herbivory for an Indo-Pacific parrotfish in Panama: implications for coral-algal competition. Coral Reefs 27: 397–405.

Sponaugle, S. 2009. Daily otolith increments in the early stages of tropical fish. pp. 93–132. *In*: B. Green, B. Mapstone, G. Carlos and G. Begg (eds.). Tropical Fish Otoliths: Information for Assessment, Management and Ecology. Springer, Netherlands.

Srinivasan, M. and G. Jones. 2006. Extended breeding and recruitment periods of fishes on a low latitude coral reef. Coral Reefs 25: 673–682.

St John, J. 1995. Feeding ecology of the coral trout, *Plectropomus leopardus* (Serranidae) on the Great Barrier Reef, Australia. Ph.D Thesis, James Cook University, Australia.

Steinberg, P.D. 1989. Biogeographical variation in brown algal polyphenolics and other secondary metabolites: comparison between temperate Australasia and North America. Oecologia 78: 373–382.

Steneck, R.S. 1988. Herbivory on coral reefs: a synthesis. Proceedings of the 6th International Coral Reef Symposium, Townsville, Australia.

Stephens, D.W. and J.R. Krebs. 1986. Foraging theory. Princeton University Press, Princeton.

Stimson, J. 2013. Consumption by herbivorous fishes of macroalgae exported from coral reef flat refuges to the reef slope. Mar. Ecol. Prog. Ser. 472: 87–99.

Thyrel, M., I. Berglund, S. Larsson and I. Näslund. 1999. Upper thermal limits for feeding and growth of 0+ Arctic charr. J. Fish Biol. 55: 199–210.

Tootell, J. and M. Steele. 2015. Distribution, behavior, and condition of herbivorous fishes on coral reefs track algal resources. Oecologia. 181(1): 13–24.

Tyler, W.A. and F.G. Stanton. 1995. Potential influence of food abundance on spawning patterns in a damselfish, *Abudefduf abdominalis*. Bull. Mar. Sci. 57: 610–623.

Unsworth, R.K.F., J.J. Bell and D.J. Smith. 2007a. Tidal fish connectivity of reef and sea grass habitats in the Indo-Pacific. J. Mar. Biol. Assoc. U. K. 87: 1287–1296.

Unsworth, R.K.F., J.D. Taylor, A. Powell, J.J. Bell and D.J. Smith. 2007b. The contribution of scarid herbivory to seagrass ecosystem dynamics in the Indo-Pacific. Estuarine Coastal Shelf Sci. 74: 53–62.

Unsworth, R.K., S.L. Garrard, P.S. De Leon, L.C. Cullen, D.J. Smith, K.A. Sloman and J.J. Bell. 2009. Structuring of Indo-Pacific fish assemblages along the mangrove-seagrass continuum. Aquat. Biol. 5: 85–95.

Valentine, J.F., K.L. Heck, D. Blackmon, M.E. Goecker, J. Christian, R.M. Kroutil, K.D. Kirsch, B.J. Peterson, M. Beck and M.A. Vanderklift. 2007. Food web interactions along seagrass-coral reef boundaries: effects of piscivore reductions on cross-habitat energy exchange. Mar. Ecol. Prog. Ser. 333: 37–50.

Valles, H., D.L. Kramer and W. Hunte. 2008. Temporal and spatial patterns in the recruitment of coral-reef fishes in Barbados. Mar. Ecol. Prog. Ser. 363: 257–272.

van Marken Lichtenbelt, W.D. 1993. Optimal foraging of a herbivorous lizard, the green iguana in a seasonal environment. Oecologia 95: 246–256.

van Rooij, J.M. 1996. Behavioural energetics of the parrotfish *Sparisoma viride*: Flexibility in a coral reef setting. PhD Thesis, University of Groningen, The Netherlands.

van Rooij, J.M., F.J. Kroon and J.J. Videler. 1996. The social and mating system of the herbivorous reef fish *Sparisoma viride*: One-male versus multi-male groups. Environ. Biol. Fishes 47: 353–378.

van Rooij, J.M., J.J. Videler and J.H. Bruggemann. 1998. High biomass and production but low energy transfer efficiency of Caribbean parrotfish: implications for trophic models of coral reefs. J. Fish Biol. 53: 154–178.

van Rooij, J.M., J.H. Bruggemann, J.J. Videler and A.M. Breeman. 1995a. Ontogenic, social, spatial and seasonal variations in condition of the reef herbivore *Sparisoma viride*. Mar. Biol. 123: 269–275.

van Rooij, J.M., J.H. Bruggemann, J.J. Videler and A.M. Breeman. 1995b. Plastic growth of the herbivorous reef fish *Sparisoma viride*: field evidence for a trade-off between growth and reproduction. Mar. Ecol. Prog. Ser. 122: 93–105.

Varpe, O., C. Jorgensen, G.A. Tarling and O. Fiksen. 2009. The adaptive value of energy storage and capital breeding in seasonal environments. Oikos 118: 363–370.

Villegas-Sánchez, C.A., J.H. Lara-Arenas, J.M. Castro-Pérez and J.E. Arias-González. 2015. Patrones de reclutamiento de 4 especies ícticas en hábitats de parche y cordillera del arrecife Banco Chinchorro (Caribe mexicano). Revista Mexicana de Biodiversidad 86: 396–405.

Vivas, M., V.C. Rubio, F.J. Sánchez-Vázquez, C. Mena, B. García García and J.A. Madrid. 2006. Dietary self-selection in sharpsnout seabream (*Diplodus puntazzo*) fed paired macronutrient feeds and challenged with protein dilution. Aquaculture 251: 430–437.

Vuki, V.C. and I.R. Price. 1994. Seasonal changes in the *Sargassum* populations on a fringing coral reef, Magnetic Island, Great barrier reef region, Australia. Aquat. Bot. 48: 153–166.

Wanders, J.B.W. 1977. Role of benthic algae in shallow reef of Curacao (Netherlands Antilles) III. The significance of grazing. Aquat. Bot. 3: 357–390.

Watabe, N., K. Tanaka, J. Yamada and J.M. Dean. 1982. Scanning electron microscope observations of the organic matrix in the otolith of the teleost fish *Fundulus heteroclitus* (Linnaeus) and *Tilapia nilotica* (Linnaeus). J. Exp. Mar. Biol. Ecol. 58: 127–134.

Wernberg, T., M.A. Vanderklift, J. How and P.S. Lavery. 2006. Export of detached macroalgae from reefs to adjacent seagrass beds. Oecologia 147: 692–701.

Wilkinson, C.R., P.W. Sammarco and L.A. Trott. 1985. Seasonal and fish grazing effects on rates of nitrogen fixation on coral reefs (Great Barrier Reef). Proceedings of the 5th International Coral Reef Congress, Tahiti.

Wilkinson, C.R. and P.W. Sammarco. 1983. Effects of fish grazing and damselfish territoriality on coral reef algae. II Nitrogen fixation. Mar. Ecol. Prog. Ser. 13: 15–19.

Williams, A.J., C.R. Davies and B.D. Mapstone. 2006. Regional patterns in reproductive biology of *Lethrinus miniatus* on the Great Barrier Reef. Mar. Freshwater Res. 57: 403–414.

Williams, A., C. Davies and B. Mapstone. 2005. Variation in the periodicity and timing of increment formation in red throat emperor (*Lethrinus miniatus*) otoliths. Mar. Freshwater Res. 56: 529–538.

Williams, S.L. and R.C. Carpenter. 1997. Grazing effects on nitrogen fixation in coral reef algal turfs. Mar. Biol. 130: 223–231.

Wilson, S.K., D.R. Bellwood, J.H. Choat and M.J. Furnas. 2003. Detritus in the epilithic algal matrix and its use by coral reef fishes. Oceanogr. Mar. Biol. Annu. Rev. 41: 279–309.

Wilson, S.K., K. Burns and S. Codi. 2001. Sources of dietary lipids in the coral reef blenny *Salarias patzneri*. Mar. Ecol. Prog. Ser. 222: 291–296.

Wilson, S.K., M. Depczynski, R. Fisher, T.H. Holmes, R.A. O'Leary and P. Tinkler. 2010. Habitat associations of juvenile fish at Ningaloo Reef, western Australia: the importance of coral and algae. PLoS One 5: e15185.

Wismer, S., A.S. Hoey and D.R. Bellwood. 2009. Cross-shelf benthic community structure on the Great Barrier Reef: relationships between macroalgal cover and herbivore biomass. Mar. Ecol. Prog. Ser. 376: 45–54.

Wright, P.J. and F.M. Gibb. 2005. Selection for birth date in North Sea haddock and its relation to maternal age. J. Anim. Ecol. 74: 303–312.

Yogo, Y., A. Nakazono and H. Tsukahara. 1980. Ecological studies on the spawning of the parrotfish, *Scarus sordidus* Forsskal. Sci. Bull. Fac. Agric. Kyushu Univ. 34: 105–114.

CHAPTER
12

The Ecology of Parrotfishes in Marginal Reef Systems

Andrew S. Hoey[1], Michael L. Berumen[2], Roberta M. Bonaldo[3], John A. Burt[4], David A. Feary[5], Carlos E.L. Ferreira[6], Sergio R. Floeter[7] and Yohei Nakamura[8]

[1] ARC Centre of Excellence for Coral Reef Studies, James Cook University,
Townsville, Queensland 4811, Australia
Email: Andrew.hoey1@jcu.edu.au

[2] Red Sea Research Center, Division of Biological and Environmental Science and Engineering,
King Abdullah University of Science and Technology, Thuwal, 23955, Saudi Arabia
Email: michael.berumen@kaust.edu.sa

[3] Grupo de História Natural de Vertebrados Museu de Zoologia,
Universidade Estadual de Campinas, Campinas, SP, Brazil, 13083–863
Email: robertabonaldo@gmail.com

[4] Center for Genomics and Systems Biology, New York University, Abu Dhabi,
PO Box 129188, Abu Dhabi, United Arab Emirates
Email: John.Burt@nyu.edu

[5] School of Life Sciences, University of Nottingham, Nottingham NG7 2RD, UK
Email: David.Feary@nottingham.ac.uk

[6] Laboratório de Ecologia e Conservação de Ambientes Recifais, Universidade Federal
Fluminense, Rio de Janeiro, RJ, Brazil, 24001-970
Email: carlosferreira@id.uff.br

[7] Laboratório de Biogeografia e Macroecologia Marinha, Universidade Federal de Santa Catarina,
Florianópolis, Santa Catarina, Brazil, 88010-970
Email: sergio.floeter@ufsc.br

[8] Graduate School of Kuroshio Science, Kochi University 200 Monobe,
Nankoku, Kochi, Japan 783-8502
Email: ynakamura@kochi-u.ac.jp

Introduction

Parrotfishes (Labridae, Scarinae) are a ubiquitous group of reef fishes that are primarily distributed across the world's tropical oceans (Fig. 1). Currently, 100 species of parrotfishes are recognized (Parenti and Randall 2011), with most occupying shallow (0–50 m) tropical marine habitats, primarily coral reefs, although the distribution of several species extends into subtropical and temperate latitudes (Bonaldo et al. 2014). Indeed, the first parrotfish to be described to science was the Mediterranean parrotfish *Sparisoma cretense* (Linnaeus, 1758) based on a specimen collected in Crete (35°N). Given their distribution, it is perhaps not surprising that the majority of research on the ecology of parrotfishes has focused on low latitude, or tropical, coral-dominated habitats (e.g., Ogden and Buckman 1973, Robertson et al. 1982, Bellwood and Choat 1990, Bruggemann et al. 1994, Hoey and Bellwood 2008, Bonaldo et al. 2011), with comparatively little research on the ecology of parrotfishes on marginal, or high-latitude, reefs. There is, however, growing interest in the ecology of marginal reefs, largely due to the poleward range expansions of organisms and the tropicalization of marginal reef communities (e.g., Greenstein and Pandolfi 2008, Yamano et al. 2011, Baird et al. 2012, Verges et al. 2014, 2016), and the potential of marginal reefs to act as refugia from climate change (e.g., Riegl and Piller 2003, Lybolt et al. 2011). So what conditions define a 'marginal' reef?

Kleypas et al. (1999) define marginal reefs as those reefs that occur near or beyond the 'normal' environmental limits or reef distribution. Through the analysis of environmental data for ca. 1,000 reef locations, Kleypas et al. (1999) conclude that reef distribution is limited by temperature (weekly sea surface temperature: 18.0–31.5°C), salinity (monthly average 30–40 PSU), nutrients (nitrate: < 2 µmol litre^{-1}; phosphate: < 0.4 µmol litre^{-1}), aragonite saturation (Ω-arag > 3.5), and light penetration, such that any reefs that exist outside these limits are considered marginal. While the relative importance of each of these environmental conditions in limiting reef development remains unclear (e.g., temperature, aragonite saturation, and light penetration often covary), temperature has been suggested to be a useful proxy for the limits of reef development, in particular along latitudinal gradients (Jokiel and Coles 1977, Johannes et al. 1983, Kleypas et al. 1999). Indeed, the majority of reefs considered to be 'marginal' are those that occur at high latitudes and hence outside the normal temperature limits for reef development (e.g. Lord Howe Island, Arabian/Persian Gulf, mainland Japan, Gulf of California, southern Brazil, see Fig. 2). We acknowledge that marginal reefs exist outside of high latitude environments (e.g., low salinity and low light environments in the Gulf of Thailand or inshore Great Barrier Reef), however they fall outside of the scope of this chapter.

In this chapter, we provide an overview of what is currently known of the ecology of parrotfishes on marginal, high latitude reefs. We start by describing the general biotic and abiotic conditions that characterize high latitude reefs, and how these may influence various aspects of parrotfish ecology within these systems. We then use case studies from four marginal reef systems (the Arabian Peninsula, eastern Australia, southern Japan, and Brazil) to explore the ecology of parrotfishes in these unique environments in greater detail. For each of these reef systems, we describe how changes in benthic communities and topographic complexity along latitudinal gradients relate to the distribution, abundance, and species richness of parrotfishes. We also consider how environmental conditions influence the functional impact of parrotfishes within these systems, and how these vary among taxonomic groups. In doing so, we seek to identify any generalities in the response of parrotfishes to the biotic and abiotic conditions experienced within these marginal reef systems.

Characteristics of Marginal Reefs

Marginal, or high latitude, reefs by definition occur beyond the normal environmental limits for coral reef development (see above), and as such one may expect marginal reefs to have lower coral cover than those of low latitude reefs. This is often, but not always the case. For example, the cover of live coral is relatively low in the Gulf of Oman (16 percent, Bento et al. 2016), Middleton Reef (19 percent, Hoey et al. 2014), Northwest Hawaiian Islands (10–20 percent, Vroom 2011), and southern Florida (< 6 percent, Moyer et al. 2003), whereas other marginal reefs have coral cover that is directly comparable or even greater

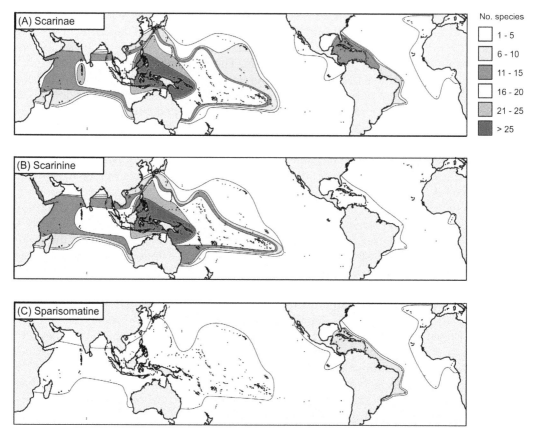

Fig. 1. Geographic distribution of species richness within the parrotfishes (A) Scarinae (all genera), and the two parrotfishes clades, (B) Scarinine (*Bolbometopon, Chlorurus, Cetoscarus, Hipposcarus,* and *Scarus*), and (C) Sparisomatine (*Calotomous, Cryptotomus, Leptoscarus, Nicholsina,* and *Sparisoma*). The distributions highlight the major biogeographic patterns with the Indo-Pacific supporting a higher diversity of scarinine species, while the Atlantic supports a higher diversity of sparisomatine species. Strong latitudinal gradients in species richness are evident in both the Indo-Pacific and Atlantic Oceans.

than that of low latitude reefs (e.g., southern Arabian Gulf: 56 percent, Bento et al. 2016; Sodwana Bay: 59 percent, Schleyer et al. 2008; Lord Howe Island: 37 percent, Hoey et al. 2011). These differences in coral cover likely reflect variation in the recent disturbance histories of each location (e.g., Bento et al. 2016), rather than suppressed rates of coral growth and/or recruitment due to lower water temperatures and geographic isolation. Moving further poleward from these reefs corals cease to produce enough calcium carbonate for reef accretion (Buddemeier and Smith 1999), but often exist, together with other organisms, as a thin veneer over rocky substrata.

Apart from potential differences in the cover of live corals, there are predictable differences in the taxonomic and morphological composition of corals between marginal and low latitude reefs. Several studies have shown that coral communities on high latitude reefs have fewer species, and are dominated by species with larger depth ranges, more robust morphologies, greater tolerance of turbid waters, and in some locations a greater proportion of species that brood larvae compared to those on lower latitude reefs (Coles 2003, Sommer et al. 2014, Keith et al. 2015, Mizerek et al. 2016). Species with robust

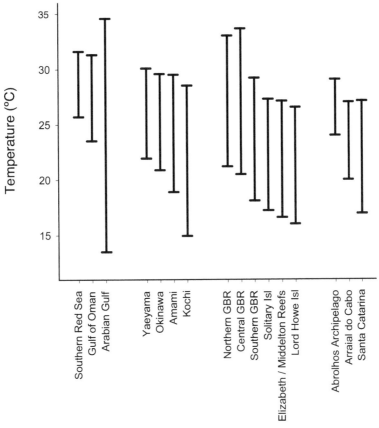

Fig. 2. Annual temperature ranges (minimum and maximum sea surface temperatures) experienced by several marginal reef systems around the Arabian peninsula, southern Japan, Australia's east coast, and Brazil. Temperature ranges for adjacent low latitude reefs are included for comparison.

morphologies (i.e., encrusting and massive growth forms), while less likely to be damaged or dislodged by oceanic swells, provide less complex physical structure than branching and tabular growth forms that are common on low latitude reefs. The physical structure of reef habitats is generally positively related to the abundance and/or diversity of associated reef fish assemblages (Graham and Nash 2013, Richardson et al. 2017), and as such any reductions in topographic complexity are likely to have adverse effects on fish assemblages, including parrotfishes (e.g. Heenan et al. 2016), and the delivery of their functional impact (Cvitanovic and Hoey 2010).

Macroalgae, although often viewed as a sign of degradation on low latitude reefs, is typically abundant on marginal, high latitude reefs, including those in isolated locations (e.g., Hoey et al. 2011, Vroom 2011, Dalton and Roff 2013). This elevated abundance and biomass of macroalgae has several implications for the parrotfishes, and is covered in detail elsewhere in this volume (Fox Chapter 13). Briefly, the abundance and/or biomass of parrotfishes on tropical reefs has been shown to be negatively related to macroalgal cover (e.g. Williams and Polunin 2001, Fox and Bellwood 2007, Wismer et al. 2009, Rasher et al. 2013), with this relationship often viewed as evidence for the top-down control of algal communities. Conversely, macroalgae may influence parrotfish populations through the provision of dietary resources or physical structure (i.e., bottom-up processes).

There has been some uncertainty regarding the nutritional targets of parrotfishes, however, recent evidence demonstrates scarinine parrotfishes target protein-rich epilithic, endolithic and epiphytic microscopic phototrophs, primarily cyanobacteria (Clements et al. 2017, Clements and Choat Chapter 3). Given these nutritional targets it is not surprising that scarinine parrotfishes are rarely observed feeding on fleshy macroalgae (e.g., Hoey and Bellwood 2009, Loffler et al. 2015, Plass-Johnson et al. 2015). The few exceptions include records of some species feeding on calcified algae, such as *Halimeda* and *Amphiroa* (Mantyka and Bellwood 2007, Rasher et al. 2013), however it is possible they are targeting small epiphytic phototrophs on the surface of the algae rather than the alga itself. Similarly, although sparisomatine parrotfishes are widely regarded as browsers of macroalgae (Streelman et al. 2002, Bonaldo et al. 2014), it has been hypothesized that the majority of species are targeting protein rich epiphytes on the surfaces of macroalgae and seagrass (Clements and Choat Chapter 3). An exception to this may be the Indo-Pacific sparisomatine *Leptoscarus vaigiensis* that appears to predominantly feed on macroalgae and seagrass (Ohta and Tachihara 2004, Gullström et al. 2011).

The structure provided by stands of tall fleshy macroalgae has been shown to suppress feeding by scarinine parrotfishes on an inshore reef on the Great Barrier Reef, and it was suggested that the avoidance of dense stands was related to inability of these fishes to visually detect potential predators (Hoey and Bellwood 2011). Yet macroalgal beds appear to be an important juvenile habitat for several species of coral reef fish, including parrotfishes (Wilson et al. 2010, Hoey et al. 2013, Evans et al. 2014), as well as an important adult habitat for *L. vaigiensis* (Lim et al. 2016, Tano et al. 2017).

Together with these changes in benthic composition and topographic complexity, the environmental conditions of marginal reefs (e.g., colder and more variable water temperatures, higher wave energy, higher productivity and reduced light penetration) may pose physiological challenges for tropical parrotfishes inhabiting these areas. Fishes are ectotherms and as such their metabolism and energy requirements are largely governed by environmental temperature (Pörtner and Farrell 2008). Any changes in environmental temperature will, therefore, change the rates of biochemical and cellular processes required for homeostasis, and the energetic cost of growth, activity, and reproduction. These changes in metabolic demands may be met through increased intake of energy, reduced energy expenditure, or a combination of both. To date, the majority of research investigating the effects of temperature on the physiological performance of reef fish has focused on small, site-attached species, primarily pomacentrids and apogonids. This body of work has shown that increasing water temperature generally increases the oxygen consumption of fishes at rest, decreases their aerobic scope (a proxy for surplus energy available for physiological or ecological activities), and has variable effects on activity (reviewed in Hoey et al. 2016a). Similar responses have been recorded for larger-bodied species, such as the coral trout *Plectropomus leopardus*, with increasing temperature leading to a reduction in activity, and an increase in food consumption and oxygen consumption (Johansen et al. 2014, 2015, Clark et al. 2017). We are unaware of any studies that have examined the physiological responses of parrotfishes to changing temperature, however it would seem reasonable to expect that they would respond in a similar manner.

Marginal Reefs of the Arabian Region

The marine region of the north-western Indian Ocean, comprising the Red Sea, Arabian Sea, and Arabian/Persian Gulf, is characterized by a high level of endemism (DiBattista et al. 2016) and forms a biogeographic unit that is distinct from the rest of the western

Indian Ocean (Keith et al. 2013, Kulbicki et al. 2013). Large spatial and temporal variation in environmental conditions makes this one of the world's most variable tropical marine environments (Bauman et al. 2013). The Red Sea has relatively stable environmental conditions and is isolated from the Indian Ocean by the narrow and relatively shallow (140 m) Strait of Bab al Mandab. The area outside the Red Sea (i.e., Arabian Sea and Gulf of Oman) is characterized by monsoonal-driven upwelling events that cause major fluctuations in nutrients and water temperature (Fig. 2), resulting in limited coral reef development (i.e. rocky reefs with sparse coral cover) (McIlwain et al. 2011). The Arabian (or Persian) Gulf is possible one of the most extreme and variable environments in which extant coral reefs occur. Its shallow depth, restricted water exchange through the Strait of Hormuz, and high latitude location contribute to extreme variation in salinity and temperature (ca. 12–36°C, Fig. 2), high levels of turbidity and sedimentation (Riegl 1999) and low nutrient levels (Bauman et al. 2013). As a consequence of these extreme and variable conditions, coral reefs within the southern Arabian Gulf are largely two-dimensional structures with little vertical relief, and dominated by massive and submassive corals (Burt et al. 2011, Bauman et al. 2013).

Ecology of Parrotfishes on Marginal Reefs of the Arabian Region

Twenty species of parrotfish (18 scarinine, 2 sparisomatine) have been reported from the Arabian region, with the vast majority (17 species) occurring within the Red Sea. Many of these species are regional or local endemics. For example, *Scarus arabicus* and *Scarus zufar* are only known from the Arabian Sea and southern coast of Oman, respectively, while *Scarus persicus* is restricted to the Arabian Gulf and coast of Oman (Choat et al. 2012). Consequently, there is a strong subregional structure in the taxonomic composition of parrotfishes between the Arabian Gulf, Arabian Sea, and Red Sea (Hoey et al. 2016b). Reefs in the Arabian Gulf are characterized by two species (*Sc. persicus*, *Scarus ghobban*), in the Arabian Sea by the same two species and *Scarus fuscopurpureus* in the north, and *Sc. arabicus* and *Sc. zufar* in the south, and the Red Sea by a diversity of scraping and excavating species (Hoey et al. 2016b). Although variation in the abundance and/or biomass of parrotfishes among habitats in the Red Sea (e.g. Bonaldo et al. 2014, Khalil et al. 2017) resembles that of other regions (Burkepile et al. Chapter 7), the lack of reef development and hence clearly defined reef zones precludes among-habitat comparisons for the Arabian Sea or Arabian Gulf.

The species richness of parrotfish and the abundance of excavating parrotfishes on shallow reef slopes (ca. 6m depth) decline from the Red Sea to the Arabian Sea and Arabian Gulf (Fig. 3). The decrease in species richness may be expected given the biogeography of the region (e.g., DiBattista et al. 2016), however the near-complete absence of excavating parrotfishes in the Arabian Sea is striking. Two species of excavating parrotfishes (*Chlorurus sordidus* and *Chlorurus strongylocephalus*) have been recorded along the coast of Oman, but they appear to be extremely rare. The lack of excavating parrotfishes on reefs in the Arabian Sea is in stark contrast to the relatively high abundance of scraping parrotfishes in this region (Fig. 3), and suggests that some aspect of their feeding or nutritional ecology may be contributing to these differences. Scraping and excavating parrotfishes typically feed from similar surfaces (dead coral or carbonates covered by epilithic algal matrix, EAM) but differ in the depth of their bite. Scraping parrotfishes typical take shallow (< 1 mm) bites, while excavating parrotfishes take deeper bites and remove greater quantities of the underlying substrata with each bite (Hoey and Bellwood 2008). This difference in feeding mode and the hard underlying substrata (i.e. rocky reef) in this area may have contributed

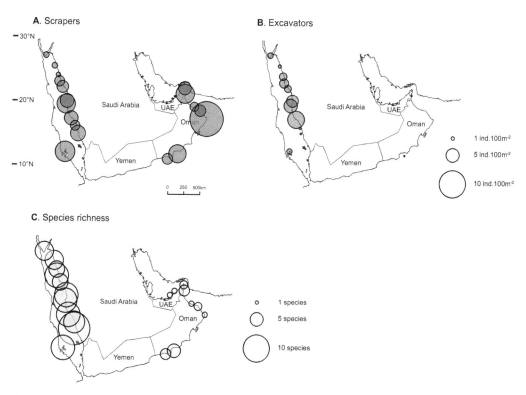

Fig. 3. Spatial variation in the abundance and species richness of scarinine parrotfishes on shallow reef slopes in the Arabian Peninsula. (A) Abundance of scraping parrotfishes (*Scarus, Hipposcarus*), (B) abundance of excavating parrotfishes (*Bolbometopon, Cetoscarus, Chlorurus*) and (C) species richness of parrotfishes. Circles are proportional to mean abundances or total number of species recorded in each location. Data are from Alwany et al. 2009, Afeworki et al. 2013, Hoey et al. 2016b. Note, no data for the abundance or species richness of parrotfishes were available for the coast of Yemen.

to the lack of excavating parrotfishes. Feeding on hard surfaces has been reported to cause dentition damage in large-bodied Atlantic parrotfishes (Bonaldo et al. 2007).

Five species of parrotfish have been reported from the Arabian Gulf, however all appear to be rare, with only two species (*Sc. persicus* and *Sc. ghobban*) recorded during extensive surveys of reefs in the southern Arabian Gulf, with densities of less than one individual per 1000 m² for both species combined (Feary et al. 2010, Hoey et al. 2016b). Other herbivorous fishes (i.e., Siganidae, Acanthuridae) are equally rare on these reefs (Feary et al. 2010, Burt et al. 2011), suggesting the low densities may be a consequence of the extreme temperatures on the physiology of these fishes, reductions in the quantity and/ or quality of dietary resources, or fishing. Although fishing has led to marked declines in parrotfish populations in many locations (e.g., Bellwood et al. 2012, Taylor et al. 2014), they are not primary targets of fishers within the Arabian Gulf (Grandcourt 2012).

Tropical organisms, including reef fishes, have evolved in relatively stable thermal environments and typically have a narrower thermal tolerance than temperate species, and are therefore potentially more sensitive to changes in temperature (Tewksbury et al. 2008). The extreme temperature variation within the Arabian Gulf (annual temperature range ≥ 20°C) may preclude many tropical species from inhabiting this region. Indeed, the reef fish fauna of the Arabian Gulf is depauperate (Burt et al. 2011). Further, although the favored feeding surfaces of parrotfishes (i.e. EAM covered surfaces) are abundant within the southern Arabian Gulf (Bauman et al. 2013, Bento et al. 2016), the high rates of sedimentation (Riegl 1999) reduce their nutritional value to herbivorous fishes (e.g., Bellwood and Fulton 2008, Gordon et al. 2016).

Irrespective of the underlying mechanism(s), these changes in the abundance and composition of parrotfishes has important implications for the spatial distribution of their functional impact. Using bite rates for each species and the area or volume of bite scars of closely related species, Hoey et al. (2016a) estimated that the area of reef surface grazed by parrotfishes decreased from an average of 210 and 150 percent year^{-1} in the Red Sea and Arabian Sea, respectively, to only 4 percent year^{-1} in the Arabian Gulf. Similarly, estimated erosion rates decreased from an average of 1.57 kg m^{-2}year^{-1} in the Red Sea to 0.43 and 0.01 kg m^{-2}year^{-1} in the Arabian Sea and Arabian Gulf, respectively. Importantly, rates of grazing and erosion within the Arabian Gulf, and to a lesser extent the Arabian Sea, may be even lower than these estimates due to the suppression of feeding during the colder months. Preliminary data for *Sc. persicus* from the northern Arabian Sea indicate that a 6°C decrease in water temperature (from 29°C to 23°C) led to a 50–60 percent decrease in bite rate (Hoey unpublished data). While changes in the nutritional quality of food items cannot be discounted, the reduction in feeding is consistent with predicted temperature-induced changes in metabolic demands.

Marginal Reefs of Eastern Australia

There is extensive coral reef formation along Australia's east coast, from the Great Barrier Reef and reefs of the Coral Sea in the north, to the high latitude oceanic reefs of Lord Howe Island, Elizabeth and Middleton Reefs, and several non-accreting coral communities along the New South Wales coast (Harriott et al. 1999, Mizerek et al. 2016). Lord Howe Island (31°33'S), approximately 630 km east of mainland Australia, is the world's southernmost coral reef, and Elizabeth (30°S) and Middleton (29°30'S) reefs, approximately 200–260 km to the north, are considered the world's southernmost platform reefs. These isolated reef systems receive warm tropical waters from the East Australian Current (EAC) that originates between 17–19°S in the Coral Sea and tracks largely southward along the east Australian coast until 30–32°S where it bifurcates, with one part flowing eastward across the Tasman Sea delivering warm water to these southern offshore reefs, and the other part continuing largely along the coast (Ridgway and Dunn 2003).

Over 100 coral species have been recorded from the offshore reefs of Lord Howe Island, and Elizabeth and Middleton Reefs, however there is little consensus as to which species are present (see Baird et al. 2017 for discussion). This is considerably lower than the > 300 coral species recorded on the GBR, and markedly higher than most reefs between 28°S and 31°S on the east Australian coast, the only exception being the Solitary Islands where as many as 90 coral species have been reported (Harriott et al. 1999, Harriott and Banks 2002). Warming of global sea surface temperatures and the strengthening of the EAC is leading to poleward expansions of both corals (e.g. Baird et al. 2012) and fish (e.g., Feary et al. 2014) down Australia's east coast. Such tropicalization of marine organisms is leading to novel interactions as tropical species are exposed to temperate and subtropical species for the first time, and can have a dramatic and lasting impact of marine ecosystems (Verges et al. 2016).

Together with these differences in species richness there are considerable differences in the cover of live coral and macroalgae, and the composition of coral communities both among and within these marginal reefs (e.g., Dalton and Roff 2013). Coral cover is relatively high on Lord Howe Island (mean: 37 percent; range: 2–57 percent), moderate on Elizabeth Reef (mean: 29 percent; range: 15–37 percent), and low on Middleton Reef (mean: 19 percent; range: 8–26 percent) (Hoey et al. 2011, 2014). Coral cover tends to be lower on the subtropical coastal reefs of eastern Australia (< 1–25 percent, Harriott et al. 1999, Harriott and Banks 2002) and dominated by encrusting and submassive corals attached to the underlying rocky substratum (Harriott and Banks 2002, Dalton and Roff 2013). One unifying feature of most of the high latitude reefs in the region, both coastal and offshore, is the relatively high cover (ca. 20–30 percent) of fleshy macroalgae (Hoey et al. 2011, 2014, Dalton and Roff 2013).

Ecology of Parrotfishes on Marginal Reefs of Eastern Australia

The offshore high latitude reefs of Lord Howe Island, Elizabeth and Middleton Reef have a distinctive marine fauna, with several species of endemic fishes (Francis 1993). However, the parrotfish fauna of these reefs are primarily a subset of the 27 species that occur on GBR, with 14 species reported from Lord Howe Island, and 22 from Elizabeth and Middleton reefs (Hoey et al. 2014). Interestingly, the excavating species *Chlorurus frontalis* is one of the most common parrotfish species on Elizabeth and Middleton reefs, but is rare on the GBR. Relatively few parrotfish species (2–3 species) are present on the coastal high latitude reefs in the region, although large individuals of *Sc. ghobban* and *Scarus altipinnis* (>30 cm total length) have been observed as far south as Sydney (A. Hoey pers. obs.).

Comparisons of the abundances and species richness of parrotfishes along Australia's east coast reveal clear spatial variation (Fig. 4). The abundances of scraping parrotfishes were broadly comparable between mid-shelf reef crests of the GBR and Elizabeth and Middleton Reefs (ca. 10 ind 100m^{-2}), but decreased markedly at Lord Howe Island (2.2 ind 100m^{-2}), and were almost completely absent on the coastal reefs of the Solitary Islands (< 0.1 ind 100m^{-2}; Fig. 4). Similar spatial variation was apparent in the abundances of excavating parrotfishes, ranging from 0 ind 100m^{-2} in the Solitary Islands to 4.1 ind 100m^{-2} on the GBR. The decline in the abundances of both groups of scarinine parrotfishes from Elizabeth and Middleton reefs to Lord Howe Island is difficult to resolve as they are both isolated offshore reefs, with similar temperature ranges (Fig. 2), are separated by only 200–260 km, have low human population densities (Elizabeth and Middleton reefs are unpopulated, Lord Howe Island's population is less than 800), and parrotfishes are not targeted by fishers. The differences in parrotfish populations may be related to reef structure. A recent study investigating the drivers of parrotfish biomass in the central and western Pacific showed that low islands and atolls (such as those of Elizabeth and Middleton reefs) support higher biomass of excavating parrotfishes than reefs surrounding high islands (such as Lord Howe Island) (Heenan et al. 2016). Similarly, island geomorphology was an important predictor of parrotfish assemblages across Micronesia, and was suggested to be due to the broad-scale habitat diversity associated with different geomorphologies (Taylor et al. 2015).

Perhaps the most striking pattern is the almost complete absence of parrotfishes within the Solitary Islands, despite similar latitude and temperature ranges as the offshore reefs, and moderate coral cover. Only five individual parrotfishes (two *Sc. altipinnis* and three *Sc. ghobban*) were recorded in surveys that covered over 10 hectares of shallow reef, with all individuals being observed on reefs surrounding the northernmost island, North Solitary Island (Hoey unpublished). It is likely that the lack of parrotfishes within the Solitary Islands is related, at least in part, to the lack of accreting reef in this area, and hence the

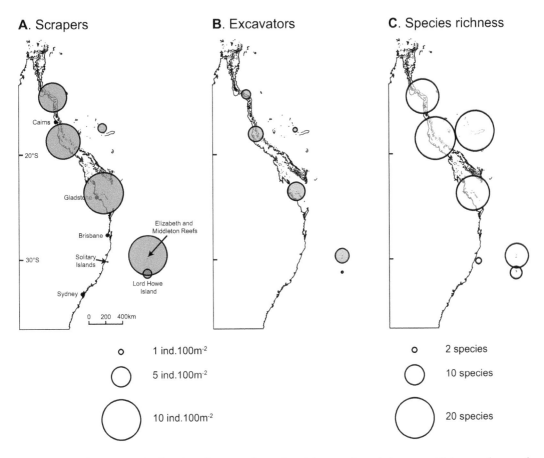

Fig. 4. Spatial variation in the abundance and species richness of scarine parrotfishes on low and high latitude reefs of eastern Australia. (A) Abundance of scraping parrotfishes (*Scarus, Hipposcarus*), (B) abundance of excavating parrotfishes (*Bolbometopon, Cetoscarus, Chlorurus*) and (C) species richness of parrotfishes. Circles are proportional to mean abundances or total number of species recorded in each location. Data are based on replicate 50 × 5 m belt transects on shallow reef crests in each location (GBR: Trapon et al. 2013; Coral Sea: Ceccarelli et al. 2008; Lord Howe Island: Hoey et al. 2011; Elizabeth and Middleton reefs: Hoey et al. 2014; Solitary Islands: Hoey unpublished data – March 2013).

impact of the hard underlying substratum (i.e., rock) on their feeding (as discussed above for the Arabian Sea).

The decreased abundance and altered taxonomic composition of parrotfishes on these high latitude reefs will lead to decreased rates of grazing and erosion. This may be further accentuated by any reductions in metabolic rates, and hence feeding rates at higher latitudes. Comparisons of feeding rates of adult individuals of three species of scarine parrotfishes (*Sc. ghobban, Scarus psittacus, Scarus schlegeli*) show declines of 39 to 71 percent between the GBR (Lizard Island 14°30′S; Heron Island 23°30′S) and Lord Howe Island (Fig. 5). It should be noted that these data are based on limited sample sizes, did not account for potential changes in the nutritional quality of dietary resources or the quantity consumed per bite, however the consistency of the declines in feeding rate suggest they may be a consequence of lower water temperatures on Lord Howe Island. If this is the case, then

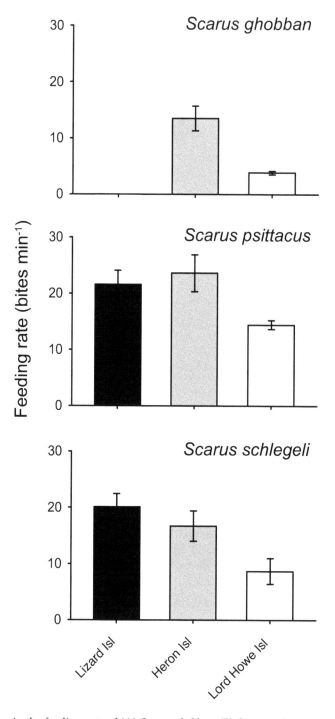

Fig. 5. Variation in the feeding rate of (A) *Scarus ghobban,* (B) *Scarus psittacus,* and (C) *Scarus schlegeli* among three locations spanning 17° of latitude on Australia's east coast (Lizard Island: 14°30'S, Heron Island: 23°30'S, Lord Howe Island: 31°30'S). Feeding rates for Heron Island and Lizard Island are from Bellwood and Choat 1990. Feeding rates for Lord Howe Island are based on replicate 3-minute focal individual observations.

such reductions in feeding rate of parrotfishes may be widespread across high latitude reefs, and together with decreases in the abundance of parrotfishes will reduce the amount of material they ingest, the volume of carbonates they erode and sediments they rework. Further, increases in other herbivorous taxa, such as subtropical acanthurids (i.e., *Prionurus* spp) and urchins on both coastal and offshore high latitude reefs (e.g., Hoey et al. 2011, Verges et al. 2016) compared to low latitude reefs (e.g., Cheal et al. 2012) may indicate a reduced importance of parrotfishes in the functioning of these reefs.

Marginal Reefs of Southern Japan

Japanese coral reefs extend from the southern Ryukyu Islands (24°N) to Tanegashima Island (31°N), and in the Ogasawara (Bonin) Islands south of 27°N (Ministry of Environment and Japanese Coral Reef Society, 2004). Reef formation transitions from continuous fringing reefs with extensive reef flats (up to 1 km wide) in the southern Ryukyu Islands (south of 27°N), to patchily distributed fringing reefs with narrow reef flats in the northern Ryukyu Islands (27–31°N), and non-accreting coral communities on rocky substrates in the area north of 31°N to the central mainland of Japan.

Approximately 415 scleractinian species belonging to 78 genera have been recorded from Japanese reefs, with the vast majority of species occurring in the Ryukyu Islands (Ministry of Environment and Japanese Coral Reef Society, 2004). This species-rich coral fauna is due, at least in part, to the northward flowing Kuroshio Current, which brings warm tropical waters from the Philippines along the Ryukyu Islands to the Pacific coast of the Japanese mainland. Declines in sea surface temperature (SST) associated with the Kuroshio Current have been related to the latitudinal gradient in coral species richness. The greatest number of coral species occur in the southern Ryukyu Islands (ca. 380 species: Yaeyama region, 24°N) and decrease with increasing latitude (340 species: Okinawa region, 26°N; 220 species: Amami region, 28°N; 200 species: western and central Japanese mainland 32°N; 50 species: eastern Japanese mainland, 34–35°N). Over the last few decades, however, rising SSTs and a strengthening of the Kuroshio Current has contributed to the range expansion of several tropical coral and fish species onto the temperate Japanese coasts (Yamano et al. 2011, Verges et al. 2014).

Ecology of Parrotfishes on Marginal Reefs of Southern Japan

Up to 36 parrotfish species belonging to seven genera have been reported from Japanese waters (Nakabo 2013), although several of these on marginal reefs appear to be tropical vagrants and the number of species with established populations is likely to be considerable lower. Of these 36 species, the vast majority (35 spp.) have been recorded in the Ryukyu Islands, the only exception being *Scarus obishime*, which is endemic to the Ogasawara Islands. Parrotfishes tend to display clear among-habitat distributions on reefs within the Ryukyu Islands, as have been reported in many other locations (Burkepile et al. Chapter 7). Among scraping and excavating (i.e., scarinine) parrotfishes, species such as *Chlorurus bowersi*, *Scarus niger*, *Scarus festivus*, and *Scarus forsteni* occur primarily on the outer reef slopes, whereas *Chlorurus spilurus*, *Scarus rivulatus*, and *Scarus schlegeli* occur in both the inner reef flats and the outer reef slopes (Shibuno et al. 2008). Among the browsing (i.e., sparisomatine) species, *Calotomus carolinus* occurs in coral reef habitats, whereas *Leptoscarus vaigiensis* and *Calotomus spinidens* are restricted to seagrass and macroalgal beds (Sano 2001, Shibuno et al. 2008).

Species richness and abundance of parrotfishes within the Ryukyu Islands decrease with increasing latitude and is largely driven by changes in the abundance of scraping and excavating (i.e., scarinine) parrotfishes (Fig. 6). In particular, significant decreases in the abundance of parrotfishes are evident between subtropical (Ryukyu Islands) and temperate (Japan mainland) regions, and are likely related to the lack of coral reef habitat as well as the low winter water temperatures (<18°C) in temperate regions. Contrary to the latitudinal distribution patterns of most parrotfish species, *Calotomus japonicus* and *Scarus ovifrons* are primarily distributed from the northern Ryukyu Islands to the central Japanese mainland (Nakabo 2013), and as such are regarded as subtropical species. *Calotomus japonicus* occur in rocky habitats and/or seaweed beds where they have been observed to feed on both red and brown macroalgae, including *Sargassum* (Terazono et al. 2012). The feeding activity of *Ca. japonicus* is positively related with seawater temperature, and increasing coastal SST during autumn and winter over the last 30 years have extended their period of intensive feeding. The increase in the feeding of *Ca. japonicus*, as well as other herbivorous fishes, has been implicated in the loss kelp forests in southern Japan (Serizawa et al. 2004, Verges et al. 2014). *Scarus ovifrons* is a large-bodied species (up to 80 cm total length, TL) that occurs in rocky and coral-dominated reefs, however its feeding ecology and life history characteristics are poorly understood.

Species richness and abundance of tropical parrotfish species in temperate Japanese waters are approximately two and five times greater, respectively, in coral-rich than rocky reef habitats, and seasonal changes in these numbers closely match those of SST (Nakamura et al. 2013). The majority of parrotfishes on these temperate reefs are juveniles (< 10 cm TL);

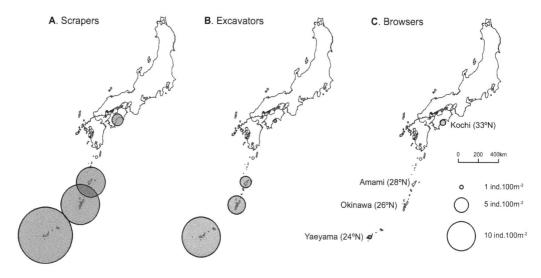

Fig. 6. Spatial variation in the abundance of parrotfishes in southern Japan. (A) Abundance of scarinine scraping parrotfishes (*Scarus, Hipposcarus*), (B) abundance of scarinine excavating parrotfishes (*Cetoscarus, Chlorurus*), and (C) abundance of sparisomatine 'browsing' parrotfishes. Circles are proportional to mean abundances or total number of species recorded in each location. Data are from 56–60 replicate 20 m² transects along inner reef flats (within 1–3 m depth) and shallow reef slopes (2–10 m) in the Ryukyu Islands, and coral-rich habitats (2–10 m) in Kochi (*n* = 56 for each region in the Ryukyu Islands and *n* = 60 for Kochi). Data were collected during spring (April–May), summer (August), and late autumn (November–December) in each region in 2004–2005 for the Ryukyu Islands and in 2009–2010 for Kochi.

the lack of larger-bodied individuals suggests they may not be surviving the colder water temperatures during winter. The only exception to this is *Sc. ghobban* which comprises approximately 75 percent of the total abundance of tropical parrotfishes at Kochi (33°N), southern Japan, with individuals ranging in size from 4 to 30 cm TL. The higher abundance coupled with the presence of larger individuals suggest *Sc. ghobban* may have established populations on the temperate Japanese reefs (Nakamura et al. 2013).

The mechanism/s that have allowed *Sc. ghobban* to establish on these temperate reefs, while other parrotfish species appear unable to overwinter, remains unclear but may be related to differences in thermal tolerances or feeding habitats among species. Interestingly, small-bodied (< 10 cm TL) *Sc. ghobban* have been observed to feed almost exclusively from the surfaces of dead corals, while larger individuals (10–30 cm TL) tend to feed equally from the surfaces of dead coral, boulders, and rocky reefs (Nakamura unpublished). The feeding rates of *Sc. ghobban* (> 10 cm TL) on these temperate Japanese reefs during summer (ca. 15 bites min^{-1}) are directly comparable to those reported from other tropical reefs (GBR: 13.9 bites min^{-1}; Red Sea 10.9 bites min^{-1}; Bonaldo et al. 2014), but in late autumn and winter the feeding rates decrease substantially (ca. 1–5 bites min^{-1}). Nevertheless, the expansion of coral habitat coupled with increasing winter SSTs on Japanese temperate reefs may lead to an increase in suitable feeding substrates for tropical scraping and excavating parrotfish species.

Marginal Reefs of Brazil

The Brazilian coast and oceanic islands encompass unique reef formations and marine life, with high rates of endemism for a number of groups of reef organisms (Maida and Ferreira 1997, Floeter et al. 2001, Rocha 2003). The region hosts a large number of endemic species, and consequently represents a biogeographic province in the Western Atlantic (Gilbert 1972, Briggs 1974, Floeter and Gasparini 2000), separated from the Caribbean by the mouths of the Amazon and Orinoco rivers, and by the sediment-rich coastline of the Guyanas. Larval exchange between the Brazilian and the Caribbean provinces is limited, as connectivity between these areas has been discontinuously maintained along geological time by eustatic processes that weaken the Amazon influence (Floeter et al. 2001, Rocha 2003).

The Brazilian province congregates a number of geological landscape features that constrain the establishment of coral reef ecosystems, such as the high rainfall and massive riverine input, convergence of subtropical currents, and the relatively narrow continental platform (Maida and Ferreira 1997, Leão et al. 2003). Growth of biogenic reefs, including coral reefs, is limited to north-eastern Brazil, an area characterized by seasonal inputs of freshwater and sediments, and the resuspension of sediments by strong winds and currents (Maida and Ferreira 1997). Eighteen species of scleractinian corals have been recorded from the Brazilian province, and not surprisingly all have a massive or encrusting morphology that are resistant to sedimentation and high turbidity (Maida and Ferreira 1997). There is a complete absence of branching scleractinian corals and *Acropora* species (Maida and Ferreira 1997, Leão et al. 2003). The richest region in the province in terms of the number of coral species and reef area size is the Abrolhos bank, off north-eastern Brazil, where coral formations have a particular mushroom shape and reefs do not display the typical zonation pattern observed in other Atlantic coral reefs (Maida and Ferreira 1997, Leão et al. 2003).

Water temperature constrains the development of shallow coral reefs in the south and south-eastern Brazilian coast. As a result, benthic communities in these regions are usually composed of massive corals and hydrocorals, together with algae, sponges and other benthic organisms that form a veneer on profuse rocky shores. As temperature decreases towards the south of the province, rocky shores become more dominant in relation to biogenic reefs and there is a marked reduction in the taxonomic and functional diversity of tropical reef fish communities (Ferreira et al. 2004, Floeter et al. 2001, 2005).

Reef fish fauna in the Brazilian province mirrors the particular physical conditions and benthic fauna of the system. Approximately 20 percent of the Brazilian reef fishes are endemic to the region, with a number of species closely related to the fauna in the Caribbean (Moura and Sazima 2000, Floeter et al. 2008). Ten parrotfish species have been recorded within the Brazilian province: two *Scarus*, six *Sparisoma*, *Cryptotomus roseus* and *Nicholsina ulsta* (Robertson et al. 2006, Floeter et al. 2008), with assemblages dominated by sparisomatine species (Fig. 7). Sparisomatine parrotfishes (i.e., *Sparisoma* or seagrass clade) are regarded as having a weaker association to coral reef habitats, especially in comparison to scarinine species (i.e., *Scarus*, or reef, clade; Streelman et al. 2002). Among the sparisomatine parrotfishes *Sparisoma* spp. are dominant on most reefs along the Brazilian coast and oceanic islands (Ferreira et al. 2004, Floeter et al. 2005), while *Cr. roseus*, *N. ulsta* and *Sparisoma radians* inhabit shallow vegetated areas (i.e., seagrass and macroalgae) adjacent to reefs (C.E.L. Ferreira pers. obs.). Although scarinine parrotfishes originated in reef habitats and have been suggested to have closer, and in some cases obligate, association with coral reefs (Streelman et al. 2002), both Brazilian *Scarus* species appear to be doing well on both the tropical coral-rich north-eastern and subtropical coral-poor south-eastern coasts (Ferreira et al. 2001, Cordeiro et al. 2016).

Five of the parrotfish species in the Brazilian province are recently described endemics that were previously confounded with their sister species in the Caribbean (e.g. Moura et al. 2001, Gasparini et al. 2003). Besides morphological resemblances, pairs of sister species in the Caribbean and Brazil often differ in abundance in their respective regions. *Sparisoma viride*, for example, is present in high densities throughout much of the Caribbean (Bruggemann et al. 1994, McAfee and Morgan 1996, Mumby and Harborne 2010), yet its sister taxon, *Sparisoma amplum*, is relatively rare on Brazilian reefs and only occurs in higher densities on the north-eastern coast and oceanic islands (Ferreira et al. 2004, Francini-Filho et al. 2010). The clade formed by *Sp. amplum* and *Sp. viride* may thus be considered as central for coral reefs in the Caribbean, but peripheral for Brazilian reefs. In contrast, while *Sp. rubripinne* and *Sp. chrysopterum* are generally found in low densities on Caribbean reefs (McAfee and Morgan 1996), usually near seagrass beds and other marginal habits, their sister species, *Sp. axillare* and *Sp. frondosum*, respectively, are the most abundant parrotfish species on Brazilian reefs (Fig. 8; Ferreira et al. 2001, Ferreira et al. 2004, Floeter et al. 2007, Francini-Filho et al. 2010, Cordeiro et al. 2016).

Differences in the relative abundances of sister taxa between the Caribbean and Brazil likely reflects variations in habitat requirements of these species. The sparisomatine parrotfishes *Sp. amplum* and *Sp. viride* possess a suite of features more characteristic of scarinine parrotfishes, namely robust jaws with simple and strong articulations among elements and well-developed musculature to generate a forceful bite (Bellwood 1994). Consequently, *Sp. amplum* and *Sp. viride* are the only sparisomatine parrotfishes considered to have an excavating feeding mode (Rotjan and Lewis 2005, 2006, Francini-Filho et al. 2008), and appear to represent an example functional convergence with scarinine parrotfishes (Streelman et al. 2002, Robertson et al. 2006). These features may explain the higher abundance of these species in coral reefs in the Caribbean and on tropical reefs of Brazil. In contrast, *Sp. rubripinne* / *Sp. axillare* and *Sp. chrysopterum* / *Sp. frondosum* are

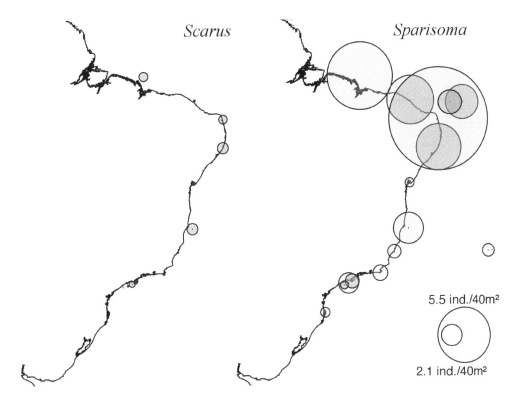

Fig. 7. Spatial variation in the density of scarinine and sparisomatine parrotfishes along the Brazilian coast. Circles are proportional to mean abundances recorded in each location. Data of species abundances based on 20 × 2 m belt transects within shallow (3–15 m) rocky or biogenic (coral/coralline algae) reefs (see details at Floeter et al. 2005).

more representative of typical sparisomatine parrotfishes, feeding on macroalgae and/or associated epiphytic material (McAfee and Morgan 1996, Bonaldo et al. 2006, Ferreira and Gonçalves 2006). Such feeding habits may reduce their reliance on conditions usually associated with coral-rich habitats in favour of the higher abundance of macroalgae on marginal reefs in Brazil.

In addition to differences in parrotfish assemblages between the Caribbean and Brazil, the abundance of parrotfishes in relation to other nominally roving herbivorous fishes differs between these two provinces. While parrotfishes typically dominate herbivorous fish assemblages on Caribbean reefs, surgeonfishes dominate herbivorous fish assemblages on most reefs in Brazil (Floeter et al. 2005). Furthermore, there is a clear decrease in the abundance of parrotfishes and surgeonfishes with increasing latitude along the Brazilian coast. This decline is hypothesized to be related to the physiological constraints of these species to low temperatures, although nutritional hypotheses need to be considered (Ferreira et al. 2004, Floeter et al. 2005). Importantly, feeding pressure by these groups is also reduced at higher latitudes and lower temperatures (Longo et al. 2014, Fig. 9). For example, feeding pressure of *Scarus zelindae* and *Scarus trispinosus* decrease by 95–100 percent from the Abrolhos Archipelago to Arraial do Cabo (Fig. 9). In Santa Catarina, the southernmost region for reefs in Brazil, parrotfishes become less abundant and surgeonfishes vagrant (Ferreira et al. 2004). In this region, species with subtropical-temperate affinities, such

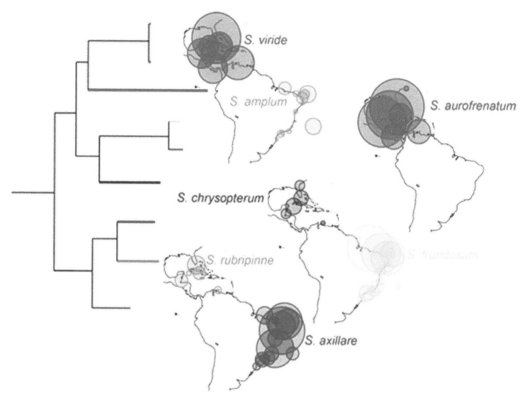

Fig. 8. Phylogeny of *Sparisoma* (modified from Robertson et al. 2006) showing relationships among species and their abundances along the Brazilian coast. Circle sizes representing the mean abundance (square root) of species in the tropical Western Atlantic. Data of species abundances based on 20 × 2 m belt transects within shallow (3–15 m) rocky or biogenic (coral/coralline algae) reefs (see details at Floeter et al. 2005).

as browsing *Kyphosus* spp. and the omnivorous *Diplodus argenteus*, are found in higher densities (Ferreira et al. 2004).

Studies on behavior and ecology of reef fishes, including parrotfishes, are relatively recent in Brazil (e.g. Ferreira et al. 1998, Floeter et al. 2005, Bonaldo et al. 2006, Francini-Filho et al. 2008, Francini-Filho et al. 2010), and several aspects of the ecology of parrotfishes are still to be elucidated, especially regarding the impact of these species on benthic communities and the influence of benthic communities in shaping parrotfish populations. On coral reefs, the balance between live coral colonies and benthic algae is fundamentally important, as corals provide the basis of the structure and dynamics of these systems. In this scenario, parrotfishes along with other reef herbivores are considered key in mediating coral-algal interactions since they can directly influence the structure of benthic communities (Lewis 1986, Bellwood et al. 2006, Burkepile and Hay 2008). On Brazilian reefs where low coral cover and high abundance of other benthic organisms naturally characterize the benthic community, the importance of parrotfishes in the structure and dynamics the reef benthos remains to be evaluated. Conversely, there is emerging evidence of the importance of bottom-up processes in shaping parrotfish populations (e.g., Heenan et al. 2016, Hamilton et al. 2017). Given the differences in the feeding ecology among parrotfishes and the geographical distribution of species in the group, the extrapolation of

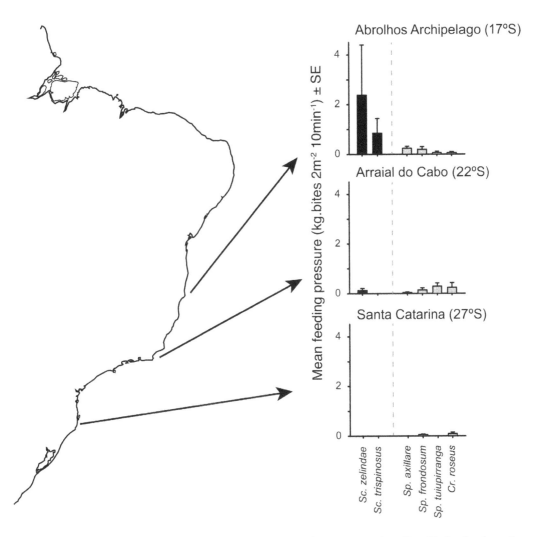

Fig. 9. Mean feeding pressure of parrotfishes at three study sites spanning 10° of latitude along the Brazilian coast (Abrolhos Archipelago, 17°S, Arraial do Cabo, 22°S, and Santa Catarina, 27°S). Data collected with remote filming of 2 m² plots (79–121 10–min videos per site) (see Longo et al. 2014).

patterns and processes derived from coral reefs in the Caribbean or Indo-Pacific to other areas should be done with care.

In spite of the unique nature and high endemism of reef fishes of the Brazilian province, reef ecosystems in this area are threatened by several anthropogenic activities, such as overfishing, pollution and habitat destruction. In the specific case for parrotfishes, a number of species have been suffering severe declines in their abundance and body size (Floeter et al. 2006, 2007, Bender et al. 2014). The largest endemic parrotfish in Brazil, *Sc. trispinosus* (maximum length 70 cm), for instance, used to be abundant on reefs along the north-east and south-east coast, but has been heavily fished during the last two decades. As a consequence, the species has suffered marked reductions and local extinctions throughout most of its historic range (Floeter et al. 2007, Bender et al. 2014) and has been recently classified as endangered by the IUCN Red List of species (Ferreira et al. 2012). Other large parrotfishes are facing similar declines, especially as populations of former targets of spearfishers, such as groupers (Serranidae) and snappers (Lutjanidae), have collapsed on Brazilian reefs. Studies on population dynamics and ecology of parrotfishes in the Brazilian Province, especially on demography, nutritional ecology, and the role of this group in shaping local benthic communities, are fundamentally important to promote management and conservation of these unique species and habitats.

Summary

This chapter has explored how the variable and sometimes extreme environmental conditions of marginal reefs influence the structure of parrotfish assemblages and their interactions with the reef environment, and the challenges such conditions pose. In doing so, this chapter has raised many questions but answered few. Declines in the abundance and diversity of parrotfishes with latitude are widespread, and pronounced among scarinine, especially excavating, parrotfishes. It has been previously suggested that declines in tropical herbivorous fishes, including parrotfishes, with latitude are related to the effects of temperature on digestive physiology, or reductions in the quantity and quality of algal and detrital food sources in marginal environments (reviewed by Choat 1991). We do not discount the potential importance of nutritional ecology in shaping these patterns, rather we offer an additional mechanism for the near absence of parrotfishes in some marginal environments; the inability to penetrate the hard substratum of marginal reefs when feeding. Scarinine parrotfishes scrape and/or excavate portions of the reef substratum when feeding, presumably targeting endolithic phototrophs (Clements et al. 2017). In marginal environments with limited coral reef development benthic communities often occur as a veneer over harder granitic or basaltic substratum, essentially a rocky reef. The feeding apparatus of scraping and excavating parrotfishes, while effective at gouging into calcareous substrata on coral reefs does not have the capacity to penetrate these rocky substrata, and attempts to do so often result in damage to the oral jaws.

Sparisomatine parrotfishes appear to be better suited to marginal coral reef environments, are abundant on rocky reefs along the Brazilian coast, and one species, *Sp. cretense*, is largely restricted to temperate reef environments. Sparisomatines are widely regarded as browsers of macroalgae (e.g., Streelman et al. 2002) and are often observed taking bites from macroalgae and seagrass (e.g., Adam et al. 2015), however, it has recently been hypothesized that the majority of species are targeting protein rich epiphytic phototrophs on the surfaces of macrophytes (Clements and Choat Chapter 3). Such a feeding mode alleviates the need to gouge hard surfaces when feeding and may explain the less pronounced decline in this group at higher latitudes. Although it should be noted that the excavating *Sp. amplum* and larger-bodied individuals of other *Sparisoma* spp. regularly takes bites on hard substrata, and such a feeding habit has been shown to cause dentition damage on the basaltic reefs of Fernando de Noronha Archipelago, north-eastern Brazil (Bonaldo et al. 2007).

Perhaps the biggest challenges for parrotfishes inhabiting marginal reefs are the extreme and variable environmental conditions, namely temperature. Cellular processes, metabolism, energy requirements, and individual performance (activity, growth, reproduction) of ectotherms, including parrotfishes, are all linked to environmental temperature. In turn, these effects of temperature on individuals will manifest as changes in parrotfish populations and communities, and ultimately their functional impact. For example, the mean maximum of length and longevity of *Ch. spilurus* varies across latitude, with individuals at low latitude sites being smaller and having shorter life spans than those at high latitudes (Taylor et al. Chapter 4). It is unclear if such relationships will hold for other parrotfish species, or in regions that experience both extremely high and low temperatures (i.e., the Arabian Gulf). The limited data on latitudinal and seasonal differences in feeding presented in this chapter also point toward the potential importance of environmental temperature in influencing the performance of parrotfishes. Any latitudinal or seasonal patterns, however, are not solely attributable to temperature. Many other factors, including productivity and the nutritional quality of dietary resources, are likely to covary with temperature and need also to be considered. Notwithstanding, a greater understanding of the effect of temperature on the physiology and performance of parrotfishes will not only allow us to better understand the factors that shape parrotfish assemblages on marginal reefs, but also enable us to better predict likely changes in the distribution and ecology of parrotfishes under climate change.

Throughout this chapter we compare the abundance of parrotfishes among locations and latitudes. One problem of such an approach is that it doesn't account for any differences in body size or longevity. If the latitudinal patterns in maximum body size and longevity of *Ch. spilurus* are representative of other parrotfish species, then the shorter life spans at lower latitudes suggest that these populations would have higher rates of turnover. Conversely, those at higher latitudes would have lower turnover, and populations would be composed of individuals from numerous cohorts, and may be extremely susceptible to disturbances or fishing.

Based on the phylogenetic reconstruction of the evolutionary history of 61 species of scarinine parrotfish, Choat et al. (2012) suggest that the ancestral habitats of *Scarus* were rocky reefs (Choat et al. 2012). Despite this, however, it appears that marginal reefs are not the realm of extant parrotfishes. Although some species are endemic to regions of marginal reef (Arabian Sea: *Sc. zufar, Sc. arabicus*; Eastern Pacific: *Scarus perrico, Scarus hoefleri*; Mediterranean: *Sp. cretense*), and others (i.e., *Sc. ghobban* species complex) are regularly found on marginal reefs, the majority of parrotfish species are rare or absent from marginal environments. Parrotfishes often dominate herbivore biomass on low latitude reefs, especially in areas of high turf algal cover and low topographic complexity (Fox Chapter 13), and are viewed as critical to maintaining the balance between corals and algae. Within high latitude marginal reef systems, however, parrotfishes appear to be relatively minor players, with subtropical acanthurids and urchins dominating herbivore assemblages.

Acknowledgments

We thank A. Bauman, D. Barneche, B. Busteed, C. Cvitanovic, J. Hoey, G. Longo, R. Morais, M. Pratchett, G. Vaughan for logistical support and/or discussions on the ideas presented in this chapter, and the Australian Research Council, CNPq (Conselho Nacional de Desenvolvimento Científico e Tecnológico), and FAPESP (Fundação de Amparo à Pesquisa do Estado de São Paulo) for financial support.

References Cited

Adam, T.C., M. Kelley, B.I. Ruttenberg and D.E. Burkepile. 2015. Resource partitioning along multiple niche axes drives functional diversity in parrotfishes on Caribbean coral reefs. Oecologia. 179: 1173–1185.

Alwany, M.A., E. Thaler and M. Stachowitsch. 2009. Parrotfish bioerosion on Egyptian red sea reefs. J. Exp. Mar. Biol. Ecol. 371: 170–176.

Afeworki, Y., J.J. Videler and J.H. Bruggemann. 2013. Seasonally changing habitat use patterns among roving herbivorous fishes in the southern Red Sea: the role of temperature and algal community structure. Coral Reefs 32: 475–485.

Baird, A.H., B. Sommer and J.S. Madin. 2012. Pole-ward range expansion of *Acropora* spp. along the east coast of Australia. Coral Reefs 31: 1063–1063.

Baird, A.H., M.O. Hoogenboom and D. Huang. 2017. *Cyphastrea salae*, a new species of hard coral from Lord Howe Island, Australia (Scleractinia, Merulinidae). Zoo Keys 662: 49–66.

Bauman, A.G., D.A. Feary, S.F. Heron, M.S. Pratchett and J.A. Burt. 2013. Multiple environmental factors influence the spatial distribution and structure of reef communities in the northeastern Arabian Peninsula. Mar. Poll. Bull. 72: 302–312.

Bellwood, D.R. 1994. A phylogenetic study of the parrotfishes family Scaridae (Pisces: Labroidei), with a revision of the genera. Rec. Austr. Mus. 20: 1–86.

Bellwood, D.R. and J.H. Choat. 1990. A functional analysis of grazing in parrotfishes (family Scaridae): the ecological implications. Environ. Biol. Fishes 28: 189–214.

Bellwood, D.R. and C.J. Fulton. 2008. Sediment-mediated suppression of herbivory on coral reefs: decreasing resilience to rising sea-levels and climate change. Limnol. Oceanogr. 53: 2695–2701.

Bellwood, D.R., T.P. Hughes and A.S. Hoey. 2006. Sleeping functional group drives coral-reef recovery. Curr. Biol. 16: 2434–2439.

Bellwood, D.R., A.S. Hoey and T.P. Hughes. 2012. Human activity selectively impacts the ecosystem roles of parrotfishes on coral reefs. Proc. R. Soc. B 271: 1621–1629.

Bender, M.G., G.R. Machado, P.J.A. Silva, S.R. Floeter, C. Monteiro-Neto, O.J. Luiz and C.E.L. Ferreira. 2014. Local ecological knowledge and scientific data reveal overexploitation by multigear artisanal fisheries in the Southwestern Atlantic. PLoS One 9: e110332.

Bento, R., A.S. Hoey, A.G. Bauman, D.A. Feary and J.A. Burt. 2016. The implications of recurrent disturbances within the world's hottest coral reef. Mar. Poll. Bull. 105: 466–472.

Bonaldo, R.M., J.P. Krajewski, C. Sazima and I. Sazima. 2006. Foraging activity and resource use by three parrotfish species at Fernando de Noronha Archipelago, tropical West Atlantic. Mar. Biol. 149: 423–433.

Bonaldo, R.M., J.P. Krajewski, C. Sazima and I. Sazima. 2007. Dentition damage in parrotfishes feeding on hard surfaces at Fernando de Noronha Archipelago, southwest Atlantic Ocean. Mar. Ecol. Prog. Ser. 342: 249–254.

Bonaldo, R.M., J.P. Krajewski and D.R. Bellwood. 2011. Relative impact of parrotfish grazing scars on massive *Porites* corals at Lizard Island, Great Barrier Reef. Mar. Ecol. Prog. Ser. 423: 223–233.

Bonaldo, R.M., A.S. Hoey and D.R. Bellwood. 2014. The ecosystem roles of parrotfishes on tropical reefs. Oceanogr. Mar. Biol. Annu. Rev. 52: 81–132.

Briggs, J.C. 1974. Marine Zoogeography. McGraw-Hill, New York.

Bruggemann, J.H., M.J.H. van Oppen and A.M. Breeman. 1994. Foraging by the stoplight parrotfish *Sparisoma viride* I. Food selection in different, socially determined habitats. Mar. Ecol. Prog. Ser. 106: 41–55.

Buddemeier, R.W. and S.V. Smith. 1999. Coral adaptation and acclimatization: a most ingenious paradox. Amer. Zool. 39: 1–9.

Burkepile, D.E. and M.E. Hay. 2008. Herbivore species richness and feeding complementarily affect community structure and function: the case for Caribbean reefs. Proc. Natl. Acad. Sci. USA. 105: 16201–16206.

Burt, J., S. Al-Harthi and A. Al-Cibahy 2011. Long-term impacts of bleaching events on the world's warmest reefs. Mar. Environ. Res. 72: 225–229.

Burt, J., D. Feary, A. Bauman, P. Usseglio, G. Cavalcante and P. Sale 2011. Biogeographic patterns of reef fish community structure in the northeastern Arabian Peninsula. ICES J. Mar. Sci. 68: 1875–1883.

Ceccarelli, D., J.H. Choat, A.M. Ayling, Z.T. Richards, L. van Herwerden, G.D. Ewels, J.P. Hobbs and B. Cuff. 2008. Coringa-Herald National Nature Reserve Marine Survey 2007. Report for the Department of the Environment, Water, Heritage and the Arts, Canberra, Australia.

Cheal, A., M. Emslie, I. Miller and H. Sweatman. 2012. The distribution of herbivorous fishes on the Great Barrier Reef. Mar. Biol. 159: 1143–1154.

Choat, J.H. 1991. The biology of herbivorous fishes on coral reefs. pp. 120–155. *In*: P.F. Sale (ed.). The Ecology of Fishes on Coral Reefs. Academic Press, San Diego.

Choat, J.H., L. Herwerden, D.R. Robertson and K.D. Clements. 2012. Patterns and processes in the evolutionary history of parrotfishes (Family Labridae). Biol. J. Linnean Soc. 107: 529–557.

Clark, T.D., V. Messmer, A.J. Tobin, A.S. Hoey and M.S. Pratchett. 2017. Rising temperatures may drive fishing-induced selection of low-performance phenotypes. Sci. Rep. 7: 40571.

Clements, K.D., D.P. German, J. Piché, A. Tribollet and J.H. Choat. 2017. Integrating ecological roles and trophic diversification on coral reefs: multiple lines of evidence identify parrotfishes as microphages. Biol. J. Linn. Soc. 120: 729–751.

Coles, S. 2003. Coral species diversity and environmental factors in the Arabian Gulf and the Gulf of Oman: a comparison to the Indo-Pacific region. Atoll Res. Bull. 507: 1–19.

Cordeiro, C.A.M.M., T.C. Mendes, A.R. Harbone and C.E.L. Ferreira. 2016. Spatial distribution of nominally herbivorous fishes across environmental gradients on Brazilian rocky reefs. J. Fish Biol. 89: 939–958.

Cvitanovic, C. and A.S. Hoey 2010. Benthic community composition influences within-habitat variation in macroalgal browsing on the Great Barrier Reef. Mar. Freshw. Res. 61: 999–1005.

Dalton, S.J. and G. Roff. 2013. Spatial and temporal patterns of eastern Australia subtropical coral communities. PloS One 8: e75873.

DiBattista, J.D., J.H. Choat, M.R. Gaither, J.P.A. Hobbs, D.F. Lozano-Cortés, R.F. Myers, G. Paulay, L.A. Rocha, R.J. Toonen, M.W. Westneat and M.L. Berumen. 2016. On the origin of endemic species in the Red Sea. J. Biogeogr. 43: 13–30.

Evans, R.D., S.K. Wilson, S.N. Field and J.A.Y. Moore. 2014. Importance of macroalgal fields as coral reef fish nursery habitat in north-west Australia. Mar. Biol. 161: 599–607.

Feary, D.A., J.A. Burt, A.G. Bauman, P. Usseglio, P.F. Sale and G.H. Cavalcante. 2010. Fish communities on the world's warmest reefs: what can they tell us about the effects of climate change in the future? J. Fish Biol. 77: 1931–1947.

Feary, D.A., M.S. Pratchett, M.J. Emslie, A.M. Fowler, W.F. Figueira, O.J. Luiz, Y. Nakamura and D.J. Booth. 2014. Latitudinal shifts in coral reef fishes: why some species do and others do not shift? Fish Fish. 15: 593–615.

Ferreira, B.P., S.R. Floeter, L.A. Rocha, C.E.L. Ferreira, R. Francini-Filho, R. Moura, A.L. Gaspar and C. Feitosa. 2012. *Scarus trispinosus*. IUCN 2014. IUCN Red List of Threatened Species. Version 2014.1. http://www.iucnredlist.org.

Ferreira, C.E.L., S.R. Floeter, J.L. Gasparini, B.P. Ferreira and J.C. Joyeux. 2004. Trophic structure patterns of Brazilian reef fishes: a latitudinal comparison. J. Biogeogr. 31: 1093–1106.

Ferreira, C.E.L. and J.E.A. Gonçalves. 2006. Community structure and diet of roving herbivo rous reef fishes in the Abrolhos Archipelago, south-western Atlantic. J. Fish Biol. 69: 1533–1551.

Ferreira, C.E.L., J.E.A. Gonçalves and R. Coutinho. 2001. Community structure of fishes and habitat complexity on a tropical rocky shore. Environ. Biol. Fishes 61: 353–369.

Ferreira, C.E.L., A.C. Peret and R. Coutinho. 1998. Seasonal grazing rates and food processing by tropical herbivorous fishes. J. Fish Biol. 53: 222–235.

Floeter, S.R., M.D. Behrens, C.E.L. Ferreira, M.J. Paddack and M.H. Horn. 2005. Geographical gradients of marine herbivorous fishes: patterns and processes. Mar. Biol. 147: 1435–1447.

Floeter, S.R., C.E.L. Ferreira and J.L. Gasparini. 2007. Os efeitos da pesca e da proteção através de UCs Marinhas: Três estudos de caso e implicações para os grupos funcionais do Brasil. Ministério do Meio Ambiente, Brasília.

Floeter, S.R. and J.L. Gasparini. 2000. The southwestern Atlantic reef fish fauna: composition and zoogeographic patterns. J. Fish Biol. 56: 1099–1114.

Floeter, S.R., R.Z.P. Guimarães, L.A. Rocha, C.E.L. Ferreira, C.A. Rangel and J.L. Gasparini. 2001. Geographic variation in reef-fish assemblages along the Brazilian coast. Global Ecol. Biogeogr. 10: 423–431.

Floeter, S.R., B.S. Halpern and C.E.L. Ferreira. 2006. Effects of fishing and protection on Brazilian reef fishes. Biol. Cons. 128: 391–402.

Floeter, S.R., L.A. Rocha, D.R. Robertson, J.C. Joyeux, W.F. Smith-Vaniz, P. Wirtz, A.J. Edwards, J.P. Barreiros, C.E.L. Ferreira, J.L. Gasparini, A. Brito, J.M. Falcón, B.W. Bowen and G. Bernardi. 2008. Atlantic reef fish biogeography and evolution. J. Biogeogr. 35: 22–47.

Fox, R.J. and D.R. Bellwood. 2007. Quantifying herbivory across a coral reef depth gradient. Mar. Ecol. Progr. Ser. 339: 49–59.

Francis, M.P. 1993. Checklist of the coastal fishes of Lord Howe, Norfolk, and Kermadec Islands, southwest Pacific Ocean. Pac. Sci. 47: 136–170.

Francini-Filho, R.B., C.M. Ferreira, E. Oliveira, C. Coni, R.L. Moura and L. Kaufman. 2010. Foraging activity of roving herbivorous reef fish (Acanthuridae and Scaridae) in eastern Brazil: influence of resource availability and interference competition. J. Mar. Biol. Assoc. U.K. 90: 481–492.

Francini-Filho, R.B., R.L. Moura, C.M. Ferreira and E.O.C. Coni. 2008. Live coral predation by parrotfishes (Perciformes: Scaridae) in the Abrolhos Bank, eastern Brazil, with comments on the classification of species into functional groups. Neotrop. Ichthyol. 6: 191–200.

Gasparini, J.L., J.C. Joyeux and S.R. Floeter. 2003. *Sparisoma tuiupiranga*, a new species of parrotfish (Perciformes: Labroidei: Scaridae) from Brazil, with comments on the evolution of the genus. Zootaxa. 384: 1–14.

Gilbert, C.R. 1972. Characteristics of the western Atlantic reef-fish fauna. Quarterly Journal of the Florida Academy of Sciences, Gainesville 35: 130–144.

Gordon, S.E., C.H. Goatley and D.R. Bellwood. 2016. Low-quality sediments deter grazing by the parrotfish *Scarus rivulatus* on inner-shelf reefs. Coral Reefs 35: 285–291.

Graham, N.A.J. and K.L. Nash. 2013. The importance of structural complexity in coral reef ecosystems. Coral Reefs 32: 315–326.

Grandcourt, E. 2012. Reef fish and fisheries in the Gulf. pp. 127–161. *In*: B.M. Riegl and S.J. Purkis (eds.). Coral Reefs of the Gulf: Adaptations to Climatic Extremes. Springer, Dordrecht.

Greenstein, B.J. and J.M. Pandolfi. 2008. Escaping the heat: range shifts of reef coral taxa in coastal Western Australia. Glob. Change Biol. 14: 513–528.

Gullström, M., C. Berkström, M.C. Öhman, M. Bodin and M. Dahlberg. 2011. Scale-dependent patterns of variability of a grazing parrotfish (*Leptoscarus vaigiensis*) in a tropical seagrass-dominated seascape. Mar. Biol. 158: 1483–1495.

Hamilton, R.J., G.R. Almany, C.J. Brown, J. Pita, N.A. Peterson and J.H. Choat. 2017. Logging degrades nursery habitat for an iconic coral reef fish. Biol. Conserv. 210: 273–280.

Harriott, V. and S. Banks. 2002. Latitudinal variation in coral communities in eastern Australia: a qualitative biophysical model of factors regulating coral reefs. Coral Reefs 21: 83–94.

Harriott, V.J., S.A. Banks, R.L. Mau, D. Richardson and L.G. Roberts. 1999. Ecological and conservation significance of the subtidal rocky reef communities of northern New South Wales, Australia. Mar. Freshw. Res. 50: 299–306.

Heenan, A., A.S. Hoey, G.J. Williams and I.D. Williams. 2016. Natural bounds on herbivorous coral reef fishes. Proc. R. Soc. B 283: 20161716.

Hoey, A.S. and D.R. Bellwood. 2008. Cross-shelf variation in the role of parrotfishes on the Great Barrier Reef. Coral Reefs 27: 37–47.

Hoey, A.S. and D.R. Bellwood. 2009. Limited functional redundancy in a high diversity system: single species dominates key ecological process on coral reefs. Ecosystems 12: 1316–1328.

Hoey, A.S. and D.R. Bellwood. 2011. Suppression of herbivory by macroalgal density: a critical feedback on coral reefs? Ecol. Lett. 14: 267–273.

Hoey, A.S., M.S. Pratchett and C. Cvitanovic. 2011. High macroalgal cover and low coral recruitment undermines the potential resilience of the world's southernmost coral reef assemblages. PLoS One 6: e25824.

Hoey, A.S., S.J. Brandl and D.R. Bellwood. 2013. Diet and cross-shelf distribution of rabbitfishes (f. Siganidae) on the northern Great Barrier Reef: implications for ecosystem function. Coral Reefs 32: 973–984.

Hoey, A.S., M.S. Pratchett, J. Johansen and J. Hoey. 2014. Marine ecological survey of Elizabeth and Middleton Reefs, Lord Howe Commonwealth Marine Reserve. Report for the Department of the Environment, Canberra, Australia.

Hoey, A.S., E. Howells, J.L. Johansen, J.P.A. Hobbs, V. Messmer, D.M. McCowan, S.K. Wilson and M.S. Pratchett 2016a. Recent advances in understanding the effects of climate change on coral reefs. Diversity 8: 12.

Hoey, A.S., D.A. Feary, J.A. Burt, G. Vaughan, M.S. Pratchett and M.L. Berumen. 2016b. Regional variation in the structure and function of parrotfishes on Arabian reefs. Mar. Poll. Bull. 105: 524–531.

Johannes, R.E., W.J. Wiebe, C.J. Crossland, D.W. Rimmer and S.V. Smith. 1983. Latitudinal limits of coral reef growth. Mar. Ecol. Prog. Ser. 11: 105–111.

Johansen, J.L., V. Messmer, D.J. Coker, A.S. Hoey and M.S. Pratchett. 2014. Increasing ocean temperatures reduce activity patterns of a large commercially important coral reef fish. Glob. Change Biol. 20: 1067–1074.

Johansen, J.L., M.S. Pratchett, V. Messmer, D.J. Coker, A.J. Tobin and A.S. Hoey. 2015. Large predatory coral trout species unlikely to meet increasing energetic demands in a warming ocean. Sci. Rep. 5: 13830.

Jokiel, P.L. and S.L. Coles. 1977. Effects of temperature on the mortality and growth of Hawaiian reef corals. Mar. Biol. 43: 201–208.

Keith, S.A., A.H. Baird, T.P. Hughes, J.S. Madin and S.R. Connolly. 2013. Faunal breaks and species composition of Indo-Pacific corals: the role of plate tectonics, environment and habitat distribution. Proc. R. Soc. B 280: 20130818.

Keith, S.A., E.S. Woolsey, J.S. Madin, M. Byrne and A.H. Baird. 2015. Differential establishment potential of species predicts a shift in coral assemblage structure across a biogeographic barrier. Ecography 38: 1225–1234.

Khalil, M., J. Bouwmeester and M.L. Berumen. 2017. Spatial variation in coral reef fish and benthic communities in the central Saudi Arabian Red Sea. Peer J. 5: e3410.

Kleypas, J.A., J.W. McManus and L.A. Meñez. 1999. Environmental limits to coral reef development: where do we draw the line? Amer. Zool. 39: 146–159.

Kulbicki, M., V. Parravicini, D.R. Bellwood, E. Arias-Gonzàlez, P. Chabanet, S.R. Floeter, A. Friedlander, J. McPherson, R.E. Myers, L. Vigliola and D. Mouillot. 2013. Global biogeography of reef fishes: a hierarchical quantitative delineation of regions. PLoS One 8: e81847.

Leão, Z.M.A.N., R.K.P. Kikuchi and V. Testa. 2003. Corals and coral reefs of Brazil. pp. 9–52. In: J. Cortés (ed.). Latin American Coral Reefs. Elsevier Science, Amsterdan.

Lewis, S.M. 1986. The role of herbivorous fishes in the organization of a Caribbean reef community. Ecol. Monogr. 56: 183–200.

Lim, I.E., S.K. Wilson, T.H. Holmes, M.M. Noble and C.J. Fulton. 2016. Specialization within a shifting habitat mosaic underpins the seasonal abundance of a tropical fish. Ecosphere 7: e01212.

Loffler, Z., D.R. Bellwood and A.S. Hoey. 2015. Among-habitat algal selectivity by browsing herbivores on an inshore coral reef. Coral Reefs 34: 597–605.

Longo, G.O., C.E.L. Ferreira and S.R. Floeter. 2014. Herbivory drives large-scale spatial variation in reef fish trophic interactions. Ecol. Evol. 4: 4553–4566.

Lybolt, M., D. Neil, J. Zhao, Y. Feng, K.F. Yu and J. Pandolfi. 2011. Instability in a marginal coral reef: the shift from natural variability to a human-dominated seascape. Front. Ecol. Environ. 9: 154–160.

Maida, M. and B.P. Ferreira. 1997. Coral reefs of Brazil: an overview. Proc. 8th Int. Coral Reef Symp. 1: 263–274.

Mantyka, C.S. and D.R. Bellwood. 2007. Macroalgal grazing selectivity among herbivorous coral reef fishes. Mar. Ecol. Prog. Ser. 352: 177–185.

McAfee, S.T. and S.G. Morgan. 1996. Resource use by five sympatric parrotfishes in the San Blas Archipelago, Panama. Mar. Biol. 125: 427–437.

McIlwain, J.L., E.S. Harvey, S. Grove, G. Shiell, H. Al Oufi and N. Al Jardani. 2011. Seasonal changes in a deep-water fish assemblage in response to monsoon-generated upwelling events. Fish. Oceanogr. 20: 497–516.

Ministry of Environment and Japanese Coral Reef Society (eds.). 2004. Coral Reefs of Japan. Ministry of the Environment, Tokyo.

Moura, R.L., J.L. Figueiredo and I. Sazima. 2001. A new parrotfish (Scaridae) from Brazil, and revalidation of *Sparisoma amplum* (Ranzani, 1842), *Sparisoma frondosum* (Agassiz, 1831), *Sparisoma axillare* (Steindachner, 1878) and *Scarus trispinosus* Valenciennes, 1840. Bull. Mar. Sci. 68: 505–524.

Mizerek, T.L., A.H. Baird, L.J. Beaumont and J.S. Madin. 2016. Environmental tolerance governs the presence of reef corals at latitudes beyond reef growth. Glob. Ecol. Biogeogr. 25: 979–987.

Moura, R.L. and I. Sazima. 2000. Species richness and endemism levels of the Southwestern Atlantic reef fish fauna. Proc. 9th Int. Coral Reef Symp. 1: 23–27.

Moyer, R.P., B. Riegl, K. Banks and R.E. Dodge. 2003. Spatial patterns and ecology of benthic communities on a high-latitude South Florida (Broward County, USA) reef system. Coral Reefs 22: 447–464.

Mumby, P.J. and A.R. Harborne. 2010. Marine reserves enhance the recovery of corals on Caribbean reefs. PLoS One 51: e8657.

Nakabo, T. (ed.). 2013. Fishes of Japan with pictorial keys to the species, third edition. Tokai University Press, Tokyo.

Nakamura, Y., D.A. Feary, M. Kanda and Y. Yamaoka. 2013. Tropical fishes dominate temperate reef fish communities within western Japan. PLoS ONE 8: e81107.

Ogden, J.C. and N.S. Buckman. 1973. Movements, foraging groups, and diurnal migratons of the striped parrotfish Scarus croicensis Bloch (Scaridae). Ecology 54: 589–596.

Ohta, I. and K. Tachihara. 2004. Larval development and food habits of the marbled parrotfish, *Leptoscarus vaigiensis*, associated with drifting algae. Ichthyol. Res. 51: 63–69.

Parenti, P. and J.E. Randall. 2011. Checklist of the species of the families Labridae and Scaridae: an update. Smithiana Bull. 13: 29–44.

Plass-Johnson, J.G., S.C. Ferse, J. Jompa, C. Wild and M. Teichberg. 2015. Fish herbivory as key ecological function in a heavily degraded coral reef system. Limnol. Oceanogr. 60: 1382–1391.

Pörtner, H.O. and A.P. Farrell. 2008. Physiology and climate change. Science 322: 690–692.

Pratchett, M.S., A.S. Hoey, S.K. Wilson, V. Messmer and N.A. Graham. 2011. Changes in biodiversity and functioning of reef fish assemblages following coral bleaching and coral loss. Diversity 3: 424–452.

Rasher, D.B., A.S. Hoey and M.E. Hay. 2013. Consumer diversity interacts with prey defenses to drive ecosystem function. Ecology 94: 1347–1358.

Richardson, L.E., N.A. Graham, M.S. Pratchett and A.S. Hoey. 2017. Structural complexity mediates functional structure of reef fish assemblages among coral habitats. Environ. Biol. Fishes 100: 193–207.

Ridgway, K.R. and J.R. Dunn. 2003. Mesoscale structure of the mean East Australian Current System and its relationship with topography. Prog. Oceanogr. 56: 189–222.

Riegl, B. 1999. Coral communities in a non-reef setting in the southern Arabian Gulf (Dubai, UAE): fauna and community structure in response to recurrent mass mortality. Coral Reefs 18: 63–73.

Riegl, B. and W.E. Piller. 2003. Possible refugia for reefs in times of environmental stress. Int. J. Earth. Sci. 92: 520–531.

Robertson, R.D., R. Reinboth and R.W. Bruce. 1982. Gonochorism, protogynous sex-change and spawning in three sparisomatinine parrotfishes from the western Indian Ocean. Bull. Mar. Sci. 32: 868–879.

Robertson, D.R., F. Karg, R.L. Moura, B. Victor and G. Bernardi. 2006. Mechanisms of speciation and faunal enrichment in Atlantic parrotfishes. Mol. Phylogenet. Evol. 40: 795–807.

Rocha, L.A. 2003. Patterns of distribution and processes of speciation in Brazilian reef fishes. J. Biogeogr. 30: 1161–1171.

Rotjan, R.D. and S.M. Lewis. 2005. Selective predation by parrotfishes on the reef coral Porites astreroides. Mar. Ecol. Progr. Ser. 305: 193–201.

Rotjan, R.D. and S.M. Lewis. 2006. Parrotfish abundance and selective corallivory on a Belizean coral reef. J. Exp. Mar. Biol. Ecol. 335: 292–301.

Sano, M. 2001. Short-term responses of fishes to macroalgal overgrowth on coral rubble on a degraded reef at Iriomote Island, Japan. Bull. Mar. Sci. 68: 543–556.

Schleyer, M.H., A. Kruger and L. Celliers. 2008. Long-term community changes on a high-latitude coral reef in the Greater St Lucia Wetland Park, South Africa. Mar. Poll. Bull. 56: 493–502.

Serizawa, Y., Z. Imoto, T. Ishikawa and M. Ohno. 2004. Decline of the *Ecklonia cava* population associated with increased seawater temperature in Tosa Bay, southern Japan. Fish. Sci. 70: 189–191.

Shibuno, T., Y. Nakamura, M. Horinouchi and M. Sano. 2008. Habitat use patterns of fishes across the mangrove-seagrass-coral reef seascape at Ishigaki Island, southern Japan. Icthyol. Res. 55: 218–237.

Sommer, B., P.L. Harrison, M. Beger and J.M. Pandolfi. 2014. Trait-mediated environmental filtering drives assembly at biogeographic transition zones. Ecology 95: 1000–1009.

Streelman, J.T., M. Alfaro, M.W. Westneat, D.R. Bellwood and S.A. Karl. 2002. Evolutionary history of the parrotfishes: biogeography, ecomorphology, and comparative diversity. Evolution 56: 961–971.

Tano, S.A., M. Eggertsen, S.A. Wikström, C. Berkström, A.S. Buriyo and C. Halling. 2017. Tropical seaweed beds as important habitats for juvenile fish. Mar. Freshw. Res. doi: 10.1071/MF16153.

Taylor, B.M., P. Houk, G.R. Russ and J.H. Choat. 2014. Life histories predict vulnerability to overexploitation in parrotfishes. Coral Reefs 33: 869–878.

Taylor, B.M., S.J. Lindfield and J.H. Choat. 2015. Hierarchical and scale-dependent effects of fishing pressure and environment on the structure and size distribution of parrotfish communities. Ecography 38: 520–530.

Terazono, Y., Y. Nakamura, Z. Imoto and M. Hiraoka. 2012. Fish response to expanding tropical *Sargassum* beds on the temperate coasts of Japan. Mar. Ecol. Prog. Ser. 464: 209–220.

Tewksbury, J.J., R.B. Huey and C.A. Deutsch. 2008. Putting the heat on tropical animals. Science 320: 1296–1297.

Trapon, M.L., M.S. Pratchett and A.S. Hoey. 2013. Spatial variation in abundance, size and orientation of juvenile corals related to the biomass of parrotfishes on the Great Barrier Reef, Australia. PloS One, 8: e57788.

Vergés, A., P.D. Steinberg, M.E. Hay, A.G. Poore, A.H. Campbell, E. Ballesteros, K.L. Heck, D.J. Booth, M.A. Coleman, D.A. Feary and W. Figueira. 2014. The tropicalization of temperate marine ecosystems: climate-mediated changes in herbivory and community phase shifts. Proc. R. Soc. B 281: 20140846.

Vergés, A., C. Doropoulos, H.A. Malcolm, M. Skye, M. Garcia-Pizá, E.M. Marzinelli, A.H. Campbell, E. Ballesteros, A.S. Hoey, A. Vila-Concejo, Y.M. Bozec and P.D. Steinberg. 2016. Long-term empirical evidence of ocean warming leading to tropicalization of fish communities, increased herbivory, and loss of kelp. Proc. Natl Acad. Sci. USA 113: 13791–13796.

Vroom 2011, P.S. 2011. Coral dominance: a dangerous ecosystem misnomer? J. Mar. Biol. 2011: 164127.

Williams, I. and N. Polunin. 2001. Large-scale associations between macroalgal cover and grazer biomass on mid-depth reefs in the Caribbean. Coral Reefs 19: 358–366.

Wilson, S.K., M. Depczynski, R. Fisher, T.H. Holmes, R.A. O'Leary and P. Tinkler. 2010. Habitat associations of juvenile fish at Ningaloo Reef, Western Australia: the importance of coral and algae. PLoS One, 5: e15185.

Wismer, S., A.S. Hoey and D.R. Bellwood. 2009. Cross-shelf benthic community structure on the Great Barrier Reef: relationships between macroalgal cover and herbivore biomass. Mar. Ecol. Prog. Ser. 376: 45–54.

Yamano, H., K. Sugihara and K. Nomura. 2011. Rapid poleward range expansion of tropical reef corals in response to rising sea surface temperatures. Geogr. Res. Lett. 38: L04601.

The Ecology of Parrotfishes on Low Coral Cover Reefs

Rebecca J. Fox[1,2]

[1] School of Life Sciences, University of Technology Sydney, Ultimo, NSW 2007, Australia
[2] Division of Ecology & Evolution Research School of Biology,
   The Australian National University, Canberra, ACT 0200, Australia
   Email: rebecca.fox-1@uts.edu.au

Introduction

The benthic community of a coral reef is a complex and diverse entity, comprised not just of coral but many other benthic organisms (e.g. filamentous turf algae, macroalgae, encrusting coralline algae, sand, detritus), all of which contribute to the productivity and functioning of the reef ecosystem. Depending on the individual system, coral cover on a healthy and pristine reef may therefore only approach a maximum level of 60-70 percent (Wilkinson 2008). On Lizard Island, an island of low human habitation in the mid shelf region of the northern Great Barrier Reef (GBR), Australia, which has been under the management of the Great Barrier Reef Marine Park Authority since 1978, coral cover on exposed reef crest and reef slope habitats reaches a maximum of 42 percent and 47 percent average live coral cover respectively (Hoey and Bellwood 2010a). Is this high coral cover? And, if so, what constitutes low coral cover? Some reefs labour under biogeographic and environmental constraints such that a level of coral cover of 20 percent would be considered 'high' (Vroom 2011). What level of coral cover under those circumstances would be considered low? In short, the concept of a "low coral cover reef" poses a number of definitional challenges, which must first be navigated in order to set the context for the ecology of parrotfishes within such systems.

One of the complicating factors in defining the idea of low coral cover is that generalisations are made difficult by the sheer level of variation through space and time. A reef defined as high coral cover in one year can have low coral cover the next should an unfortunate confluence of disturbance events intervene. For example, reefs in the Seychelles were heavily impacted by a mass bleaching event in 1998 associated with the El Nino warming. In the space of one year, coral cover on these reefs declined from levels at or above 50 percent to less than 5 percent cover (Ahamada et al. 2008 and references therein). Recovery since then has been patchy, with some reefs now exhibiting coral cover of around 25 percent, but others still with levels less than 5 percent (Ledlie et al. 2007, Ahamada et al. 2008, Graham et al. 2015). As this example shows, coral cover cannot only vary temporally, but also spatially, with large differences in the level of cover between reefs within a single region. Within the system of reefs that make up the GBR, average coral cover can vary from 1-50 percent across the continental shelf from inner to outer shelf reefs (Wismer et al. 2009). There can also be large differences in coral cover between reefs within the same local area (e.g. average coral cover on the reef slope of mid-shelf reefs in the same segment of the GBR can range from 7-74 percent, Nash et al. 2012). Even within a single reef, coral cover will vary significantly between different reef zones (e.g. an average of 35 percent live coral cover on the reef slope within Pioneer Bay, GBR compared to 6 percent coral cover on the inner part of the reef flat, Fox and Bellwood 2007).

In this chapter I therefore first define the idea of a low coral cover reef, taking into account the qualitative and relative nature of the terminology, the difficulties associated with meta-analysis of coral cover data (see Hughes et al. 2010) and acknowledging the fact that shifting baselines can make one generation's "low coral cover" the next generation's "high quality reef habitat". Having defined what is meant by a low coral cover reef and the nature of these ecosystems, the remainder of the chapter deals with different aspects of parrotfish ecology within such systems. I start with a section on the distribution and abundance of parrotfishes observed on low coral cover reefs at varying spatial scales, noting the dominant geographic and local patterns. The next section takes the approach of combining multiple aspects of parrotfish ecology (trophic, behavioural and population ecology) to examine the functional impact of parrotfishes within low coral cover reef systems, with reference to region-specific case studies. Finally I draw together the various threads to summarize the key aspects of parrotfish ecology as it relates to low coral cover systems and highlight areas for future research within this field.

What is a Low Coral Cover Reef?

Low Coral Cover as a Relative versus Qualitative Concept

One of the key problems in defining low coral cover reefs is that the concept is relative, i.e. low, relative to what? Compared to the reefs of Fiji, which have a reported average coral cover of 45% (Chin et al. 2011), the subtropical reefs of the North West Hawaii Islands (NWHI) with an average coral cover of 19% (Friedlander et al. 2008) would be described as low coral cover, yet compared to reefs in the Seychelles that have an average coverage of 1-2% following the mass bleaching event of 1998 (Ledlie et al. 2007), the NWHI might be considered "high coral cover". In actual fact, the reefs of the NWHI are typical of subtropical reefs and marginal reef systems (see Hoey et al. Chapter 12) and are considered some of the most pristine in US waters. They are rated as healthy, partly because of the stability though time of this level of coral coverage (Chin et al. 2011). Similarly, for inshore reefs of the GBR, low yet stable levels of coral coverage are a natural function of the sedimentary and light regime in which these reefs have evolved (Hopley et al. 1983, Hopley and Choat 1990, Larcombe and Woolfe 1999). In short, the idea of low coral cover is relative and the devil is in the geographical and historical detail (Vroom 2011).

The other misconception concerning low coral cover reefs is a qualitative one, i.e. does low coral cover always mean low quality reef habitat? The idea of low coral cover within an ecosystem named for the word "coral" would certainly seem to have a negative connotation. However, the qualitative nature of the descriptor "low coral cover" partly depends on the historical setting and the corollary. As seen in the case of the NWHI (above), the historical context can indicate whether low coral cover is the natural state of the system. For example, compare the NWHI reefs to those of the main Hawaiian Islands (MHI) which were also reported to have average live coral cover of 19 percent in 2008 (Friedlander et al. 2008). Two systems with the same level of coral cover, yet MHI reefs were rated as being in poor health. The difference is in the dynamics and the coral that makes up that level of coverage. In the NWHI, the 19 percent coverage has remained relatively stable through time and the system has elements of resilience: factors such as the presence of old, single coral colonies and a good variety of coral growth forms, some of which are now uncommon in the MHI (Chin et al. 2011). By contrast, in the MHI the 19 percent figure represents a decline over the last 30 years from an average of 31 percent and a loss of functional diversity in terms of growth forms (Chin et al. 2011). The key message is that absolute level of coral cover does not tell the whole story, it is about the specifics of community composition and system dynamics.

Moving Targets and Shifting Baselines

Declines in coral cover are an all-too-familiar story on reefs around the globe, from the Caribbean (Gardner et al. 2003, Jackson et al. 2014) to the Western Indian Ocean (Graham et al. 2006, Ahamada et al. 2008) to the Indo-Pacific (Bruno and Selig 2007, Chin et al. 2011). Even the Great Barrier Reef, Australia, which is under the protection of management that has been hailed as world-leading, has seen average levels of coral cover on reefs decline over the past 50 years (Bellwood et al. 2004, De'ath et al. 2012, Hughes et al. 2017), linked to increased rates of sedimentation following European settlement (Larcombe et al. 1996, Wolanski and Spagnol 2000), outbreaks of crown-of thorn starfish (COTS), and coral bleaching (but see Sweatman et al. 2011). Average coral cover for the wider Caribbean is now just 14.3 percent, down from an average of 35 percent in 1970-83 (Jackson et al. 2014). Bruno and Selig (2007) found an average level of coral cover of 22.1 percent across reefs of the Indo-Pacific and Goatley and Bellwood (2011) recently calculated that the average level of coral cover on reefs globally could be just 26 percent. These declines are just part of a continuum of degradation of coral reefs (Pandolfi et al. 2003). What was once considered low coral cover is now the average.

The phenomenon is one of 'shifting baselines' (Pauly 1995, Friedlander and DeMartini 2002, Knowlton and Jackson 2008), where each generation has a different idea to the one preceding about what constitutes the 'norm' for a given ecosystem. The effect is dangerous as it allows for gradual declines in ecosystem health to pass under the radar, until it is too late and the system finds itself locked in a new, stable (typically degraded) state (Hughes et al. 2010). If low coral cover is defined in relative terms then, as the goal-posts move over generations, so does the definition. From the perspective of conservation and advocacy of management action, we should definitely be in the business of absolutes. This chapter therefore defines a low coral cover reef as one with an average of <20 percent live coral cover, although importantly, it makes no overall value judgement about the quality or health of reefs that fall into this category. Such judgements can only be made on an individual reef basis, taking into account the particular ecosystem drivers of coral cover and the overall historical and biogeographic context.

To summarize, it is clear that there are low coral cover reefs that were formerly high coral cover, but are on a downward trajectory (e.g. sites in the Caribbean; Jackson et al. 2014, and in the Seychelles; Ledlie et al. 2007) and there are low coral cover reefs that are more stable through time (e.g. those of the NWHI; Chin et al. 2011). In many cases, the distinction between these two categories can be viewed in terms of the corollary of low coral, i.e. what is the high coverage component of the benthic community? Typically (although not always) a reef that exhibits low coral cover along with high turf algal cover is more likely to be in the category of "stable and healthy", whereas a reef that exhibits low coral cover along with high macroalgal cover is more likely (although, again, not always) to be on a downward trajectory. Rather than make any value judgements as to which type of reef is higher quality, the remainder of this chapter will examine the ecology of parrotfishes within the context of both categories, looking at their distribution across low coral (<20 percent) cover reefs of each type and the interactions between parrotfishes and their environment in both.

The Distribution of Parrotfishes on Low Coral Cover Reefs

Since low coral cover reefs are not confined to one single area of the globe or one particular biogeographic region, the ecology of parrotfishes on such reefs will necessarily be underpinned by variation in their distribution at the geographic scale (Choat 1991, Thresher 1991, Bellwood and Wainwright 2002, Floeter et al. 2005, Bonaldo et al. 2014, see also Kulbicki et al. Chapter 10). The individual species protagonists will differ between regions, for example the function of reef bioerosion is performed by *Chlorurus microrhinos* on the GBR and by *Sparisoma viride* in the Caribbean (Bellwood and Wainwright 2002, Bonaldo et al. 2014, Fig. 1). Suffice to say that the ecology of parrotfishes on low coral cover reefs is governed by the fundamental dominance of *Scarus* (scrapers) and *Chlorurus* (excavators) in the Indo-West Pacific (Choat 1991, Bonaldo et al. 2014) and the dominance of *Sparisoma* (grazers and bioeroders) in the Caribbean (Choat 1991, Bonaldo et al. 2014).

In most cases, the biomass of parrotfish within low coral cover reef systems are dominated by two or three species (Fig. 1). Within the Caribbean this result is partly an artefact of the lower overall species richness of the reef fauna in the region, but even in the Indo-Pacific, with its species-rich assemblage (55 species in total; Choat 1991), parrotfish biomass on low-coral cover reefs tend to be dominated by a single species (Hoey and Bellwood 2008). On low coral cover reefs of the Caribbean, parrotfish assemblages vary according to fishing density, with lightly fished reefs of Bonaire and Saba dominated by *Sp. viride* and *Scarus vetula* (Bruggemann et al. 1996, Hawkins and Roberts 2004). By contrast, on the heavily-fished reefs of Jamaica, *Sp. viride* and *Sc. vetula* are uncommon and the smaller species *Scarus taeniopterus*, *Sc. iseri* and *Sparisoma aurofrenatum* dominate (Hawkins and Roberts 2004, Fig. 1). On reefs of the Florida Keys, *Sp. viride* and *Sc. iseri* (previously *Scarus coelestinus*) dominate the shallow reef habitats, while *Sp. aurofrenatum* and *Sc. taeniopterus* are abundant at depth (Paddack et al. 2006, Burkepile and Hay 2009). In unfished parts of Belize, *Sp. viride* and *Sc. iseri* dominate, while *S vetula* is uncommon. (Lewis and Wainwright 1985, Fig. 1). Meanwhile, on the low coral cover reefs of the Southern Red Sea (Eritrea), *Scarus ferrugineus* and *Hipposcarus harid* dominate the biomass (Afeworki 2014, Fig. 1). On inner-shelf reefs of the GBR, the parrotfish assemblage is dominated by the scraping *Scarus rivulatus* and the excavating *Ch. microrhinos* (Fox and Bellwood 2007, Hoey and Bellwood 2008, Fig. 1).

The relationship between parrotfish abundance and level of coral cover is somewhat controversial. Across the Caribbean, Newman et al. (2006) found no correlation between herbivorous fish biomass and coral cover. Stockwell et al. (2009) also found no significant relationship between herbivore (predominantly parrotfish and surgeonfish) biomass and coral cover across reefs in the Philippines. However, other studies at the regional level have shown that parrotfish biomass increases with decreasing coral cover and that low coral cover reefs can be where parrotfishes reach their maximum within-region abundance. For

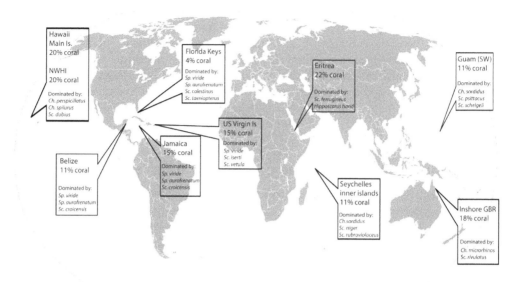

Fig. 1. Map of the world showing the dominant taxa of parrotfishes (by biomass) on selected low (≤ 20%) coral cover reefs within individual regions. Assessment of average coral cover is based on data presented in Wilkinson (2008) with the exception of Florida Keys (from Paddack et al. 2006), inner-shelf GBR (from Wismer et al. 2009) and Southern Red Sea (from Afeworki 2014). Assessment of dominance is based on data from: NWHI – Friedlander and DeMartini 2002; Belize - Lewis and Wainwright 1985; Jamaica – McGinley 2014; US Virgin Islands – Carpenter 1990a; Florida - Paddack et al. 2006, Burkepile and Hay 2009; Southern Red Sea – Afeworki et al. 2013; Seychelles – Ledlie et al. 2007; Inner-shelf GBR – Fox and Bellwood 2007, Hoey and Bellwood 2008, Bennett and Bellwood 2011; Guam – McIlwain and Taylor 2009.

example, Hoey and Bellwood (2008) showed that the density of parrotfishes on reefs across the continental shelf of the GBR reaches maximum abundance on the inner-shelf reefs where coral cover is typically lower. In a study spanning three island states of Micronesia (Palau, Guam and Pohnpei), Mumby et al. (2013) found that large-bodied parrotfish (in particular *Ch. microrhinos*) were more abundant when coral cover was high, while the smaller-bodied species (in particular *Chlorurus spilurus* (previously *Ch. sordidus*) and *Scarus psittacus*) were more abundant when coral cover was low. The latter result, however, was most likely driven by the high abundance of *Sc. psittacus* on the low coral cover reefs of Guam (average live coral cover of 23% in 2008; Burdick et al. 2008) where the herbivorous fish community structure differs significantly from the other two islands (Mumby et al. 2013). A positive correlation between parrotfish biomass and coral cover was also observed across reefs of the Coral Coast of Viti Levu, Fiji by Rasher et al. (2013), where a comparison of areas inside no-take marine reserves with adjacent fished areas revealed that a 7-17x higher level of herbivore (including parrotfish) biomass was associated with a 3-11x higher level of coral cover on reefs inside the no-take reserve areas.

Of course the relationship between parrotfish biomass and coral cover has its corollary in the relationship between parrotfish biomass and macroalgal cover. As coral cover increases so will the intensity of herbivore grazing, with feeding being confined to progressively smaller areas of algal substrate (Mumby et al. 2013). This higher level of grazing disturbance per unit area should reduce the cover of macroalgae. In the Viti

Levu study, the 7-17x higher biomass of herbivores within no-take reserve areas was associated with both a lower abundance (26-61x) and lower species richness (3-4x) of macroalgae (Rasher et al. 2013). By video recording feeding activity on seven species of macroalgae transplanted from fished areas to reserve areas of reef, Rasher et al. (2013) demonstrated that initial phase (IP) *Ch. spilurus* was responsible for the majority of bites taken off the two red algal species of assays. Macroalgal abundance has also previously been shown to be negatively related to parrotfish biomass in the Philippines (Stockwell et al. 2009), on the GBR (Fox and Bellwood 2007), in Hawaii (Friedlander et al. 2007) and across the Caribbean (Williams and Polunin 2001, Newman et al. 2006), where the average biomass of parrotfish increases from 2.9 g m^{-2} on reefs with 68 percent macroalgal cover to 8.5 g m^{-2} on reefs with 27 percent macroalgal cover (Williams and Polunin 2001, Fig. 2). Incorporating comparisons from low coral cover reefs around the globe, however, highlights the difficulty in inferring relationships between parrotfish abundance and levels of macroalgal or turf algal cover (Fig. 2). For example, a 40 percent level of macroalgal cover is associated with low levels of parrotfish biomass (5.8 g m^{-2}) in Cuba (fished site), but moderate levels (15 g m^{-2}) on the GBR (unfished site), demonstrating that the correlation between fish biomass and macroalgal abundance is not necessarily a simple one. Nevertheless, a detailed

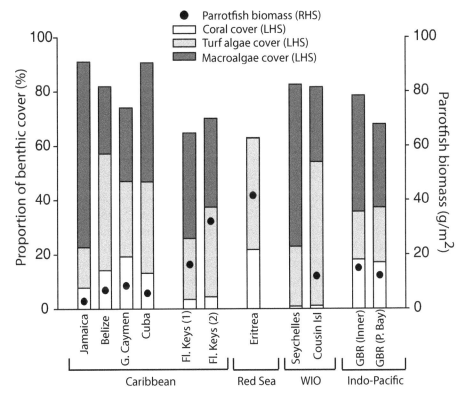

Fig. 2. Relationship between total parrotfish biomass, macroalgal cover and turf algal cover for low (< 20%) coral cover reefs around the globe. Each column and point represents biomass and benthic community composition taken from a particular study: Caribbean – Bruggemann et al. 1994a, b, 1996, Williams and Polunin 2001, Hawkins and Roberts 2004, Paddack et al. 2006, Burkepile and Hay 2009; Red Sea – Afeworki 2014; Western Indian Ocean (WIO) – Chong-Seng et al. 2012, Ledlie et al. 2007; Inner-shelf GBR – Hoey and Bellwood 2008, Fox and Bellwood 2007.

meta-analysis of the relationship between coral cover, macroalgal cover and parrotfish abundance on reefs of the Caribbean recently published by the IUCN (Jackson et al. 2014), has highlighted low parrotfish biomass as a key driver of reef degradation within that region since the 1970s. Reefs where parrotfish had been overfished suffered greater subsequent declines in coral cover and increases in macroalgae than Caribbean reefs that had intact populations of parrotfish (Jackson et al. 2014). The distribution of parrotfishes on low coral cover reefs of the Caribbean therefore appears to be negatively correlated with levels of macroalgal abundance and positively correlated to the health and resilience of individual reefs.

The Ecology of Parrotfishes on "Low Coral/High Turf Algae" Cover Reefs

High Algal Turf Cover Reefs: Parrotfishes in their Element

On many reefs, algal turfs are a significant, or even the most abundant, component of benthic cover (Wismer et al. 2009, Bruno et al. 2014). Algal turfs are a natural and important component of reef benthos, being one of the major contributors to reef primary productivity (Klumpp and Polunin 1990) and are targeted by parrotfishes, particularly when they contain a high percentage of detritus (Crossman et al. 2001, Purcell and Bellwood 2001, Wilson et al. 2003). On a reef in healthy dynamic balance, this grazing pressure typically keeps the algae well cropped, with low turf lengths and associated sediment loads (Fig. 3) and the regular disturbance impact helps to prevent the succession of the algal community to stages associated with the development of undesirable and unpalatable macroalgae (Hixon and Brostoff 1996). Reefs that combine low coral cover with high coverage of turf algae, otherwise known as the Epilithic Algal Matrix (EAM) should therefore see parrotfishes in their element.

Low Coral Cover, High Turf Algae Cover, Low Topographic Complexity: A Parrotfish Trifecta in the Southern Red Sea

Along with high algal cover, the other corollary of low coral cover tends to be low topographic complexity (Alvarez-Filip et al. 2009, Graham and Nash 2013, Komyakova et al. 2013). Based on the positive correlation typically observed between structural complexity and reef fish biomass (e.g. Jones et al. 2004, Wilson et al. 2009) and between parrotfish biomass and structural complexity (Graham and Nash 2013), it might be thought that low coral cover reefs would not be conducive to parrotfish abundance. However, as we have already seen on the GBR, within individual systems, parrotfishes can reach their greatest abundances where coral cover is lowest. Parrotfish, due to the constraints of their morphology (see Gobalet Chapter 1, Wainwright and Price Chapter 2) are comparatively limited when it comes to feeding in tight spaces associated with high topographic complexity (Fox and Bellwood 2013). They preferentially feed off convex surfaces (Bellwood and Choat 1990, Bruggemann et al. 1994b, Afeworki et al. 2011, Roff et al. 2011) and tend to graze in relatively open areas of reef (Fox and Bellwood 2013). The lack of complexity within low coral cover reefs offers the potential of high resource availability, combined with suitable grazing surfaces and therefore the potential to support a large population density of parrotfishes. It has been suggested that areas of low topographic complexity are fundamentally less attractive than their high complexity counterparts due to their increased "riskiness" and greater rates of predation associated

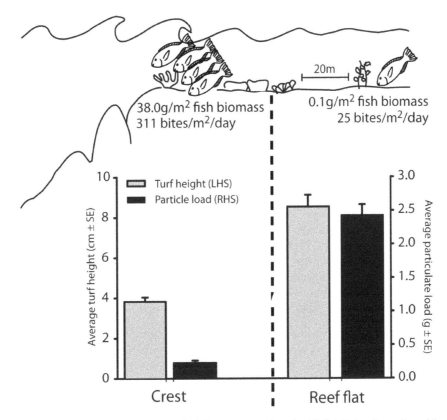

Fig. 3. Characteristics of algal turfs sampled from areas of reef with high (reef crest) and low (reef flat) parrotfish grazing intensity. Algal turf data are taken from variables presented in Bonaldo and Bellwood (2011) for an inshore reef on the GBR (Pioneer Bay, Orpheus Island) and parrotfish biomass and number of bites per day for the same reef are recalculated from R.J.F. data published in Fox and Bellwood 2007, 2008a.

with the lack of shelter (Schneider 1984, Dickman 1992). However, there is evidence that high complexity habitats can present greater risks in terms of the level of visual occlusion they encompass and several studies have now found higher feeding rates by coral reef fishes in areas of lower structural and topographic complexity (Beukers and Jones 1998, Bennett et al. 2010, Hoey and Bellwood 2011).

Low coral cover, high turf algal coverage and low topographic complexity could therefore be described as the parrotfish trifecta. The combination is typical of higher-latitude rocky reefs of the west Atlantic and seasonal reefs of the southern Red Sea. On Sheikh Said Island fringing reef in the southern Red Sea, parrotfish biomass reaches 41.7 g m^{-2} (Fig. 2), almost three times the level seen on the inner shelf of the GBR, demonstrating the extent to which parrotfishes may be in their element within these low coral/high turf algal cover/low complexity environments (topographic complexity on Sheikh Said Island fringing reef measured as the ratio of chain length to planar length ranges from an average of 1.2 on the reef flat to just under 1.35 on the reef crest; Afeworki 2014). As Afeworki (2014) points out, the high levels of parrotfish biomass cannot be attributed just to low fishing intensity as they exceed those of other unfished locations.

Instead, he highlights the high levels of primary productivity that exist in the southern Red Sea as compared to other tropical reef locations and hypothesizes that primary productivity may be the driving factor behind parrotfish abundance in this system. Within this reef system characterized by seasonality and extreme summer temperatures, *Sc. ferrugineus, H. harid* and *Ch. sordidus* predominate, with *Sc. ferrugineus* the most abundant in terms of density and biomass. This species has been shown to preferentially feed on EAM on endoliths (Afeworki et al. 2011), presumably due to their higher nutritional quality (Bruggemann et al. 1994a). Interestingly, the reduction in availability of this preferred resource during the cool season did not precipitate a switch to other, more abundant macroalgal resources, hinting at a limited flexibility in resource use for parrotfishes (Afeworki et al. 2011) which may have implications for parrotfish energetics and population dynamics under changing environmental conditions. The overall ecological importance of parrotfishes within these marginal reef systems that typically have low coral cover, high algal cover and low topographical complexity is perhaps best reflected in the fact that this book devotes a separate chapter (Hoey et al. Chapter 12) to discussion of the role of parrotfishes on marginal reefs and the reader is directed there for a full treatment of the topic.

Low Coral Cover, High Turf Algae Cover: Inner-shelf Reefs of the GBR

The reefs located along the inner part of the continental shelf that forms Australia's GBR are typically viewed as low coral cover systems in the relative sense, compared to their mid- and outer-shelf counterparts. These reefs have developed under the influence of a nearshore marine sediment wedge and have historically experienced higher levels of turbidity and sedimentation than offshore reef systems, which have acted as natural curbs on coral growth and development (Hopley and Choat 1990, Larcombe and Woolfe 1999).

Associated with the lower levels of coral cover on these inshore reefs are higher levels of algal cover (McCook and Price 1997a, b, Wismer et al. 2009), including high levels of turf algal cover (Wismer et al. 2009, Sweatman et al. 2011) and macroalgal cover (McCook 1996, 1997, Schaffelke and Klumpp 1997, Hoey and Bellwood 2010b), with evidence that this has been the case even in a historical context (Hopley et al. 1983, Hopley and Choat 1990, Larcombe and Woolfe 1999). It is therefore perhaps no coincidence that the inner-shelf reefs of the GBR are where parrotfishes reach their maximum densities on the GBR (6.7-8.5 ind. 100 m^{-2}), driven by the abundance of scraping parrotfishes (*Scarus*), in particular *Sc. rivulatus, Sc. ghobban* and *Sc. psittacus* (Hoey and Bellwood 2008). The inner shelf reefs are also where parrotfishes show their highest grazing rates (Hoey and Bellwood 2008). Together, these factors suggest that parrotfishes play a key role in the ecosystem dynamics of low coral cover reefs on the GBR.

Much of the experimental work on the ecology of parrotfishes on low coral cover reefs of the GBR has been conducted within and around Pioneer Bay, a well-developed fringing reef system located on the western, leeward side of Orpheus Island (Fig. 4). Orpheus forms part of the Palm Island group, located approximately 12 km from the east Australian mainland. Its proximity to the mainland means that it is under the influence of flood plumes from two of the region's major rivers: the Herbert and Burdekin. Terrigenous sediments carried by these flood plumes and run-off from the island itself impact on the reef system, particularly during significant wet seasons, leading to low (typically between 3-7 m) visibility. In 2007, live coral cover along the reef crest habitat averaged 22.7 percent

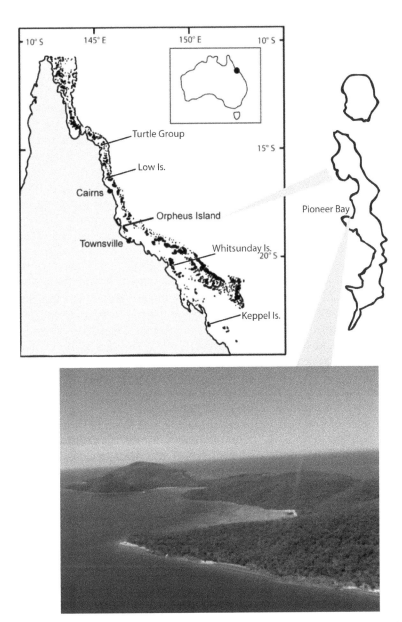

Fig. 4. Summary of the geographic location of studies investigating the functional ecology of parrotfishes on inshore reefs of the Great Barrier Reef since the early 2000s: Turtle Group and Nymph reef (see Hoey and Bellwood 2008), Low Islands, Whitsunday Islands and Keppel Islands (see Bennett and Bellwood 2011), Orpheus Island (see Bellwood et al. 2006, Fox and Bellwood 2007, 2008a, Cvitanovic and Bellwood 2009, Lefèvre and Bellwood 2010, 2011, Bonaldo and Bellwood 2011). Majority of studies at Orpheus Island have been conducted in Pioneer Bay located on the western (leeward) side of the island. Aerial photograph of Pioneer Bay is courtesy of D.R. Bellwood.

(± 2.2 s.e.; Fox and Bellwood 2007), a figure that is respectable by today's world standards, but qualifies as 'low' for the GBR ecosystem as a whole in the context of historical data presented by Bellwood et al. (2004). As expected for an inshore reef in this region, Pioneer Bay shows relatively high levels of algal (including macroalgal) cover over parts of the reef, but declining out from the shoreline to just 1.5 ± 0.6% (mean ± s.e) cover at the reef crest (Fig. 5a).

The biomass of the herbivorous fish community at Pioneer Bay is overwhelmingly dominated by three species, two of which are parrotfish: *Sc. rivulatus* and *Ch. microrhinos* (Fig. 5b). What is more, it is interesting to note that these two dominant species each represent a different feeding functional group, with *Sc. rivulatus* a "scraper" of the EAM and *Ch. microrhinos* an "excavator" that contributes to bioerosion of the reef substratum (see Burkepile et al. Chapter 7 for a full treatment of the importance of functional variation among parrotfish species). Fox and Bellwood (2008a) estimated the impact of these two species on the reef system from feeding rates observed via remote underwater video camera in 1 m^2 areas of reef across five habitat zones and found that, on average, *Sc. rivulatus* would scrape 52 percent (± 18 s.e.) of each m^2 patch of substratum on the reef crest each month (Fig. 5c), while *Ch. microrhinos* would excavate across 19 percent (±11% s.e.) of each m^2 of crest habitat per month (Fig. 5c). The distribution of parrotfish impact across the reef at Pioneer Bay was significantly negatively correlated with the abundance of macroalgae across the same habitat gradient (Fox and Bellwood 2007).

The measures of functional impact obtained by Fox and Bellwood (2008a) for individual parrotfish species were comparable with those calculated by Hoey and Bellwood (2008) for total parrotfish grazing impact across inner, mid and outer-shelf reefs of GBR (where impact was inferred from presence x feeding rate x bite size). They calculated that each m^2 of back reef and reef slope/crest would be grazed 8.5-11 times per year i.e. 70-92 percent of each m^2 grazed per month by all species of scraping parrotfishes, with *Sc. rivulatus* accounting for 70 percent of that grazing activity or 49-64 percent of each m^2. They, too, found that nearly all of the bioerosion on inner shelf reefs was performed by the excavating species *Ch. microrhinos* (Hoey and Bellwood 2008).

The importance of parrotfish grazing impact across the reef in terms of maintaining turf algal communities in an early stage of succession and preventing the encroachment of macroalgae beyond its inner reef flat refuge was demonstrated experimentally at Pioneer Bay by Bonaldo and Bellwood (2011). When all herbivores, including parrotfishes, were excluded from grazing on algal turfs at the reef crest of Pioneer Bay for a period of 4 days, the turfs within caged plots increased in length by as much as 26 percent, as well as in particulate content and in sediment load. Samples of algal turf on substratum from the inshore part of the reef flat that typically experience low grazing rates were also transplanted out to the reef crest (where high grazing pressure is the norm). These transplanted algal turfs were cropped to a shorter length and total particulate content and sediment load within the turfs was reduced over just four days (Bonaldo and Bellwood 2011). At the same time, algal turfs moved from the crest (high grazing pressure) to the reef flat (low grazing pressure) grew in length by about 25 percent and had a higher level of particulate and sediment content at the end of four days (Bonaldo and Bellwood 2011).

When High Turf Cover Becomes High Sediment Cover: Parrotfish Deterrence on the GBR

Of course, a natural side effect of the crucial process of parrotfish grazing and bioerosion is the production of sediment. Inshore reefs of the GBR that require high rates of parrotfish grazing in order to maintain low turf algal biomass will therefore also have to contend with associated rates of sediment production. Hoey and Bellwood (2008) calculated that parrotfishes would produce around 45 kg m^{-2} y^{-1} of sediment in back-reef habitats of inner shelf reefs and around 29 kg m^{-2} y^{-1} of sediment in slope/crest habitats, the vast

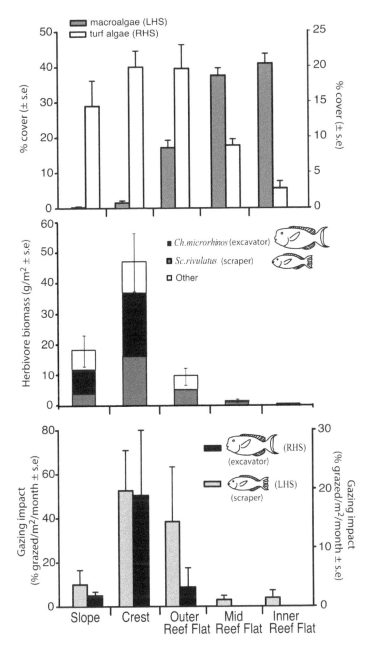

Fig. 5. (a) Distribution of turf algae and macroalgal cover across the reef at Pioneer Bay, Orpheus Island, as measured by benthic point intercept transects (b) Distribution of herbivore biomass across the reef gradient at Pioneer Bay, Orpheus Island, as measured by underwater visual censuses (10 min timed swims by divers). Total biomass is broken down into contributions from *Scarus rivulatus*, *Chlorurus microrhinos* and "other", where "other" represents the remaining 25 species of roving herbivore censused within the Bay. Functional impact of the population of (c) *Sc. rivulatus* and *Ch. microrhinos* in Pioneer Bay in terms of proportion of reef area grazed per month as measured by remotely deployed underwater video cameras. Monthly impact values were calculated by multiplying daily rates by 28. Overall standard errors were calculated using Goodman's estimator.
      Data are redrawn from: (a, b) Fox and Bellwood 2007 and (c) Fox and Bellwood 2008a.

majority (91 percent) of which was actually found to come from reworking of sediment by the scraping *Scarus* parrotfishes. This is in stark contrast to the process of production of sediment on the outer-shelf reefs which is predominantly via bioerosion activity of the bumphead parrotfish *Bolbometopon muricatum* (Hoey and Bellwood 2008). The problem is that these inner shelf reefs are already high sediment environments, by virtue of their proximity to the mainland and the impacts of increased terrigenous inputs from flood plumes, agricultural run-off and industrial development. There exists the potential for the grazing process itself to exacerbate the problem.

Recent research has found that, beyond a certain point, increased benthic sediment loads within algal turfs can deter grazing by herbivorous fishes, including parrotfishes (Bellwood and Fulton 2008, Goatley and Bellwood 2012). For example, when sediment was cleared from areas of reef crest habitat on mid-shelf reefs of the GBR, the number of bites taken by parrotfishes increased (Goatley and Bellwood 2012). This result is particularly interesting because it suggests that even for species of herbivore such as parrotfish, adapted to rework sediment intake via modifications of the pharyngeal apparatus designed to triturate food with high sediment loads (Bellwood 1996), there is a limit to the amount or types of sediment they are prepared to ingest. The reduction in grazing brought about by the increased sediment load can result in the development of longer length turfs (Goatley and Bellwood 2013), setting the benthos on a new developmental trajectory which is unlikely to favour the establishment of juvenile corals and which could therefore see low coral cover/high turf cover reefs become dominated by unpalatable and undesirable turf barrens.

Increased benthic sediment loads that become trapped within algal turfs are posing a new threat to reefs, including the low coral cover reefs of the inshore GBR. A recent study conducted by Goatley and colleagues at Pioneer Bay, Orpheus Island sought to revisit some of the earlier measures of herbivore grazing activity used by Fox and Bellwood (2008a) and Bonaldo and Bellwood (2011) with a view to providing an updated health check of the state of the reef there following sediment deposition events associated with Severe Tropical Cyclone Yasi in 2011. The results were not encouraging. The grazing intensity of parrotfishes on the reef crest, as measured by remotely deployed video cameras, had declined by 98% between 2005 and 2012 and, unlike in 2009 when turf algal assays transplanted from the reef flat to the reef crest were grazed down within four days, the length of experimental turf assays moved out to the reef crest actually increased by 16% in 2012 (Goatley et al. 2016). This was despite no apparent change in the biomass or functional composition of parrotfishes within Pioneer Bay over the seven year period from 2005 to 2012 detected in the diver visual censuses (Goatley et al. 2016). The consequences of this reduced grazing for the low coral cover reef system in Pioneer Bay are yet to be seen, but these changing ecological processes may signal that we are in that critical policy window where action can be taken to avert any potential impending regime shift in this particular system.

The Ecology of Parrotfishes on "Low Coral/High Macroalgae" Cover Reefs

No examination of low coral cover reefs would be complete without reference to the Caribbean. In this case, the low coral cover descriptor has to be viewed as a qualitative one in that reefs of this region have seen a massive decline in coral cover over the past 30 years (Hughes 1994, Gardner et al. 2003, Wilkinson 2008). The associated increase in macroalgal abundance on reefs of the Caribbean now constitutes one of the most famous examples of a phase shift in textbooks of ecological theory. The series of accidents leading up to this most notorious of regime shifts is well documented and will not be dealt with in detail here, suffice to say that historical overfishing of large predatory fishes and then herbivorous fishes (including parrotfishes) through the first half of the 20th century led to a reliance by the ecosystem on grazing by the sea-urchin *Diadema antillarium* (Hughes 1994, Lessios et al. 2001). This overreliance on a single keystone species went unnoticed through the 1960s and 70s until two significant events in the early 1980s brought the ecological imbalance to light. In August 1980 Hurricane Allen passed through the Caribbean. It was one of the strongest hurricanes in recorded history and severely damaged reefs across the region (Woodley et al. 1981), toppling corals and opening up space within the reef matrix, effectively resetting the community to an earlier stage of ecological succession. Shortly after, an outbreak of disease within the *Diadema* population in 1983 caused a crash in numbers across the entire Caribbean (Lessios et al. 1984a, b, Lessios 1988a, b, Carpenter 1988, 1990a, b) and effectively made the organism functionally extinct (Lessios 2005). With the population of herbivorous fishes depleted by overfishing and an artificially high surfeit of space on the reef effectively diluting the impact of those fish that remained, the standing stock of herbivores was not sufficient to maintain grazing rates and to compensate for the loss of *Diadema*. The algal community experienced a competitive release from grazing and the result was an increase in macroalgal cover, the most famous example being from Discovery Bay, Jamaica where algal cover increased from around 5 percent cover in 1977 to levels of 60-90 percent in the 1990s with an associated decrease in coral cover from up to 75 percent in the late 1970s to an average of around 10 percent in 1994 (Hughes et al. 1987, Hughes 1994). At sites all around Jamaica the pattern was similar. Fleshy macroalgae (predominantly *Dictyota* sp. and *Lobophora* sp.; Williams and Polunin 2001) went from being an average of 4 percent of benthic cover in 1977-80 to 92 percent coverage in 1990-93 and coral cover on Jamaican reefs fell from an average of 52 percent in 1977-80 to just 3% in 1990-93 (Hughes et al. 1987, Hughes 1994). That is low coral cover by any definition.

High Macroalgal Cover by Accident in the Caribbean: Competitive Release of Parrotfishes and the Decline and Fall of the Urchin Empire

At the same time as the algae in the Caribbean underwent competitive release from grazing pressure following the mass mortality of *Diadema*, populations of herbivorous fishes remaining on Caribbean reefs by the 1980s also experienced their own competitive release in terms of no longer having to compete with urchins for access to algal resources. Ironically, this process had just been demonstrated experimentally on reefs of St. Thomas and St. Croix, U.S. Virgin Islands, where Hay and Taylor (1985) had removed *Diadema* from a 50 m strip of reef in St. Thomas and observed an immediate (within 4 months) increase in grazing rates and densities of parrotfish at the site. Parrotfish abundances in the 50 m experimental plots increased approximately three-fold and feeding rates (measured via consumption of seagrass bioassays) went up by a similar order of magnitude (Hay and Taylor 1985). At the same time, however, the experiment was almost prophetic in foreshadowing the fact that initial increases in grazing by fishes were unlikely to be sufficient to offset the reduction in urchin grazing. Macroalgal cover increased following *Diadema* removal, despite the increase in fish feeding. Reseeding of urchin populations in the cleared sites by adjacent (still healthy) populations allowed the system to correct for the imbalance, showing the extent to which these reefs relied on a diversity of functional groups. Within 16 months of the start of the experiment Hay and Taylor (1985) found that

Diadema numbers were back to 70% of their original density and macroalgal cover had been reduced to levels equal to surrounding control plots.

When the region-wide mass mortality of *Diadema* struck just one year later, the repeatability of Hay and Taylor's results would be put to the test in the context of an uncontrolled natural experiment, but this time without the possibility of a renewed influx of urchins. Unfortunately for Caribbean reefs, their work turned out to be impeccably accurate. Examining the impacts of the *Diadema* die-off on herbivore abundances at Tague Bay Reef, St Croix, US Virgin Islands, Carpenter (1990a) found that parrotfishes increased in abundance over the four years following the urchin die-offs (Fig. 6). This was mostly due to increases in the numbers of juvenile parrotfishes, suggesting that populations had previously been resource constrained. Ironically, the resource constraint may not have been one of food availability. Paddack and Sponaugle (2008) found that recruitment of the stoplight parrotfish *Sp. viride* (a dominant component of parrotfish biomass in the region) was strongly tied to abundance of the brown macroalga *Dictyota* sp. The increase in numbers of juvenile parrotfishes recorded by Carpenter may therefore have been a

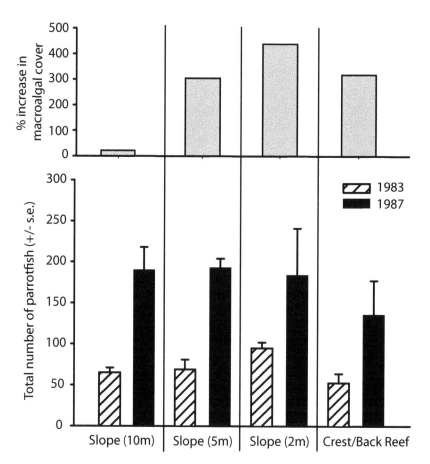

Fig. 6. Mean number of individual parrotfish (± s.e.) across four habitat zones from visual censuses conducted at Tague Bay Reef, St Croix, US Virgin Islands before and after mass mortality of the urchin *Diadema antillarum*. Species of parrotfish recorded at this location were: *Scarus taeniopterus, Sc. vetula, Sc. iseri, Sparisoma viride, Sp. aurofrenatum, Sp. chrysopterum, Sp. rubipinne.* Figure redrawn from tables presented in Carpenter 1990a and data in Carpenter 1990b.

response to the increased availability of macroalgal cover as a habitat resource, rather than to the reduced competition with urchins for the turf algae dietary resource. However, the overall increase in grazing pressure generated by these population increases, along with measured increases in fish feeding rates was not sufficient to control the biomass of algae on this reef. Nearly two years after the urchin-die off, algal biomass at Tague Bay Reef had increased between 300-400 percent along the back reef, reef crest and shallow fore-reef (Carpenter 1990b).

The increases in macroalgal cover on Caribbean reefs continued on into the 1990s and 2000s (Hughes 1994, Gardner et al. 2003, Hughes et al. 2010), suggesting that Carpenter's results from St. Croix were robust to time scales longer than two years. However, as Paddack et al. (2006) point out, almost all of the reported increases in macroalgal cover came from reefs that had been subjected to heavy overfishing of herbivores prior to the urchin die-off and which were therefore starting from artificially low levels (Jamaica: Liddell and Ohlhorst 1986, Hughes 1994; St Croix: Carpenter 1988, 1990a, b; US Virgin Islands: Hay and Taylor 1985, Levitan 1988). Expecting these low standing stocks of herbivores to provide compensatory grazing effects was therefore always going to be a tall order. The question was whether, in the case of Caribbean reefs that did have relatively healthy herbivore populations, the fishes could maintain high enough grazing rates to prevent the spread of macroalgae and maintain the benthic status quo? Williams and Polunin (2001) examined the relationship between macroalgal cover and biomass of herbivorous fishes across 19 low coral cover reefs (<25 percent cover at all but one and most <10 percent cover) in seven Caribbean countries at a standard depth of 12-15 m. They found that parrotfish biomass was higher than surgeonfish biomass at this depth at all locations (cf. Lewis and Wainwright 1985) and that herbivorous fish biomass was strongly negatively correlated with macroalgal cover and positively correlated with cropped substratum.

On the face of it, Williams and Polunin's (2001) results suggested that where parrotfish populations were intact, macroalgae coverage was lower and that parrotfishes were able to provide compensatory grazing pressure in the face of urchin loss. But, as they pointed out, this was only part of the story. Even on lightly-fished reefs, macroalgal cover at 12-15 m was still 20-50% and the fish were only able to maintain about 40-60% of the reef substratum in a cropped state. Their hypothesis was that the increase in macroalgal abundance in the Caribbean was not just a response to reduced grazing pressure of fishes compared to both fish and urchin grazing, but a result of the combination of low grazing and low coral cover leaving a significant portion of the substratum available for macroalgal settlement and development. When Williams et al. (2001) simulated the effect of increases in coral cover at sites (12 m depth) in Ambergris Caye, Belize by attaching artificial corals to the substratum, they found that, within three months, the percentage cover of macroalgae in their experimental plots had decreased. In some cases this was without any significant increase in the biomass of parrotfishes (predominantly *Sp. aurofrenatum* and *Sc. iseri*) present. It appeared that, where coral cover was high, the fish could maintain a high enough grazing rate to keep the macroalgae under control, but the high levels (70-90 percent) of available substratum on low coral cover reefs were simply too much for the fish to maintain in a cropped state. Paddack et al. (2006) did find evidence that the existing standing stock of herbivores on reefs in the Florida Keys (predominantly *Sp. viride* and *Sp. rubripinne* on high relief reefs and *Sp. viride* and *Sc. coelestinus* on low relief reefs) were able to consume the majority (55-100 percent) of algal production on offshore reefs, but only 31-51 percent of production on inshore reefs. Their results confirmed the suspicion that although parrotfishes would be able to limit the spread of macroalgae within those systems, they would not be able to exclude it.

With no signs of recovery in the *Diadema* populations to the present (Williams and Polunin 2001, Lessios 2005, although see Miller at al. 2003) and the continuing decline in coral cover on Caribbean reefs (Gardner et al. 2003), the dominance of macroalgae is now starting to have negative feedbacks on herbivore populations. Using datasets from across 20 countries and territories in the Caribbean, Paddack et al. (2009) found an average 4 percent per annum decline in herbivore abundances over the period 1996-2007. They attributed the decrease to the loss of three-dimensional reef structure associated with the decline in coral cover seen across the region. Overall, therefore, the evidence for parrotfishes being able to increase their population numbers or raise their grazing rate sufficiently to be able to exclude macroalgae from algal communities of the Caribbean is limited. The best that can be expected of parrotfishes on low coral cover reefs of the Caribbean is that they are able to limit the rate of macroalgal spread and maintain some areas of reef in a state that can allow for coral settlement and recruitment. The ecology of parrotfishes on what are now, unfortunately, low coral cover reefs of the Caribbean is essentially a story of the emergence of a new keystone functional group. To the extent that *Diadema* populations are not recovering on many reefs in the region, the buck now stops with the fish and particularly parrotfish as a significant component of the herbivorous fish biomass on reefs of the region. The issues of competitive release and trophic cascades are fundamental aspects of the ecology of parrotfishes in the Caribbean. However this keystone group is under significant pressure. The average parrotfish biomass on low coral cover reefs of the Caribbean is now calculated to be 14 g m^{-2}, compared with a historic high of 71 g m^{-2} for that region (Jackson et al. 2014). The International Union for Conservation of Nature (IUCN) has recently released a report co-authored by members of the IUCN, Global Coral Reef Monitoring Network and United Nations Environment Program calling for parrotfish to be listed as a specially protected species on the Protocol Concerning Specially Protected Areas and Wildlife (SPAW Protocol; Jackson et al. 2014). This recommendation is based on the assertion that declines in parrotfish biomass on Caribbean reefs have been a key driver of the declines in coral cover on these reefs since 1970. There is no doubt that these fishes are a fundamental component of coral reef ecosystem health and functioning in this region and require protection from over-exploitation. Whether the IUCN recommendation will be adopted remains to be seen.

High Macroalgal Cover by Experimental Design on the GBR: The Fundamental Limitations of Scarus and Chlorurus as Reversers of Regime Shifts

Although the inshore reefs of the GBR are low coral cover more by historical design than by accident, their status in this category does make them more vulnerable to drivers of ecosystem change such as bleaching events, increased sedimentation and nutrient run-off that could result in an undesirable regime shift to macroalgal dominance. For this reason, such reefs have been the focus of much research attention, with particular attention to the role of herbivorous fishes such as parrotfishes in maintaining the already delicate balance between coral and algae and in preventing the expansion of macroalgal cover.

The Fundamental Limitations of Scarus and Chlorurus as Reversers on the GBR

The large-scale caging exclusion experiment carried out by Terry Hughes and colleagues in Pioneer Bay, Orpheus Island was designed precisely to examine the question of whether an intact population of herbivorous fishes could reverse the effects of a regime shift to macroalgal dominance (Hughes et al. 2007). The experiment was significant both in its scale (5 m × 5 m cages and cage controls installed along the reef crest at Pioneer Bay for a period of 26 months) and its findings. The large 5 m × 5 m cages effectively prevented large fishes from grazing the reef substratum for a period of just over two years, in which time the benthic community in the experimental plots underwent a fundamental regime shift to macroalgal dominance (e.g. 3 m stands of the fleshy brown macroalga *Sargassum* were recorded within the caged areas; Hughes et al. 2007). When the cages were finally removed, remote underwater video cameras were deployed to monitor feeding by herbivorous fishes on the experimentally-induced macroalgal community and to document species responsible for macroalgal removal. Surprisingly, it was the batfish, *Platax pinnatus*, a species previously only described as an invertebrate feeder, that was found to be predominantly responsible for the rapid (over five days) reversal of the macroalgal-dominated system to a coral/EAM dominated state (Bellwood et al. 2006). Feeding by *Sc. rivulatus* and *Ch. microrhinos* on the macroalgae was insignificant, despite their dominance in the system in terms of biomass (see figure 3 in Bellwood et al. 2006).

Subsequent experimental work conducted at Orpheus Island in the late 2000s, provided further confirmation of the fact that although parrotfishes seemed to be critical to these inshore systems in terms of maintaining the status quo and preventing the spread of macroalgae, they could do little to reverse the situation once the algae had taken hold. Fox and Bellwood (2008a) recorded the feeding behaviour of herbivorous fishes in 1 m² plots across the reef using remotely deployed video cameras and then again in the presence of macroalgal bioassays (small quantities of the brown algae *Sargassum* sp. consisting of four single thalli artificially attached to fishing line and a piece of natural substratum with between 4-10 strands attached). This work, done at the same location as the exclusion experiment run by Hughes and colleagues, but at a much smaller scale, also showed a lack of involvement of parrotfishes when it came to the removal of fleshy, brown macroalgae. Of the 14,656 bites observed being taken from the *Sargassum* assays on the slope, crest and outer reef flat habitats of Pioneer Bay, only 172 (or 1.2%) were taken by the two dominant species of parrotfish at this site (*Sc. rivulatus* and *Ch. microrhinos*). Instead, it was the rabbitfish, *Siganus canaliculatus* that was responsible for removal of the algae (Fox and Bellwood 2008b). At the same location, Hoey and Bellwood (2011) examined the impact of varying densities of *Sargassum* on rates of macroalgal consumption and found that the kyphosid *Kyphosus vaigiensis* and the surgeonfish *Naso unicornis* were responsible for almost 97 percent of the bites taken from the *Sargassum* bioassays (but with lower browsing rates at high densities of algae). When algal deployments were repeated on reefs adjacent to Pioneer Bay (Cvitanovic and Bellwood 2009) and at inshore reefs across latitudes spanning 900 km of the GBR (Bennett and Bellwood 2011), a similar story emerged in terms of the lack of feeding by parrotfishes, although each time with a slightly different hero in terms of the actual species that was responsible for macroalgal removal. Only when assay deployments were conducted over an annual cycle, did parrotfish feature. Lefèvre and Bellwood (2011) showed that the parrotfish *Sc. rivulatus* took significantly more bites from *Sargassum* assays in the winter months compared to the summer months in which all previous experiments had been conducted (they recorded a total of 2,884 bites in winter compared to just 71 bites from *Sargassum* assays in the summer). However, this increased feeding by *Sc. rivulatus* coincided with the period when epiphyte cover on the plants was heaviest (Lefèvre and Bellwood 2010), suggesting that they were targeting the epiphytes rather than the *Sargassum* sp. and leaving the question as to whether the biomass of the algae itself was at all decreased by grazing during this winter period. It also coincided with the period during which plants were not growing and would naturally experience their annual die-back and tissue-shedding event, meaning that parrotfish feeding is unlikely to be significant in terms of reducing algal biomass from a system trapped in a macroalgal-dominated state.

Overall, therefore, this body of work has now fairly convincingly demonstrated that parrotfish do not play a significant role in the removal of large brown macroalgae from inshore reefs of the GBR and are therefore unlikely to play any significant functional role in the reversal of regime shifts from coral to macroalgal dominance. In addition, data from this work suggests that not only can macroalgae persist in the presence of parrotfishes, but that the presence of macroalgal stands can decrease the normal feeding rate of parrotfishes on clear areas of healthy algal turf immediately adjacent to algal stands. When Fox and Bellwood recorded the number of bites taken by herbivorous fishes within 1 m² areas of reef crest habitat in both the presence and absence of small quantities of macroalgal bioassays, they observed changes in the feeding behaviour of the dominant species of parrotfish, *Sc. rivulatus,* when even small densities of macroalgae were present (10 single strands of *Sargassum* sp. in linear array and clumped stand of 8-10 strands attached to live rock. Assays covered approximately 10% of the 1 m² experimental area). Compared to a "natural" feeding rate of 0.5 ± 0.1 bites per minute (mean ± s.e.) recorded within un-manipulated 1 m² plots on the reef crest, the rate of parrotfish feeding off EAM within 1 m² plots in the presence of *Sargassum* assays almost halved to 0.3 bites per minute (± 0.1 s.e.) (Fig. 7a). The interference effect of the macroalgal assays could actually be seen in the dynamics of *Sc. rivulatus* feeding behaviour. When bites rates by this species of parrotfish off the reef substratum surrounding the assays were calculated across individual 90 min segments of the total 4.5 hr deployment period, a significant pattern was observed. The rate of feeding by *S. rivulatus* on EAM within the 1 m² experimental assay plots showed a significant increase from 0.3 bites per min (mean ± 0.2 s.e.) in the first 90 mins to 1.3 bites per min (mean ± 0.4 s.e.) in the final 90 mins (repeated measures ANOVA, $F_{(2,14)}$=5.106, $p < 0.05$) (Fig. 7b). The interference effect was also visible when comparing the feeding rates of *Sc. rivulatus* on reef substratum within plots where bioassays were "removed" by the rabbitfish *Siganus canaliculatus* and those cases where the macroalgal biomass did not decrease over the 4.5 hr deployment period. In cases where bioassays were not fully removed, zero grazing was observed by *Sc. rivulatus* on the EAM-covered substratum within the experimental area in the second and third segments of the deployment period (Fig. 7c).

Again within Pioneer Bay, Hoey and Bellwood (2011) demonstrated similar negative correlations between the presence of macroalgae (as experimental bioassay deployments) and the feeding rate of scraping parrotfishes belonging to the genus *Scarus* and the excavating parrotfish *Ch. microrhinos*. The pattern of suppression of scraping and excavating feeding behaviour by parrotfishes in the presence of varying densities of *Sargassum* sp. was consistent over 4 separate daily deployments (see figures 5 and S3 in Hoey and Bellwood 2011). Such data strongly suggests the potential for a positive feedback loop based on an "interference effect" of the algae reducing the level of grazing disturbance on adjacent areas of reef substratum and allowing for neighbouring communities to develop to later stages of algal succession (usually associated with the growth of unpalatable macroalgae). The importance of these positive feedbacks and feeding "edge effects" for the establishment and persistence of macroalgae within low coral cover systems is a key area of future research that is currently being advanced by Hoey and co-workers.

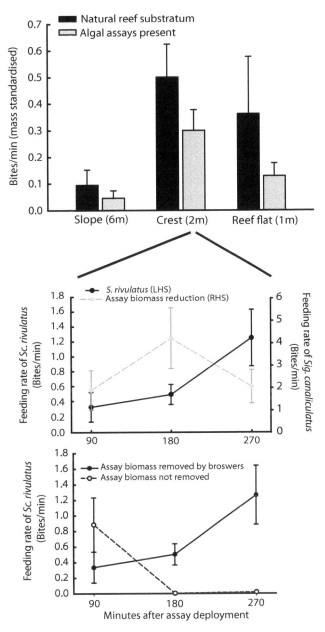

Fig. 7. Feeding rate (bites per minute) of *Scarus rivulatus* within experimental 1 m² plots on the reef crest, Pioneer Bay, Orpheus Island, GBR. All grazing rates quantified from footage obtained via remotely deployed underwater video recorders in order to minimize impact of divers on fish behaviour. (a) Grazing rate of *Sc. rivulatus* at the reef crest within 1 m² areas in the presence and absence of macroalgal assays. (b) Temporal change in grazing rate of *Sc. rivulatus* in the presence of macroalgal assays, shown as rate in the first, second and third 90 minute period following bioassay deployment. (c) Temporal change in grazing rate of *Sc. rivulatus* in the presence of macroalgal assays comparing rates where algal biomass did and did not decrease over the 4.5 hr deployment period. In (b) and (c), the feeding rate of *Sc. rivulatus* is shown alongside that of *Siganus canaliculatus* as a proxy for the extent of removal of macroalgae. (R.J. Fox unpublished data).

Summary and Future Research Directions

Which point brings me to the future and a summary of the past. This chapter has explored the interactions that parrotfishes have with their environment in the context of low coral cover reefs and the challenges and opportunities such environments present for parrotfish. We have seen that in cases where the corollary of low coral cover is high turf algal cover and low complexity, parrotfishes may be in their element. Due to their dominance of the herbivore biomass within such systems they are critical ecosystem engineers in terms of maintaining high levels of disturbance on the benthic community, in terms of maintaining the balance between coral and algae. In these well-balanced systems the main threats are reduction of parrotfish biomass via overfishing and the sudden pulse loading of the system with nutrients (via eutrophication) or increased sedimentation (either from deposition associated with natural storm and flooding events or anthropogenic loading from industrial and agricultural development). Low coral cover reefs therefore present opportunities for parrotfishes. But they can also present challenges, such as those discussed above in the context of low coral, high macroalgal cover reef systems. In these environments the challenge is to be able to maintain sufficiently high grazing rates on remaining reef substrata to prevent further spread of macroalgae, often in the context of depleted population densities, altered species (and hence functional group) composition, and size-class distributions skewed by fishing pressure on large individuals.

It is in the context of these environmental conditions that the agenda for future research on the ecology of parrotfishes on low coral cover reefs will be set. As experiments on the GBR and uncontrolled experiments in the Caribbean have shown, parrotfish are not a magic bullet solution. They can act only within the constraints of their evolutionary, morphological, dietary constraints. They are great stabilizers, but require the assistance of the cavalry once the system starts to go into decline. The next steps in terms of understanding the ecology of parrotfishes on low coral cover reefs will come from linking together aspects of their behaviour, physiology and ecology. In terms of their behaviour, understanding more about their movement patterns and habitat use within low coral cover environments will enable us to get to grips with the detail of where parrotfish exert their grazing impact, how this interacts with algal (including macroalgal) growth and life-history parameters and how these grazing patterns might change under altered behaviour caused by changing physical parameters within reef habitats under future environmental conditions. For parrotfishes, these patterns of movement and habitat use are strongly under the influence of their complex social and mating systems and work to understand the dynamics of these social systems in all of the species that have been identified as dominant components of the biomass on low coral cover reefs will be an important aspect of future research.

Of course, function and functional impact is also very much linked to physiological processes and understanding more about parrotfish physiology: their energetics, metabolism, and the hormonal regulation of their physiology will enable us to better predict likely changes in their ecology under changing environmental conditions. By tackling a research agenda on the degree of plasticity (both behavioural and physiological) that parrotfish can demonstrate, we will start to understand more about how they might either acclimate or adapt over generations to the conditions that are expected under future scenarios of global change. Advances in technology in terms of acoustic telemetry and archival tagging can now provide us with the behavioural and physiological data and advances in genetics, epigenetics and genomics, along with the power of new analysis such as are provide by the field of quantitative genetics (Kruuk 2004, Kruuk et al. 2008) that have the potential to throw new light on these issues. We now need the datasets to analyse. What many of the studies of Caribbean populations in the late 1980s and early 1990s highlighted was the variability of population densities of parrotfishes on these reef systems and just how much a long-term picture of the herbivore populations on these systems can improve our understanding of the dynamics of parrotfish grazing impact. Repeated censusing of individual locations over the long-term will not only allow us to answer questions relating to the grazing process with greater confidence, but will open up possibilities of examining the sorts of evolutionary questions about the genetic components of phenotypic diversity now being addressed via long-term studies of terrestrial taxa (Nussey et al. 2005, Wilson et al. 2006, Kruuk et al. 2008). This will give us the power to address questions relating to the degree of phenotypic plasticity within parrotfish populations and the likely response of these wild populations to environmental variation associated with the effects of climate change on low coral cover reefs.

Long-term data sets that record not just the abundance of parrotfishes on low coral cover reefs, but quantify the actual dynamics of grazing process itself will also be crucial. Such data offers the possibility of being able to detect subtle changes in this critical ecosystem process as a potential early-warning sign of changes in ecosystem state (cryptic loss of ecosystem resilience, Bellwood et al. 2004). In the context of reef governance, quantification of these early-warning signs will be crucial in the argument to convince policy-makers that critical windows for action are soon closing. At current rates of decline, there is a real risk that the epithet "low coral cover" will become applicable to more and more of the world's reef systems and that "the ecology of parrotfishes on low coral cover reefs" will simply become "the ecology of parrotfishes". I hope this fear proves to be unfounded.

Acknowledgments

I am grateful to D. Bellwood, A. Hoey, T. Hill, J.H. Choat, R. Bonaldo, C. Fulton and other members of the Reef Fish Ecology lab at James Cook University (past and present) for all their helpful advice and discussions that have informed the views expressed in this chapter and shaped its construction. All errors are, of course, my own. A very special thank-you to M. Jennions for hosting me in the J-lab during the write-up of this chapter and to A. Cockburn, S. Keogh and colleagues within the Division of Ecology & Evolution for all their incredible support and encouragement.

References Cited

Afeworki, Y. 2014. Population ecology of the rusty parrotfish *Scarus ferrugineus*, a dominant grazer on a seasonal coral reef. PhD dissertation. University of Groningen, Groningen, The Netherlands.

Afeworki, Y., J.H. Bruggemann and J.J. Videler. 2011. Limited flexibility in resource use in a coral reef grazer foraging on seasonally changing algal communities. Coral Reefs 30: 109–122.

Afeworki, Y., J.J. Videler and J.H. Bruggemann. 2013. Seasonally changing habitat use patterns among roving herbivorous fishes in the southern Red Sea: the role of temperature and algal community structure. Coral Reefs 32: 475–485.

Ahamada, S., J.P. Bijoux, B. Cauvin, A. Hagan, A. Harris, M. Koonjul, S. Meunier and J.-P. Quod. 2008. Status of the coral reefs of the South-West Indian Ocean Island States: Comoros, Madagascar, Mauritius, Reunion, Seychelles. pp. 105–118 *In*: C. Wilkinson (ed.). Status of Coral Reefs of the World: 2008. Australian Institute of Marine Science, Townsville, Australia.

Alvarez-Filip, L., N.K. Dulvy, J.A. Gill, I.M. Côté and A.R. Watkinson. 2009. Flattening of Caribbean coral reefs: region-wide declines in architectural complexity. Proc. R. Soc. B Biol. Sci. 276: 3019–3025.

Bellwood, D.R. 1996. Production and reworking of sediment by parrotfishes (family Scaridae) on the Great Barrier Reef, Australia. Mar. Biol. 125: 795–800.

Bellwood, D.R. and J.H. Choat. 1990. A functional analysis of grazing in parrotfishes (family Scaridae): the ecological implications. Environ. Biol. Fishes 28: 189–214.

Bellwood, D.R. and P.C. Wainwright. 2002. The history and biogeography of fishes on coral reefs. pp. 5–32 *In*: P.F. Sale (ed.). Coral Reef Fishes: Dynamics and Diversity in a Complex Ecosystem. Academic Press, San Diego, USA.

Bellwood, D.R. and C.J. Fulton. 2008. Sediment-mediated suppression of herbivory on coral reefs: decreasing resilience to rising sea levels and climate change? Limnol. Oceanogr. 53: 2695–2701.

Bellwood, D.R., T.P. Hughes, C. Folke and M. Nyström. 2004. Confronting the coral reef crisis. Nature 429: 827–833.

Bellwood, D.R., T.P. Hughes and A.S. Hoey. 2006. Sleeping functional group drives coral-reef recovery. Curr. Biol. 16: 2434–2439.

Bennett, S. and D.R. Bellwood. 2011. Latitudinal variation in macroalgal consumption by fishes on the Great Barrier Reef. Mar. Ecol. Prog. Ser. 426: 241–252.

Bennett, S., A. Vergés and D.R. Bellwood. 2010. Branching coral as a macroalgal refuge in a marginal coral reef system. Coral Reefs 29: 471–480.

Beukers, J.S. and G.P. Jones. 1989. Habitat complexity modifies the impact of piscivores on a coral reef fish population. Oecologia 114: 50–59.

Bonaldo, R.M. and D.R. Bellwood. 2011. Spatial variation in the effects of grazing on epilithic algal turfs on the Great Barrier Reef, Australia. Coral Reefs 30: 381–390.

Bonaldo, R.M., A.S. Hoey and D.R. Bellwood. 2014. The ecosystem role of parrotfishes on tropical reefs. Oceanogr. Mar. Biol. 52: 81–132.

Bruggemann, J.H., J. Begeman, E.M. Bosma, P. Verburg and A.M. Breeman. 1994a. Foraging by the stoplight-parrotfish *Sparisoma viride*. II. Intake and assimilation of food, protein and energy. Mar. Ecol. Prog. Ser. 106: 57–71.

Bruggemann, J.H., M.W.M. Kuyper and A.M. Breeman. 1994b. Comparative analysis of foraging and habitat use by the sympatric Caribbean parrotfish *Scarus vetula* and *Sparisoma viride* (Scaridae). Mar. Ecol. Prog. Ser. 112: 51–66.

Bruggemann, J.H., A.M. vanKessel, J.M. vanRooij and A.M. Breeman. 1996. Bioerosion and sediment ingestion by the Caribbean parrotfish *Scarus vetula* and *Sparisoma viride*: Implications of fish size, feeding mode and habitat use. Mar. Ecol. Prog. Ser. 134: 59–71.

Bruno, J.F. and E.R. Selig. 2007. Regional decline of coral cover in the Indo-Pacific: timing, extent, and subregional comparisons. PLoS One 2: e711.

Bruno, J.F., W.F. Precht, P.S. Vroom and R.B. Aronson. 2014. Coral reef baselines: How much macroalgae is natural? Mar. Pollut. Bull. 80: 24–29.

Burdick, D., V. Brown, J. Asher, C.F. Caballes, M. Gawel, L. Goldman, A. Hall, J. Kenyon, T. Leberer, E. Lundblad, J.L. McIlwain, J. Miller, D. Minton, M. Nadon, N. Pioppi, L. Raymundo, B. Richards, R. Schroeder, P.J. Schupp, E. Smith and B. Zgliczynski. 2008. Status of the coral reef ecosystems of Guam. Bureau of Statistics and Plans, Guam Coastal Management Program, Guam.

Burkepile, D.E. and M.E. Hay. 2009. Nutrient versus herbivore control of macroalgal community development and coral growth on a Caribbean reef. Mar. Ecol. Prog. Ser. 389: 71–84.

Carpenter, R.C. 1988. Mass mortality of a Caribbean sea urchin: immediate effects on community metabolism and other herbivores. Proc. Natl. Acad. Sci. USA 85: 511–514.

Carpenter, R.C. 1990a. Mass mortality of *Diadema antillarum* II. Effects on population densities and grazing intensity of parrotfishes and surgeonfishes. Mar. Biol. 104: 67–77.

Carpenter, R.C. 1990b. Mass mortality of *Diadema antillarum* I. Long term effects on sea-urchin population-dynamics and coral reef algal communities. Mar. Biol. 104: 79–86.

Chin, A., T. Lison de Loma, K. Reytar, S. Planes, K. Gerhardt, E. Clua, L. Burke and C. Wilkinson. 2011. Status of coral reefs of the Pacific and outlook: 2011. Global Coral Reef Monitoring Network.

Choat, J.H. 1991. The biology of herbivorous fishes on coral reefs. pp. 120–155. *In*: P.F. Sale (ed.). The Ecology of Fishes on Coral Reefs. Academic Press, Inc., San Diego, USA.

Chong-Seng, K.M., T.D. Mannering, M.S. Pratchett, D.R. Bellwood and N.A.J. Graham. 2012. The influence of coral reef benthic condition on associated fish assemblages. PLoS One 7: e42167.

Crossman, D.J., J.H. Choat, K.D. Clements, T. Hardy and J. McConochie. 2001. Detritus as food for grazing fishes on coral reefs. Limnol. Oceanogr. 46: 1596–1605.

Cvitanovic, C. and D.R. Bellwood. 2009. Local variation in herbivore feeding activity on an inshore reef of the Great Barrier Reef. Coral Reefs 28: 127–133.

De'ath, G., K.E. Fabricius, H.P.A. Sweatman and M. Puotinen. 2012. The 27-year decline of coral cover on the Great Barrier Reef and its causes. Proc. Natl. Acad. Sci. USA 109: 17995–17999.

Dickman, C.R. 1992. Predation and habitat shift in the house mouse, *Mus domesticus*. Ecology 313–322.

Floeter, S.R., M.D. Behrens, C.E.L. Ferreira, M.J. Paddack and M.H. Horn. 2005. Geographical gradients of marine herbivorous fishes: patterns and processes. Mar. Biol. 147: 1435–1447.

Fox, R.J. and D.R. Bellwood. 2007. Quantifying herbivory across a coral reef depth gradient. Mar. Ecol. Prog. Ser. 339: 49–59.

Fox, R.J. and D.R. Bellwood. 2008a. Direct versus indirect methods of quantifying herbivore grazing impact on a coral reef. Mar. Biol. 154: 325–334.

Fox, R.J. and D.R. Bellwood. 2008b. Remote video bioassays reveal the potential feeding impact of the rabbitfish *Siganus canaliculatus* (f: Siganidae) on an inner-shelf reef of the Great Barrier Reef. Coral Reefs 27: 605–615.

Fox, R.J. and D.R. Bellwood. 2013. Niche partitioning of feeding microhabitats produces a unique function for herbivorous rabbitfishes (Perciformes, Siganidae) on coral reefs. Coral Reefs 32: 13–23.

Friedlander, A.M. and E.E. DeMartini. 2002. Contrasts in density, size, and biomass of reef fishes between the northwestern and the main Hawaiian islands: the effects of fishing down apex predators. Mar. Ecol. Prog. Ser. 230: e264.

Friedlander, A.M., E. Brown and M.E. Monaco. 2007. Defining reef fish habitat utilization patterns in Hawaii: comparisons between marine protected areas and areas open to fishing. Mar. Ecol. Prog. Ser. 351: 221–233.

Friedlander, A.M., J.E. Maragos, R.E. Brainard, A. Clark, G. Aeby, B. Bowen, E. Brown, K. Chaston, J. Kenyon, C.P. Meyer, P. McGowan, J. Miller, T. Montgomery, R. Schroeder, C. Smith, P.S. Vroom, W.J. Walsh, I.D. Williams, W. Wiltse and J. Zamzow. 2008. Status of coral reefs in Hawai'i and United States Pacific remote island areas (Baker, Howland, Palmyra, Kingman, Jarvis, Johnstone, Wake) in 2008. pp. 213–224. *In*: C. Wilkinson (ed.). Status of Coral Reefs of the World: 2008. Australian Institute of Marine Science, Townsville, Australia.

Gardner, T.A., I.M. Côté, J.A. Gill, A. Grant and A.R. Watkinson. 2003. Long-term region-wide declines in Caribbean corals. Science 301: 958–960.

Goatley, C.H.R. and D.R. Bellwood. 2011. The roles of dimensionality, canopies and complexity in ecosystem monitoring. PLoS One 6: e27307.

Graham, N.A.J, S. Jennings, M.A. MacNeil, D. Mouillot and S.K. Wilson. 2015. Predicting climate-driven regime shifts versus rebound potential in coral reefs. Nature 518: 94–97.

Goatley, C.H.R. and D.R. Bellwood. 2012. Sediment suppresses herbivory across a coral reef depth gradient. Biol. Lett. 8: 1016–1018.

Goatley, C.H.R. and D.R. Bellwood. 2013. Ecological consequences of sediment on high-energy coral reefs. PLoS One 8: e77737.

Goatley, C.H.R., R.M. Bonaldo, R.J. Fox and D.R. Bellwood. 2016. Sediments and herbivory as sensitive indicators of coral reef degradation. Ecology and Society 21: 29.

Graham, N.A.J. and K.L. Nash. 2013. The importance of structural complexity in coral reef ecosystems. Coral Reefs 32: 315–326.

Graham, N.A.J., S.K. Wilson, S. Jennings, N.V.C. Polunin, J.P. Bijoux and J. Robinson. 2006. Dynamic fragility of oceanic coral reef ecosystems. Proc. Natl. Acad. Sci. USA 103: 8425–8429.

Hawkins, J.P. and C.M. Roberts. 2004. Effects of artisanal fishing on Caribbean coral reefs. Conserv. Biol. 18: 215–226.

Hay, M.E. and P.R. Taylor. 1985. Competition between herbivourous fishes and urchins on Caribbean reefs. Oecologia 65: 591–598.

Hixon, M.A. and W.N. Brostoff. 1996. Succession and herbivory: effects of differential fish grazing on Hawaiian coral-reef algae. Ecol. Monogr. 66: 67–90.

Hoey, A.S. and D.R. Bellwood. 2008. Cross-shelf variation in the role of parrotfishes on the Great Barrier Reef. Coral Reefs 27: 37–47.

Hoey, A.S. and D.R. Bellwood. 2010a. Among-habitat variation in herbivory on *Sargassum* spp. on a mid-shelf reef in the northern Great Barrier Reef. Mar. Biol. 157: 189–200.

Hoey, A.S. and D.R. Bellwood. 2010b. Cross-shelf variation in browsing intensity on the Great Barrier Reef. Coral Reefs 29: 499–508.

Hoey, A.S. and D.R. Bellwood. 2011. Suppression of herbivory by macroalgal density: a critical feedback on coral reefs? Ecol. Lett. 14: 267–273.

Hopley, D. and J.H. Choat. 1990. The effects of mainland land use on adjacent reef systems of the Great Barrier Reef. Agriculture and the Ecosystem in North Queensland: proceedings of a Symposium 1: 1–16.

Hopley, D., A.M. Slocombe, F. Muir and C. Grant. 1983. Nearshore fringing reefs in North Queensland. Coral Reefs 1: 151–160.

Hughes, T.P. 1994. Catastrophes, Phase Shifts, and Large-Scale Degradation of a Caribbean Coral Reef. Science 265: 1547–1551.

Hughes, T.P., D.C. Reed and M.-J. Boyle. 1987. Herbivory on coral reefs: community structure following mass mortalities of sea urchins. J. Exp. Mar. Biol. Ecol. 113: 39–59.

Hughes, T.P., M.J. Rodrigues, D.R. Bellwood, D. Ceccarelli, O. Hoegh-Guldberg, L. McCook, N. Moltschaniwskyj, M.S. Pratchett, R.S. Steneck and B.L. Willis. 2007. Phase shifts, herbivory, and the resilience of coral reefs to climate change. Curr. Biol. 17: 360–365.

Hughes, T.P., N.A.J. Graham, J.B.C. Jackson, P.J. Mumby and R.S. Steneck. 2010. Rising to the challenge of sustaining coral reef resilience. Trends Ecol. Evol. 25: 633–642.

Hughes, T.P., J.T. Kerry, M. Álvarez-Noriega, J.G. Álvarez-Romero, K.D. Anderson, A.H. Baird, R.C. Babcock, M. Beger, D.R. Bellwood, R. Berkelmans, T.C. Bridge, I.R. Butler, M. Byrne, N.E. Cantin, S. Comeau, S.R. Connolly, G.S. Cumming, S.J. Dalton, G. Diaz-Pulido, C.M. Eakin, W.F. Figueira, J.P. Gilmour, H.B. Harrison, S.F. Heron, A.S. Hoey, J.-P.A. Hobbs, M.O. Hoogenboom, E.V. Kennedy, C.-Y. Kuo, J.M. Lough, R.J. Lowe, G. Liu, M.T. McCulloch, H.A. Malcolm, M.J. McWilliam, J.M. Pandolfi, R.J. Pears, M.S. Pratchett, V. Schoepf, T. Simpson, W.J. Skirving, B. Sommer, G. Torda, D.R. Wachenfeld, B.L. Willis and S.K. Wilson. 2017. Global warming and recurrent mass bleaching of corals. Nature 543: 373–377.

Jackson, J.B.C., M.K. Donovan, K.L. Cramer and V.V. Lam. 2014. Status and Trends of Caribbean Coral Reefs: 1970–2012. Global Coral Reef Monitoring Network, IUCN, Gland, Switzerland.

Jones, G.P., M.I. McCormick, M. Srinivasan and J.V. Eagle. 2004. Coral decline threatens fish biodiversity in marine reserves. Proc. Natl. Acad. Sci. USA 101: 8251–8253.

Klumpp, D.W. and N.V.C. Polunin. 1990. Algal production, grazers and habitat partitioning on a coral reef: positive correlation between grazing rate and food availability. Trophic Relationships in the Marine Environment: Proceedings of the 24th European Marine Biological Symposium. 1: 372–388.

Knowlton, N. and J.B.C. Jackson. 2008. Shifting baselines, local impacts and global change on coral reefs. PLoS Biol. 6: e54.

Komyakova, V., P.L. Munday and G.P. Jones. 2013. Relative importance of coral cover, habitat complexity and diversity in determining the structure of reef fish communities. PLoS One 8: e83178.

Kruuk, L.E.B. 2004. Estimating genetic parameters in natural populations using the 'animal model'. Philos. Trans. R. Soc. B Biol. Sci. 359: 873–890.

Kruuk, L.E.B., J. Slate and A.J. Wilson. 2008. New answers for old questions: the evolutionary quantitative genetics of wild animal populations. Annu. Rev. Ecol. Evol. Syst. 39: 525–548.

Larcombe, P. and K.J. Woolfe. 1999. Terrigenous sediments as influences upon Holocene nearshore coral reefs, central Great Barrier Reef, Australia. Aust. J. Earth Sci. 46: 141–154.

Larcombe, P., K.J. Woolfe and R.G. Purdon. 1996. Terrigenous sediment fluxes and human impacts. CRC Reef Research Centre, Townsville, Australia.

Ledlie, M., N.A.J. Graham, J. Bythell, S.K. Wilson, S. Jennings, N.V.C. Polunin and J. Hardcastle. 2007. Phase shifts and the role of herbivory in the resilience of coral reefs. Coral Reefs 26: 641–653.

Lefèvre, C.D. and D.R. Bellwood. 2010. Seasonality and dynamics in coral reef macroalgae: variation in condition and susceptibility to herbivory. Mar. Biol. 157: 955–965.

Lefèvre, C.D. and D.R. Bellwood. Temporal variation in coral reef ecosystem processes: herbivory of macroalgae by fishes. Mar. Ecol. Prog. Ser. 422: 239–251.

Lessios, H.A. 1988a. Mass mortality of *Diadema antillarum* in the Caribbean: what have we learned? Annu. Rev. Ecol. Syst. 371–393.

Lessios, H.A. 1988b. Population dynamics of *Diadema antillarum* (Echinodermata: Echinoidea) following mass mortality in Panama. Mar. Biol. 99: 515–526.

Lessios, H.A. 2005. *Diadema antillarum* populations in Panama twenty years following mass mortality. Coral Reefs 24: 125–127.

Lessios, H.A., D.R. Robertson and J.D. Cubit. 1984a. Spread of *Diadema* mass mortality through the Caribbean. Science 226: 335–337.

Lessios, H.A., J.D. Cubit, D.R. Robertson, M.J. Shulman, M.R. Parker, S.D. Garrity and S.C. Levings. 1984b. Mass mortality of *Diadema antillarum* on the Caribbean coast of Panama. Coral Reefs 3: 173–182.

Lessios, H.A., M.J. Garrido and B.D. Kessing. 2001. Demographic history of *Diadema antillarum*, a keystone herbivore on Caribbean reefs. Proc. R. Soc. B Biol. Sci. 268: 2347–2353.

Levitan, D.R. 1988. Algal-urchin biomass responses following mass mortality of *Diadema antillarum* Philippi at Saint John, US Virgin Islands. J. Exp. Mar. Biol. Ecol. 119: 167–178.

Lewis, S.M. and P.C. Wainwright. 1985. Herbivore abundance and grazing intensity on a Caribbean coral reef. J. Exp. Mar. Biol. Ecol. 87: 215–228.

Liddell, W.D. and S.L. Ohlhorst. 1986. Changes in benthic community composition following the mass mortality of *Diadema* at Jamaica. J. Exp. Mar. Biol. Ecol. 95: 271–278.

McCook, L.J. 1996. Effects of herbivores and water quality on *Sargassum* distribution on the central Great Barrier Reef: cross-shelf transplants. Mar. Ecol. Prog. Ser. 139: 179–192.

McCook, L.J. 1997. Effects of herbivory on zonation of *Sargassum* spp. within fringing reefs of the central Great Barrier Reef. Mar. Biol. 129: 713–722.

McCook, L.J. and I.R. Price. 1997a. The state of the algae of the Great Barrier Reef: what do we know? pp. 194–204. *In*: D.R. Wachenfeld, J.K. Oliver and K. Davis (eds.). State of the Great Barrier Reef World Heritage Area Workshop. Great Barrier Reef Marine Park Authority, Townsville, Australia.

McCook, L.J. and I.R. Price. 1997b. Macroalgal distributions on the Great Barrier Reef: a review of patterns and causes. The Great Barrier Reef: Science, Use and Management: A National Conference. 2: 37–46.

McGinley, M. 2014. Common coral reef fishes of Jamaica. The Encyclopedia of Earth. http://www.eoearth.org/view/article/151353 (accessed 18 July 2014)

McIlwain, J.L. and B.M. Taylor. 2009. Parrotfish population dynamics from the Marianas Islands, with a description of the demographic and reproductive characteristics of *Chlorurus sordidus*. Final Report to the Western Pacific Regional Fishery Management Council. University of Guam, Mangilao, Guam.

Miller, R.J., A.J. Adams, N.B. Ogden, J.C. Ogden and J.P. Ebersole. 2003. *Diadema antillarum* 17 years after mass mortality: is recovery beginning on St. Croix? Coral Reefs 22: 181–187.

Mumby, P.J., S. Bejarano, Y. Golbuu, R.S. Steneck, S.N. Arnold, R. Van Woesik and A.M. Friedlander. 2013. Empirical relationships among resilience indicators on Micronesian reefs. Coral Reefs 32: 213–226.

Nash, K.L., N.A.J. Graham, F.A. Januchowski-Hartley and D.R. Bellwood. 2012. Influence of habitat condition and competition on foraging behaviour of parrotfishes. Mar. Ecol. Prog. Ser. 457: 113–124.

Newman, M.J.H., G.A. Paredes, E. Sala and J.B.C. Jackson. 2006. Structure of Caribbean coral reef communities across a large gradient of fish biomass. Ecol. Lett. 9: 1216–1227.

Nussey, D.H., T.H. Clutton-Brock, S.D. Albon, J. Pemberton and L.E.B. Kruuk. 2005. Constraints on plastic responses to climate variation in red deer. Biol. Lett. 1: 457–460.

Paddack, M.J. and S. Sponaugle. 2008. Recruitment and habitat selection of newly settled *Sparisoma viride* to reefs with low coral cover. Mar. Ecol. Prog. Ser. 369: 205–212.

Paddack, M.J., R.K. Cowen and S. Sponaugle. 2006. Grazing pressure of herbivorous coral reef fishes on low coral-cover reefs. Coral Reefs 25: 461–472.

Paddack, M.J., J.D. Reynolds, C. Aguilar, R.S. Appeldoorn, J. Beets, E.W. Burkett, P.M. Chittaro, K. Clarke, R. Esteves, A.C. Fonseca, G.E. Forrester, A.M. Friedlander, J. Garcia-Sais, G. Gonzalez-Sanson, L.K. Jordan, D.B. McClellan, M.W. Miller, P.P. Molloy, P.J. Mumby, I. Nagelkerken, M. Nemeth, R. Navas-Camacho, J. Pitt, N.V.C. Polunin, M.C. Reyes-Nivia, D.R. Robertson, A. Rodriguez-Ramirez, E. Salas, S.R. Smith, R.E. Spieler, M.A. Steele, I.D. Williams, C.L. Wormald, A.R. Watkinson and I.M. Côté. 2009. Recent region-wide declines in Caribbean reef fish abundance. Curr. Biol. 19: 590–595.

Pandolfi, J.M., R.H. Bradbury, E. Sala, T.P. Hughes, K.A. Bjorndal, R.G. Cooke, D. McArdle, L. McClenachan, M.J.H. Newman, G. Paredes, R.R. Warner and J.B.C. Jackson. 2003. Global trajectories of the long-term decline of coral reef ecosystems. Science 301: 955–958.

Pauly, D. 1995. Anecdotes and the shifting baseline syndrome of fisheries. Trends Ecol. Evol. 10: 430.

Purcell, S.W. and D.R. Bellwood. 2001. Spatial patterns of epilithic algal and detrital resources on a windward coral reef. Coral Reefs 20: 117–125.

Rasher, D.B., A.S. Hoey and M.E. Hay. 2013. Consumer diversity interacts with prey defenses to drive ecosystem function. Ecology 94: 1347–1358.

Roff, G., M.H. Ledlie, J.C. Ortiz and P.J. Mumby. 2011. Spatial patterns of parrotfish corallivory in the Caribbean: the importance of coral taxa, density and size. PLoS One 6: e29133.

Schaffelke, B. and D.W. Klumpp. 1997. Biomass and productivity of tropical macroalgae on three nearshore fringing reefs in the central Great Barrier Reef, Australia. Bot. Mar. 40: 373–384.

Schneider, K.J. 1984. Dominance, predation, and optimal foraging in white-throated sparrow flocks. Ecology: 1820–1827.

Stockwell, B., C.R.L. Jadloc, R.A. Abesamis, A.C. Alcala and G.R. Russ. 2009. Trophic and benthic responses to no-take marine reserve protection in the Philippines. Mar. Ecol. Prog. Ser. 389: 1–15.

Sweatman, H.P.A., S. Delean and C. Syms. 2011. Assessing loss of coral cover on Australia's Great Barrier Reef over two decades, with implications for longer-term trends. Coral Reefs 30: 521–531.

Thresher, R.E. 1991. Geographic variability in the ecology of coral reef fishes: evidence, evolution and possible implications. pp. 401–435. *In*: P.F. Sale (ed.). The Ecology of Fishes on Coral Reefs. Academic Press, Inc, San Diego, USA.

Vroom, P.S. 2011. "Coral Dominance": a dangerous ecosystem misnomer? J. Mar. Biol. 2011: 164127.

Wilkinson, C. (ed.). 2008. Status of Coral Reefs of the World: 2008. Global Coral Reef Monitoring Network and Rainforest Research Centre, Townsville, Australia.

Williams, I.D. and N.V.C. Polunin. 2001. Large-scale associations between macroalgal cover and grazer biomass on mid-depth reefs in the Caribbean. Coral Reefs 19: 358–366.

Williams, I.D., N.V. Polunin and V.J. Hendrick. 2001. Limits to grazing by herbivorous fishes and the impact of low coral cover on macroalgal abundance on a coral reef in Belize. Mar. Ecol. Prog. Ser. 222: 187–196.

Wilson, A.J., J.M. Pemberton, J.G. Pilkington, D.W. Coltman, D.V. Mifsud, T.H. Clutton-Brock and L.E.B. Kruuk. 2006. Environmental coupling of selection and heritability limits evolution. PLoS Biol. 4: e216.

Wilson, S.K., D.R. Bellwood, J.H. Choat and M.J. Furnas. 2003. Detritus in the epilithic algal matrix and its use by coral reef fishes. Oceanogr. Mar. Biol. 41: 279–310.

Wilson, S.K., A.M. Dolman, A.J. Cheal, M.J. Emslie, M.S. Pratchett and H.P.A. Sweatman. 2009. Maintenance of fish diversity on disturbed coral reefs. Coral Reefs 28: 3–14.

Wismer, S., A.S. Hoey and D.R. Bellwood. 2009. Cross-shelf benthic community structure on the Great Barrier Reef: relationships between macroalgal cover and herbivore biomass. Mar. Ecol. Prog. Ser. 376: 45–54.

Wolanski, E. and S. Spagnol. 2000. Pollution by mud of Great Barrier Reef coastal waters. J. Coast. Res. 16: 1151–1156.

Woodley, J., E. Chornesky, P. Cliffo, J.B.C. Jackson, L.S. Kaufman, N. Knowlton, J.C. La, M.P. Pearson, J.W. Porter and M.C. Rooney. 1981. Hurricane Allen's impact on a Jamaican coral reef. Science 214: 749–754.

No-take Marine Reserve Effects on Parrotfish and Parrotfish-benthos Interactions

Sarah-Lee A. Questel[1] and Garry R. Russ[2]

[1] College of Science and Engineering, James Cook University, Townsville,
  Queensland 4811, Australia
  Email: sarah.questel@gmail.com
[2] College of Science and Engineering and ARC Centre of Excellence for Coral Reef Studies,
  James Cook University, Townsville, Queensland 4811, Australia
  Email: garry.russ@jcu.edu.au

Introduction

Parrotfish and their Role on Coral Reefs

Parrotfish (Labridae, Scarinae) are conspicuous, often abundant herbivorous/detritivorous fishes on coral reefs that are recognized by their fused-beak like jaws and often bright coloration of terminal phase males (Bonaldo et al. 2014). Although parrotfish are known to graze primarily on algal covered surfaces (Bellwood and Choat 1990, Choat et al. 2002, Bellwood 2003, Clements et al. 2009), parrotfish obtain substantial nutrition from detritus (Crossman et al. 2005) and protein-rich epilithic, endolithic and epiphytic microscopic phototrophs (Clements and Choat Chapter 3). Some parrotfish species are also known to consume live corals (Bellwood et al. 2003, Rotjan and Lewis 2005, Bonaldo and Rotjan Chapter 9). Their powerful oral jaws enable them to remove the epilithic algal matrix (EAM) and underlying calcium carbonate substratum, while their pharyngeal jaws grind ingested material into a fine paste (Bellwood et al. 2003). This feeding mode and mechanical trituration sets parrotfish apart from other herbivores in their use of, and potential effect on, the substratum (Bellwood 2003, Bellwood et al. 2003, Bonaldo et al. 2014). Their abundance and feeding behaviour result in parrotfish playing critical roles on coral reefs (Bellwood et al. 2003, Bonaldo et al. 2014), including the top-down control of algal communities (Hughes et al. 2007a, Adam et al. 2015), the production and transport of reef sediments (Bellwood et al. 2003, Hoey and Bellwood 2008, Mallela and Fox Chapter 8) and acting as a key trophic link between primary producers (algae) and primary consumers (Hatcher 1981, 1988).

Based on their morphology and feeding behaviour, parrotfish have been categorized into three functional groups: excavators, scrapers and browsers (Bellwood and Choat 1990, Bellwood et al. 2003, Bonaldo et al. 2014). Both excavators and scrapers generally feed from EAM covered surfaces, but differ in the amount of material that is removed when feeding. Excavators remove large portions of limestone and dead corals when feeding, whereas scrapers only take shallow bites and have minimal impact on the benthic substratum (Bellwood and Choat 1990, Bellwood et al. 2012). Browsers feed primarily on macroalgae and associated epiphytic material and remove it without affecting the substratum (Bellwood and Choat 1990). All three functional groups can differ markedly in their contribution to removal of benthic algae, bioerosion, modification of the benthos, replenishment and recovery of corals and the dynamics of tropical reefs (Bellwood and Choat 1990, Bellwood et al. 2003, 2012, Bonaldo et al. 2014).

Overfishing and Modified Food Webs

On many tropical reefs, overfishing has been associated with reduction in top-down control of lower trophic levels, resulting in modified food webs and increased vulnerability to physical disturbances (Pandolfi et al. 2003, Hughes et al. 2007b, 2010, McClanahan et al. 2011a, Bellwood et al. 2012, Adam et al. 2015). A classic example of this occurred in Jamaica where a fishing-induced decline of herbivorous fishes (including parrotfishes), declines of herbivorous urchins due to a disease, and hurricane damage, ultimately led to a benthic phase-shift from a coral- to a macroalgal-dominated state (Hughes 1994). In many tropical regions, parrotfish are important fishery targets, especially in developing nations (Bellwood et al. 2003, Hawkins and Roberts 2004a, McClanahan et al. 2007, Comeros-Raynal et al. 2012, Edwards et al. 2014). Increased fishing pressure has resulted in widespread declines of these fishes in the last few decades (Bellwood et al. 2012, Comeros-Raynal et al. 2012, Edwards et al. 2014, Jackson et al. 2014). Although parrotfish may not be as vulnerable to fishing as other groups of reef fish of similar size in different or even the same trophic levels (Abesamis et al. 2014, Taylor et al. 2014a), they still possess particular life-history and behavioural traits that make them susceptible to exploitation (Dulvy and Polunin 2004, Hawkins and Roberts 2004b, Bellwood et al. 2012, Taylor 2014a). For example, they are conspicuous, have shallow depth distributions and small home ranges, and are easily targeted at night (Taylor et al. 2014a). Thus, there is concern that the loss of parrotfishes that exert top-down control on primary producers will eventually result in detrimental macroalgal-dominated states on coral reefs (Hughes 1994, Bellwood et al. 2012, Hughes et al. 2007a, Jackson et al. 2014, Adam et al. 2015).

An extensive review of the status and trends of Caribbean coral reefs from 1970 to 2012 concluded that overfishing of herbivores, particularly parrotfish, was a major driver of decline of coral cover in the Caribbean (Jackson et al. 2014). The review recommended reduction of fishing, bans on spearfishing and fish traps and the possibility of total bans on fishing for parrotfish on Caribbean coral reefs. Similarly, an extensive review of Indian Ocean coral reef fisheries suggested that the proportion of herbivorous fish in the fishable biomass could decline to a point that had major consequences to the coral reef ecosystem (McClanahan et al. 2011a). In contrast, a growing body of research suggests that trophic cascades resulting from overfishing of herbivorous fish, causing a shift from coral to macroalgal-dominance, may be an over-simplication (Aronson and Precht 2006, Cote et al. 2013, Toth et al. 2014). A global review of the incidence of benthic phase shifts from coral to algal dominance concluded that such events were still relatively rare, particularly on Indo-West Pacific coral reefs (Bruno et al. 2009). Roff and Mumby (2012) also concluded that evidence for such benthic phase shifts was more common on Caribbean than Indo-Pacific coral reefs. Further, a global meta-analysis showed that marine protected areas did not affect either coral recovery nor macroalgal development following major environmental disturbances, and led to the recommendation that management aimed specifically at protecting herbivorous reef fish in New Caledonia would do little to alleviate the vulnerability of coral reefs to major environmental disturbances (Carassou et al. 2013).

No-take Marine Reserve Effects

No-take marine reserves (NTMRs), areas that have been permanently closed to fishing, generally increase body size, density and biomass of organisms targeted by fishing (Halpern 2003, Lester et al. 2009, Molloy et al. 2009, Graham et al. 2011, Edgar et al. 2014). Such effects of NTMRs can vary with trophic level, with NTMR effects usually stronger on large-bodied predatory fish, and sometimes less obvious for smaller–bodied herbivores, omnivores and planktivores (Lester et al. 2009, Molloy et al. 2009, Graham et al. 2011, Abesamis et al. 2014, Edgar et al. 2014). Effects of NTMRs on target and non-target species always need to be interpreted in relation to the fishery in which they are embedded. The higher the fishing mortality on target species (Edwards et al. 2014), and the better the NTMR compliance (Campbell et al. 2012, Bergseth et al. 2015), the stronger the direct NTMR effect should be. These points are especially important when considering effects of NTMRs on parrotfish, a group that are not universally fished on coral reefs. For example, fishing mortality rates on parrotfish are generally much higher on coral reefs in developing than developed nations (Edwards et al. 2014). Furthermore, not all parrotfish are targeted equally by fisheries, with greater fishing effort often directed towards larger-bodied than smaller-bodied species (Dulvy and Polunin 2004, Bellwood et al. 2012, Edwards et al. 2014, Taylor et al. 2014a, Heenan et al. 2016). Different life-histories among parrotfish species may also make certain species more or less vulnerable to fishing. For example, Taylor et al. (2014a) showed that higher fishing pressure on coral reefs in Guam led to a higher proportion of smaller-bodied, early maturing species of parrotfish. Furthermore, fishing may have different effects on parrotfishes that perform different functional roles (e.g. scraping vs bioerosion - Bellwood et al. 2012). Many of these factors suggest that direct effects of NTMRs on parrotfish density, biomass and size structure might be more varied, and perhaps less common in the literature, compared to effects on heavily targeted, large predatory reef fish.

In addition to direct NTMR effects on targeted species, there are expectations that NTMRs will also have indirect effects on organisms, usually via trophic interactions (Babcock et al. 2010). Several studies have reported changes in species composition and trophic interactions as a result of recovery of predators in NTMRs (McClananhan 1994, Graham et al. 2003, Mumby et al. 2006; Edgar et al. 2009, Boaden and Kingsford 2015). Babcock et al. (2010) showed that NTMRs have both direct effects (increase in body size, density and biomass of fish targeted by fishing) after ~five years and indirect effects (secondary, food-web interactions) after ~13 years. The best examples of NTMRs altering trophic interactions and thus benthic structure come from kelp beds in Tasmania (Edgar et al. 2009, Babcock et al. 2010), New Zealand (Babcock et al. 1999, 2010) and California (Behrens and Lafferty 2004, Babcock et al. 2010). In these studies, NTMRs increased abundance of targeted predatory lobsters and reef fish, decreasing density of their herbivorous urchin prey and thus facilitating recovery of kelps. It is notable that these examples come from systems with substantially lower species diversity, lower trophic diversity and thus lower functional redundancy than is typical of most coral reefs. Nevertheless, such indirect effects of NTMRs on trophic interactions can and do impact benthos on coral reefs (McClanahan 1994, Mumby et al. 2006, Babcock et al. 2010,

Rasher et al. 2013, Adam et al. 2015, Bonaldo et al. 2017). For example, Kenyan NTMRs have been shown to increase biomass of herbivorous fish (including parrotfish) and cover of crustose coralline algae, while decreasing urchin densities and maintaining or reducing cover of fleshy algae (McClanahan 1994, 2014, Babcock et al. 2010, Humphries et al. 2014). A Caribbean NTMR had higher biomass of, and grazing rates by, large parrotfish inside it relative to fished areas, resulting in lower cover of macroalgae and enhanced coral cover in the NTMR relative to fished areas (Mumby 2006, Mumby et al. 2006, 2007, Mumby and Harborne 2010). In contrast, an emerging literature questions the ability of NTMRs to alter trophic interactions, the benthos and ecosystem structure on coral reefs (e.g. Aronson and Precht 2006, Kramer and Heck 2007, Bruno et al. 2009, Carassou et al. 2013, Russ et al. 2015) and points to the limited role that NTMRs may play in alleviating the effects of major environmental disturbances on coral reefs (Jones et al. 2004, Coelho and Manfrino 2007, Huntington et al. 2011, Carassou et al. 2013, Russ et al. 2015, Hughes et al. 2017).

This chapter will firstly examine the evidence for both direct and indirect effects of NTMRs on parrotfish. A direct effect of a NTMR is defined as the removal of fishing directly targeting parrotfish that subsequently affects size structure, density and biomass of the target fish. Indirect effects of NTMRs are defined as removal of fishing that affects either predators or competitors of parrotfish, but specifically excludes NTMR effects on the benthos that may subsequently affect parrotfish. Secondly, if the parrotfish are affected directly or indirectly by NTMRs, this may modify the ways in which parrotfish then affect the benthos or are themselves affected by changes in the benthos occurring either due to NTMR protection or independent of NTMR protection.

Direct and Indirect Effects of NTMRs on Parrotfish

Direct Effects

Removal of Fishing Pressure

Properly enforced NTMRs should generally increase mean body size, density and biomass of targeted, site-attached species (Fig. 1) (Lester et al. 2009, Molloy et al. 2009, Graham et al. 2011, Edgar et al. 2014). Abesamis et al. (2014) identified four approaches used to detect direct NTMR effects on target species: single-point-in-time comparisons of fish variates inside and outside NTMRs, meta-analyses of NTMR studies, space-for-time substitutions, and Before-After-Control-Impact-Pair (BACIP) studies incorporating long-term (decadal-scale) monitoring. Here, we examine all four of these as they relate to parrotfish size structure, density, and biomass. To obtain information on the direct and indirect effects of NTMRs on parrotfish, and subsequent interactions of parrotfish with benthos, a literature search of peer-reviewed studies (published between 1980 and 2014) was performed using the terms 'parrotfish', 'scaridae', 'reserve', 'abundance', 'biomass', 'size' and 'species richness' in Google Scholar. Relevant references in studies obtained from the literature search were also selected using the 'snowball sampling' technique. That is, a systematic examination was made of the literature cited in the papers from the literature search.

Many studies of NTMR effects simply compare fish variables inside and outside reserves at a single-point-in-time; our literature search identified fourteen such studies that specifically did this for parrotfish (Table 1). Density, biomass, and body size of parrotfishes was higher inside NTMRs for most studies, consistent with a very positive direct response

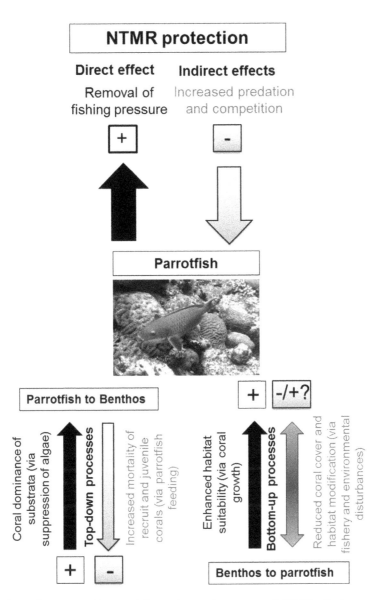

Fig. 1. Summary of potential no-take marine reserve (NTMR) effects on parrotfish and subsequent parrotfish interactions with the benthos.

of parrotfish to NTMR protection. In contrast, biomass of herbivorous coral reef fish, including parrotfish, did not show strong relationships with fishing pressure on some reefs in Fiji (Jennings and Polunin 1996, 1997) and the Seychelles (Jennings et al. 1995, 1996) measured at one point in time. Single-point-in-time studies of NTMRs have been criticized since differences in fish biomass between fished and NTMR sites can be due to a number of factors that may have existed prior to reserve establishment (e.g. differences in habitat, history and larval supply, biased choice of NTMR or control site, variable NTMR compliance; Russ 1985, 2002, Jones et al. 1992, Edgar et al. 2004, Osenberg et al. 2006, Miller and Russ 2014). Furthermore, inferring recovery rates inside NTMRs from such studies is problematic (Russ et al. 2005, Abesamis et al. 2014).

Table 1. Single-point-in-time spatial comparisons of parrotfish density, biomass and body size in NTMR and fished sites (+ = higher in NTMR relative to fished area; - = lower in NTMR relative to fished area, 0 = no difference).

Location	Years of NTMR protection	Biomass	Density	Body size	Authors
Caribbean and Florida Keys					
Exuma Cays, Bahamas	46 (19 enforced)	+ (2-fold)	+ (2-fold)	-(small bodied species<23cm)	Mumby et al. 2006
Saba	4	+	+		Polunin and Roberts 1993
Belize	4	+ (for 1 out of 2 species)	+ (for 1 out of 5 species)	+ (for 1 out of 2 species)	Polunin and Roberts 1993
U.S. Virgin Islands	3 (Monument)	- (43%)			Monaco et al. 2007
Anse Chastanet, Saint Lucia	2	+			Roberts and Hawkins 1997
Jamaica, Belize, Grand Cayman	5-12	0			Williams and Polunin 2001
Upper and lower Florida Keys	4 and 6		+ (large-bodied reef-associated species in lower Keys)		Valentine et al. 2007
Florida Keys	7		+ (adults) - (juveniles)		Kramer & Heck 2007
Indian Ocean					
Kenya	18-23		+ ~ 10 times (15 spp.)		McClanahan 1994
Kenya	28		- (for all 3 species)		McClanahan et al. 2010
East Africa	less than 10		+		McClanahan and Arthur 2001
South America					
Abrohlos reefs, Brazil	19 (16 enforced)		-	+	Floeter et al. 2006
Pacific Ocean					
Western Solomon Islands	2		+	+	Aswani and Sabetian 2010
Fiji	9-10	+ (herbivorous fishes including parrotfish: 7-17 fold)			Rasher et al. 2013

Meta-analyses provided broader inferences of NTMR effects, even when the majority of studies analysed were single-point-in-time or short-term (Cote et al. 2001, Halpern and Warner 2002, Halpern 2003, Lester et al. 2009, Maliao et al. 2009, Molloy et al. 2009, Graham et al. 2011, Abesamis et al. 2014, Edgar et al. 2014). However, few of these meta-analyses provided specific information on parrotfish. Both Halpern and Warner (2002) and Maliao et al. (2009) found greater density of herbivores inside NTMRs than in fished controls and Halpern and Warner (2002) also found greater biomass of herbivores in NTMRs, suggesting a positive direct response to NTMR protection. The only meta-analysis that presented parrotfish-specific data indicated that parrotfish had a very positive direct response to NTMR protection across the 12 studies (10-20 fold higher density in NTMR than fished sites; Mosquera et al. 2000) (Fig. 2). Meta-analyses thus also suggested a positive direct response of parrotfish to NTMR protection, but such conclusions are still subject to limitations of site choice of NTMRs and controls (Edgar et al. 2004), habitat heterogeneity among NTMR and control sites, lack of 'before' NTMR data (Osenberg et al. 2006, Miller and Russ 2014), and often a general lack of information about the nature of the surrounding fishery (intensity, targets) and the level of compliance to NTMR protection. Publication bias also means that NTMR 'success stories' are more often published than non-significant NTMR results (Huntington 2011). Note also that data quality can be problematic in meta-analyses, with variations in management effectiveness or confounding ecological factors common (Hedges and Olkin 1985, Edgar et al. 2004, Osenberg et al. 2006, Abesamis et al. 2014, Miller and Russ 2014).

A third approach, 'space-for-time' substitution, examines single-point-in-time data inside and outside NTMRs of different age (e.g., Halpern and Warner 2002, Stockwell et al. 2009). Alternatively, the approach can combine 'space-for-time' substitution with temporal monitoring inside and outside NTMRs of different age (e.g., McClanahan and Graham 2005, McClanahan et al. 2007, Humphries et al. 2014, McClanahan 2014). Such approaches effectively assume that reserve age is far more important in determining size structure, density, biomass and assemblage structure of fish than any inherent differences in such things as general environmental conditions, habitat, history and larval supply among the NTMRs, and that NTMR compliance is equally effective among locations. Space-for-time substitutions documenting recovery of parrotfish inside NTMRs have been made in Kenya (Indian Ocean) and in the Philippines (Pacific Ocean) (Fig. 3). A study of 15 inshore (adjacent to mainland islands) Philippine NTMRs suggested a rapid increase of parrotfish biomass (21 species combined) during the first five years of NTMR protection, followed by a slight

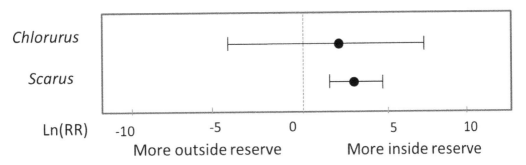

Fig. 2. Response ratio (RR= Ratio of NTMR: Fished density for *Chlorurus* and *Scarus* parrotfish (four and 20 species, respectively) with bootstrap-generated confidence interval (Redrawn from Mosquera et al. 2000)

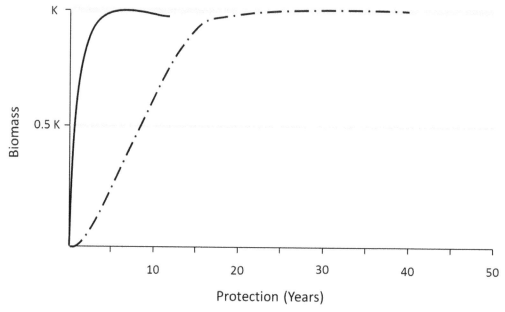

Fig. 3. Recovery trends for parrotfish biomass inside Philippine (black line) and Kenyan (dashed line) NTMRs. Lines indicate general trends of recovery based on models fitted to field data from the original studies. Original data rescaled according to probable local carrying capacity (K) [Redrawn from Abesamis et al. (2014)].

decrease in biomass in older NTMRs (Stockwell et al. 2009; Fig. 4). Parrotfish biomass inside four Kenyan NTMRs continued to increase for 40 years, but parrotfish density peaked at between 18 and 28 years and then declined (McClanahan et al. 2007, Babcock et al. 2010, McClanahan 2014). However, both Kenyan and Philippine studies suggested relatively rapid initial biomass recovery (within ~ 5-10 years) (Fig. 3). Interestingly in the Philippine example, parrotfish biomass asymptoted after 11 years of NTMR protection, whereas surgeonfish biomass increased exponentially over this time-scale (Stockwell et al. 2009; Fig. 4).

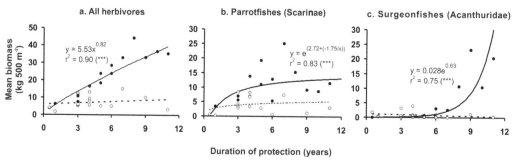

Fig. 4. Mean biomass of (a) all herbivorous fishes (b) parrotfishes and (c) surgeonfishes against duration of protection on inshore Philippine coral reefs (black dots: NTMRs sites; white dots: fished sites). Solid and broken lines represent trends in NTMRs and fished sites, respectively (Stockwell et al. 2009). Duration of protection refers to the NTMRs only. The data for fished sites were sampled at the same time as the NTMRs were sampled and thus are, for the most-part, 'paired'.

Fourthly, BACIP (Before-After-Control-Impact-Pairs) designs combined with long-term monitoring are viewed as the most effective approach to assessing rates and patterns of recovery of exploited species following fishery closure (Jones et al. 1992, Russ 2002, Abesamis et al. 2014). Wantiez et al. (1997) used two estimates of abundance, one before and one five years after NTMR establishment in New Caledonia, and showed that density, biomass and species richness of parrotfish increased significantly in the NTMR relative to the fished control. Hawkins et al. (2006) in a seven year BACIP study in St. Lucia, Caribbean reported a strong NTMR effect on parrotfish biomass at one NTMR and a more rapid increase in biomass inside than outside NTMRs for three other NTMRs. O'Farrell et al. (2015) reported a logistic recovery of parrotfish biomass, asymptoting at seven to nine years, following a ban on trap fishing in Bermuda. In contrast, Huntington et al. (2011) estimated parrotfish abundance at two points in time after NTMR establishment (1998–1999 and 2008–2009) in Belize and found that density of parrotfish had declined both inside and outside of the NTMR. Samoilys et al. (2007) did not detect clear NTMR effects on density of parrotfish in a six-year BACIP study of five inshore NTMRs in the Philippines. In a decade long BACIP study of NTMRs in the Florida Keys National Marine Sanctuary (FKNMS), Bohnsack et al. (2009) and Smith et al. (2011) showed that densities of two species of parrotfish (*Scarus iseri* and *Sparisoma viride*), which were not targeted by fishing in this area, showed no significant differences in density between NTMR and fished sites. This latter study included annual monitoring from 1998-2007 and provided baseline (pre-NTMR) data collected from 1994-1997. Long-term (decade or more) and spatially extensive BACIP-type studies have also demonstrated no significant effects of NTMRs on parrotfish density on Australia's Great Barrier Reef (GBR), where fisheries do not target parrotfish (Williamson et al. 2014, Emslie et al. 2015).

Russ and Alcala (1998a, b), in a BACIP-type study of two offshore Philippine NTMRs over a decade, reported that the direct effects of NTMR protection on density and species richness of parrotfish were weak, despite the fact that they are targeted by fishing, although they did record a distinct drop in species richness of parrotfish when fishing resumed in Sumilon Reserve in 1991. In a longer study of the effects of NTMR protection on a parrotfish at Sumilon and Apo Islands, Philippines, Babcock et al. (2010) reported that density of *Scarus tricolor* increased inside the Apo NTMR relative to the fished control after 8-12 years of NTMR protection but then decreased thereafter to be not significant after up to 26 years of protection. In contrast, density of *S. tricolor* increased rapidly inside the Sumilon NTMR relative to the control in the first 5 years of protection, then asymptoted and remained high in the NTMR for up to 15 years of protection (Babcock et al. 2010). This rapid increase in the first five years of NTMR protection on an offshore island was similar to the results reported by Stockwell et al. (2009) using space-for-time substitution on inshore Philippine coral reefs (Fig. 4). More recently, Russ et al. (2015) showed that NTMR protection could increase parrotfish density by factors of 2-3 at three Philippine islands over periods of protection up to 8-14 years.

Overall, single-point-in-time studies, meta-analyses and studies using space-for-time substitutions suggest a positive direct response of parrotfish density and biomass to NTMR protection where fisheries target parrotfish, and NTMR compliance is presumably high. However, the most convincing direct NTMR effects on parrotfish come from BACIP studies with long-term monitoring in places where fishing targets parrotfish (e.g., The Caribbean, Kenya, the Philippines). It is worthy of note that long-term BACIP-type studies in places where parrotfish are not targeted by fishing (e.g., Australia's GBR, Florida Keys) fail to demonstrate direct NTMR effects on density and biomass.

Indirect Effects

Increasing Abundance of Parrotfish Predators in NTMRs

The build-up of predators inside NTMRs is hypothesized to cause a decline in the abundance of their prey (McClanahan 1994, Graham et al. 2003, Micheli et al. 2004, Babcock et al. 2010). On coral reefs, increased predation by large piscivores inside NTMRs is believed to negatively impact herbivorous fish populations (McClanahan et al. 2007, 2011a, b, Mumby et al. 2006, 2007, Graham et al. 2003, Boaden and Kingsford 2015). Such indirect effects were observed within the Exuma Cays Land and Sea Park (ECLSP) in the Caribbean where size and grazing rates of small-bodied parrotfish were inferred to have declined due to increased predation by piscivores, particularly the grouper, *Epinephelus striatus* (Mumby et al. 2006). Nevertheless, the study also suggested that large-bodied parrotfish were able to escape predation by exceeding the predator's gape size and, as a result, parrotfish biomass was twice as high inside the ECLSP than in fished areas. Reduction in density of parrotfish in NTMRs relative to fished areas due to higher levels of predation by piscivores was also suggested on Australia's GBR (Graham et al. 2003). Biomass of the predator *Plectropomus leopardus* (Serranidae) was 2-3 times higher inside NTMRs and the density of six prey species (including one parrotfish species) were significantly lower inside NTMRs relative to fished areas of inshore islands of the GBR (Graham et al. 2003). Boaden and Kingsford (2015) reported lower densities of piscivores, and higher densities of herbivorous fish (including parrotfish), in fished areas than NTMRs on the GBR. In contrast to these studies, Rizzari et al. (2015) reported that shark abundance and meso-predator biomass had no discernible effect on density and biomass of different functional groups of herbivorous fish (including parrotfish) in different no-fishing zones of the GBR. Similarly, Soler et al. (2015) showed that the biomass of all trophic groups increased significantly in effective no-take marine protected areas worldwide (n = 79 MPAs) relative to fished control sites. Soler et al. (2015) concluded that the direct effects of fishing on trophic structure appeared stronger than any top-down effects on lower trophic levels that would be imposed by intact predator populations. Note that all of these studies were single-point-in-time comparisons of fished and NTMR sites.

Babcock et al. (2010) argued that the recovery of large predatory reef fishes inside Apo reserve, Philippines possibly caused the lack of recovery of the parrotfish *Sc. tricolor* in the reserve. The authors also suggested that because predators of parrotfish typically take a long time to grow and often target juvenile parrotfish, a time lag may be expected before any effect on parrotfish populations is detected. This was also suggested by McClanahan et al. (2007) in a space-for-time substitution study that incorporated long-term monitoring, where parrotfish populations declined within Kenyan NTMRs after decades of protection, possibly due to increased predation over time on juvenile parrotfish. Alternatively, declines in parrotfish abundance with time in well-enforced NTMRs could be related to reductions in food availability. If coral cover increases over time in reserves there should be consequent decreases in cover of dead substrata covered with algal turfs. The evidence that NTMRs indirectly affect parrotfish populations through restoration of parrotfish predators remains limited.

Increasing Abundance of Parrotfish Competitors

If competitors of parrotfish are the target of reef fishing, their density and biomass may increase over time in NTMRs, thus potentially having a detrimental effect on parrotfish density and biomass. This review could find no convincing evidence for such a process, despite suggestions that it may occur (Babcock et al. 2010). Extensive studies on Kenyan coral reefs have shown that NTMRs can increase density of predators of sea urchins (triggerfish), resulting in higher predation rates on urchins in NTMRs (e.g. McClanahan 1994, 2014, McClanahan and Shafir 1990, McClanahan et al. 2011a). It has also been suggested that this process mediates competition between herbivorous urchins and herbivorous fish like parrotfish (McClanahan 1994, 2014, McClanahan and Shafir 1990, Bellwood et al. 2004).

Effects of Parrotfish on Benthos and Benthos on Parrotfish

Effects of Parrotfish on the Benthos

Parrotfish Feeding Suppresses Algae, Facilitating Coral Dominance of Benthic Substrata

Phase-shifts from coral- to macroalgal-dominance have largely been attributed to, or have been underpinned by, fishing-induced reductions in herbivores (e.g. Hughes 1994, Bellwood et al. 2004, Mumby et al. 2006, Jackson et al. 2014). Thus, NTMRs are hypothesized to prevent, or reverse, shifts to undesirable macroalgal-dominated states by maintaining natural densities of herbivorous fish and natural grazing rates (Mumby et al. 2006, 2007, Hughes et al. 2007b, 2010, Mumby and Harborne 2010, Jackson et al. 2014). Such effects typically involve the recovery of depleted herbivores such as parrotfish (McClanahan et al. 2007, 2011a, Rasher et al. 2013, McClanahan 2014, Adam et al. 2015), which in turn leads to an increase in grazing intensity on turf and fleshy algae (Mumby and Steneck 2008, Humphries et al. 2014). Macroalgae are known to harm coral populations by inhibiting coral recruitment (Hughes et al 2007a, Diaz-Pulido et al. 2010) or by overgrowing and out-competing corals for space (McCook et al. 2001, Jompa and McCook 2003, Nugues and Bak 2006). MacNeil et al. (2015) estimated recovery rates from fishing of reef fish on 800 coral reefs globally, arranged along a 'space-for-time' exploitation gradient. Their analysis suggested that returns of functions like browsing, scraping/excavating and grazing were relatively rapid, and that key herbivore functions could be fulfilled at intermediate, as opposed to pristine, levels of biomass of fish like parrotfish. Preventing proliferation of macroalgae on coral reefs should benefit coral communities, and it is often suggested that NTMRs could play a key role in such prevention (Fig. 1).

Studies providing clear evidence of herbivores reducing macroalgal abundance over time inside NTMRs on coral reefs are surprisingly rare (e.g., Mumby et al. 2006, McClanahan et al. 2007, Stockwell et al. 2009, Babcock et al. 2010, Mumby and Harborne 2010). Even fewer studies have looked specifically at the effects of parrotfish assemblages inside and outside NTMRs on coral reef algae over time (Adam et al. 2015). Hughes et al. (2007a), in an experimental study at Orpheus Island, GBR showed that reducing grazing pressure with fish exclusion cages resulted in a dramatic increase in macroalgal cover. Mumby et al. (2006) attributed four-times less macroalgal cover inside the ECSLP in the Bahamas to higher grazing pressure by large parrotfish. However, Harborne et al. (2008) showed that indirect effects of the ECSLP on macroalgae varied by habitat. Stockwell et al. (2009) studied 15 inshore Philippine NTMRs and inferred a significant herbivore-macroalgal interaction that was modified by NTMR protection (Fig. 5). They found that biomass of herbivorous fish was 15 times higher, and macroalgal cover 13 times less in the oldest (11 yr) reserve, relative to fished areas. Stockwell et al. (2009) also suggested that this pattern may have been driven mainly by parrotfishes, as their recovery was more rapid (within five yr of NTMR protection) than other grazers (surgeonfishes) (Fig. 4). However, Stockwell et al. (2009) stressed that coral cover did not differ significantly between NTMR and fished sites, macroalgal cover was still generally low, even in fished sites, and that

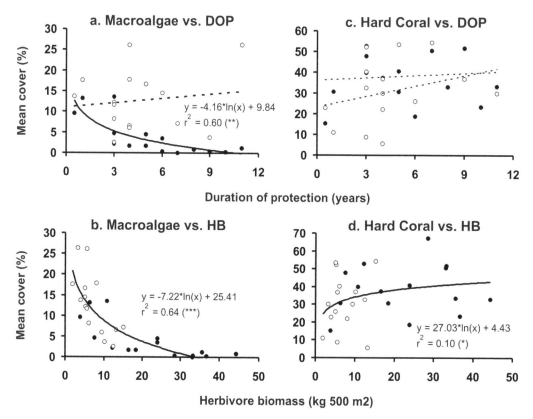

Fig. 5. Mean percent cover of macroalgae (a,b) and hard coral (c,d) versus either duration of protection (DOP) or biomass of herbivorous fish (HB) (black dots: NTMRs sites; white dots: fished sites) (Stockwell et al. 2009).

there was no evidence of 'benthic phase shifts' (i.e., shifts in the dominant form of benthic cover) between NTMRs and fished sites in their study.

Williams and Polunin (2001) examined NTMRs and fished sites in the Caribbean, and reported a negative correlation between biomass of herbivorous fish (parrotfish, surgeonfish) and macroalgal cover. Kopp et al. (2010) reported that macroalgal cover was much lower in old NTMRs in Guadeloupe and attributed this to the grazing activities of four species of herbivorous fish (three surgeonfishes and one parrotfish) that had greater biomass in NTMRs. Similar relationships have been reported for small NTMRs in Fiji (Rasher et al. 2013, Bonaldo et al. 2017). Rasher et al. (2013) reported the biomass of herbivorous fish was 7-17 times higher, and macroalgal cover 27–61 times lower, in NTMRs than fished sites. Four species of herbivorous fish (one of which was a parrotfish) were responsible for 97 percent of macroalgal consumption in the NTMRs (Rasher et al. 2013). However, all of these studies are single-point-in-time comparisons (Mumby et al. 2006, Stockwell et al. 2009, Kopp et al. 2010, Rasher et al. 2013) and thus they can only infer indirect effects of NTMR protection. In contrast, a 14-year study in Belize reported very weak relationships between herbivore abundance and macroalgal cover inside NTMRs (McClanahan et al. 2011b).

An extensive and long-term study of NTMRs in Kenya using space-for-time substitutions combined with long-term (decadal scale) monitoring (e.g., McClanahan and

Graham 2005, McClanahan et al. 2007, Babcock et al. 2010, McClanahan 2014, Humphries et al. 2014) has demonstrated increases in biomass of herbivorous fish (including parrotfish) and decreases in abundance of herbivorous urchins (due to increases in urchin predators, which are fished) over almost 40 years of NTMR protection. These studies have also reported maintenance of low cover of fleshy erect algae inside NTMRs relative to fished areas (Babcock et al. 2010) and that cover of fleshy algae did not change with duration of NTMR closure up to 40 years (McClanahan 2014). A very clear result of NTMR protection over time was an increase in cover of crustose coralline algae (CCA) inside NTMRs relative to fished areas (Babcock et al. 2010, McClanahan 2014), where grazing rates of herbivorous fish are higher in NTMRs than in fished areas (Humphries et al. 2014). Humphries et al. (2014) suggested a more complex effect of duration of NTMR closure on cover of macroalgae. In fished areas, high densities of urchins, and in old (20-40 year) government-managed NTMRs, high biomass of herbivorous fish, are each suggested to keep algal assemblages in early successional stages dominated by turfs (and crustose coralline algae in old NTMRs). In community-managed NTMRs of intermediate age (<10 years), a herbivore assemblage of urchins and small fish allow macroalgae to quickly develop from turf into early and then late successional algal stages. These reefs in intermediate aged NTMRs are suggested to represent an intermediate or transitional system of herbivore dominance, with benthos subsequently characterized by macroalgae. Note also that duration of protection of Kenyan NTMRs had no clear effect on coral cover (Babcock et al. 2010, Humphries et al. 2014, McClanahan 2014).

Several studies have reported positive relationships between abundance of CCA and coral recruitment inside NTMRs and related this to the presence or absence of herbivorous fishes (Sammarco 1985, Hughes et al. 2007a, Mumby et al. 2007, O'Leary and McClanahan 2010, Mumby and Harborne 2010, Smith et al. 2010, Bonaldo et al. 2017). An experimental study in a GBR NTMR was able to demonstrate a positive effect of grazing by herbivorous fish on coral recruitment by manipulating density of herbivorous fishes with cages (Hughes et al. 2007a). However, the direct effect of grazing by parrotfish was confounded by the fact that all large fish were excluded by cages. Mumby et al. (2007) observed a strong negative correlation between macroalgal cover (itself controlled by parrotfish grazing) and the density of coral recruits within the Exuma Cays SLP. This provides some of the best evidence that parrotfish assemblages protected inside NTMRs can indirectly enhance coral recruitment. A more recent study at Exuma Cays SLP showed that corals were able to recover faster inside the NTMR as a result of parrotfish reducing macroalgae inside the NTMR (Mumby and Harbone 2010), but this study spanned only 2.5 years. In contrast, a study of Philippine NTMRs on inshore reefs showed no significant indirect effects of grazer biomass on coral cover (Stockwell et al. 2009) (Fig. 5). Coral cover and cover of fleshy algae also did not increase with duration of NTMR protection on Kenyan coral reefs (Babcock et al. 2010, McClanahan 2014). Several recent studies have questioned the ability of NTMRs to maintain or possibly increase coral cover in the Caribbean (Coelho and Manfrino 2007, Kramer and Heck 2007, Huntington et al. 2011), the Florida Keys (Toth et al. 2014) and in New Caledonia (Carassou et al. 2013), especially in the face of environmental disturbances.

A useful area of future research is how fishing, and protection from fishing in NTMRs, affects the differential functional impacts of parrotfish. Fishing may have different effects on parrotfishes that perform different functional roles (Bellwood et al. 2012, Heenan et al. 2016). NTMRs embedded in fisheries that have greater direct impact on browsing, as opposed to scraping and excavating species of parrotfish (or herbivorous fish in general), may play a greater role in preventing, or reversing, phase-shifts from coral to macroalgal dominance. For example, there are clear differences in taxonomic and functional composition of parrotfish on coral reefs of the Caribbean and Indo-Pacific (Bonaldo et al. 2014). Caribbean taxa are dominated by the 'seagrass' or *Sparisoma* clade (Sparisomatine parrotfishes), with relatively higher proportions of browsers that feed on seagrass and macroalgae, while Indo-Pacific taxa are dominated by the 'reef' or *Scarus* clade (Scarinine parrotfishes), with relatively higher proportions of grazers and excavators (Bonaldo et al. 2014). This suggests that NTMRs in the Caribbean may be more effective in preventing, or reversing, coral-macroalgal phase-shifts than NTMRs in the Indo-Pacific. On a smaller scale, NTMRs may be particularly effective at preventing or reversing coral-macroalgal phase shifts on the coral reefs of Lord Howe Island, where assemblages of large herbivorous fish are composed of 84 percent macroalgal browsing species (Hoey et al. 2011).

Evidence for strong herbivore-macroalgal-coral interactions inside NTMRs over time is rare and somewhat equivocal. Bascompte et al. (2005) studied Caribbean reefs and suggested that strong interactions between adjacent trophic levels were uncommon, even when fishing pressure was reduced. Similarly, Carassou et al. (2013) questioned if protection of herbivores inside NTMRs in New Caledonia had an overall positive effect on coral reef condition. Furthermore, a review of coral-macroalgal phase-shifts found little evidence of wide-spread benthic phase shifts that could be related unequivocally to overfishing of herbivorous fish, and reported weak negative relationships between coral and macroalgal cover (Bruno et al. 2009). Evidence that coral survival and recruitment is enhanced inside NTMRs via algal removal following the recovery of parrotfish assemblages remains relatively scarce and comes only from single-point-in-time or short-time monitoring studies (e.g., Mumby et al. 2006, Mumby and Harborne 2010).

Effects of Parrotfish on Recently-settled and Juvenile Corals

Actual or incidental predation by grazing fishes can be an important source of mortality for recently-settled and juvenile corals (Sammarco 1985, Christiansen et al. 2009, Mumby 2009, Penin et al. 2010, Trapon et al. 2013a, b). The relationship between survivorship of newly settled or juvenile corals and the grazing activities of fishes is generally believed to be positive since the grazing reduces coral-overgrowth by algae (Sammarco 1985, Mumby et al. 2006). However, the association between increased levels of incidental mortality of young corals and the greater abundance of grazers observed by Penin et al. (2010) on French Polynesian reefs suggests that the grazing fish-coral relationship may be more complex.

Parrotfishes, particularly scrapers and excavators, have the ability to remove portions of the underlying carbonate substratum when feeding and are therefore presumed the most likely source of incidental mortality of young corals (Sammarco 1985, Bellwood et al. 2003, Trapon et al. 2013a, 2013b, Bonaldo et al. 2014). Thus, if NTMR protection increases density of parrotfish, a higher mortality of coral recruits and juveniles may result (Fig. 1). A recent study showed that mortality of coral recruits was strongly associated with the abundance of grazing parrotfish and that incidental predation on early post-settlement corals by parrotfish was the likely mechanism causing this higher coral mortality (Penin et al. 2010). Another study found that the density of juvenile corals was negatively correlated with the biomass of parrotfish (grazers and scrapers) across nine reefs spanning over 1000 km on the GBR, suggesting that high biomass of these fishes could also affect corals during their juvenile stage (Trapon et al. 2013a). Conversely, in the Exuma Cays Park, , increased grazing activity was negatively correlated to macroalgal cover and caused a two-fold increase in the density of juvenile corals, suggesting parrotfish have a positive effect on coral recruitment (Mumby et al. 2007). Hoey et al. (2011) also reported a positive relationship between biomass of grazing fish and density of juvenile corals at Lord Howe Island. However, with the exception of Mumby et al. (2007), studies investigating the relationship between parrotfish abundance and survivorship of juvenile corals inside NTMRs are rare.

Some large-bodied parrotfish eat corals directly as opposed to incidentally (Trapon et al. 2013a, Bonaldo et al. 2014, Bonaldo and Rotjan Chapter 9). These include the bumphead parrotfish *Bolbometopon muricatum,* which frequently feeds on living corals (Bellwood et al. 2003, 2012, Hoey and Bellwood 2008). McClanahan et al. (2005) transplanted coral fragments inside and outside of Kenyan NTMRs and found that parrotfish were amongst the dominant coral predators, negatively impacting on adult corals in NTMRs. This pattern did not apply to coral recruits. Grazing by corallivorous parrotfish was also found to negatively affect the fitness and recovery of adult corals following a bleaching event in Belize (Rotjan et al. 2006). A review by Mumby (2009) concluded that corallivory by parrotfish was not offsetting the benefits of parrotfish removing macroalgae and enhancing coral recruitment in the Caribbean. Nevertheless, the possibility that higher density and biomass of parrotfish inside NTMRs may directly affect coral recruits and juveniles and subsequently affect adult coral populations has not been fully explored despite recent evidence that grazing parrotfish may play a more complex role in shaping coral populations (Penin et al. 2010, Trapon et al. 2013a, b, Bonaldo et al. 2014).

Effects of Benthos on Parrotfish

Enhanced Structural Complexity of the Benthos Providing Refuges from Predation for Parrotfish

NTMRs have been hypothesised to benefit coral communities, directly by reducing fishing techniques destructive to corals, or indirectly by maintaining herbivory (Selig and Bruno 2010). Graham et al. (2011) reviewed this subject and concluded that coral cover increased inside NTMRs relative to fished areas, but the magnitude of this effect was influenced by the history of fishing (e.g., presence of destructive fishing techniques, whether herbivores were heavily targeted) and the frequency and intensity of environmental disturbances (e.g., storms, coral bleaching). Few studies have considered the possibility that recovery of corals in NTMRs may affect parrotfish density and biomass in a positive manner (Fig.1).

Bozec et al. (2013) showed that increased structural complexity of the benthos (via coral growth) in an NTMR in Belize increased parrotfish density. Furthermore, overall grazing intensity was also enhanced in the NTMR, despite the reduced surface area available for parrotfish grazing due to the increases in coral cover. Tzadik and Appeldorn (2013) showed that reef structure was the most important variable accounting for parrotfish density in Puerto Rico. Structurally complex (rugose) habitats that have high coral cover may provide shelter from predators and provide night-time resting sites (Shephard 1994). Several studies have reported strong positive links between parrotfish density and benthic habitat characteristics (Russ 1984, Sluka and Miller 2001, Mellin et al. 2007, Taylor et al. 2014b, Heenan et al. 2016). In a recent study of the parrotfish assemblage across eight Micronesian islands, Taylor et al. (2014b) concluded that, despite widespread fishing pressure, the structure of parrotfish assemblages at broad spatial scales was largely driven by benthic habitat, not fishing. Other studies have suggested that parrotfishes are benthic habitat generalists, lacking strict associations with particular types of benthos (Gust 2002, Jones et al. 2004, Pinca et al. 2012).

Environmental Disturbances that Reduce Coral Cover may Increase or Decrease Availability of Habitat Suitable for Parrotfish Feeding

Environmental disturbances are important processes structuring coral reef communities (Connell 1978, Syms and Jones 2000, Jones et al. 2004). Physical disturbances to coral reef benthos are generally not related to fishing (e.g., storms, crown-of-thorns outbreaks, coral bleaching, coral diseases), although they can be if fishing techniques like explosives and drive-nets reduce coral cover (e.g., Russ and Alcala 1989). NTMRs have been promoted as a potential solution to enhance the recovery of benthic communities following environmental disturbances (Mumby et al. 2006, Hughes et al. 2007b, 2010, Selig and Bruno 2010, Wilson et al. 2012). The rationale for this is that removal of extractive activities by humans may result in a more natural coral reef ecosystem which may be more resistant to environmental disturbances, or be able to recover more quickly from such disturbances. Environmental disturbances have often been associated with substantial declines in coral cover and structural complexity (Halford et al. 2004, Myers and Ambrose 2009, Wilson et al. 2006, De'ath et al. 2012). Many studies report NTMRs do not prevent the effects of major environmental disturbances such as coral bleaching, COTS outbreaks and typhoons (e.g., Jones et al. 2004, Coelho and Manfrino 2007, Myers and Ambrose 2009, Graham et al. 2011, Huntington et al. 2011, Carassou et al. 2013, Toth et al. 2014, Emslie et al. 2015, Hughes et al. 2017) and there is increasing evidence that these disturbances to reef structure can also alter the abundance of coral reef-associated fish (Jones et al. 2004, Pratchett et al. 2008, Wilson et al. 2008). For example, Halford et al. (2004) showed that the 1998 mass coral bleaching event was responsible for 75 percent of the decline in species richness of parrotfishes in the Capricorn Bunker Group on the GBR. Furthermore, a decline in the density of parrotfish was also observed, with 24 percent of the variation in abundance explained by temporal trends in the benthos. However, Halford et al. (2004) pointed out that the separate effects of habitat complexity and coral cover *per se* on fish abundance could not be partitioned effectively (see also Emslie et al. 2014).

Although the difference is generally small, coral cover is often hypothesized to be higher inside than outside NTMRs (Selig and Bruno 2010; Graham et al. 2011). If parrotfish are more abundant in NTMRs, then habitat loss via environmental disturbance may have a greater impact on parrotfish assemblages in NTMRs (Fig. 1). However, very few studies suggest that this is the case. For instance, a study in the Seychelles showed almost equal declines in smaller sized herbivorous fish (mainly parrotfishes and surgeonfishes) both inside and outside NTMRs following coral bleaching and a loss of structural complexity (Graham et al. 2007). The authors attributed habitat loss as the main factor responsible for these declines.

Williamson et al. (2014) showed that a freshwater flood plume led to a decrease in density of grazers, including parrotfish, in the Keppel Islands, GBR, but the opposite effect was noted for a coral bleaching event. A meta-analysis by Wilson et al. (2006) showed that parrotfish density and biomass tended to increase as a result of coral loss caused by environmental disturbances to the benthos. They suggested that environmental disturbances affecting the benthos may affect parrotfish density and biomass positively, due to an increased availability of dead surfaces on which their algal and detrital food is abundant. Furthermore, Russ and Alcala (1989) showed that, when Sumilon NTMR was subjected to destructive fishing practices in 1984, parrotfish increased dramatically in density. The authors suggested that this increase was most likely due to these fish being driven off the reef flat (where parrotfish had been abundant previously) onto the reef slope where surveys were conducted. However, they also suggested that the increase in parrotfish density was possibly related to the loss of coral cover and an increase in cover of rubble and dead substratum on which parrotfish like to feed. In a subsequent study spanning a decade, Russ and Alcala (1998a, b) reported that the pulse fishing event at Sumilon reserve in 1984 reduced coral cover and increased rubble, resulting in increased density and species richness of parrotfish over the next eight years.

The study of parrotfish density in relation to NTMR protection and benthic habitat at Sumilon Island has now extended to 30 years (Russ et al. 2015). Figure 6 provides an example of the results for the excavating parrotfish *Chlorurus spilurus*. There is some evidence that NTMR protection over 19 years maintained density of this parrotfish inside the NTMR as density declined in the fished area (Fig. 6A). There was a far stronger effect on density of *Ch. spilurus* caused by environmental disturbances that affected the benthos in Sumilon Reserve over 30 years (1983-2013) of regular monitoring (Fig. 6B). The percentage cover of dead substratum, a proxy for food availability, as this is a favoured feeding substratum of this species, increased sharply following the use of explosives and drive nets in Sumilon Reserve in 1984. This resulted in a sharp and rapid increase in density of *Ch. spilurus* (Fig. 6B). Coral recovered from 1988 to 1997, and this led to a decline in both cover of dead substratum and density of *Ch. spilurus* (Fig. 6B). Coral cover again declined sharply in 1998 due to coral bleaching and a COTS outbreak, resulting in an increase in cover of dead substratum and a concomitant increase in density of *Ch. spilurus* from 1998 to 2008 (Fig. 6B). As coral recovered again (2009-2013), cover of dead substratum declined slightly, associated with a slight decline in density of the parrotfish (Fig. 6B). This concordance of parrotfish density with cover of dead substratum provides a very clear, long-term natural experiment demonstrating how the density of the parrotfish, although affected by the top-down process of fishing (Fig. 6A), is driven strongly by bottom-up availability of benthic habitat suitable for feeding, with benthic habitat itself driven by environmental disturbances (Fig. 6B, Russ et al. 2015).

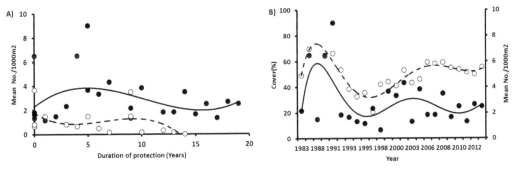

Fig. 6. A. Density of *Chlorurus spilurus* in Sumilon No-Take Marine Reserve (black data points and solid line) for up to 19 years of NTMR protection and in a nearby fished site (white data points, dashed line). B. Percentage cover of dead substratum (dead hard coral substratum, rubble and sand combined) (white data points, dashed line) and density of *Ch. spilurus* (black data points, solid line) in Sumilon NTMR for 30 years from 1983 to 2013. Dead substratum here is a preferred feeding substratum for this fish. Lines in A and B are polynomials fitted to emphasize temporal trends in parrotfish density or cover of dead substratum. See also Russ et al. (2015).

Conclusions and Suggestions for Future Research

While most studies reported a positive direct response of NTMR protection on parrotfish size structure, density and biomass, few provided unequivocal evidence of this. Space-for-Time substitutions, sometimes supplemented with long-term monitoring (e.g., McClanahan et al. 2007, Stockwell et al. 2009, McClanahan 2014) and BACIP studies combined with long-term monitoring (e.g., Hawkins et al. 2006, Babcock et al. 2010, Russ et al. 2015) best demonstrated the recovery of parrotfish inside NTMRs, particularly in terms of biomass. However, such studies were from a limited number of locations (Kenya, Philippines, Caribbean). It is worthy of note that long-term BACIP-type studies in places where parrotfish are not targeted by fishing (e.g., Australia's GBR, Florida Keys: Smith et al. 2011, Emslie et al. 2015) fail to demonstrate direct positive NTMR effects on parrotfish. A particular example where non-detection of NTMR effects on a species of parrotfish might be expected, is the case of the giant bumphead parrotfish *B. muricatum* (Dulvy and Polunin 2004). In this case the species is regionally so rare, likely related to high fishing pressure, that there may be little capacity for recovery of the species simply by implementing NTMRs. Clearly, future studies of NTMR effects on parrotfish should make more effort to document the nature of the surrounding fishery (target species, mortality rates, life history data, fishing gears etc.) and the degree of compliance to NTMR protection. Without detailed knowledge of the surrounding fishery and NTMR compliance, there will always be uncertainty associated with detection and non-detection of NTMR effects, both direct and indirect.

Studies indicating possible indirect effects of NTMRs on parrotfish assemblages were rare, probably because such indirect effects often take much longer to develop than direct effects (Babcock et al. 2010). Those studies that showed negative effects of predator restoration on parrotfish density or biomass (e.g., Mumby et al. 2006) were mostly single-point-in-time comparisons between fished and NTMR sites, and thus inferential. Only one long- term (several decades) study suggested that predation on juvenile parrotfish was potentially impeding their recovery (see Babcock et al. 2010). Furthermore, the evidence that NTMRs may enhance the abundance of parrotfish competitors is rare (McClanahan 2014). BACIP studies of NTMR effects combined with long-term monitoring are required to elucidate any such patterns and processes.

Evidence that NTMRs can modify the ways in which parrotfish (or herbivorous fish in general) exert top-down control on the benthos remains equivocal, coming mainly from single-point-in-time or short-term studies. The evidence for wide-spread benthic phase shifts on coral reefs caused by overfishing of parrotfish, comes mostly from the Caribbean, and seems far less common on Indo-Pacific coral reefs (Bruno et al. 2009, Roff and Mumby 2012, Bonaldo et al. 2014, Adam et al. 2015). Thus, it is perhaps not surprising that unequivocal demonstrations of the prevention or reversal of such benthic phase shifts in NTMRs (i.e., using spatially replicated BACIP studies monitored over decades) are non-existent. Given the very long history of coral reef research in the Caribbean (e.g. Jackson et al. 2014) and the level of debate on the significance of fishing-induced trophic cascades/benthic phase shifts and how NTMRs may prevent or reverse these (e.g. Hughes 1994, Aronson and Precht 2006, Mumby et al. 2006, 2007, Hughes et al. 2007b, 2010, Jackson et al. 2014, Toth et al. 2014), the paucity of well-designed, decadal-scale studies of Caribbean NTMRs is unfortunate. A very notable exception is the Florida Keys National Marine Sanctuary where decadal-scale monitoring of NTMR protection has indicated little effect on either parrotfish density (Bohnsack et al. 2009, Smith et al. 2011) nor coral cover (Toth et al. 2014).

Note also that the reversal of benthic phase shifts caused by recovery of parrotfish abundance in NTMRs, or by other means, may be problematic if parrotfish do not eat macroalgae (Choat et al. 2002, 2004, Clements et al. 2009, Hoey and Bellwood 2009, Adam et al. 2015, Clements and Choat Chapter 3), or avoid areas of abundant macroalgae (Hoey and

Bellwood 2011). Two rare, long-term studies of benthic phase-shifts from coral to macroalgal-dominance provide contrasting results. Such a phase-shift occurred independent of the abundance of scraping and excavating parrotfish on inshore coral reefs of the GBR, and showed no evidence of reversing (Cheal et al. 2010). Contrarily, recovery from algal to coral dominance in the Seychelles (following coral bleaching) was more likely to occur at locations where densities of herbivorous fish were high and nutrients low (Graham et al. 2015). Whether recovery of parrotfish abundance can reverse benthic phase shifts will depend substantially on what parrotfish eat. We agree with calls for greater understanding of the dietary composition of, and nutrient extraction by, marine herbivorous fishes and how these factors affect the impact of such fish on coral reefs (Choat et al. 2002, 2004, Crossman et al. 2005, Clements et al. 2009, Cheal et al. 2013, Clements and Choat Chapter 3).

When assessing parrotfish interactions with the benthos, and how NTMRs may modify these interactions, most studies have focused on top-down processes like the role of fishing in shifts from coral to macroalgal dominance (e.g., Hughes 1994, Bellwood et al. 2004, 2012, Mumby et al. 2006, McClanahan 2014). Few studies have considered bottom-up processes, with dead substratum considered a proxy for a feeding substratum for many species of parrotfish, despite emerging evidence that they may be important in shaping parrotfish assemblages (e.g. Bozec et al. 2013, Tzadik and Appeldorn 2013, Taylor et al. 2014b, Russ et al. 2015; Fig. 6). It is clear that future studies should adopt a more bi-directional approach when assessing parrotfish interactions with the benthos inside and outside NTMRs. Furthermore, it is still unclear whether environmental disturbances, resulting in a loss of coral cover and/or structural complexity of the substratum, increase or decrease local parrotfish density and biomass, if such effects are short-term or long-term, and how NTMR protection may affect responses of parrotfish. More long-term monitoring studies that intercept a range of environmental disturbances both inside NTMR and at fished sites could possibly help resolve this issue.

References Cited

Abesamis, R.A., A.L. Green, G.R. Russ and C.R.L. Jadloc. 2014. The intrinsic vulnerability to fishing of coral reef fishes and their differential recovery in fishery closures. Rev. Fish. Biol Fisheries 24: 1033–1063.

Adam, T.C., D.E. Burkepile, B.I. Ruttenberg, M.J. Paddack. 2015. Herbivory and the resilience of Caribbean coral reefs: knowledge gaps and implications for management. Mar. Ecol. Prog. Ser. 520: 1–20.

Aronson, R.B. and W.F. Precht. 2006. Conservation, precaution, and Caribbean reefs. Coral Reefs 25: 441–450.

Aswani, S. and A. Sabetian. 2010. Implications of urbanization for artisanal parrotfish fisheries in the Western Solomon Islands. Conserv. Biol. 24: 520–530.

Babcock, R.C., S. Kelly, N.T Shears, J.W. Walker and T.J. Willis. 1999. Changes in community structure in temperate marine reserves. Mar. Ecol. Prog. Ser. 189: 125–134.

Babcock, R.C., N.T. Shears, A.C. Alcala, N.S. Barrett, G.J. Edgar, K.D. Lafferty, T.R. McClanahan and G.R. Russ. 2010. Decadal trends in marine reserves reveal differential rates of change in direct and indirect effects. Proc. Natl. Acad. Sci. USA.107: 18256–18261.

Bascompte, J., C.J. Melián and E. Sala. 2005. Interaction strength combinations and the overfishing of a marine food web. Proc. Natl. Acad. Sci. USA.102: 5443–5447.

Behrens, M.D. and K.D. Lafferty. 2004. Effects of marine reserves and urchin disease on southern California rocky reef communities. Mar. Ecol. Prog. Ser. 279: 129–139.

Bellwood, D.R. 2003. Origins and escalation of herbivory in fishes: a functional perspective. Paleobiology. 29: 71–83.

Bellwood D.R. and J.H. Choat. 1990. A functional analysis of grazing in parrotfishes (family Scaridae): the ecological implications. Environ. Biol. Fish. 28: 189–214.

Bellwood, D.R., A.S. Hoey and J.H. Choat. 2003. Limited functional redundancy in high diversity systems: resilience and ecosystem function on coral reefs. Ecol. Lett. 6: 281–285.

Bellwood, D.R., T.P. Hughes, C. Folke and M. Nyström. 2004. Confronting the coral reef crisis. Nature 429: 827–833.

Bellwood, D.R., A.S. Hoey and T.P. Hughes. 2012. Human activity selectively impacts the ecosystem roles of parrotfishes on coral reefs. Proc. R. Soc. B. 279: 1621–1629.

Bergseth, B.J., G.R. Russ and J.E. Cinner. 2015. Measuring and monitoring compliance in no-take marine reserves. Fish. Fish. 16: 240–258.

Boaden, A.E. and M.J. Kingsford. 2015. Predators drive community structure in coral reef fish assemblages. Ecosphere 6: 1–33.

Bohnsack, J.A., D.E. Harper, D.B. McClellan, G.T. Kellison, J.S. Ault, S.G. Smith and N. Zurcher. 2009. Coral reef fish response to FKNMS management zones: the first ten years (1997–2007). Progress report to the Florida Keys National Marine Sanctuary. PRBD 08/09-10. pp. 30.

Bonaldo, R.M., A.S. Hoey and D.R. Bellwood. 2014. The ecosystem roles of parrotfishes on tropical reefs. Oceanogr. Mar. Biol. Annu. Rev. 52: 81–132.

Bonaldo, R.M., M.M. Pires, P.R.G. Junior, A.S. Hoey and M.E. Hay. 2017. Small Marine Protected Areas in Fiji Provide Refuge for Reef Fish Assemblages, Feeding Groups, and Corals. PLoS ONE. 12: e0170638.

Bozec, Y.M., L. Yakob, S. Bejarano and P.J. Mumby. 2013. Reciprocal facilitation and non-linearity maintain habitat engineering on coral reefs. Oikos. 122: 428–440.

Bruno, J.F., H. Sweatman, W.F. Precht, E.R. Selig and V.G. Schutte. 2009. Assessing evidence of phase shifts from coral to macroalgal dominance on coral reefs. Ecology. 90: 1478–1484.

Campbell, S.J., A.S. Hoey, J. Maynard, T. Kartawijaya, J. Cinner, N.A.J. Graham and A.H. Baird. 2012. Weak compliance undermines the success of no-take zones in a large government-controlled marine protected area. PLoS ONE 7: e50074.

Carassou, L., M. Léopold, N. Guillemot, L. Wantiez and M. Kulbicki. 2013. Does herbivorous fish protection really improve coral reef resilience? Study from New Caledonia (South Pacific). PLoS ONE 8: e60564.

Cheal, A.J., M.A. MacNeil, E. Cripps, M.J. Emslie, M. Jonker, B. Schaffelke and H.P.A. Sweatman. 2010. Coral-macroalgal phase shifts or reef resilience: links with diversity and functional roles of herbivorous fishes on the Great Barrier Reef. Coral Reefs 29: 1005–1015.

Cheal, A.J., M. Emslie, M.A. MacNeil, I. Miller and H. Sweatman. 2013. Spatial variation in the functional characteristics of herbivorous fish communities and the resilience of coral reefs. Ecol. Appl. 23: 174–188

Choat, J.H., K.D. Clements and W.D. Robbins. 2002. The trophic status of herbivorous fishes on coral reefs. I: dietary analyses. Mar. Biol. 140: 613–623.

Choat, J.H., W.D. Robbins and K.D. Clements. 2004. The trophic status of herbivorous fishes on coral reefs. II: food processing modes and trophodynamics. Mar. Biol. 145: 445–454.

Christiansen, N.A., S. Ward, S. Harii and I.R. Tibbetts. 2009. Grazing by a small fish affects the early stages of a post-settlement stony coral. Coral Reefs 28: 47–51.

Clements, K.D., D. Raubenheimer and J.H. Choat. 2009. Nutritional ecology of marine herbivorous fishes: ten years on. Funct. Ecol. 23: 79–92.

Coelho, V.R. and C. Manfrino. 2007. Coral community decline at a remote Caribbean island: marine no-take reserves are not enough. Aquat. Conserv. 17: 666–685.

Comeros-Raynal, M.T., J.H. Choat, B.A. Polidoro, K.D. Clements, R. Abesamis, M.T. Craig, M.E. Lazuardi, J. McIlwain, A. Muljadi and R.F. Myers. 2012 The likelihood of extinction of iconic and dominant herbivores and detritivores of coral reefs: the parrotfishes and surgeonfishes. PLoS One 7: e39825.

Connell, J.H. 1978. Diversity in tropical rain forests and coral reefs. Science 199: 1302–1310.

Cote, I.M., I. Mosqueira and J.D. Reynolds. 2001. Effects of marine reserve characteristics on the protection of fish populations: a meta-analysis. J. Fish. Biol. 59: 178–189.

Cote, I.M., W.F. Precht, R.B. Aronson and T.A. Gardner. 2013. Is Jamaica a good model for understanding Caribbean coral reef dynamics? Mar. Pollut. Bull. 76: 28–31.

Crossman, D.J., J.H. Choat and K.D. Clements. 2005. Nutritional ecology of nominally herbivorous fishes on coral reefs. Mar. Ecol. Prog. Ser. 296: 129–142.

De'ath, G., K.E. Fabricius, H. Sweatman and M. Puotinen. 2012. The 27-year decline of coral cover on the Great Barrier Reef and its causes. Proc. Nat. Acad. Sci. USA 109: 17995–17999.

Diaz-Pulido, G., S. Harii, L. McCook and O. Hoegh-Guldberg. 2010. The impact of benthic algae on the settlement of a reef-building coral. Coral Reefs 29: 203–208.

Dulvy, N.K. and N.V.C. Polunin. 2004. Using informal knowledge to infer human-induced rarity of a conspicuous reef fish. Anim. Conserv. 7: 365–374.

Edgar, G.J., R. Bustamante, J.-M. Farina, M. Calvopina, C. Martinez and M. Toral-Granda. 2004. Bias in evaluating the effects of marine protected areas: the importance of baseline data for the Galapagos Marine Reserve. Environ. Conserv. 31: 212–218.

Edgar, G.J., N.S. Barrett and R.D. Stuart-Smith. 2009. Exploited reefs protected from fishing transform over decades into conservation features otherwise absent from seascapes. Ecol. Appl. 19: 1967–1974.

Edgar, G.J., R.D. Stuart-Smith, T.J. Willis, S. Kininmonth, S.C. Baker, S. Banks, N.S. Barrett, M.A. Becerro, A.T. Bernard, J. Berkhout, C.D. Buxton, S.J. Campbell, A.T. Cooper, M. Davey, S.C. Edgar, G. Försterra, D.E. Galván, A.J. Irigoyen, D.J. Kushner, R. Moura, P.E. Parnell, N.T. Shears, G. Soler, E.M.A. Strain and R.J. Thomson. 2014. Global conservation outcomes depend on marine protected areas with five key features. Nature. 506: 216–220.

Edwards, C.B., A. Friedlander, A. Green, M. Hardt, E. Sala, H. Sweatman, I. Williams, B. Zgliczynski, S. Sandin and J. Smith. 2014. Global assessment of the status of coral reef herbivorous fishes: evidence for fishing effects. Proc. R. Soc. B 281: 20131835.

Emslie, M.J., A.J. Cheal and K.A. Johns. 2014. Retention of habitat complexity minimizes disassembly of reef fish communities following disturbance: a large-scale natural experiment. PLoS One 9: e105384.

Emslie, M.J., M. Logan, D.H. Williamson, A.M. Ayling, M.A. MacNeil, D. Ceccarelli, A.J. Cheal, R.D. Evans, K.A. Johns, M.J. Jonker, I.R. Miller, K. Osborne, G.R. Russ and H.P.A. Sweatman. 2015. Expectations and outcomes of reserve network performance following re-zoning of the Great Barrier Reef Marine Park. Curr. Biol. 25(8): 983–992.

Floeter, S., B.S. Halpern and C. Ferreira. 2006. Effects of fishing and protection on Brazilian reef fishes. Biol. Conserv. 128: 391–402.

Graham, N.A., R.D. Evans and G.R. Russ. 2003. The effects of marine reserve protection on the trophic relationships of reef fishes on the Great Barrier Reef. Environ. Conserv. 30: 200–208.

Graham, N.A., S.K. Wilson, S. Jennings, N.V. Polunin, J. Robinson, J.P. Bijoux and T.M. Daw. 2007. Lag effects in the impacts of mass coral bleaching on coral reef fish, fisheries, and ecosystems. Conserv. Biol. 21: 1291–1300.

Graham, N.A., T.D. Ainsworth, A.H. Baird, N.C. Ban, L.K. Bay, J.E. Cinner, D.M. De Freitas, G. Diaz-Pulido, M. Dornelas, S.R. Dunn, P.I.J. Fidelman, S. Foret, T.C. Good, J. Kool, J. Mallela, L. Penin, M.S. Pratchett, D.H. Williamson. 2011. From microbes to people: tractable benefits of no-take areas for coral reefs. Oceanogr. Mar. Biol. Annu. Rev. 49: 105–136.

Graham, N.A.J., S. Jennings, M.A. MacNeil, D. Mouillot and S.K. Wilson. 2015. Predicting climate-driven regime shifts versus rebound potential in coral reefs. Nature 518: 94–97.

Gust, N. 2002. Scarid biomass on the northern Great Barrier Reef: the influence of exposure, depth and substrata. Environ. Biol. Fish. 64: 353–366.

Halford, A., A. Cheal, D. Ryan and D.M. Williams. 2004. Resilience to large-scale disturbance in coral and fish assemblages on the Great Barrier Reef. Ecology 85: 1892–1905.

Halpern, B.S. 2003. The impact of marine reserves: do reserves work and does reserve size matter? Ecol. Appl. 13: S117–S137.

Halpern, B.S. and R.R. Warner. 2002. Marine reserves have rapid and lasting effects. Ecol. Lett. 5: 361–366.

Harborne, A.R., P.J. Mumby, C.V. Kappel, C.P. Dahlgren, F. Micheli, K.E. Holmes, J.N. Sanchirico, K. Broad, I.A. Elliott and D.R. Brumbaugh. 2008. Reserve effects and natural variation in coral reef communities. J. Appl. Ecol. 45: 1010–1018.

Hatcher, B.G. 1981. The interaction between grazing organisms and the epilithic algal community of a coral reef: a quantitative assessment. Proc. 4th. Int. Coral. Reef. Symp. 2: 515–524.

Hatcher, B.G. 1988. Coral reef primary productivity: a beggar's banquet. Trends. Ecol. Evol. 3: 106–111.

Hawkins, J.P. and C.M. Roberts. 2004a. Effects of artisanal fishing on Caribbean coral reefs. Conserv. Biol. 18: 215–226.

Hawkins, J.P. and C.M. Roberts. 2004b. Effects of fishing on sex-changing Caribbean parrotfishes. Biol. Conserv. 115: 213–226.

Hawkins, J.P., C.M. Roberts, C. Dytham, C. Schelten and M.M. Nugues. 2006. Effects of habitat characteristics and sedimentation on performance of marine reserves in St. Lucia. Biol. Conserv. 127: 487–499.

Hedges, L. and I. Olkin. 1985. Statistical methods for meta-analysis. Academic, Orlando, FL.

Heenan, A., A.S. Hoey, G.J. Williams and I.D. Williams. 2016. Natural bounds on herbivorous coral reef fishes. Proc. R. Soc. B 283: rspb20161716.

Hoey, A.S. and D.R. Bellwood. 2008. Cross-shelf variation in the role of parrotfishes on the Great Barrier Reef. Coral Reefs 27: 37–47.

Hoey, A.S. and D.R. Bellwood. 2009. Limited functional redundancy in a high diversity system: single species dominates key ecological process on coral reefs. Ecosystems 12: 1316–1328.

Hoey, A.S. and D.R. Bellwood. 2011. Suppression of herbivory by macroalgal density: a critical feedback on coral reefs. Ecol. Lett. 14: 267–273.

Hoey, A.S., M.S. Pratchett and C. Cvitanovic. 2011. High Macroalgal Cover and Low Coral Recruitment Undermines the Potential Resilience of the World's Southernmost Coral Reef Assemblages. PLoS One 6(10): e25824.

Hughes, T.P. 1994. Catastrophes, phase-shifts, and large-scale degradation of a Caribbean coral reef. Science 265: 1547–1551.

Hughes, T.P., M.J. Rodrigues, D.R. Bellwood, D. Ceccarelli, O. Hoegh-Guldberg, L. McCook, N. Moltschaniwskyj, M.S. Pratchett, R.S. Steneck and B. Willis. 2007a. Phase shifts, herbivory, and the resilience of coral reefs to climate change. Curr. Biol. 17: 360–365.

Hughes, T.P., D.R. Bellwood, C.S. Folke, L.J. McCook and J.M. Pandolfi. 2007b. No-take areas, herbivory and coral reef resilience. Trends. Ecol. Evol. 22: 1–3.

Hughes, T.P., N.A. Graham, J.B. Jackson, P.J. Mumby and R.S. Steneck. 2010. Rising to the challenge of sustaining coral reef resilience. Trends. Ecol. Evol. 25: 633–642.

Hughes, T.P., J.T. Kerry, M. Álvarez-Noriega, J.G. Álvarez-Romero, K.D. Anderson, A.H. Baird, R.C. Babcock, M. Beger, D.R. Bellwood, R. Berkelmans, T.C. Bridge, I.R. Butler, M. Byrne, N.E. Cantin, S. Comeau, S.R. Connolly, G.S. Cumming, S.J. Dalton, G. Diaz-Pulido, C.M. Eakin, W.F. Figueira, J.P. Gilmour, H.B. Harrison, S.F. Heron, A.S. Hoey, J.-P.A. Hobbs, M.O. Hoogenboom, E.V. Kennedy, C.-Y. Kuo, J.M. Lough, R.J. Lowe, G. Liu, M.T. McCulloch, H.A. Malcolm, M.J. McWilliam, J.M. Pandolfi, R.J. Pears, M.S. Pratchett, V. Schoepf, T. Simpson, W.J. Skirving, B. Sommer, G. Torda, D.R. Wachenfeld, B.L. Willis and S.K. Wilson. 2017. Global warming and recurrent mass bleaching of corals. Nature 543: 373–377.

Humphries, A., T. McClanahan and C. McQuaid. 2014. Differential impacts of coral reef herbivores on algal succession in Kenya. Mar. Ecol. Prog. Ser. 504: 119–132.

Huntington, B.E. 2011. Confronting publication bias in marine reserve meta-analyses. Front. Ecol. Environ. 9: 375–376.

Huntington, B.E., M. Karnauskas and D. Lirman. 2011. Corals fail to recover at a Caribbean marine reserve despite ten years of reserve designation. Coral Reefs 30: 1077–1085.

Jackson, J.B.C., M.K. Donovan, K.L. Cramer and V.V. Lam. 2014. Status and Trends of Caribbean Coral Reefs: 1970–2012. Global Coral Reef Monitoring Network, IUCN, Gland, Switzerland.

Jennings, S. and N.V.C. Polunin. 1996. Effects of fishing effort and catch rate upon the structure and biomass of Fijian reef fish communities. J. Appl. Ecol. 33: 400–412.

Jennings, S. and N.V.C. Polunin. 1997. Impacts of predator depletion by fishing on the biomass and diversity of non-target reef fish communities. Coral Reefs. 16: 71–82.

Jennings, S., E.M. Grandcourt and N.V. C. Polunin. 1995. The effects of fishing on the diversity, biomass and trophic structure of Seychelles' reef fish communities. Coral Reefs. 14: 225–235.

Jennings, S., S.S. Marshall and N.V.C. Polunin. 1996. Seychelles' marine protected areas: comparative structure and status of reef fish communities. Biol. Conserv. 75: 201–209.

Jompa, J. and L.J. McCook. 2003. Coral-algal competition: macroalgae with different properties have different effects on corals. Mar. Ecol. Prog. Ser. 258: 87–95.

Jones, G.P., R.C. Cole and C.N. Battershill. 1992. Marine reserves: do they work? Proc. 2nd Int. Temperate Reef Symp. Auckland 1: 7–10.

Jones, G.P., M.I. McCormick, M. Srinivasan and J.V. Eagle. 2004. Coral decline threatens fish biodiversity in marine reserves. Proc. Natl. Acad. Sci. USA 101: 8251–8253.

Kopp, D., Y. Bouchon-Navaro, S. Cordonnier, A. Haouisee, M. Louis and C. Bouchon. 2010. Evaluation of algal regulation by herbivorous fishes on Caribbean coral reefs. Helgoland Marine Research 64: 181–190.

Kramer, K.L. and K.L. Jr. Heck. 2007. Top-down trophic shifts in Florida Keys patch reef marine protected areas. Mar. Ecol. Prog. Ser. 349: 111–123.

Lester, S.E., B.S. Halpern, K. Grorud-Colvert, J. Lubchenco, B.I. Ruttenberg, S.D. Gaines, S. Airame and R.R. Warner. 2009. Biological effects within no-take marine reserves: a global synthesis. Mar. Ecol. Prog. Ser. 384: 33–46.

MacNeil, M.A., N.A.J. Graham, J.E. Cinner, S.K. Wilson, I.D. Williams, J. Maina, S. Newman, A.M. Friedlander, S. Jupiter, N.V.C. Polunin and T.R. McClanahan. 2015. Recovery potential of the world's coral reef fishes. Nature 520: 341–344.

Maliao, R., A. White, A. Maypa and R. Turingan. 2009. Trajectories and magnitude of change in coral reef fish populations in Philippine marine reserves: a meta-analysis. Coral Reefs. 28: 809–822.

McClanahan, T.R. 1994. Kenyan coral reef lagoon fish: effects of fishing, substrate complexity, and sea urchins. Coral Reefs. 13: 231–241.

McClanahan, T.R. 2014. Recovery of functional groups and trophic relationships in tropical fisheries closures. Mar. Ecol. Prog. Ser. 497: 12–23.

McClanahan, T.R. and S.H. Shafir. 1990. Causes and consequences of sea urchin abundance and diversity in Kenyan coral reef lagoons. Oecol. 83: 362–370.

McClanahan, T.R. and R. Arthur. 2001. The effect of marine reserves and habitat on populations of East African coral reef fishes. Ecol. Appl. 11: 559–569.

McClanahan, T.R. and N.A.J. Graham. 2005. Recovery trajectories of coral reef fish assemblages within Kenyan marine protected areas. Mar. Ecol. Prog. Ser. 294: 241–248.

McClanahan, T.R., J. Maina, C.J. Starger, P. Herron-Perez and E. Dusek. 2005. Detriments to post-bleaching recovery of corals. Coral Reefs 24: 230–246.

McClanahan, T.R., N.A.J. Graham, J.M. Calnan and M.A. MacNeil. 2007. Toward pristine biomass: Reef fish recovery in coral reef marine protected areas in Kenya. Ecol. Appl. 17: 1055–1067.

McClanahan, T.R., B. Kaunda-Arara and J.O. Omukoto. 2010. Composition and diversity of fish and fish catches in closures and open-access fisheries of Kenya. Fisheries. Manag. Ecol. 17: 63–76.

McClanahan, T.R., N.A.J. Graham, M.A. MacNeil, M.A. Muthiga, J.E. Cinner, J.H. Bruggerman and S.K. Wilson. 2011a. Critical thresholds and tangible targets for ecosystem-based management of coral reef fisheries. Proc. Natl. Acad. Sci. USA 108: 17230–17233.

McClanahan, T.R., N.A. Muthiga and R. Coleman. 2011b. Testing for top-down control: can post-disturbance fisheries closures reverse algal dominance? Aquat. Conserv. 21: 658–675.

McCook, L., J. Jompa and G. Diaz-Pulido. 2001. Competition between corals and algae on coral reefs: a review of evidence and mechanisms. Coral Reefs 19: 400–417.

Mellin, C., M. Kulbicki and D. Ponton. 2007. Seasonal and ontogenetic patterns of habitat use in coral reef fish juveniles. Estuar. Coast. Shelf Sci. 75: 481–491.

Micheli, F., B.S. Halpern, L.W. Botsford and R.R. Warner. 2004. Trajectories and correlates of community change in no-take marine reserves. Ecol. Appl. 14: 1709–1723.

Miller, K.I. and G.R. Russ. 2014. Studies of no-take marine reserves: methods for differentiating reserve and habitat effects. Ocean. Coast. Manage. 96: 51–60.

Molloy, P.P., I.B. McLean and I.M. Cote. 2009. Effects of marine reserve age on fish populations: a global meta-analysis. J. Appl. Ecol. 46: 743–751.

Monaco, M., A. Friedlander, C. Caldow, J. Christensen, C. Rogers, J. Beets, J. Miller and R. Boulon. 2007. Characterising reef fish populations and habitats within and outside the US Virgin Islands

Coral Reef National Monument: a lesson in marine protected area design. Fisheries Manag. Ecol. 14: 33–40.

Mosquera, I., I.M. Cote, S. Jennings and J.D. Reynolds. 2000. Conservation benefits of marine reserves for fish populations. Anim. Conserv. 3: 321–332.

Mumby, P.J. 2006. The impact of exploiting grazers (Scaridae) on the dynamics of Caribbean coral reefs. Ecol. Appl. 16: 747–769.

Mumby, P.J. 2009. Herbivory versus corallivory: are parrotfish good or bad for Caribbean coral reefs? Coral Reefs 28: 683–690.

Mumby, P.J., C.P. Dahlgren, A.R. Harborne, C.V. Kappel, F. Micheli, D.R. Brumbaugh, K.E. Holmes, J.M. Mendes, K. Broad, J.N. Sanchirico, K. Buch, S. Box, R.W. Stoffle and A.B. Gill. 2006. Fishing, trophic cascades, and the process of grazing on coral reefs. Science 311: 98–101.

Mumby, P.J., A.R. Harborne, J. Williams, C.V. Kappel, D.R. Brumbaugh, F. Micheli, K.E. Holmes, C.P. Dahlgren, C.B. Paris and P.G. Blackwell. 2007. Trophic cascade facilitates coral recruitment in a marine reserve. Proc. Natl. Acad. Sci. USA 104: 8362–8367.

Mumby, P.J. and R.S. Steneck. 2008. Coral reef management and conservation in light of rapidly evolving ecological paradigms. Trends. Ecol. Evol. 23: 555–563.

Mumby, P.J. and A.R. Harborne. 2010. Marine reserves enhance the recovery of corals on Caribbean reefs. PLoS One 5: e8657.

Myers, M.R. and R.F. Ambrose. 2009. Differences in benthic cover inside and outside marine protected areas on the Great Barrier Reef: influence of protection or disturbance history. Aquat. Conserv. 19: 736–747.

Nugues, M.M. and R.P. Bak. 2006. Differential competitive abilities between Caribbean coral species and a brown alga: a year of experiments and a long-term perspective. Mar. Ecol. Prog. Ser. 315: 75–86.

O'Farrell, S., A.R. Harborne, Y.M. Bozec, B.E. Luckhurst and P.J. Mumby. 2015. Protection of functionally important parrotfishes increases their biomass but fails to deliver enhanced recruitment. Mar. Ecol. Prog. Ser. 522: 245–254.

O'Leary, J.K. and T.R. McClanahan. 2010. Trophic cascades result in large-scale coralline algae loss through differential grazer effects. Ecology 91: 3584–3597.

Osenberg, C.W., B.M. Bolker, J. White, C.M.S. Mary and J.S. Shima. 2006. Statistical issues and study design in ecological restorations: lessons learned from marine reserves. Foundations of restoration ecology, Island Press, Washington, DC 24: 280–302.

Pandolfi, J.M., R.H. Bradbury, E. Sala, T.P. Hughes, K.A. Bjorndal, R.G. Cooke, D. McArdle, L. McClenachan, M.J. Newman and G. Paredes. 2003. Global trajectories of the long-term decline of coral reef ecosystems. Science 301: 955–958.

Penin, L., F. Michonneau, A.H. Baird, S.R. Connolly, M.S. Pratchett, M. Kayal and M. Adjeroud. 2010. Early post-settlement mortality and the structure of coral assemblages. Mar. Ecol. Prog. Ser. 408: 55–64.

Pinca, S., M. Kronen, F. Magron, B. McArdle, L. Vigliola, M. Kulbicki and S. Andrefouet. 2012. Relative importance of habitat and fishing in influencing reef fish communities across seventeen Pacific Island Countries and Territories. Fish Fish. 13: 361–379.

Polunin, N. and C. Roberts. 1993. Greater biomass and value of target coral-reef fishes in two small Caribbean marine reserves. Mar Ecol. Prog. Ser. 100: 167–176.

Pratchett, M.S., P. Munday, S.K. Wilson, N.A. Graham, J.E. Cinner, D.R. Bellwood, G.P. Jones, N.V. Polunin and T.R. McClanahan. 2008. Effects of climate-induced coral bleaching on coral-reef fishes. Ecological and economic consequences. Oceanography and Mar. Biol. Annu. Rev. 46: 251–296.

Rasher, D.B., A.S. Hoey and M.E. Hay. 2013. Consumer diversity interacts with prey defenses to drive ecosystem function. Ecology 94: 1347–1358.

Rizzari, J.R., B.J. Bergseth and A.J. Frisch. 2015. Impact of conservation areas on trophic interactions between apex predators and herbivores on coral reefs. Conserv. Biol. 29: 418–429.

Roberts, C.M. and J.P. Hawkins. 1997. How small can a marine reserve be and still be effective? Coral Reefs 16: 150.

Roff, G. and P.J. Mumby. 2012. Global disparity in the resilience of coral reefs. Trends. Ecol. Evol. 27: 404–413.

Rotjan, R.D. and S.M. Lewis. 2005. Selective predation by parrotfishes on the reef coral *Porites astreoides*. Mar. Ecol. Prog. Ser. 305: 193–201.

Rotjan, R.D., J.L. Dimond, D.J. Thornhill, J.J. Leichter, B. Helmuth, D.W. Kemp and S.M. Lewis. 2006. Chronic parrotfish grazing impedes coral recovery after bleaching. Coral Reefs 25: 361–368.

Russ, G. 1984. Distribution and abundance of herbivorous grazing fishes in the central Great Barrier Reef. II: patterns of zonation of mid-shelf and outer-shelf reefs. Mar. Ecol-Prog. Ser. 20: 35–44.

Russ, G.R. 1985. Effects of protective management on coral reef fishes in the central Philippines. Proc. 5th Int. Coral Reef Congress, Tahiti. 4: 219–224.

Russ, G.R. 2002. Yet another review of marine reserves as reef fishery management tools. Coral reef fishes: dynamics and diversity in a complex ecosystem. Academic Press, New York, 421–444.

Russ, G.R. and A.C. Alcala. 1989. Effects of intense fishing pressure on an assemblage of coral reef fishes. Mar. Ecol. Prog. Ser. 56: 13–27.

Russ, G.R. and A.C. Alcala. 1998a. Natural fishing experiments in marine reserves 1983–1993: community and trophic responses. Coral Reefs 17: 383–397.

Russ, G.R. and A.C. Alcala. 1998b. Natural fishing experiments in marine reserves 1983–1993: roles of life history and fishing intensity in family responses. Coral Reefs 17: 399–416.

Russ, G.R., B. Stockwell and A.C. Alcala. 2005. Inferring versus measuring rates of recovery in no-take marine reserves. Mar. Ecol. Prog. Ser. 292: 1–12.

Russ, G.R., S.A. Questel, J.R. Rizzari and A.C. Alcala. 2015. The parrotfish–coral relationship: refuting the ubiquity of a prevailing paradigm. Mar. Biol. 162: 2029–2045.

Sammarco, P.P. 1985. The Great Barrier Reef vs. the Caribbean: comparisons of grazers, coral recruitment patterns and reef recovery. Proceedings of the 5th International Coral Reef Congress, Tahiti, 4: 391–398.

Samoilys, M.A., K.M. Martin-Smith, B.G. Giles, B. Cabrera, J.A. Anticamara, E.O. Brunio and A.C. Vincent. 2007. Effectiveness of five small Philippines' coral reef reserves for fish populations depends on site-specific factors, particularly enforcement history. Biol. Conserv. 136: 584–601.

Selig, E.R. and J.F. Bruno. 2010. A global analysis of the effectiveness of marine protected areas in preventing coral loss. PLoS One 5: e9278.

Shephard, K.L. 1994. Functions for fish mucus. Rev. Fish. Biol. Fisheries 4: 401–429.

Sluka, R.D. and M.W. Miller. 2001. Herbivorous fish assemblages and herbivory pressure on Laamu Atoll, Republic of Maldives. Coral Reefs. 20: 255–262.

Smith, J.E., C.L. Hunter and C.M. Smith. 2010. The effects of top-down versus bottom-up control on benthic coral reef community structure. Oecologia 163: 497–507.

Smith, S.G., J.S. Ault, J.A. Bohnsack, D.E. Harper, J. Luo and D.B. McClellan. 2011. Multispecies survey design for assessing reef-fish stocks, spatially explicit management performance, and ecosystem condition. Fish. Res. 109: 25–41.

Soler, G.A., G.J. Edgar, R.J. Thomson, S. Kininmonth, S.J. Campbell, T.P. Dawson, N.S. Barrett, A.T.F. Bernard, D.E. Galvan, T.J. Willis, T.J. Alexander and R.D. Stuart-Smith. 2015. Reef fishes at all trophic levels respond positively to effective marine protected areas. PLoS One 10: e0140270.

Stockwell, B., C.R.L. Jadloc, R.A. Abesamis, A.C. Alcala and G.R. Russ. 2009. Trophic and benthic responses to no-take marine reserve protection in the Philippines. Mar. Ecol. Prog. Ser. 389: 1–15.

Syms, C. and G.P. Jones. 2000. Disturbance, habitat structure, and the dynamics of a coral-reef fish community. Ecology 81: 2714–2729.

Taylor, B.M. 2014. Drivers of protogynous sex change differ across spatial scales. Proc. R. Soc. B 281: rspb20132423.

Taylor, B.M., P. Houk, G.R. Russ and J.H. Choat. 2014a. Life histories predict vulnerability to overexploitation in parrotfishes. Coral Reefs 33: 869–878.

Taylor, B.M., S.J. Lindfield and J.H. Choat. 2014b. Hierarchical and scale-dependent effects of fishing pressure and environment on the structure and size distribution of parrotfish communities Ecography 38: 520–530.

Toth, L.T., R. van Woesik, T.J.T. Murdoch, S.R. Smith, J.C. Ogden, W.F. Precht and R.B. Aronson. 2014. Do no-take reserves benefit Florida's corals? 14 years of change and stasis in the Florida Keys National Marine Sanctuary. Coral Reefs 33: 565–577.

Trapon, M.L., M.S. Pratchett and A.S. Hoey. 2013a. Spatial variation in abundance, size and orientation of juvenile corals related to the biomass of parrotfishes on the Great Barrier Reef, Australia. PloS One 8: e57788.

Trapon, M.L., M.S. Pratchett, A.S. Hoey and A.H. Baird. 2013b. Influence of fish grazing and sedimentation on the early post-settlement survival of the tabular coral Acropora cytherea. Coral Reefs 32: 1051–1059.

Tzadik, O.E. and R.S. Appeldoorn. 2013. Reef structure drives parrotfish species composition on shelf edge reefs in La Parguera, Puerto Rico. Cont. Shelf. Res. 54: 14–23.

Valentine, J.F., K.L. Jr. Heck, D. Blackmon, M.E. Goecker, J. Christian, R.M. Kroutil, K.D. Kirsch, B.J. Peterson, M. Beck and M.A. Vanderklift. 2007. Food web interactions along seagrass-coral reef boundaries: effects of piscivore reductions on cross-habitat energy exchange. Mar. Ecol. Prog. Ser. 333: 37–50.

Wantiez, L., P. Thollot and M. Kulbicki. 1997. Effects of marine reserves on coral reef fish communities from five islands in New Caledonia. Coral Reefs 16: 215–224.

Williams, I.D. and N.V.C. Polunin. 2001. Large-scale associations between macroalgal cover and grazer biomass on mid-depth reefs in the Caribbean. Coral Reefs 19: 358–366.

Williamson, D.H., D.M. Ceccarelli, R.D. Evans, G.P. Jones and G.R. Russ. 2014. Habitat dynamics, marine reserve status, and the decline and recovery of coral reef fish communities. Ecol. Evol. 4: 337–354.

Wilson, S.K., N.A. Graham, M.S. Pratchett, G.P. Jones and N.V. Polunin. 2006. Multiple disturbances and the global degradation of coral reefs: are reef fishes at risk or resilient? Glob. Change. Biol. 12: 2220–2234.

Wilson, S.K., R. Fisher, M.S. Pratchett, N.A. Graham, N.K. Dulvy, R.A. Turner, A. Cakacaka, N.V. Polunin and S.P. Rushton. 2008. Exploitation and habitat degradation as agents of change within coral reef fish communities. Glob. Change. Biol. 14: 2796–2809.

Wilson, S.K., N.A. Graham, R. Fisher, J. Robinson, K. Nash, K. Chong-Seng, N.V. Polunin, R. Aumeeruddy and R. Quatre. 2012. Effect of macroalgal expansion and Marine Protected Areas on coral recovery following a climatic disturbance. Conserv. Biol. 26: 995–1004.

CHAPTER 15

Differential Vulnerabilities of Parrotfishes to Habitat Degradation

Michael J. Emslie[1] and Morgan S. Pratchett[2]

[1] Australian Institute of Marine Science, PMB No 3, Townsville MC,
 Townsville, Queensland, Australia
 Email: m.emslie@aims.gov.au
[2] ARC Centre of Excellence for Coral Reef Studies, James Cook University, Townsville, Australia
 Email: morgan.pratchett@jcu.edu.au

Introduction

Habitat degradation has a devastating influence on the structure and dynamics of ecological assemblages (Vitousek, 1997, Fahrig, 2001) and is increasingly recognised as the major contributor to global biodiversity loss (Tilman et al. 1994, Brooks et al. 2002, Hoekstra et al. 2005, Mantyka-Pringle et al. 2012). Losses of marine diversity are highest in shallow coastal habitats (Gray 1997), representing the confluence of land-based perturbations (e.g. sedimentation, eutrophication and pollution) with high levels of human use and exploitation, all set against a backdrop of naturally occurring disturbances. Globally, 65 percent of coastal habitat has been significantly degraded (Lotze et al. 2006), resulting in significant declines in biodiversity within these habitats. Moreover, declines in biodiversity within coastal habitats compromise ecosystem functioning and services (Worm et al. 2006), further accelerating the degradation of coastal habitats. For example, major losses of suspension feeders and coastal vegetation, which filter and contain toxins, have contributed to declining water quality (Worm et al. 2006).

On coral reefs, habitat degradation is manifest as declines in the cover of habitat-forming scleractinian corals (e.g. Gardner et al. 2003, De'ath et al. 2012), often accompanied by corresponding shifts in the composition of the benthic community, exemplified by increases in the cover of fleshy macroalgae, or seaweeds (e.g. Hughes et al. 2010, Cheal et al. 2010). The major cause(s) or contributors to coral reef degradation vary geographically (e.g. Pandolfi et al. 2003, Bruno and Selig 2007). However, it is generally those areas closest to urban centres and large human populations that exhibit the most pronounced degradation (Jackson et al. 2001; Pandolfi et al. 2003; Wilkinson 2008); a disproportionate number of coral reefs have been significantly degraded in east Africa, south-east Asia, and the central and southern Caribbean (Wilkinson 2004), where anthropogenic pressure on coral reefs is greatest, leading to chronic pollution, eutrophication, sedimentation, overfishing and/ or destructive fishing. These direct anthropogenic disturbances are also contributing to coral bleaching, outbreaks of crown-of-thorns starfish (*Acanthaster* cf. *solaris*), and coral disease, which all add to widespread degradation of coral reefs (Sano et al. 1987, Graham et al. 2006, 2007, Osborne et al. 2011, Emslie et al. 2011, 2012, De'ath et al. 2012, Hughes et al. 2017). On the Great Barrier Reef (GBR), for example, recent coastal runoff now contains 5.7 times higher concentrations of nitrogen and 5.5 times the total suspended solids than prior to European settlement (Brodie et al. 2012, Kroon et al. 2012) and these elevated nutrient concentrations may cause or exacerbate outbreaks of crown-of-thorns starfish (Fabricius et al. 2010, Pratchett et al. 2014). Similarly, high levels of coral disease have been correlated with increased nutrient levels in both the Caribbean and Pacific (Bruno et al. 2003, Aeby et al. 2011). The long-term and increasing effects of both anthropogenic and natural disturbances are also now being compounded by global climate change, further accelerating coral loss (Hughes et al. 2003, 2017, Hoegh-Guldberg et al. 2007).

Significant and widespread degradation of coral reef ecosystems is of critical concern given strong reliance on coral reef ecosystems in many tropical maritime nations. Millions of people (many in some of the world's poorest countries) are directly reliant on coral reefs for food and livelihoods, while sustained and ongoing habitat degradation is threatening goods and services derived from these ecosystems (Bell et al. 2009, 2013), undermining fisheries productivity, tourism, ecosystem function and biodiversity. Degraded coral reef ecosystems support lower abundance, biomass and diversity of coral reef fishes (reviewed by Wilson et al. 2006, Pratchett et al. 2008), which has a direct effect on fisheries productivity (Pratchett et al. 2011a). Declines in the abundance and diversity of fishes are of even greater concern where the species affected perform critical functional roles like herbivory (Bellwood et al. 2004). If for example, degradation of reef ecosystems leads to declines in the abundance or function of fishes that could otherwise prevent or reverse phase shifts (Hoey and Bellwood 2011), this will result in a feedback that further enhances or accelerates habitat degradation.

Parrotfishes (Labridae, Scarinae) are a unique and conspicuous component of the fish fauna in shallow tropical marine habitats, especially coral reefs and seagrass meadows (Bellwood 1994). Along with rabbitfishes (Siganidae), surgeonfishes (Acanthuridae) and rudderfishes (Kyphosidae), parrotfishes are one of the major groups of herbivorous fishes on coral reefs (Bellwood et al. 2004, Cheal et al. 2010, but see Choat et al. 2002, 2004) and are considered fundamental to the resilience of these ecosystems. Characteristic phase shifts from coral- to seaweed-dominance are inherently linked to declines in the abundance or function of herbivorous fishes (Hughes 1994, Graham et al. 2006, Rasher et al. 2013), often due to overexploitation. In Discovery Bay, Jamaica, for example, catastrophic phase shifts that occurred following Hurricane Allen and the die-off of *Diadema* urchins, were ultimately attributed to the historical depletion of herbivorous fishes (Hughes 1994). Moreover, subsequent implementation of no-take areas in the Caribbean has led to increased biomass of parrotfishes and associated reductions in macroalgae (Mumby et al. 2006). In another example, inter-reef variation in abundance of herbivorous fishes has been implicated in the differential occurrence of coral-macroalgae phase shifts following multiple disturbances on the inshore GBR, Australia (Cheal et al. 2010).

One of the most powerful demonstrations of the functional importance of parrotfishes (and other herbivorous reef fishes) was a large-scale fish exclusion experiment undertaken by Hughes et al. (2007) at Orpheus Island on the inshore GBR, Australia. Large (5 x 5 x 4m) replicate cages, designed to simulate localised depletion of large herbivorous fishes, caused by chronic overfishing, were installed to investigate the role of herbivory in the recovery of coral assemblages following mass bleaching. Importantly, the biomass of herbivorous fishes inside cages was seven to ten times lower than partial and uncaged plots, and benthic communities diverged widely among the treatments. Most notably, there were sustained increases in cover and biomass of seaweed over two years, which in turn contributed to reduced recruitment and survivorship of corals inside cages (Hughes et al. 2007). This experiment showed that exclusion of herbivores and proliferation of seaweeds does not in itself necessarily contribute to coral loss, but profoundly erodes the resilience of reefs and the ability of coral assemblages to regenerate after acute disturbances (e.g., following bleaching).

While parrotfishes are functionally important, there are conflicting reports about their vulnerability to the sustained and ongoing degradation of coral reef ecosystems (Comeros-Raynal et al. 2012). This largely reflects the disparate nature of studies that report the effects of different disturbance types from wide-ranging geographic localities, often using different methodologies. In most instances, parrotfishes seem resilient to habitat degradation from natural disturbances with numerous studies reporting increases or no change in the short-term (Walsh 1983, Hart et al. 1996, Lewis 1997, Cheal et al. 2002, Sano 2004, Lindahl et al. 2001, Graham et al. 2006, Garpe et al. 2006, Yahya et al. 2011, Khalil et al. 2013, Adam et al. 2015). However, they do appear vulnerable to disturbances which degrade habitat complexity, over either the long-term, tracking the slow gradual erosion of dead coral skeletons following bleaching induced coral mortality (Graham et al. 2006, Garpe et al. 2006) or immediately following large storm swells which scour reef surfaces clean (Emslie et al. 2014). Inherent to the variety of responses recorded by studies to date is that parrotfish communities generally exhibit high response diversity to disturbances.

Marked discrepancies in reported responses of parrotfishes to localised coral loss has led to recent attempts to synthesise the existing information. Comeros-Raynal et al. (2012) relied heavily on expert opinion, and suggested that some reef-associated parrotfishes (e.g. *Bolbometopon muricatum*, *Cetoscarus bicolor* and *Chlorurus* spp.) are reliant on corals for food and/ or settlement, and will therefore, be sensitive to widespread coral loss and degradation of reef habitats. It is unknown however, to what extent those parrotfishes (e.g., *B. muricatum*) that are facultative corallivores (reviewed by Cole et al. 2008, Bonaldo et al. 2014) are obligately dependent on live coral, or will be negatively affected if they have little or no access to live corals. Fisheries exploitation can result in numerous impacts to parrotfish populations. Aside from the obvious mortality associated with these extractive activities, fisheries exploitation can also result in sub-lethal effects which can shift population structure, alter sex ratios, diminish reproductive output and alter their ability to perform necessary ecosystem functions (Taylor 2014). Graham et al. (2011) established the specific vulnerability of 24 species of parrotfishes (as well as surgeonfishes, butterflyfishes and wrasses) to fisheries exploitation versus climate-induced changes in benthic habitats. It was concluded that abundance and function of parrotfishes (unlike many other reef fish families) are generally insensitive to climate-induced degradation of coral reef habitats. These fishes are however, considered to be highly vulnerable to fisheries exploitation, such that over-fishing would rapidly deplete local assemblages of parrotfishes thereby leading to habitat degradation, rather than *vice versa* (Graham et al. 2011).

The purpose of this chapter was to explore species-specific vulnerability of parrotfishes to different disturbances, considering the responses of parrotfishes to acute disturbances (e.g. cyclones, bleaching, and outbreaks of crown-of-thorns starfish) that cause coral loss and reef degradation. This review is important because parrotfishes, perhaps more than any other group of coral reef fishes, play a critical role in maintaining the structure and function of their specific habitats (Bellwood et al. 2003, Bonaldo and Bellwood 2009). To explore species-specific vulnerability of parrotfishes, we compiled published data on changes in the abundance of parrotfishes during major disturbances (mainly severe tropical storms and climate-induced coral bleaching) that caused localised coral loss (e.g.

Cheal et al. 2002, Rousseau et al. 2010). However, given the paucity of published data, we complemented this meta-data with new analyses of data from long-term and spatially extensive surveys across the Great Barrier Reef (see Cheal et al. 2012) to explicitly examine responses of parrotfishes to temporally and spatially discrete disturbance events.

Resource Requirements

The vulnerability of species to habitat degradation is fundamentally dependent upon their resource requirements, including both the specific range of resources that they reportedly use (e.g., corals and other benthic substrata on which they are seen to feed), as well as the complete range of alternative resources they could use, should preferred resources become depleted. Species that use a generally narrower range of different resources (ecological specialists) are expected to be more constrained by resource availability and much more susceptible to disturbances that reduce the local availability of essential resources ("the specialisation-disturbance hypothesis", Vázquez and Simberloff 2002). However, it is important to realise that many seemingly specialist species (realised specialists) are actually capable of using a much wider range of different resources, but tend to use only restricted range of resources when available or abundant (e.g. Pratchett et al. 2004). A species' actual resource requirements and flexibility in resource use can only really be established through manipulative experiments, testing an individual's use and performance across a range of different resources, independent of vagaries in resource access or availability (Devictor et al. 2010). However, these experiments are very complex and rarely consider the full range of possible resources (e.g. Berumen and Pratchett 2008). Testing for intra-specific variation in dietary composition across broad geographic locations with known differences in resource availability (e.g. Lawton et al. 2011) may therefore, be the best way to assess the extent to which patterns of prey use are constrained by extrinsic (mostly prey availability and competition for resources) or intrinsic factors (specific adaptations or individualistic feeding preferences). For highly specialised species, patterns of resource use are expected to be invariant, such that specialist species can only exist in habitats and locations that provide adequate access to necessary resources (Devictor et al. 2010). Conversely, when species are shown to use only a very restricted range of resources at a single time and location, this may be due to inherent prey preferences, high abundance of a particular prey type, or strong constraints on prey use due to environmental factors or biological interactions (Fox and Morrow 1981).

While patterns of resource-use are key to predicting effects of habitat degradation on motile species (Feary et al. 2007, Wilson et al. 2008a), there is surprisingly limited data on critical resource (diet or habitat) requirements for most coral reef fishes. Notably, the importance of corals for many coral reef fishes is assumed based on loose associations between individual fishes and specific coral types (e.g. Coker et al. 2014), while there is little to no evidence about whether these fishes are obligatory dependent upon live coral. Coker et al. (2014) lists 320 species of coral reef fishes (eight percent of species), which have been recorded to inhabit live corals, including 11 species of parrotfishes (e.g. *Chlorurus spirilus*, *Scarus oviceps*, and *Sparisoma aurofrenatum*). However, this is considered to be a very conservative estimate of the number of fishes that will be negatively affected by severe coral depletion, mainly because it fails to account for fishes that are indirectly affected by live coral loss and/ or associated declines in structural complexity that often accompanies extensive coral loss (e.g. Wilson et al. 2006, Emslie et al. 2014). There are also at least 30 species of parrotfishes that are have been recorded to feed on live corals (Bonaldo et al. 2014), which may also be adversely affected by extensive coral loss.

Habitat-use

Broad habitat types used by extant parrotfishes are coral reefs, seagrass meadows, and/ or rocky reefs (Bellwood 1994). Coral reef parrotfishes (*Scarus* spp., *Hipposcarus* spp., *Chlorurus* spp., *B. muricatum*, *Cetoscarus* spp., *Calotomus* spp. and some *Sparisoma* spp.) are generally restricted to shallow reef environments with extensive carbonate matrix, as well as immediately adjacent sand or rubble habitats. Parrotfishes exhibit clear distinctions in habitat preferences across small spatial scales (100 m, e.g. among habitats within individual reefs; Russ 1984), but also broader spatial scales (10 km) such as preferences for sheltered embayments or inshore reefs, versus exposed and offshore reef habitats (e.g., Gust et al. 2001, Hoey and Bellwood 2008, Cheal et al. 2012). Most notably, *B. muricatum*, the largest parrotfish species, is often most abundant on shallow habitats on highly exposed offshore reefs (Bellwood and Choat 2011, Taylor et al. 2014). On Australia's GBR, large adult *B. muricatum* are most abundant on outer reefs (Hoey and Bellwood, 2008) and mainly in the northern GBR (Bellwood and Choat 2011). In contrast, three species of *Scarus* (*Sc. rivulatus*, *Sc. ghobban* and *Sc. psittacus*) are found predominantly on inner-shelf reefs (Hoey and Bellwood 2008; see also Gust et al. 2001). Such patterns of broad habitat preferences have resulted in clear spatial patterns in the distribution and abundance of parrotfishes on the GBR, including a general increase in diversity and abundance from inshore coastal reefs to offshore reefs (Cheal et al. 2012).

Differential preferences of parrotfishes for sheltered versus exposed habitats may be attributable to differential swimming ability (e.g. Bellwood et al. 2002), or spatial variation in abundance of habitat or dietary resources (Cheal et al. 2012). Importantly, large herbivorous fishes may favour habitats with highest algal productivity (Russ 2003). Some species are also reported to be restricted to coral-rich habitats (e.g. *Scarus maculipinna*, Comeros-Raynal et al. 2012) or vary in abundance among sites in direct accordance with live coral cover (e.g. *Scarus niger* and *Scarus tricolor*; Jennings et al. 1996). Overall, it is clear that habitat structure (biological and physical) exerts a major influence on the structure and diversity of parrotfish assemblages (e.g. Taylor et al. 2014, Heenan et al. 2016) as has been shown for virtually all coral reef fishes (Friedlander and Parrish 1998, Friedlander et al. 2010, Messmer et al. 2011). To date, there are at least 17 species of parrotfishes reported to preferentially utilise coral rich habitats at settlement and/ or as juveniles (Table 1), but this is likely to be an underestimate and there remains definitive need to better understand specific habitat requirements of parrotfishes (Bellwood and Choat 2011).

Reliance on specific corals or coral-rich habitats by coral reef fishes is often highest for small or juvenile individuals (Munday and Jones 1998). For example, Jones et al. (2004) found 65 percent of reef fish species in Kimbe Bay (New Britain, Papua New Guinea) preferentially settle into live corals. Settlement preferences of most coral reef parrotfishes are very poorly known. However, Adam et al. (2011) showed that 92 percent of newly settled (<5 cm TL) *Ch. spirilus* and *Sc. psittacus* in Moorea, French Polynesia were associated with live coral, specifically *Porites rus*. Moreover, juvenile *Ch. spirilus* and *Sc. psittacus* were five times more abundant at sites dominated by *P. rus* compared to similar sites with little *P. rus* (Adam et al. 2011). Similarly, in St. Croix (U.S. Virgin Islands), Tolimieri (1998) reported *Sparisoma viride* settles preferentially to, and has higher post-settlement persistence within, the common branching coral, *Porites porites*. On the GBR, Bellwood and Choat (1989) reported juveniles of *Scarus oviceps* and *Scarus spinus* are predominantly found in areas with abundant live coral. Johnson et al. (2011) went further to suggest these two species preferentially use specific coral species, namely *Acropora pulchra*. In contrast, many other *Scarus* species (*Sc. chameleon*, *Sc. ghobban*, *Sc. globiceps*, *Sc. niger*, *Sc. schlegeli*,

Table 1. Parrotfishes reported to settle into specific corals and/ or coral rich habitats, or strongly associated with such habitats during their early life-history

No.	Group/Species	Preferred habitat/ micro-habitat	Principal source
Excavators			
1	*Bolbometopon muricatum*	Branching corals	Hamilton et al. 2017
2	*Chlorurus perspicillatus*	Branching corals	DeMartini et al. 2010
3	*Chlorurus microrhinos*	Coral-rich habitats	Wilson et al. 2010
4	*Chlorurus sordidus*	Coral-rich habitats	Wilson et al. 2010
5	*Chlorurus spirilus*	Branching corals	DeMartini et al. 2010
		Porites rus	Adam et al. 2011
Scrapers			
6	*Scarus atrilunula*	Coral-rich habitats	Jennings et al. 1996
7	*Scarus dimidiatus*	Coral-rich habitats	Lieske and Meyers, 2001
8	*Scaurus dubius*	Branching corals	DeMartini et al. 2010
9	*Scarus frenatus*	Coral-rich habitats	Wilson et al. 2010
10	*Scarus prasiognathos*	Coral-rich habitats	Wilson et al. 2010
11	*Scarus niger*	Coral-rich habitats	Jennings et al. 1996
12	*Scarus maculipinna*	Coral-rich habitats	Comeros-Raynal et al. 2012
13	*Scarus oviceps*	Coral-rich habitats	Bellwood and Choat 1989
		Acropora pulchra	Johnson et al. 2011
14	*Scarus spinus*	Coral-rich habitats	Bellwood and Choat 1989
		Acropora pulchra	Johnson et al. 2011
15	*Scarus psittacus*	*Acropora pulchra*	Johnson et al. 2011
		Porites rus	Adam et al. 2011
16	*Scarus tricolor*	Coral-rich habitats	Jennings et al. 1996
17	*Sparisoma aurofrenatum*	*Montastera* sp.	Danilowicz et al. 2001
18	*Sparisoma viride*	*Porites porites*	Tolimeiri 1998

and *Sc. rubroviolaceus*) are predominantly found in areas with low coral cover, such as near reef habitats with high cover of coral rubble and/or sand (Bellwood and Choat 1989) or associated with stands of macroalgae, such as within damselfish territories (Green 1992, 1998).

Several studies (Feary et al. 2007, Wilson et al. 2010) have shown that many reef fishes are closely associated with live corals during the early stages of their benthic life history even though they rarely associate with corals as adults. Wilson et al. (2010) quantified juvenile habitat use for 56 reef fish species (including five parrotfish species) at Ningaloo Reef, Western Australia, explicitly testing for preferential use of algal meadows versus live and recently dead coral microhabitats. A significant proportion (>20 percent) of juveniles for four of the five parrotfish species (all but *Leptoscarus vaigiensis*, a macroalgal browser) were strongly associated with live corals. Although an equal or greater number of juveniles were found in dead, but intact coral microhabitats. This suggests that juveniles of the four species (*Chlorurus microrhinos*, *Ch. spirilus*, *Scarus frenatus*, and *Scarus prasiognathos*) are reliant on the structure provided by live corals, more so than live coral *per se* (Wilson et al. 2010; Table 1). In many reef systems the availability of suitably complex microhabitats is nonetheless, fundamentally dependent on live hard corals (e.g. Graham et al. 2006) because dead branching corals become quickly eroded and decomposed and only provide effective habitat for one to four years after they die (Pratchett et al. 2009).

While knowledge of species-specific settlement preferences is obviously important, the small size and cryptic behaviour of newly settled parrotfishes poses considerable challenges to this research. One of the largest and most iconic of all coral reef parrotfishes, *B. muricatum*, has distinct habitat requirements at different stages during its' life (Aswani and Hamilton 2004) such that it is only found in locations where recruit/juvenile habitats and adult habitats occur within relatively close proximity (Comeros-Raynal et al. 2012, Taylor et al. 2014). However, the settlement habitats of *B. muricatum* have only very recently started to become apparent. On the GBR, for example, newly settled (<50 mm TL) individuals of *B. muricatum* are very rarely sighted (Bellwood and Choat 2011). In the Solomon Islands, however, high densities of newly recruited *B. muricatum* are found in coral rich habitats in inshore and sheltered lagoonal environments (Hamilton et al. 2017). Importantly, comparable inshore habitats where corals have been lost due to excessive sediment following extensive logging of coastal environments contain little or no recruits. These data are being used to highlight the long-term negative impacts of coastal degradation on local populations of *B. muricatum* (Hamilton et al. 2017).

Dietary Composition

In terms of feeding, parrotfishes are largely distinguished by the impact they have on the benthos while feeding (e.g. Bellwood and Choat 1990), rather than the specific range of food items ingested or digested (but see Clements and Choat Chapter 3). The key difference being, the extent to which they excavate the substratum; excavating parrotfishes (*Cetoscarus* spp., *B. muricatum*, *Sp. viride*, *Sparisoma amplum* and *Chlorurus* spp.) have robust jaws, taking relatively slow, but deep bites, and feed predominantly on convex surfaces, presumably targeting endolithic food items (Clements and Choat Chapter 3). Scraping parrotfishes (*Hipposcarus* spp., *Scarus* spp. and *Sparisoma aurofrenatum*) have comparatively weak jaws, taking rapid, shallow bites, and feed mainly on flat (rather than convex or concave) surfaces, where they ingest mainly epilithic food particles (Bellwood and Choat 1990). Both scraping and excavating parrotfishes are benthic-feeding fishes that predominantly graze on carbonate substrates that are not obviously occupied by sessile invertebrates, but are generally covered by the epilithic algal matrix (EAM: *sensu* Wilson et al. 2003). However, many parrotfishes (both scraping and excavating species) have also been recorded to feed from the surface of live scleractinian corals (e.g. Cole et al. 2008, Bonaldo et al. 2014, Bonaldo and Rotjan Chapter 9). A third grouping also exists, comprising 'browsing' parrotfishes (e.g. *L. vaigiensis*, *Calotomus* spp. and most *Sparisoma* spp.), which bite off distinct pieces of epilithic algae, macroalgae or seagrasses, and are generally associated with rocky reefs or seagrass habitats, rather than coral reefs (Bellwood 1994). The greatest diversity and abundance of browsing parrotfishes is in the Caribbean, whereas scraping and excavating species dominate in the Indo-Pacific and their centre of diversity is the coral triangle (Bonaldo et al. 2014).

While parrotfishes are typically categorised as herbivorous, detailed analyses of gut contents and short-chain fatty acid profiles for representative species (*Ch. microrhinos* and *Sc. schlegeli*) suggest that they feed primarily on detritus and protein-rich epilithic, endolithic and epiphytic microscopic phototrophs (Choat et al. 2004, Clements and Choat Chapter 3). There is ongoing debate about the extent to which parrotfishes (especially, scraping and excavating species) derive nutrition from plant matter versus non-plant material, but it is clear that parrotfishes do ingest large amounts of detritus, meiofauna and bacteria whilst grazing on EAM (Bonaldo et al. 2014), much of which, is far more nutritious than the algae itself (Wilson et al. 2003, Crossman et al. 2005, Kramer et al. 2013). Even if not targeting plant material (e.g. Choat et al. 2004), parrotfishes still play critical functional roles in structuring coral reef habitats, contributing to the periodic clearance of epilithic algae from carbonate substrata (Ogden and Lobel 1978, Mumby et al. 2007, Bonaldo and Bellwood 2009), as well bioerosion, re-working and transport of sediments (Bellwood 1995, 1996, Bellwood et al. 2003). Recent work has highlighted that parrotfishes may play an indirect role in coral recovery following disturbances as their scraping and excavating feeding activities increase the amount of bare substratum for settlement of corals, and influence post-settlement survivorship of corals through incidental mortality, particularly in areas of intense grazing by larger parrotfishes (Trapon et al. 2013). There is also increasing realisation that parrotfishes are very diverse (even among scaping versus excavating species) in terms of their ecology and biology (Bellwood and Choat 1990, Bellwood et al. 2003, 2006a, Hoey and Bellwood 2008, Bonaldo et al. 2014, Hoey et al. 2016a). More research is required to understand individual differences in resource use (Clements and Choat Chapter 3), which is fundamental to understanding their functional importance, as well as their vulnerability to widespread and increasing degradation of coral reef ecosystems.

Parrotfish species expected to be most at risk from significant and widespread degradation of coral reef habitats are those species that specifically graze on the live tissues of scleractinian corals (Comeros-Raynal et al. 2012). Notably, corallivorous fishes are consistently among the first and worst-affected fishes following coral depletion (e.g. Williams 1986, Wilson et al. 2006, Pratchett et al. 2008), though responses vary depending on; i) their dependence on corals for food, ii) the severity and extent of coral loss, and iii) the extent to which live coral (versus other factors such as recruitment) may be limiting their abundance (Pratchett et al. 2008, Emslie et al. 2011). The primary factor that determines the extent to which coral-feeding fishes will be sensitive to coral loss is the extent to which they actually feed on corals versus other non-coral prey. Among butterflyfishes, obligate coral-feeders (for which corals represent >90 percent of their diet: *sensu* Cole et al. 2008) are more affected by coral loss compared to facultative coral feeders (Pratchett et al. 2008, Emslie et al. 2011), which only occasionally eat live corals. For some species however, corals may represent an important and necessary, albeit minor component of their diet.

At least 30 different species of parrotfishes have been reported to feed on live corals (Table 2), but all are facultative corallivores and most take <10 percent of bites from live corals. The only parrotfish that can reasonably be considered to be a major coral predator is *B. muricatum* (Bellwood et al. 2003, Hoey and Bellwood 2008). On the outer GBR, Hoey and Bellwood (2008) showed that *B. muricatum* took close to half its' bites on live corals, feeding mainly on *Acropora*, *Isopora* and *Pocillopora* corals (see also Bellwood and Choat 1990). Other parrotfishes, meanwhile, only very occasionally feed on the surface of live corals (Table 2), grazing mostly on EAM covering non-coral substrates (Bonaldo et al. 2014). Even for *B. muricatum,* it is not clear whether individuals are selectively targeting live corals, because high rates of feeding (≥ 50 percent) on live corals have been recorded in areas with high coral cover (Hoey and Bellwood 2008), such that high rates of grazing

Table 2. Proportional use of live corals by benthic-feeding parrotfishes, based on the proportion of bites taken from the surface of living corals as opposed to bites on EAM or other non-coral substrates

No.	Group/Species	% coral use	Principal source
Excavators			
1	*Bolbometopon muricatum*	48	Hoey and Bellwood 2008
2	*Cetoscarus bicolor*	1	Bellwood and Choat 1990
3	*Cetoscarus ocellatus*	-	Bellwood 1985
4	*Chlorurus gibbus*	-	Alwany et al. 2009
5	*Chlorurus perspicillatus*	1	Ong and Holland 2010
6	*Chlorurus microrhinos*	2	Bellwood and Choat 1990
7	*Chlorurus sordidus*	-	Rotjan and Lewis 2008
8	*Chlorurus spirilus*	7	Lokranatz et al. 2008
9	*Chlorurus strongylocephalus*	27	Lokranatz et al. 2008
10	*Sparisoma amplum*	8	Francini-Filho et al. 2010
11	*Sparisoma viride*	6	Cardoso et al. 2009
Scrapers			
11	*Scarus coelestinus*	-	Randall 1974
12	*Scarus ferrugineus*	8	Afeworki et al. 2011
13	*Scarus flavipectoralis*	-	Bonaldo and Bellwood 2011
14	*Scarus frenatus*	1	Bellwood and Choat 1990
15	*Scarus ghobban*	-	Alwany et al. 2009
16	*Scarus guacamaia*	-	Rotjan and Lewis 2006
17	*Scarus iseri*	<3	Cardoso et al. 2009
18	*Scarus niger*	9	Lokranatz et al. 2008
19	*Scarus perrico*	-	Glynn et al. 1972
20	*Scarus perspicillatus*	-	Ong and Holland 2010
21	*Scarus rivulatus*	1	Bellwood and Choat 1990
22	*Scarus rubroviolaceus*	1	Ong and Holland 2010
23	*Scarus taeniopterus*	5	Cardoso et al. 2009
24	*Scarus trispinosus*	1	Francini-Filho et al. 2010
25	*Scarus vetula*	3	Cardoso et al. 2009
26	*Scarus viridifucatus*	-	McClannahan et al. 2005
27	*Sparisoma aurofrenatum*	1	Cardoso et al. 2009
28	*Sparisoma rubrippinne*	2	Cardoso et al. 2009
29	*Sparisoma chrysopterum*	<3	Cardoso et al. 2009
Browser			
30	*Calatomus carolinus*	-	McClanahan et al. 2005

on live corals may occur even if these fishes are biting randomly on coral reef substrates. These parrotfishes do however, tend to avoid biting from the surface of certain coral types, such as encrusting *Montipora* spp. (Hoey and Bellwood 2008), but this may be attributable to the physical structure of such corals (specifically, a lack of convex surfaces), rather than any specific limitations in food quality. Previous studies of coral-feeding by parrotfishes tend to focus on the ecological effects of parrotfishes on coral assemblages (e.g. Rotjan and Lewis 2006, 2008, Bonaldo and Bellwood 2011, Bonaldo et al. 2014), such that it is still unclear whether corals are an important and potentially limiting component of their diet. While *B. muricatum* exerts a major influence on the structure and dynamics of corals and coral reef habitats (Bellwood et al. 2003, Hoey and Bellwood 2008), it is unclear whether the abundance and composition of scleractinian corals also influence the distribution, abundance and fitness of these coral-feeding parrotfish.

Vulnerability to Disturbance

The two most pervasive threats to marine fishes are fisheries exploitation and habitat degradation (Wilson et al. 2008b, Graham et al. 2011), collectively accounting for >90 percent of recorded marine population and species extinctions (Dulvy et al. 2003). Importantly, both fisheries exploitation and habitat degradation continue to worsen with burgeoning human populations across many tropical island nations (Newton et al. 2007, Bell et al. 2013). These are leading to increased fisheries demand, coastal development, sedimentation, eutrophication and pollution, which are being further compounded by emerging effects of global climate change (e.g. Hughes et al. 2003, 2017, Hoegh-Guldberg and Bruno 2010). Vulnerability of parrotfishes to habitat degradation and loss will be largely dictated by their individual resource requirements, but it is important to consider that any effects of habitat degradation may be compounded by targeted fisheries exploitation of some parrotfish species.

Vulnerability to Habitat Degradation and Loss

Coral reef ecosystems are particularly vulnerable to habitat degradation owing to the sensitivity of key habitat forming organisms (namely, scleractinian corals) to a wide range of natural and anthropogenic disturbances. Most acute disturbances (e.g. cyclones, flood events, and climate-induced coral bleaching) cause elevated rates of whole colony mortality (Bellwood et al. 2004), adding to significant background rates of coral mortality (Madin et al. 2014, Pisapia et al. 2014). Significant and recurrent disturbances can also lead to loss of ecosystem resilience, defined as "the capacity of a system to absorb disturbance and reorganize while undergoing change so as to still retain essentially the same function, structure, identity and feedbacks" (Walker et al. 2004, p2). This can lead to subsequent long-term degradation if the capacity of reefs to recover from such disturbances is diminished. The capacity for recovery depends in part on the feeding activities of parrotfishes for the provision of suitable substrate allowing successful settlement and recruitment of coral larvae. Disturbances which dramatically reduce the biomass of parrotfishes may ultimately modify the composition of algal communities, with the loss of species desirable for coral settlement (turf and coralline algae; Harrington et al. 2004, Birrell et al. 2005) and the replacement by those which may inhibit settlement or increase post settlement mortality (such as young macroalgae and cyanobacteria; Tanner 1995, Kuffner et al. 2006). In such scenarios, algae or other benthic organisms can then replace corals as the dominant benthic life form (Done 1992, Scheffer et al. 2001, Nyström et al. 2008, Norström et al. 2009, Cheal et al. 2010) in an alternate stable state, which may persist for years.

Disturbances can vary in their intensity, spatial extent, duration and frequency and their effects can range from negligible to severe and can affect not only hard coral cover, but also habitat complexity (Adjeroud et al. 2002, Emslie et al. 2011, 2012). For example, physical disturbances (e.g. severe tropical storms) can kill corals, but also breakdown the structural habitat previously afforded by their skeletons (Wilson et al. 2006, Halford et al. 2004). Even if corals die but are left intact (e.g., during outbreaks of crown-of-thorns starfish and coral bleaching) the skeletons of dead corals are highly susceptible to biological and physical erosion (Glynn 1997, Hutchings 2011). Over time, skeletons of erect branching corals (e.g., *Acropora* and *Pocillopora*) break down into coral rubble (Sheppard et al. 2002, Graham et al. 2006), whereas more robust skeletons of massive corals (e.g. *Porites*) may become dislodged or gradually eroded *in situ* (Sheppard et al. 2002). There is mounting evidence that habitat complexity plays a pivotal role in structuring reef fish communities, and reductions in complexity can cause reductions in the abundance and diversity of reef fishes and in extreme cases, local extirpations (Jones et al. 2004, Bellwood 2006b, 2012b, Graham et al. 2006, Wilson et al. 2006, Emslie et al. 2014, Hoey et al. 2016b).

Data on parrotfish responses to acute episodes of coral loss associated with major disturbances (e.g., cyclones, bleaching episodes, or outbreaks of crown-of-thorns starfish) were compiled initially from 105 published studies. To standardise responses according to the severity of coral loss, proportional changes in the individual abundance of each species are expressed as a ratio of the proportional loss of coral cover, following Wilson et al. (2006) and Pratchett et al. (2008, 2011b). Standardized responses are then averaged across all studies that consider each fish species (Fig. 1). Individual responses of parrotfishes to acute episodes of coral loss are likely to vary depending on the overall level of coral loss, corresponding changes in the productivity or cover of macroalgae (Mumby et al. 2007), and/or the extent to which structural complexity is also affected (Pratchett et al. 2008). However, these analyses look only at relationships between changes in live coral cover versus individual abundance of different species.

Of the 29 species of parrotfishes considered in these analyses, only nine species exhibited consistent declines in abundance following episodes of coral depletion (Fig. 1). Of these, three species (*Scarus caudofasciatus*, *Sc. schlegeli*, and *Scarus iseri*) exhibited declines in abundance that were disproportionate to (up to 3.7 times) the loss of live coral cover. However, based on the available data, none of these species appear to be heavily reliant on live coral cover (Tables 1 and 2). Interestingly, a recent study found that the abundance of scraping, but not excavating, parrotfishes across 33 islands in the central and western Pacific were positively related to habitat complexity (Heenan et al. 2016). Most parrotfishes were either unaffected or increased in abundance following episodes of coral depletion (see also Hart et al. 1996). Increases in abundance of these species may relate to increased abundance of algae, which often occurs following localised corals loss (e.g. Pratchett et al. 2008, Cheal et al. 2010), or the increased opportunities to occupy habitats that had been vacated by species that did decline following coral depletion.

While meta-analyses are useful in drawing out generalities in parrotfish response to disturbance from the literature, they must be interpreted with caution, as data is drawn from widely disparate studies that can utilise various methodologies and generally were not designed to examine the response of parrotfishes per se, but usually looked at the whole or part of the fish community. Another issue is that the length of time after the disturbance occurred varies among studies. A bleaching event may cause mortality among the coral community; however the skeletal structure remains in place providing structure and refuge among which reef fishes can shelter. Over time, this structure may erode as coral skeletons are broken down through wave action and bioeroders. Here we

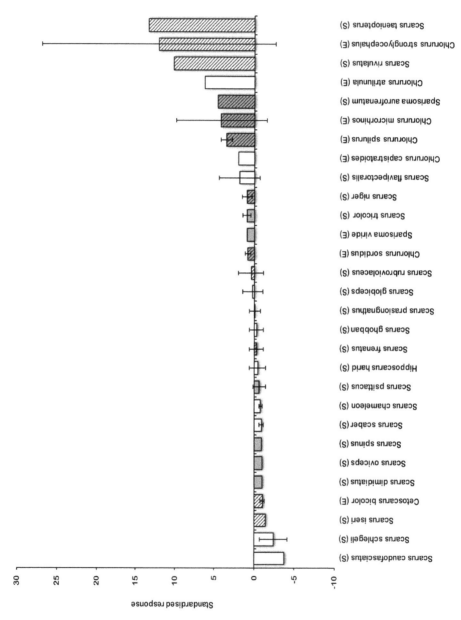

Fig. 1. Standardized responses of 32 species of parrotfishes to declines in coral cover. Responses are calculated based on the proportional changes in the local abundance of each species relative to declines in total live coral cover, following Wilson et al. (2006). Where possible, standard error is calculated to show variation in responses recorded among different studies. Species are coded according to their functional group (scraping – S, or excavating – E), association with live corals (grey shading) and reported grazing on live corals (cross-hatching).

use a long-term (~20 year) dataset collected using standardized methods to investigate changes in parrotfish assemblages following declines in hard coral cover. We then partition the changes among disturbances of various types. The Long Term Monitoring Program (LTMP) at the Australian Institute of Marine Science (AIMS) has surveyed fish and benthic communities on 47 GBR reefs between 14°S and 24°S annually from 1995 to 2005, and biennially thereafter. In the intervening years after 2005, an additional 45 reefs between 16°S and 24°S were surveyed as part of the GBR rezoning monitoring program, yielding a total of 101 reefs available for examination.

Overall, 123 natural disturbances were recorded on reefs surveyed during the course of the LTMP including; nine bleaching, 21 crown-of-thorns starfish outbreaks, nine coral disease episodes, 25 multiple disturbances (more than one disturbance occurring concurrently e.g. bleaching and storm) and 55 storms and cyclones. These disturbances caused declines in absolute hard coral cover ranging from 4.9 to 65.9 percent. In order to compare the results from the LTMP to the meta-analysis of the general literature described above, we conducted a preliminary analysis to consider how changes in hard coral cover affected parrotfish assemblages irrespective of disturbance type. The change in relative abundance (percent) of each species was calculated and standardized by coral loss as described above. Most parrotfishes were unaffected, or increased in abundance, following localised coral depletion (Fig. 2); only two species of parrotfishes, *Scarus dimidiatus* and *Sc. forsteni*, exhibited negative responses to disturbance. These results were similar to the meta-analysis of the literature and indicate that on the whole, parrotfishes appear very resilient to coral loss.

For a more thorough examination of parrotfish response to disturbance, we examined the LTMP data and extracted data from 31 reefs where a single disturbance occurred. This enabled the clearest picture of disturbance effects on parrotfishes without the confounding effects of preceding or subsequent disturbances. We assessed the short-term effects of disturbances by modelling the change in individual species abundance from the year immediately preceding and following the disturbance. The impacts of individual disturbances resulted in few negative responses from GBR parrotfish assemblages. Overall parrotfish abundance and diversity either increased or remained unchanged (Fig. 3). There was only one species (*Sc. rivulatus*) whose abundance declined significantly and this was in response to multiple disturbances impacting reefs concurrently (Fig. 4). The vast majority of species were unaffected by disturbances and only *Ch. microrhinos* and *Ch. spilurus* had significant increases in abundance, which occurred following the passage of tropical cyclones (Fig. 4).

While most parrotfishes seem largely resilient to the effects of habitat degradation facilitated by the loss of live scleractinian coral cover, they appear much more susceptible to disturbances that significantly reduce habitat complexity (Graham et al. 2006, Wilson et al. 2006, Emslie et al. 2014). A clear example comes from the GBR, where reefs of the Capricorn-Bunker sector on the southern GBR, have undergone dramatic reductions in hard coral cover following storms in 2008/2009. As the underlying substrate on these sites consisted of a flat terrace, with little in the way of vertical relief or substrata rugosity, much of the habitat complexity that was afforded by the skeletons of live coral colonies, was ultimately lost following storm impacts (Emslie et al. 2014). Of the 10 species that occurred in sufficient abundance to be analysed, eight declined by more than 50 percent of their pre-disturbance abundance and one species, *Ch. microrhinos* became locally extirpated on the sites, although there was little commonality among species that declined in abundance in

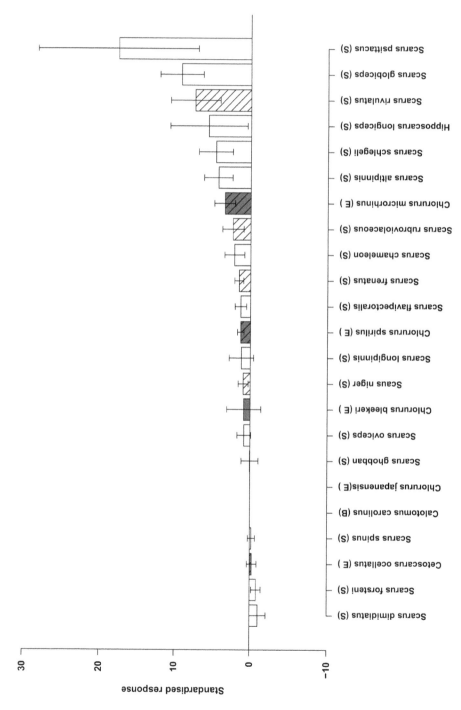

Fig. 2. Standardized response of 22 species of parrotfishes to declines in coral cover. Data are taken from the Australian Institute of Marine Science Long Term Monitoring database. Responses are calculated based on the proportional changes in the local abundance of each species relative to declines in total live coral cover, averaged across 38 individual disturbances. Standard error is calculated to show variation in responses recorded among different disturbances. Species are coded as per Fig. 1.

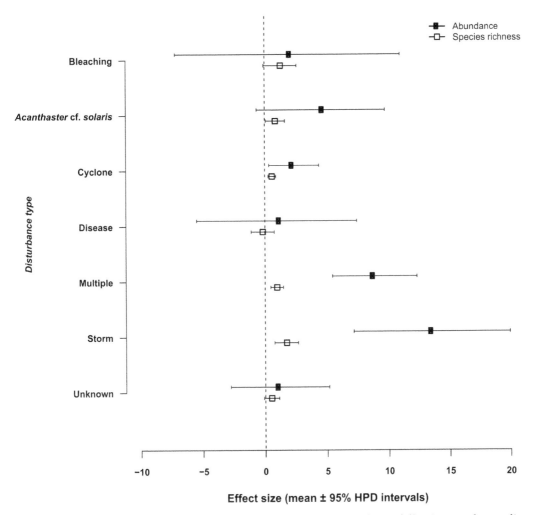

Fig. 3. Overall change in parrotfish total abundance and species richness following coral mortality caused by: bleaching (n = 1), *Acanthaster* cf. *solaris* outbreaks (n = 1), cyclones (n = 20), coral disease (n = 2), multiple disturbances (n = 7), sub-cyclonic storms (n = 2), unknown disturbances (n = 5). Data are taken from reefs which had only a single disturbance.

terms of their functional role or dependence on live hard coral. This compares to reefs that underwent similar reductions in coral cover but retained much of the habitat complexity where only one species (*Sc. flavipectoralis*) declined in abundance and then by only 20 percent. Similarly, on reefs where there was no disturbance and no concomitant change in coral cover or complexity only one species declined in abundance by less than 10 percent. This suggests that parrotfishes are largely invariant in losses of hard coral cover per se, but are particularly susceptible to losses of habitat complexity presumably through the loss of provision of shelter (Emslie et al. 2014). Similar results are reported following the erosion of coral skeleton some years after bleaching in the Seychelles, where sites that lost complexity through the erosion of coral skeletons had large reductions in the abundance of small bodied parrotfishes (Graham et al. 2006).

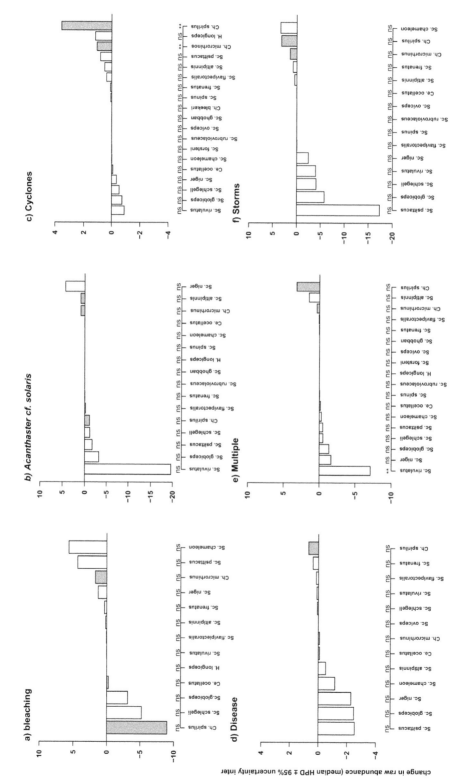

Fig. 4. Variation in responses of parrotfishes to declines in coral cover caused by bleaching ($n = 1$), *Acanthaster* cf. *solaris* outbreaks ($n = 1$), cyclones ($n = 20$), coral disease ($n = 2$), multiple disturbances ($n = 7$), sub-cyclonic storms ($n = 2$). Responses are changes in abundance as calculated from a Bayesian hierarchical linear model. Data are taken from reefs which had only a single disturbance. NS = non-significant, ** = significant decline/increase in abundance.

Vulnerability to Exploitation

Overfishing is arguably one of the most pressing human impacts in shallow coastal environments (Hughes 1994, Jackson et al. 2001, Bellwood et al. 2012a), contributing to the serial collapse of globally significant fisheries, including the Peruvian anchovy fishery (Walsh et al. 1980) and Atlantic cod fishery (Myers et al. 1996). As human populations expand, ever increasing pressure is placed on fishery resources. Increases in demand and the consequences of fisheries depletion are most pronounced in developing nations (e.g. Bell et al. 2009). A ubiquitous development among the world's fishery harvest has been a shift to taxa from lower trophic levels, as stocks of top order predators are depleted (Pauly et al. 1998). While parrotfishes have historically been exposed to only low levels of exploitation in most parts of the Indo-Pacific (e.g. Russ and Alcala 1998, Jennings et al. 1999), they are now becoming increasingly utilised in many parts of the world, forming an important component of the artisanal, subsistence and commercial fisheries (Dalzell et al. 1996, Houk et al. 2012, Bejarano et al. 2013, Rasher et al. 2013), particularly in developing nations. Recent fisheries data suggests that parrotfishes represent a significant component of the reef-associated harvest, accounting for between 40 and >75 percent in some locations in the central Pacific (Rhodes et al. 2008, Houk et al. 2012, Bejarano et al. 2013). Across Polynesia, for example, parrotfishes are among the most heavily harvested group of coral reef fishes (Pratchett et al. 2011c), accounting for up to 36.8 percent of the biomass of fishes harvested from coral reef environments. The schooling behaviour of some parrotfishes, coupled with their tendency to aggregate for spawning means they can be easily speared or caught in barrier nets. Due to their unique sleeping behaviour they are also particularly vulnerable to nocturnal fishing (Johannes 1981, Donaldson and Dulvy 2004).

The majority of parrotfishes are relatively short-lived, compared with other taxa of reef fishes (Choat et al. 1996, 2003, Choat and Robertson 2002, Taylor and Choat 2014, Taylor et al. Chapter 4). This coupled with flexible life history traits suggests they should be reasonably resilient to fishery exploitation. There have been few detailed investigations of parrotfish vulnerability to overexploitation. While some early studies suggest a weak response to fishing pressure (Russ and Alcala 1998, Jennings et al. 1998), recent investigations have revealed that some species of parrotfishes are highly vulnerable to fisheries exploitation (Bejarano et al. 2013, Taylor and Choat 2014), with changes to community structure (Hawkins and Roberts 2003, Clua and Legendre 2008) resulting from extractive activities. Furthermore, several studies have revealed strong responses of some parrotfishes to protection from no-take marine reserves (e.g. Stockwell et al. 2009, Rasher et al. 2013), all of which implies strong species-specific reactions to fishing. There's a reasonable expectation that larger-bodied species would be most heavily targeted by fishers so as to maximise their returns, however species from a wide variety of sizes, life histories and functional roles are harvested (Dalzell et al. 1996, Gillet and Moy 2006, Rhodes et al. 2008, Bellwood et al. 2012a).

The consequences of fishing exploitation may manifest themselves in a number of ways; reductions in abundance/biomass and body sizes, reductions in age at maturity, changes in community structure, recruitment failure, reduced functional capacity and exposure to extinction risk. Despite the increasing fishery focus on parrotfishes in many parts of the world, thorough evaluations of their vulnerability are uncommon and those that have been conducted describe the majority of species as having a low risk of local extinction (e.g. Comeros-Raynal et al. 2012). There is however, a growing literature documenting clear responses to exploitation and protection, which can cause changes in parrotfish communities impacting on ecological functions (Russ and Alcala 1989, Polunin and Roberts 1993, Hawkins and Roberts 2003, Clua and Legendre 2008, Bellwood et al.

2003, 2012a, Mumby et al. 2013, Edwards et al. 2014, Taylor et al. 2014). Most notably, fishing exploitation reduces the abundance and biomass of target species, and is commonly manifest in greater abundance/biomass inside areas closed to fishing, such as no take marine reserves, compared to areas where fishing occurs (Russ and Alcala 1989, 1998, Polunin and Roberts 1993, Rakitin and Kramer 1996, Friedlander and DeMartini 2002, Hawkins and Roberts 2003, Floeter et al. 2006, Heenan and Williams 2013, Rasher et al. 2013).

Direct exploitation of fish stocks can also result in a decrease in the average size of fishes in areas of heavy exploitation from fishing, as the removal of larger individuals occurs more quickly than new cohorts can grow (Taylor 2014). Where target species are protogynous hermaphrodites (female to male sex transition) as are parrotfishes, there is inherent vulnerability to size-selective exploitation and heavy levels of exploitation can lead to selective removal of large individuals (males), a reduction in the size at reproductive maturation and an overall decline in in reproductive output to levels where it may be compromised (Taylor 2014). In extreme examples, terminal phase (TP) males may be completely absent from populations, as was found to be the case in Jamaica, or severely reduced in average size and the proportion of TP males in the populations (Hawkins and Roberts 2003).

Exploitation of parrotfishes can also have serious implications for ecosystem function and resilience of coral reef ecosystems, as declines in the overall abundance of parrotfishes, but also selective removal of larger individuals greatly affects rates of bioerosion, coral predation and redistribution of sediments (Bellwood et al. 2012a), as well as reducing local rates of grazing (Lokrantz et al. 2010). Importantly, large bodied individuals play a disproportionate role in ecosystem functioning (Lokrantz et al. 2008) and fishing can therefore, have deleterious ecosystem-level effects even when harvesting is within limits that can sustain local populations of target species. Most notably, the reduction in overall grazing rates may compromise the capacity of benthic communities to recover following extensive coral loss (Hughes et al. 2007, Bejarano et al. 2013, but see Questel and Russ Chapter 14), resulting persistent high levels of macroalgal cover (Steneck 1988, Hughes 1994, Mumby et al. 2006) which can inhibit coral settlement and post-settlement survivorship.

Conclusions

While there is little doubt that parrotfishes have an important functional role (especially, within coral reef environments), the extent to which these species are vulnerable to increasing anthropogenic disturbances is equivocal. Parrotfishes are generally resilient to coral depletion and habitat degradation caused by major disturbances affecting contemporary reef environments (e.g., cyclones, coral bleaching, and outbreaks of crown-of-thorns starfish), though parrotfishes (like most reef fishes) are sensitive to widespread or significant loss of topographic complexity, which may occur due to extensive coral loss. Currently however, the major threat to parrotfishes is fisheries exploitation, with over-exploitation of key species (e.g. *B. muricatum*) already apparent at many locations (Bellwood et al. 2012a). This excessive exploitation of functionally important parrotfishes has significant consequences for the structure and function of coral reef ecosystems (e.g. Bellwood et al. 2003), such that management of these fisheries must focus not only on the sustainability of fisheries impacts, but also the larger and longer-term consequences of fisheries impacts. Given the pivotal role that parrotfishes perform in reef resilience, their continued exploitation is of considerable concern, especially given the imperative to maximise ecosystem-level resilience to emerging effects of global climate change and other direct anthropogenic disturbances.

In high diversity systems, such as coral reefs, it is generally assumed that there would be high levels of functional redundancy, whereby there are many different species capable of fulfilling essentially the same ecological role and function. Moreover, the loss of some species may be compensated for by increases in the abundance or function of other ecologically-equivalent species. There is some evidence of response diversity (Elmqvist et al. 2003) among broadly defined groups of herbivorous fishes (Pratchett et al. 2011b, Emslie et al. 2014), whereby the median response to coral loss is negative, but many species actually increase in abundance immediately following major disturbances. Examination of the responses of parrotfishes to disturbances in this chapter, both through a meta-analysis and an analysis of the LTMP dataset support this notion, showing broadly differentiated responses across different parrotfish species (Figs. 1 and 2). However, we must be cautious in assuming functional-equivalency among different parrotfishes. Importantly, there is increasing evidence that sympatric species have highly specialised and often complementary diets (e.g. Burkepile and Hay 2008, Mantyka and Bellwood 2007, Rasher et al. 2013, Loffler et al. 2015). Moreover, different species often perform unique and critical functions (Bellwood et al. 2003, 2006c, Fox et al. 2009, Hoey and Bellwood 2009, 2011, Streit et al. 2015).

Acknowledgments

We thank all past and present members of the Long Term Monitoring team at the Australian Institute of Marine Science (AIMS) for collection of the benthic and parrotfish data. The monitoring program was largely funded by AIMS, with additional funding provided by the CRC Reef Research Centre, the Australian Government's Marine and Tropical Sciences Research Facility (MTSRF) and the National Environmental Research Program (NERP). We also thank Nick Graham and Andrew Hoey for helpful comments which improved the chapter.

References Cited

Adam, T.C., R.J. Schmitt, S.J. Holbrook, A.J. Brooks, P.J. Edmunds, R.C. Carpenter and G. Bernardi. 2011. Herbivory, connectivity, and ecosystem resilience: response of a coral reef to a large-scale perturbation. PLoS One 6: e23717.

Adam, T.C., D.E. Burkepile, B.I. Ruttenberg and M.J. Paddack. 2015. Herbivory and the resilience of Caribbean coral reefs: knowledge gaps and implications for management. Mar. Ecol. Prog. Ser. 520: 1–20.

Adjeroud, M., D. Augustin, R. Galzin and B. Salvat 2002. Natural disturbances and interannual variability of coral reef communities on the outer slope of Tiahura (Moorea, French Polynesia): 1991 to 1997. Mar. Ecol. Prog. Ser. 237: 121–131.

Aeby, G.S., G.J. Williams, E.C. Franklin, J. Haapkyla, C.D. Harvell, S. Neale, C.A. Page, L. Raymundo, B. Vargas-Angel, B.L. Willis, T.M. Work, S.K. Davy. 2011. Growth anomalies on the coral genera *Acropora* and *Porites* are strongly associated with host density and human population size across the Indo-Pacific. PLoS One 6: e16887.

Afeworki, Y., J.H. Bruggemann and J.J. Videler. 2011. Limited flexibility in resource use in a coral reef grazer foraging on seasonally changing algal communities. Coral Reefs 30: 109–122.

Alwany, M.A., E. Thaler and M. Stachowitsch. 2009. Parrotfish bioerosion on Egyptian Red Sea reefs. J. Exp. Mar. Biol. Ecol. 371: 170–176.

Aswani, S. and R.J. Hamilton. 2004. Integrating indigenous ecological knowledge and customary sea tenure with marine and social science for conservation of bumphead parrotfish (*Bolbometopon muricatum*) in the Roviana Lagoon, Solomon Islands. Environ. Cons. 31: 69–83.

Bejarano, S., Y. Galbuu, T. Sapolu and P.J. Mumby. 2013. Ecological risk and the exploitation of herbivorous reef fish across Micronesia. Mar. Ecol. Prog. Ser. 482: 197–215.

Bell, J.D., M. Kronen, A. Vunisea, W.J. Nash, G. Keeble, A. Demmke, S. Pontifex and S. Andréfouët. 2009. Planning the use of fish for food security in the Pacific. Mar. Pol. 33: 64–76.

Bell, J.D., A. Ganachaud, P.C. Gehrke, S.P. Griffiths, A.J. Hobday, O. Hoegh-Guldberg, J.E. Jonhson, R. Le Borgne, P. Lehodey, J.M. Lough, R.J. Matear, T.D. Pickering, M.S. Pratchett, A. Sen Gupta, I. Senina and M. Waycott. 2013. Tropical Pacific fisheries and aquaculture have mixed responses to climate change. Nat. Climate Change 3: 591–599.

Bellwood, D.R. 1985. The functional morphology, systematic and behavioural ecology of parrotfishes (Family Scaridae). PhD thesis, James Cook University, Australia.

Bellwood, D.R. 1994. A phylogenetic study of the parrotfishes family Scaridae (Pisces: Labroidei) with a revision of genera. Rec. Aust. Mus. Suppl. 20: 1–86

Bellwood, D.R. 1995. Direct estimate of bioerosion by two parrotfish species, *Chlorurus gibbus* and *C. sordidus*, on the Great Barrier Reef, Australia. Mar. Biol. 121: 419–429.

Bellwood, D.R. 1996. Production and reworking of sediment by parrotfishes (family Scaridae) on the Great Barrier Reef, Australia. Mar. Biol. 125: 795–800.

Bellwood, D.R. and J.H. Choat. 1989. A description of the juvenile phase colour patterns of 24 parrotfish species (family Scaridae) from the Great Barrier Reef, Australia. Rec. Aust. Mus. 41: 1–41.

Bellwood, D.R. and J.H. Choat. 1990. A functional analysis of grazing in parrotfishes (family Scaridae): the ecological implications. Environ. Biol. Fish. 28: 189–214.

Bellwood, D.R. and J.H. Choat. 2011. Dangerous demographics: the lack of juvenile humphead parrotfishes *Bolbometopon muricatum* on the Great Barrier Reef. Coral Reefs 30: 549–554.

Bellwood, D.R., P.C. Wainwright, C.F. Fulton and A. Hoey. 2002. Assembly rules and functional groups at global biogeographical scales. Func. Ecol. 16: 557–562.

Bellwood, D.R., A.S. Hoey and J.H. Choat. 2003. Limited functional redundancy in high diversity systems: resilience and ecosystem function on coral reefs. Ecol. Lett. 6: 281–285.

Bellwood, D.R., T.P. Hughes, C. Folke and M. Nyström. 2004. Confronting the coral reef crisis. Nature 429: 827–833.

Bellwood, D.R., P.C. Wainwright, C.J. Fulton and A.S. Hoey. 2006a. Functional versatility supports coral reef biodiversity. Proc. R. Soc. B Biol. Sci. 273: 101–107.

Bellwood, D.R., A.S. Hoey, J.L. Ackerman and M. Depczynski. 2006b. Coral bleaching, reef fish community phase shifts and the resilience of coral reefs. Glob. Change Biol. 12: 1587–1594.

Bellwood, D.R., T.P. Hughes and A.S. Hoey. 2006c. Sleeping functional group drives coral-reef recovery. Curr. Biol. 16: 2434–2439.

Bellwood, D. R., A.S. Hoey and T.P. Hughes. 2012a. Human activity selectively impacts the ecosystem roles of parrotfishes on coral reefs. Proc. R. Soc. Lond. B Biol. Sci. 279: 1621–1629.

Bellwood, D.R., A.H. Baird, M. Depczynski, A. González-Cabello, A.S. Hoey, C.D. Lefèvre and J.K. Tanner. 2012b. Coral recovery may not herald the return of fishes on damaged coral reefs. Oecologia 170: 567–573.

Berumen, M.L. and M.S. Pratchett. 2008. Trade-offs associated with dietary specialisation in corallivorous butterflyfishes (Chaetodontidae: *Chaetodon*). Behav. Ecol. Sociobiol 62: 989–994.

Birrell, C.L., L.J. McCook and B.L. Willis. 2005. Effects of algal turfs and sediment on coral settlement. Mar. Poll. Bull. 51: 408–414.

Bonaldo, R.M. and D.R. Bellwood. 2009. Dynamics of parrotfish grazing scars. Mar. Biol. 156: 771–777.

Bonaldo, R.M. and D.R. Bellwood. 2011. Parrotfish predation on massive *Porites* on the Great Barrier Reef. Coral Reef 30: 259–269

Bonaldo, R.M., A.S. Hoey and D.R. Bellwood. 2014. The ecosystem roles of parrotfishes on tropical reefs. Ocean. Mar. Biol. Annu. Rev. 52: 81–132

Brodie, J.E., F.J. Kroon, B. Schaffelke, E.C. Wolanski, S.E. Lewis, M.J. Devlin, I.C. Bohnet, Z.T. Bainbridge, J. Waterhouse and A.M. Davis. 2012. Terrestrial pollutant runoff to the Great Barrier Reef: an update of issues, priorities and management response. Mar. Poll. Bull. 65: 81–100.

Brooks, T.M., R.A. Mittermeier, C.G. Mittermeier, G.A. Da Fonseca, A.B. Rylands, W.R. Konstant, P. Flick, J. Pilgrim, S. Oldfield, G. Magin and C. Hilton-Taylor. 2002. Habitat loss and extinction in the hotspots of biodiversity. Cons. Biol. 16: 909–923.

Bruno J.F., L.E. Petes, C.D. Harvell and A. Hettinger. 2003. Nutrient enrichment can increase the severity of coral diseases. Ecol. Lett. 6: 1056–1061.

Bruno, J.F. and E.R. Selig. 2007. Regional decline of coral cover in the Indo-Pacific: timing, extent, and subregional comparisons. PLoS One 2: e711.

Burkepile, D.E. and M.E. Hay. 2008. Herbivore species richness and feeding complementarity affect community structure and function on a coral reef. Proc. Natl. Acad. Sci. USA. 105: 16201–16206.

Cardoso, S.C., M.C. Soares, H.A. Oxenford and I.M. Côté. 2009. Interspecific differences in foraging behaviour and functional role of Caribbean parrotfish. Mar. Biodiv. Rec. 2:e148.

Cheal, A.J., G. Coleman, S. Delean, I. Miller, K. Osborne and H. Sweatman. 2002. Responses of coral and fish assemblages to a severe but short-lived tropical cyclone on the Great Barrier Reef, Australia. Coral Reefs 21: 131–142.

Cheal, A.J., M.A. MacNeil, E. Cripps, M.J. Emslie, M. Jonker, B. Schaffelke and H.P.A. Sweatman. 2010. Coral-macroalgal phase shifts or reef resilience: links with diversity and functional roles of herbivorous fishes on the Great Barrier Reef. Coral Reefs 29: 1005–1015.

Cheal, A., M. Emslie, I. Miller and H. Sweatman. 2012. The distribution of herbivorous fishes on the Great Barrier Reef. Mar. Biol. 159: 1143–1154.

Choat, J.H. and D.R. Robertson. 2002. Age-based studies. pp. 57–80. *In*: P.F. Sale (ed.). Dynamics and Diversity in a Complex Ecosystem. Academic Press, San Diego.

Choat, J.H., L.M. Axe and D.C. Lou. 1996. Growth and longevity in fishes of the family Scaridae. Mar. Ecol. Prog. Ser. 145: 33–41.

Choat, J., K. Clements and W. Robbins. 2002. The trophic status of herbivorous fishes on coral reefs. Mar. Biol. 140: 613–623.

Choat, J.H., D.R. Robertson, J.L. Ackerman and J.M. Posada. 2003. An age-based demographic analysis of the Caribbean stoplight parrotfish *Sparisoma viride*. Mar. Ecol. Prog. Ser. 246: 265–277.

Choat, J.H., W.D. Robbins and K.D. Clements. 2004. The trophic status of herbivorous fishes on coral reefs. Mar. Biol. 145: 445–454.

Clua, E. and P. Legendre. 2008. Shifting dominance among Scarid species on reefs representing a gradient of fishing pressure. Aquat. Living Resour. 21: 339–348.

Coker, D.J., S.K. Wilson and M.S. Pratchett. 2014. Importance of live coral habitat for reef fishes. Rev. Fish. Biol. Fisheries 24: 89–126.

Cole, A.J., M.S. Pratchett and G.P. Jones. 2008. Diversity and functional importance of coral-feeding fishes on tropical coral reefs. Fish Fish. 9: 286–307.

Comeros-Raynal, M.T., J.H. Choat, B.A. Polidoro, K.D. Clements, R. Abesamis, M.T. Craig, M.E. Lazuardi, J. McIlwain, A. Muljadi, R.F. Myers, C.L. Nanola Jr., S. Pardede, L.A. Rocha, B. Russell, J.C. Sanciangco, B. Stockwell, H. Harwell and K.E. Carpenter. 2012. The likelihood of extinction of iconic and dominant herbivores and detritivores of coral reefs: The parrotfishes and surgeonfishes. PLoS One 7(7): e39825.

Crossman, D.J., J.H. Choat and K.D. Clements. 2005. Nutritional ecology of nominally herbivorous fishes on coral reefs. Mar. Ecol. Prog. Ser. 296: 129–142.

Danilowicz, B.S., N. Tolimieri and P.F. Sale. 2001. Meso-scale habitat features affect recruitment of reef fishes in St. Croix, US Virgin Islands. Bull. Mar. Sci. 69: 1223–1232.

Dalzell P., T.J.H. Adams and N.V.C. Polunin. 1996. Coastal fisheries in the Pacific islands. Oceanogr. Mar. Biol. Annu. Rev. 34: 395–531.

De'ath, G., K.E. Fabricius, H.P.A. Sweatman and M. Puotinen. 2012. The 27-year decline of coral cover on the Great Barrier Reef and its causes. Proc. Nat. Acad. Sci. USA 109: 17995–17999.

DeMartini, E.E., T.W. Anderson, J.C. Kenyon, J.P. Beets and A.M. Friedlander. 2010. Management implications of juvenile reef fish habitat preferences and coral susceptibility to stressors. Mar. Freshw. Res. 61: 532–540.

Devictor, V., J. Clavel, R. Julliard, S. Lavergne, D. Mouillot, W. Thuiller, P. Venail, S. Villéger and N. Mouquet. 2010. Defining and measuring ecological specialisation. J. Appl. Ecol. 47: 15–25.

Donaldson, T.J. and N.K. Dulvy. 2004. Threatened fishes of the world: Bolbometapon muricatum (Valenciennes 1840) (Scaridae). Environ. Biol. Fish. 70: 373.

Done, T.J. 1992. Phase shifts in coral reef communities and their ecological significance. Hydrobiologia 247: 121–132.

Dulvy, N.K., Y. Sadovy and J.D. Reynolds. 2003. Extinction vulnerability in marine populations. Fish Fish. 4: 25–64.

Edwards, C.B., A.M. Friedlander, A.G. Green, M.J. Hardt, E. Sala, H.P. Sweatman, I.D. Williams, B. Zgliczynski, S.A. Sandin and J.E. Smith. 2014. Global assessment of the status of coral reef herbivorous fishes: evidence for fishing effects. Proc. R. Soc. B. 281: rspb20131835.

Elmqvist, T., C. Folke, M. Nyström, G. Peterson, J. Bengtsson, B. Walker and J. Norberg. 2003. Response diversity, ecosystem change, and resilience. Front Ecol. Environ. 1: 488–494.

Emslie, M.J., M.S. Pratchett and A.J. Cheal. 2011. Effects of different disturbance types on butterflyfish communities of Australia's Great Barrier Reef. Coral Reefs 30: 461–471.

Emslie, M.J., M. Logan, D.M. Ceccarelli, A.J. Cheal, A.S. Hoey, I. Miller and H.P.A. Sweatman. 2012. Regional-scale variation in the distribution and abundance of farming damselfish on Australia's Great Barrier Reef. Mar. Biol. 159: 1293–1304.

Emslie, M.J., A.J. Cheal and K.A. Johns. 2014. Retention of habitat complexity minimizes disassembly of reef fish communities following disturbance: a large-scale natural experiment. PLoS One. 9: e105384.

Fabricius, K., K. Okaji and G. De'ath. 2010. Three lines of evidence to link outbreaks of the crown-of-thorns seastar Acanthaster planci to the release of larval food limitation. Coral Reefs 29: 593–605.

Fahrig, L. 2001. How much habitat is enough? Biol. Cons. 100: 65–74.

Feary, D.A., G.R. Almany, M.I. McCormick and G.P. Jones. 2007. Habitat choice, recruitment and the response of coral reef fishes to coral degradation. Oecologia 153: 727–737.

Floeter, S.R., B.S. Halpern and C.E.L. Ferreira. 2006. Effects of fishing and protection on Brazilian reef fishes. Biol. Conserv. 128: 391–402.

Fox, L.R. and P.A. Morrow. 1981. Specialization – species property or local phenomenon. Science 211: 887–893.

Fox, R.J., T.L. Sunderland, A.S. Hoey and D.R. Bellwood. 2009. Estimating ecosystem function: contrasting roles of closely related herbivorous rabbitfishes (Siganidae) on coral reefs. Mar. Ecol. Prog. Ser. 385: 261–269.

Francini-Filho, R.B., C.M. Ferreira, E. Oliveira, C. Coni, R.L. De Moura and L. Kaufman. 2010. Foraging activity of roving herbivorous reef fish (Acanthuridae and Scaridae) in eastern Brazil: influence of resource availability and interference competition. J. Mar. Biol. Assoc. U.K. 90: 481–492.

Friedlander, A.M. and J.D. Parrish. 1998. Habitat characteristics affecting fish assemblages on a Hawaiian coral reef. J. Exp. Mar. Biol. Ecol. 224: 1–30.

Friedlander, A.M. and E.E. DeMartini. 2002. Contrasts in density, size, and biomass of reef fishes between the northwestern and the main Hawaiian islands: the effects of fishing down apex predators. Mar. Ecol. Prog. Ser. 230: 253–264.

Friedlander, A.M., S.A. Sandin, E.E. DeMartini and E. Sala. 2010. Spatial patterns of the structure of reef fish assemblages at a pristine atoll in the central Pacific. Mar. Ecol. Prog. Ser. 410: 219–231.

Gardner, T.A., I.M. Côté, J.A. Gill, A. Grant and A.R. Watkinson. 2003. Long-term region-wide declines in Caribbean corals. Science 301: 958–960.

Garpe, K.C., S.A.S. Yahya, U. Lindahl and M.C. Ohman. 2006. Long-term effects of the 1998 coral bleaching event on reef fish assemblages. Mar. Ecol. Prog. Ser. 315: 237–247.

Gillet, R. and W. Moy. 2006. Spearfishing in the Pacific Islands: current status and management issues. Secretariat of the Pacific Community, Noumea.

Glynn, P.W. 1997. Bioerosion and coral reef growth: a dynamic balance. pp. 68–95. *In*: C. Birkeland (ed.). Life and death of coral reefs. Chapman and Hall, New York.

Glynn, P.W., R.H. Stewart and J.E. McCosker. 1972. Pacific coral reefs of Panama: structure, distribution and predators. Geol. Rundsch. 61: 483–519.

Graham, N.A.J., S.K. Wilson, S. Jennings, N.V.C. Polunin, J.P. Bijoux and J. Robinson. 2006. Dynamic fragility of oceanic coral reef ecosystems. Proc. Natl. Acad. Sci. USA 103: 8425–8429.

Graham N.A.J., S.K. Wilson, S. Jennings, N.V.C. Polunin, J. Robinson, J.P. Bijoux and T.M. Daw. 2007. Lag effects in the impacts of mass coral bleaching on coral reef fish, fisheries, and ecosystems. Conserv. Biol. 21: 1291–1300.

Graham, N.A.J., P. Chabanet, R.D. Evans, S. Jennings, Y. Letourneur, M.A. MacNeil, T.R. McClanahan, M.C. Öhman, N.V.C. Polunin and S.K. Wilson. 2011. Extinction vulnerability of coral reef fishes. Ecol. Lett. 14: 341–348.

Gray, J.S. 1997. Marine biodiversity: patterns, threats and conservation needs. Biodivers. Conserv. 6: 153–175.

Green A.L. 1992. Damselfish territories: focal sites for studies of the early life history of Labroid fishes. Proc. 7th Int. Coral Reef Symp. Guam. 1: 601–605.

Green, A.L. 1998. Spatio-temporal patterns of recruitment of labroid fishes (Pisces: Labridae and Scaridae) to damselfish territories. Environ. Biol. Fish. 51: 235–244.

Gust, N., J.H. Choat and M.I. McCormick. 2001. Spatial variability in reef fish distribution, abundance, size, and biomass: a multi-scale analysis. Mar. Ecol. Prog. Ser. 214: 237–251.

Halford, A., A.J. Cheal, D. Ryan and D.M. Williams. 2004. Resilience to large-scale disturbance in coral and fish assemblages on the Great Barrier Reef. Ecology 85: 1892–1905.

Hamilton, R.J., G.R. Almany, C.J. Brown, J. Pita, N.A. Peterson and J.H. Choat. 2017. Logging degrades nursery habitat for an iconic coral reef fish. Biol. Conserv. 210: 273–280.

Harrington, L., K. Fabricius, G. De'ath and A. Negri. 2004. Recognition and selection of settlement substrata determine post-settlement survival in corals. Ecology 85: 3428–3437.

Hart, A., D.W. Klumpp and G.R. Russ. 1996. Response of herbivorous fishes to crown-of-thorns starfish *Acanthaster planci* outbreaks. II. Density and biomass of selected species of herbivorous fish and fish-habitat correlations. Mar. Ecol. Prog. Ser. 132: 21–30.

Hawkins, J.P. and C.M. Roberts. 2003. Effects of fishing on sex-changing Caribbean parrotfishes. Biol. Conserv. 115: 213–226.

Heenan, A. and I.D. Williams. 2013. Monitoring herbivorous fishes as indicators of coral reef resilience in American Samoa. PLoS One 8(11): e79604.

Heenan, A., A.S. Hoey, G.J. Williams and I.D. Williams. 2016. Natural bounds on herbivorous coral reef fishes. Proc. R. Soc. B. 283: rspb20161716.

Hoegh-Guldberg, O. and J.F. Bruno. 2010. The impact of climate change on the world's marine ecosystems. Science 328: 1523–1528.

Hoegh-Guldberg, O., P.J. Mumby, A.J. Hooten, R.S. Steneck, P. Greenfield, E. Gomez, C.D. Harvell, P.F. Sale, A.J. Edwards, K. Caldeira, N. Knowlton, C.M. Eakin, R. Iglesias-Prieto, N. Muthiga, R.H. Bradbury, A. Dubi and M.E. Hatziolos. 2007. Coral reefs under rapid climate change and ocean acidification. Science 318: 1737–1742.

Hoekstra, J.M., T.M. Boucher, T.H. Ricketts and C. Roberts. 2005. Confronting a biome crisis: global disparities of habitat loss and protection. Ecol. Lett. 8: 23–29.

Hoey, A.S. and D.R. Bellwood. 2008. Cross-shelf variation in the role of parrotfishes on the Great Barrier Reef. Coral Reefs 27: 37–47.

Hoey, A.S. and D.R. Bellwood. 2009. Limited functional redundancy in a high diversity system: single species dominates key ecological process on coral reefs. Ecosystems 12: 1316–1328.

Hoey, A.S. and D.R. Bellwood. 2011. Suppression of herbivory by macroalgal density: a critical feedback on coral reefs. Ecol. Lett. 14: 267–273.

Hoey, A.S., D.A. Feary, J.A. Burt, G. Vaughan, M.S. Pratchett and M.L. Berumen. 2016a. Regional variation in the structure and function of parrotfishes on Arabian reefs. Mar. Poll. Bull. 105: 524–531.

Hoey, A.S., E. Howells, J.L. Johansen, J.P.A. Hobbs, V. Messmer, D.M. McCowan, S.K. Wilson and M.S. Pratchett. 2016b. Recent advances in understanding the effects of climate change on coral reefs. Diversity 8: 1–12.

Houk, P., K. Rhodes, J. Cuetos-Bueno, S. Lindfield, V. Fread and J.L. McIlwain. 2012. Commercial coral-reef fisheries across Micronesia: a need for improving management. Coral Reefs 31: 13–26.

Hughes, T.P. 1994. Catastrophes, phase shifts, and large-scale degradation of a Caribbean coral reef. Science 265: 1547–1551.

Hughes, T.P., A.H. Baird, D.R. Bellwood, M. Card, S.R. Connolly, C. Folke, R. Grosberg O. Hoegh-Guldberg, J.B.C. Jackson, J. Kleypas, J.M. Lough, P. Marshall, M. Nyström, S.R. Palumbi, J.M. Pandolfi, B. Rosen and J. Roughgarden. 2003. Climate change, human impacts, and the resilience of coral reefs. Science 301: 929–933.

Hughes, T.P., M.J. Rodrigues, D.R. Bellwood, D. Ceccarelli, O. Hoegh-Guldberg, L. McCook, N. Moltchaniwskyj, M.S. Pratchett, R.S. Steneck and B. Willis. 2007. Phase shifts, herbivory, and the resilience of coral reefs to climate change. Curr. Biol. 17: 360–365.

Hughes, T.P., N.A. Graham, J.B. Jackson, P.J. Mumby and R.S. Steneck. 2010. Rising to the challenge of sustaining coral reef resilience. Trends. Ecol. Evol. 25: 633–642.

Hughes, T.P., A.H. Baird, E.A. Dinsdale, N.A. Moltschaniwskyj, M.S. Pratchett, J.E. Tanner and B.L. Willis. 2012 Assembly rules of reef corals are flexible along a climatic gradient. Curr. Biol. 22: 736–741.

Hughes, T.P., J.T. Kerry, M. Álvarez-Noriega, J.G. Álvarez-Romero, K.D. Anderson, A.H. Baird, A.H., R.C. Babcock, M. Beger, D.R. Bellwood, R. Berkelmans, T.C. Bridge, I.R. Butler, M. Byrne, N.E. Cantin, S. Comeau, S.R. Connolly, G.S. Cumming, S.J. Dalton, G. Diaz-Pulido, C.M. Eakin, W.F. Figueira, J.P. Gilmour, H.B. Harrison, S.F. Heron, A.S. Hoey, J.-P.A. Hobbs, M.O. Hoogenboom, E.V. Kennedy, C.-Y. Kuo, J.M. Lough, R.J. Lowe, G. Liu, M.T. McCulloch, H.A. Malcolm, M.J. McWilliam, J.M. Pandolfi, R.J. Pears, M.S. Pratchett, V. Schoepf, T. Simpson, W.J. Skirving, B. Sommer, G. Torda, D.R. Wachenfeld, B.L. Willis and S.K. Wilson. 2017. Global warming and recurrent mass bleaching of corals. Nature 543: 373–377.

Hutchings, P. 2011. Bioerosion. pp. 139–156. *In*: D. Hopley (ed.). Encyclopedia of Modern Coral Reefs. Springer, Netherlands.

Jackson, J.B.C., M.X. Kirby, W.H. Berger, K.A. Bjorndal, L.W. Botsford, B.J. Bourque, R.H. Bradbury, R. Cooke, J. Erlandson, J.A. Estes, T.P. Hughes, S. Kidwell, C.B. Lange, H.S. Lenihan, J.M. Pandolfi, C.H. Peterson, R.S. Steneck, M.J. Tegner and R.R. Warner. 2001. Historical overfishing and the recent collapse of coastal ecosystems. Science 293: 629–638.

Jennings, S., D.P. Boullé and N.V. Polunin. 1996. Habitat correlates of the distribution and biomass of Seychelles' reef fishes. Environ. Biol. Fish. 46: 15–25.

Jennings, S., J.D. Reynolds and S.C. Mills. 1998. Life history correlates of responses to fisheries exploitation. Proc. R. Soc. B. 265: 333–339.

Jennings, S., J.D. Reynolds and N.V.C. Polunin. 1999. Predicting the vulnerability of tropical reef fishes to exploitation with phylogenies and life histories. Conserv. Biol. 13: 1466–1475.

Johannes, R. 1981. Words of the lagoon: Fishing and marine lore in the Palau district of Micronesia. University of California Press, Berkeley, CA.

Johnson, M.K., S.J. Holbrook, R.J. Schmitt and A.J. Brooks. 2011. Fish communities on staghorn coral: effects of habitat characteristics and resident farmerfishes. Environ. Biol. Fish. 91: 429–448.

Jones, G.P., M.I. McCormick, M. Srinivasan and J.V Eagle. 2004. Coral decline threatens fish biodiversity in marine reserves Proc. Natl. Acad. Sci. USA 101: 8251–8253.

Khalil, M.T., J.E.M. Cochran and M.L. Berumen. 2013. The abundance of herbivorous fish on an inshore Red Sea reef following a mass coral bleaching event. Environ. Biol. Fish. 96: 1065–1072.

Kramer, M.J., O. Bellwood and D.R. Bellwood. 2013. The trophic importance of algal turfs for coral reef fishes: the crustacean link. Coral Reefs 32: 575–583.

Kroon, F.J., K.M. Kuhnert, B.L. Henderson, S.N. Wilkinson, A. Kinsey-Henderson, J.E. Brodie and R.D.R. Turner. 2012. River loads of suspended solids, nitrogen, phosphorus and herbicides delivered to the Great Barrier Reef lagoon. Mar. Poll. Bull. 65: 167–181.

Kuffner, I.B., L.J. Walters, M.A. Becerro, V.J. Paul, R. Ritson-Williams and K.S. Beach. 2006. Inhibition of coral recruitment by macroalgae and cyanobacteria. Mar. Ecol. Prog. Ser. 323: 107–117.

Lawton, R.J., V. Messmer, M.S. Pratchett and L.K. Bay. 2011. High gene flow across large geographic scales reduces extinction risk for a highly specialised coral feeding butterflyfish. Mol. Ecol. 20: 3584–3598.

Lewis, A.R. 1997. Effects of experimental coral disturbance on the structure of fish communities on large patch reefs. Mar. Ecol. Prog. Ser. 161: 37–50.

Lieske, E. and R. Myers. 2001. Coral reef fishes: Indo-pacific and Caribbean. Harper Collins, London.

Lindahl, U., M.C. Ohman and C.K. Schelten. 2001. The 1997/1998 mass mortality of corals: effects on fish communities on a Tanzanian coral reef. Mar. Poll. Bull. 42: 127–131.

Loffler, Z., D.R. Bellwood and A.S. Hoey. 2015. Among-habitat algal selectivity by browsing herbivores on an inshore coral reef. Coral Reefs 34: 597–605.

Lokrantz, J., M. Nyström, M. Thyresson and C. Johansson. 2008. The non-linear relationship between body size and function in parrotfishes. Coral Reefs 27: 967–974.

Lokrantz, J., M. Nyström, A.V. Norström, C. Folke and J.E. Cinner. 2010. Impacts of artisanal fishing on key functional groups and the potential vulnerability of coral reefs. Environ. Conserv. 36: 327–337.

Lotze, H.K., H.S. Lenihan, B.J. Bourque, R.H. Bradbury, R.G. Cooke, M.C. Kay, S.M. Kidwell, M.X. Kirby, C.H. Peterson and J.B.C. Jackson. 2006. Depletion, degradation, and recovery potential of estuaries and coastal seas. Science 312: 1806–1809.

Madin, J.S., A.H. Baird, M. Dornelas and S.R. Connolly. 2014. Mechanical vulnerability explains size-dependent mortality of reef corals. Ecol. Lett. 17: 1008–1015.

Mantyka, C.S. and D.R. Bellwood. 2007. Macroalgal grazing selectivity among herbivorous coral reef fishes. Mar. Ecol-Prog. Ser. 352: 177–185.

Mantyka-Pringle, C.S., T.G. Martin and J.R. Rhodes. 2012. Interactions between climate and habitat loss effects on biodiversity: a systematic review and meta-analysis. Glob. Change Biol. 18: 1239–1252.

McClanahan, T., J. Maina, C.J. Starger, P. Herron-Perez and E. Dusek. 2005. Detriments to post-bleaching recovery of corals. Coral Reefs 24: 230–246.

Messmer, V., G.P. Jones, P.L. Munday, S.J. Holbrook, R.J. Schmitt and A.J. Brooks. 2011. Habitat biodiversity as a determinant of fish community structure on coral reefs. Ecology 92: 2285–2298.

Mumby, P.J., C.P. Dahlgren, A.R. Harborne, C.V. Kappel, F. Micheli, D.R. Brumbaugh, K.E. Holmes, J.M. Mendes, K. Borad, J.N. Sanchirico, K. Bunch, S. Box, R.W. Stoffle and A.B. Gill. 2006. Fishing, trophic cascades, and the process of grazing on coral reefs. Science 311: 98–101.

Mumby P.J., A.R. Harborne, J. Williams, C.V. Kappel, D.R. Brumbaugh, F. Micheli, K.E. Holmes, C.P. Dahlgren, C.B. Paris and P.G. Blackwell. 2007. Trophic cascade facilitates coral recruitment in a marine reserve. Proc. Nat. Acad. Sci. USA. 104: 8362–8367.

Mumby, P.J., S. Bejarano, Y. Golbuu, R.S. Steneck, S.N. Arnold, R. Van Woesik and A.M. Friedlander. 2013. Empirical relationships among resilience indicators on Micronesian reefs. Coral Reefs. 32: 213–226.

Munday, P.L. and G.P. Jones. 1998. The ecological implications of small body size among coral-reef fishes. Oceanogr. Mar. Biol. Annu. Rev. 36: 373–411.

Myers, R.A., N.J. Barrowman, J.M. Hoenig and Z. Qu. 1996. The collapse of cod in Eastern Canada: the evidence from tagging data. ICES J. Mar. Sci. 53: 629–640.

Newton, K., I.M. Cote, G.M. Pilling, S. Jennings and N.K. Dulvy. 2007. Current and future sustainability of island coral reef fisheries. Curr. Biol. 17: 655–658.

Norström, A.V., M. Nyström, J. Lokrantz and C. Folke. 2009. Alternative states on coral reefs: beyond coral-macroalgal phase shifts. Mar. Ecol. Prog. Ser. 376: 295–306.

Nyström, M., N.A.J. Graham, J. Lokrantz and A.V. Norström. 2008. Capturing the cornerstones of coral resilience: linking theory to practice. Coral Reefs 27: 795–809.

Ogden, J.C. and P.S. Lobel. 1978. The role of herbivorous fishes and urchins in coral reef communities. Environ. Biol. Fish. 3: 49–63.

Ong, L. and K.N. Holland. 2010. Bioerosion of coral reefs by two Hawaiian parrotfishes: species, size differences and fishery implications. Mar. Biol. 157: 1313–1323.

Osborne, K., A.M. Dolman, S.C. Burgess and K.A. Johns. 2011. Disturbance and the dynamics of coral cover on the Great Barrier Reef (1995–2009). PLoS One 6: e17516.

Pandolfi, J.M., R.H. Bradbury, E. Sala, T.P. Hughes, K.A. Bjorndal, R.G. Cooke, D. McArdle, L. McClenachan, M.J.H. Newman, G. Paredes, R.R. Warner and J.B.C. Jackson. 2003. Global trajectories of the long-term decline of coral reef ecosystems. Science 301: 955–958.

Pauly, D., V. Christensen, J. Dalsgaard, R. Froese and F. Torres Jr. 1998. Fishing down marine food webs. Science 279: 860–863.

Pisapia, C., K. Anderson and M.S. Pratchett. 2014. Intraspecific variation in physiological condition of reef-building corals associated with differential levels of chronic disturbance. PLoS One 9: e91529.

Polunin, N.V.C. and C.M. Roberts. 1993. Greater biomass and value of target coral-reef fishes in two small Caribbean marine reserves. Mar. Ecol. Prog. Ser. 100: 167–176.

Pratchett, M.S., S.K. Wilson, M.L. Berumen and M.I. McCormick. 2004. Sub-lethal effects of coral bleaching on an obligate coral feeding butterflyfish. Coral Reefs 23: 352–356.

Pratchett, M.S., P.L. Munday, S.K. Wilson, N.A.J. Graham, J.E. Cinner and D.R. Bellwood. 2008. Effects of climate-induced coral bleaching on coral-reef fishes – ecological and economic consequences. Ocean. Mar. Biol. Annu. Rev. 46: 251–296.

Pratchett, M.S., S.K. Wilson, N.A.J. Graham, P.L. Munday, G.P. Jones and N.V.C. Polunin. 2009. Coral bleaching and consequences for motile reef organisms: past, present and uncertain future effects. pp. 139–158. *In*: M. van Oppen and J. Lough (eds.). Coral Bleaching: Patterns and Processes, Causes and Consequences. Springer, Heidelberg.

Pratchett, M.S., L.K. Bay, P.C. Gehrke, J.D. Koehn, K. Osborne, R.L. Pressey, H.P.A. Sweatman and D. Wachenfeld. 2011a. Contribution of climate change to degradation and loss of critical fish habitats in Australian marine and freshwater environments. Mar. Freshw. Res. 62: 1062–1081.

Pratchett, M.S., A.S. Hoey, S.K. Wilson, V. Messmer and N.A.J. Graham. 2011b. Changes in the biodiversity and functioning of reef fish assemblages following coral bleaching and coral loss. Diversity 3: 424–452.

Pratchett, M.S., P.L. Munday, N.A.J. Graham, M. Kronen, S. Pinica, K. Friedman, T. Brewer, J.D. Bell, S.K. Wilson, J.E. Cinner, J.P. Kinch, R.J. Lawton, A.J. Williams, L. Chapman, F. Magron and A. Webb. 2011c. Vulnerability of coastal fisheries in the tropical Pacific to climate change. pp. 493–576. *In*: J.D. Bell, J.E. Johnson and A.J. Hobday (eds.). Vulnerability of Tropical Pacific Fisheries and Aquaculture to Climate Change. Secretariat for the Pacific Community, Noumea, New Caledonia.

Pratchett, M.S., C. Caballes, J.A. Rivera-Posada and H.P.A. Sweatman. 2014. Limits to understanding and managing outbreaks of crown-of-thorns starfish (*Acanthaster* spp.). Oceanogr. Mar. Biol. Annu. Rev. 52: 133–200.

Rakitin, A. and D.L. Kramer. 1996. Effect of a marine reserve on the distribution of coral reef fishes in Barbados. Mar. Ecol. Prog. Ser. 131: 97–113.

Randall, J.E. 1974. The effect of fishes on coral reefs. Proc. 2nd Int. Coral Reef Symp, Brisbane. 1: 159–166.

Rasher, D.B., A.S. Hoey and M.E. Hay. 2013. Consumer diversity interacts with prey defenses to drive ecosystem function. Ecology 94: 1347–1358.

Rhodes, K.L., M.H. Tupper and C.B. Wichilmel. 2008. Characterization and management of the commercial sector of the Pohnpei coral reef fishery, Micronesia. Coral Reefs 27: 443–454.

Rotjan, R.D. and S.M. Lewis. 2006. Parrotfish abundance and selective corallivory on a Belizean coral reef. J. Exp. Mar. Biol. Ecol. 335: 292–301.

Rotjan, R.D. and S.M. Lewis. 2008. The impact of coral predators on tropical reefs. Mar. Ecol. Prog. Ser. 367: 73–91.

Rousseau, Y., R. Galzin and J.P. Marechal. 2010. Impact of hurricane Dean on coral reef benthic and fish structure of Martinique, French West Indies. Cybium 34: 243–256.

Russ, G.R. 1984. Distribution and abundance of herbivorous grazing fishes in the central Great Barrier Reef. II: patterns of zonation of mid-shelf and outer-shelf reefs. Mar. Ecol. Prog. Ser. 20: 35–44.

Russ, G.R. 2003. Grazer biomass correlates more strongly with production than with biomass of algal turfs on a coral reef. Coral Reefs 22: 63–67.

Russ, G.R. and A.C. Alcala. 1989. Effects of intense fishing pressure on an assemblage of coral reef fishes. Mar. Ecol. Prog. Ser. 56: 13–27.

Russ, G.R. and A.C. Alcala. 1998. Natural fishing experiments in marine reserves 1983–1993: community and trophic responses. Coral Reefs 17: 383–397.

Sano, M. 2004. Short-term effects of a mass coral bleaching event on a reef fish assemblage at Iriomote Island, Japan. Fish. Sci. 70: 41–46.

Sano, M., M. Shimizu and Y. Nose. 1987. Long-term effects of destruction of hermatypic corals by *Acanthaster planci* infestation on reef fish communities at Iriomote Island, Japan. Mar. Ecol. Prog. Ser. 37: 191–199.

Scheffer, M., S. Carpenter, J.A. Foley, C. Folke and B. Walker. 2001. Catastrophic shifts in ecosystems. Nature 413: 591–596.

Sheppard, C.R.C., M. Spalding, C. Bradshaw and S. Wilson. 2002. Erosion vs. recovery of coral reefs after 1998 El Niño: Chagos reefs, Indian Ocean. Ambio. 31: 40–48.

Steneck, R. 1988. Herbivory on coral reefs: a synthesis. Proc. 6th Int. Coral Reef Symp. Townsville 1: 37–49

Stockwell, B., C.R.L. Jadloc, R.A. Abesamis, A.C. Alcala and G.R. Russ. 2009. Trophic and benthic responses to no-take marine reserve protection in the Philippines. Mar. Ecol. Prog. Ser. 389: 1–15.

Streit, R.P., A.S. Hoey and D.R. Bellwood. 2015. Feeding characteristics reveal functional distinctions among browsing herbivorous fishes on coral reefs. Coral Reefs. 34: 1037–1047.

Tanner, J.E. 1995. Competition between scleractinian corals and macroalgae: an experimental investigation of coral growth, survival and reproduction. J. Exp. Mar. Biol. Ecol. 190: 151–168.

Taylor, B.M. 2014. Drivers of protogynous sex change differ across spatial scales. Proc. R. Soc. B. 281: rspb20132423.

Taylor, B.M. and J.H. Choat 2014. Comparative demography of commercially important parrotfish species from Micronesia. J. Fish Biol. 84: 383–402.

Taylor, B.M, P. Houk, G.R. Russ and J.H. Choat. 2014. Life histories predict vulnerability to overexploitation in parrotfishes. Coral Reefs 33: 869–878.

Tilman, D., R.M. May, C.L. Lehman and M.A. Nowak. 1994. Habitat destruction and the extinction debt. Nature 371: 65–66.

Tolimieri, N. 1998. Effects of substrata, resident conspecifics and damselfish on the settlement and recruitment of the stoplight parrotfish, *Sparisoma viride*. Environ. Biol. Fish. 53: 393–404.

Trapon, M.L., M.S. Pratchett, A.S. Hoey and A.H. Baird. 2013. Influence of fish grazing and sedimentation on the early post-settlement survival of the tabular coral *Acropora cytherea*. Coral Reefs 32: 1051–1059.

Vázquez, D.P. and D. Simberloff. 2002. Ecological specialisation and susceptibility to disturbance: conjectures and refutations. Am. Nat. 159: 606–623.

Vitousek, P.M. 1997. Human-domination of Earth's ecosystems. Science 275: 494–499.

Walker, B., C.S. Holling, S.R. Carpenter and A. Kinzig. 2004. Resilience, adaptability and transformability in social-ecological systems. Ecol. Soc. 9: 5.

Walsh, J.J., T.E. Whitledge, W.E. Esaias, R.L. Smith, S.A. Huntsman, H. Santander and B.R. De Mendiola. 1980. The spawning habitat of the Peruvian anchovy, *Engraulis ringens*. Deep Sea Res. 27: 1–27.

Walsh, W.J. 1983. Stability of a coral reef fish community following a catastrophic storm. Coral Reefs 2: 49–63.

Wilkinson, C.R. (ed.). 2004. Status of coral reefs of the world. Australian Institute of Marine Science, Townsville.

Wilkinson, C. (ed.). 2008. Status of coral reefs of the world. Global Coral Reef Monitoring Network, Townsville.

Williams, D.M. 1986. Temporal variation in the structure of reef slope fish communities (central Great Barrier Reef): short-term effects of *Acanthaster planci* infestation. Mar. Ecol. Prog. Ser. 28: 157–164.

Wilson, S.K., D.R. Bellwood, J.H. Choat and M.J. Furnas. 2003. Detritus in the epilithic algal matrix and its use by coral reef fishes. Ocean. Mar. Biol. Annu. Rev. 41: 279–310.

Wilson, S.K., N.A. Graham, M.S. Pratchett, G.P. Jones and N.V.C. Polunin. 2006. Multiple disturbances and the global degradation of coral reefs: are reef fishes at risk or resilient? Glob. Change Biol. 12: 2220–2234.

Wilson, S.K., S.C. Burgess, A.J. Cheal, M.J. Emslie, R. Fisher, I. Miller, N.V.C. Polunin and H.P.A. Sweatman. 2008a. Habitat utilization by coral reef fish, implications for specialists vs. generalists in a changing environment. J. Anim. Ecol. 77: 220–228.

Wilson, S.K., R. Fisher, M.S. Pratchett, N.A.J. Graham, N.K. Dulvy, R.A. Turner, A. Cakacaka, N.V.C. Polunin and S.P. Rushton. 2008b. Exploitation and habitat degradation as agents of change within coral reef fish communities. Glob. Change. Biol. 14: 2796–2809.

Wilson, S.K., M. Depczynski, R. Fisher, T.H. Holmes, R.A. O'Leary and P. Tinkler. 2010. The importance of coral in the habitat use of juvenile reef fish. PLoS One 5: e15185.

Worm, B., E.B. Barbier, N. Beaumont, J.E. Duffy, C. Folke, B.S. Halpern, J.B.C. Jackson, H.K. Lotze, F. Micheli, S.R. Palumbi, E. Sala, K.A. Selkoe, J.J. Stachowicz and R. Watson. 2006. Impacts of biodiversity loss on ocean ecosystem services. Science 314: 787–790.

Yahya, S.A.S., M. Gullström, M.C. Öhman, N.S. Jiddawi, M.H. Andersson, Y.D. Mgaya and U. Lindahl. 2011. Coral bleaching and habitat effects on colonisation of reef fish assemblages: An experimental study. Estuar. Coast. Shelf Sci. 94: 16–23.

CHAPTER

16

FAQs about Caribbean Parrotfish Management and their Role in Reef Resilience

Alastair R. Harborne[1] and Peter J. Mumby[2]

[1] Tropical Fish Ecology Lab, Department of Biological Sciences, Florida International University, Marine Science Building, 3000 NE 151 Street, North Miami, Florida 33181, USA
   Email: alastair.harborne@fiu.edu
[2] Marine Spatial Ecology Laboratory and Australian Research Council Centre of Excellence for Coral Reef Studies, School of Biological Sciences, Goddard Building, The University of Queensland, Brisbane, QLD 4072, Australia
   Email: p.j.mumby@uq.edu.au

Introduction

Research on the biology of coral reef fishes has accelerated rapidly since the development of SCUBA, and has been driven by their diversity and potential use as a model system for testing general ecological concepts (Hixon 2011). More recently, research has embraced concerns about how fish populations respond to stressors such as fishing pressure, habitat degradation, increasing ocean temperatures and acidification, and the introduction of invasive species, and there is an increasing focus on how conservation initiatives may address these threats. This wealth of research has elucidated the functional role of many species, but the role of herbivorous fish has received perhaps the greatest attention. Within the guild of herbivorous fishes, parrotfishes are the best studied taxa and, as can be seen from this book, there is a rapidly growing literature covering all aspects of their biology, management, and importance to coral reefs.

One of the most widely studied functional roles of parrotfishes is their removal of algae that might otherwise compete with corals (McCook et al. 2001). In principle, grazers such as parrotfishes benefit coral populations by facilitating recruitment and reducing the frequency and intensity of competitive interactions with algae. However, as is so often the case in ecological systems, this apparently straightforward concept belies a multitude of complex questions including the relative importance of top-down and bottom-up controls of benthic dynamics, phase shifts and alternative stable states, the degree of functional redundancy amongst herbivores, and resolving the functional versus nutritional aspects of parrotfish feeding (Adam et al. 2015a). In addition, many questions exist concerning the management of parrotfish management, including how harvesting influences the biomass of fish on reefs and their functional role, and how marine reserves affect trophic cascades within food webs.

Research into the nutritional biology of Caribbean parrotfishes has lagged behind that in the Pacific. For example, the emerging picture in the Pacific is that parrotfishes are microphages and detritivores (Crossman et al. 2001, Choat et al. 2004, Clements et al. 2017), which likely also applies to Caribbean species of the genus *Scarus*. However, the nutritional biology of the other major – and endemic – Atlantic genus, *Sparisoma*, has received little recent study. Species of *Sparisoma* differ from *Scarus* in taking a large proportion of bites from fleshy macroalgae (Bruggemann et al. 1994c), and their alkaline intestine helps dissociate protein-tannin complexes in brown algae (Lobel 1981). Furthermore, brown algal secondary metabolites do not seem to affect nutritional assimilation (Targett and Arnold 1998). The degree to which *Sparisoma* derives nutrition from detritus and microalgae remains unclear (but see Bruggemann et al. 1994c). Despite the uncertainty of the nutritional biology of Caribbean parrotfish, many studies have examined the impact of their grazing on coral reef algae, which seems to be strongly negative (e.g. Williams and Polunin 2001, Kramer 2003, Mumby et al. 2006b, Burkepile et al. 2013).

Parrotfishes are among the most abundant and conspicuous of Caribbean coral reef fish, yet comprise just 16 species, and only species from the genera *Scarus* and *Sparisoma* are functionally important grazers on reefs. Furthermore, the functional importance of these genera has increased since the mass mortality of the herbivorous urchin *Diadema antillarum* in the early 1980s (Lessios 1988), so that they are now the major grazers on most reefs in the region. The relative simplicity of this system has allowed the functional role of parrotfishes to be sufficiently well understood that it can be built into predictive models that provide realistic insights into future reef dynamics (Mumby 2006a). Such models have provided a range of new insights into coral reef resilience, which is the probability that a reef will still be able to maintain a trajectory of coral recover after some prescribed period of time during which disturbances occur (Mumby et al. 2007a, 2014).

Through our work on the biology, functional role, and management of Caribbean grazers, we have been exposed to a wide range of questions about the role of parrotfishes. Some of these queries stem from an understandable inability to keep abreast of a diverse literature, but some are driven by misunderstandings over what particular papers actually demonstrate (and we include the wider scientific literature here, not just our own). Indeed, some of the confusion is caused by apparent disagreements in the literature about the roles parrotfishes might play in reef dynamics and resilience. For example, a recent high-profile analysis of coral cover trends in the Caribbean highlighted the importance of overfishing of parrotfishes, and strongly recommended reductions in herbivore fishing (Jackson et al. 2014). This led to comments on international fora (e.g., coral-list) such as "we have parrotfishes on our reefs but coral cover is still declining", "we have a marine reserve but coral cover is still low", and "banning parrotfish fishing doesn't address the threats of climate change". Such comments reveal several misconceptions about the role of parrotfish in coral population dynamics, and if left unchecked could lead to perceived failures of management that result from unrealistic expectations.

This chapter aims to address some of the most frequently asked questions (FAQs) posed to us by researchers, reviewers, managers, and the general public. It is not intended to be a comprehensive review of Caribbean parrotfishes, but rather attempts to summarise the key literature required to answer specific questions. We begin by addressing questions regarding the basic biology of parrotfishes as this builds a foundation from which to understand their role in reef resilience, and the implications for management.

Parrotfish Biology

FAQ 1. What Controls the Diversity and Abundance of Parrotfishes on Reefs?

The Western Atlantic supports a greater diversity of parrotfishes than the Eastern Atlantic (Floeter et al. 2008, Bonaldo et al. 2014) and in the Western Atlantic the density of herbivorous fishes decreases from tropical to temperate latitudes, possibly because of the physiological challenges of utilising a relatively low-quality food in cooler water (Floeter et al. 2004, but see Clements et al. 2009). The major evolutionary radiation of *Sparisoma* was driven by allopatric processes caused by geographic separation from the Atlantic and riverine barriers (Robertson et al. 2006). Species of the genus *Scarus* likely arrived in the Caribbean through migration from an Indian Ocean source via South Africa and primarily exhibited sympatric radiation on reaching the Western Atlantic (Choat et al. 2012). Within the Caribbean, the biogeography of parrotfishes is not well documented, although there is little evidence that any species have restricted ranges within the region.

In contrast to their biogeography, there are major differences in parrotfish diversity and densities among habitat types. For example, one of the most abundant species is *Sparisoma viride*, which is absent in sandy habitat, has densities of <1 fish 100 m^{-2} in mangroves, seagrass beds and escarpments, is more common (1–2 fish 100 m^{-2}) on gorgonian-dominated pavements and deep *Orbicella*-dominated reefs, and is most common (>2 fish 100 m^{-2}) on patch reefs, back reefs, reef crests and shallow *Orbicella*-dominated reefs (reviewed by Harborne et al. 2006). The major abiotic and biotic drivers of this inter-habitat variability are relatively well understood, and reef complexity is typically positively correlated with parrotfish abundance. Refuges within the reef have a range of functions for parrotfishes, including predator avoidance and nocturnal sleeping sites (Tzadik and Appeldoorn 2013). Consequently, across a range of sites and habitats in Belize, the density and biomass of the commonest species (*Scarus iseri*, *Sparisoma aurofrenatum* and *Sp. viride*) were positively correlated with reef complexity, although the abundance of *Sparisoma chrysopterum* did not appear to be linked to rugosity (Bejarano et al. 2011). Such habitat preferences appear to be established during the settlement and recruitment period of some species (Tolimieri 1998b). Within coral-rich habitats, patches of reef with larger mean heights of coral colonies also appear to support greater biomasses of parrotfishes (Harborne et al. 2012). Critically, although increasing reef complexity increases the grazeable area on reefs, the benefits to parrotfish abundances from increasing rugosity are sufficient to drive higher grazing intensities on rugose reefs (Bozec et al. 2013). Therefore, positive feedbacks are established with increasing rugosity increasing parrotfish abundance and grazing intensity, which reduces macroalgal abundances, facilitating coral settlement and the maintenance and enhancement of processes underpinning high reef complexity (Bozec et al. 2013). Such feedbacks do not occur on flat, hard-bottom habitats where parrotfishes are less abundant and benthic dynamics are largely controlled by physical processes (Mumby 2016).

In addition to reef complexity, other controls of parrotfish populations include decreasing abundances with increasing depth, predominantly because of its effect on algal productivity, but also because of predator abundance and the density of herbivorous competitors (Lewis and Wainwright 1985, Nemeth and Appeldoorn 2009). Wave exposure also has an important influence on the composition of coral reef fish assemblages because varying water velocities favour different fin morphologies (Fulton et al. 2005), although there are limited data available for Caribbean parrotfishes (but see Bellwood et al. 2002). There are also limited data on whether pre- or post-settlement processes are most important for parrotfish demographics, although factors such as predation risk and refuge availability are more important than larval supply for regulating populations of other territorial species (Hixon et al. 2012). Populations of species such as *Sc. iseri* and *Scarus guacamaia* on reefs are enhanced by nursery habitat availability, particularly mangroves and dense seagrass beds (see FAQ 5). Finally, populations of adult parrotfishes are highly influenced by fishing pressure, but this is outlined in more detail in the Section on 'Parrotfish fisheries management' below.

FAQ 2. Are all Parrotfishes Functionally Equivalent?

When considering the conservation of parrotfishes, an important question is whether the sole target should be increasing fish biomass, or whether herbivore diversity should also be considered to maintain grazing pressure. Answering this question requires an understanding of whether all parrotfishes are functionally equivalent, or whether some species have specialised roles on reefs. Despite the uncertainties concerning their nutritional biology, all parrotfishes remove algae and there is a well-established distinction between the morphology of species that 'excavate' (grazing also removes pieces of the substratum) and those that 'scrape' (food is removed from the surface of the substratum with a non-excavating bite) (Bellwood and Choat 1990). In the Caribbean, *Scarus* species generally target algal turf assemblages, crustose coralline algae, and endolithic algae, while *Sparisoma* species generally target macroalgae (Adam et al. 2015b). More specifically, seven common species have been categorised based on the degree of removal of the substratum and major food sources (Cardoso et al. 2009), and we add our own observations to this list here. *Scarus taeniopterus* and *Sc. iseri* are categorised as 'scrapers' (leave superficial bite marks but remove more turf algae than other species). *Sp. aurofrenatum*, *Sparisoma rubripinne* and *Sp. chrysopterum* are considered 'grazers' (the term grazer is used because these species are not obligate browsers and also scrape the epilithic algal matrix), although *Sp. aurofrenatum* does frequently bite live corals (Miller and Hay 1998) and in some classifications is considered a scraper (Bernardi et al. 2000, Streelman et al. 2002). *Sp. viride* is a 'bioeroder' or 'excavator' (removes both coralline rock and live coral when feeding but feeds extensively on algal turfs and several macroalgae including *Dictyota* spp., Mumby 2006a), and *Scarus vetula* is a 'bioeroder/scraper' (removes coralline rock but feeds primarily on turf algae). These results suggest significant functional diversity among Caribbean parrotfish (Cardoso et al. 2009, Adam et al. 2015b), especially since the frequently over-fished, large-bodied *Scarus guacamaia* is not included in the classification. A single large-bodied parrotfish species can have key functional roles on reefs, even in the more diverse Indo-Pacific (Bellwood et al. 2003, 2012), although there is some evidence that *Scarus guacamaia* feeds on similar foods to *Sp. viride* (Burkepile and Hay 2011).

The suggestion that there are multiple functional roles within the parrotfish assemblage is supported by experimental evidence that individual species alone may not be able to supress successional processes resulting in macroalgae that reduce coral growth, or remove established macroalgae that have similar effects (Burkepile and Hay 2008, 2010). These functional roles of individual parrotfish species are further supported by feeding of surgeonfishes (Burkepile and Hay 2008, 2010) and, before their mass mortality, would have complemented grazing by urchins such as *Diadema antillarum* (reviewed by Lessios 1988). However, a study examining the response of the herbivore assemblage to different algal assemblages indicates that while there is complementarity between *Sparisoma* spp. and both the surgeonfishes *Acanthurus* spp. and *Scarus* spp., there may be some redundancy within the latter genera (Burkepile and Hay 2011). Finally, the lack of apparent competition and aggressive interactions among parrotfish species (Mumby and Wabnitz 2002), also suggests limited niche overlaps and the use of different feeding resources (Fig. 1). This general lack of functional redundancy in Caribbean parrotfishes is consistent with global patterns of functional diversity being highly vulnerable to losses of reef fish species (Mouillot et al. 2014). Therefore, while the absolute biomass of parrotfishes is clearly important to maintain the ecological process of grazing, there is a growing understanding that maintaining the diversity of parrotfish assemblages, and the entire herbivore guild, is also important for the benthic dynamics of reefs.

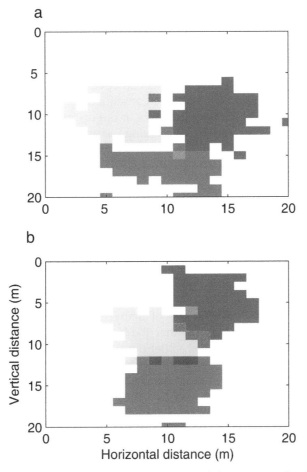

Fig. 1. Species stacking showing three adjacent parrotfish territories for *Sparisoma viride* (a) and *Sparisoma aurofrenatum* (b). Individuals of each species distinguished using red, green, and blue. Intraspecific spatial overlap among territories is infrequent but denoted using purple, olive and turquoise. Note that the territories of different species overlap (stack) in space.

FAQ 3. Are There Key Spawning Sites that Should be Conserved?

Mass spawning aggregations for species such as groupers and snappers are key sites for protection (Claydon 2004), but most Caribbean parrotfishes do not use the same reproductive strategy (though the bumphead parrotfish in the Indo-Pacific can form aggregations of up to one thousand individuals, Roff et al. 2017). In contrast, terminal phase parrotfishes typically defend territories and spawn daily with a harem of females

(Bruggemann et al. 1994a, van Rooij et al. 1996). However, in shallow water, *Sp. viride* may form non-territorial mixed groups of males and females that have only limited sexual activity, including with territorial males in deeper water (van Rooij et al. 1996). Parrotfishes have long been recognised as undertaking predictable diurnal migrations between resting and feeding sites (Ogden and Buckman 1973), but these movements do not seem to be related to reproductive behaviour. Therefore, protecting parrotfish populations within marine reserves that are sufficiently large to capture diurnal movements are likely to also incorporate spawning sites. However, one exception may be populations of species such as *Sp. rubripinne* that occur at low densities on extensive low complexity hard-bottom and bank areas (Mumby 2016). We have heard anecdotal reports of parrotfishes undertaking spawning migrations to the edge of these habitats (see also the classification of *Sp. rubripinne* as forming spawning aggregations in Nemeth 2012), presumably to release their gametes into more favourable oceanographic conditions at the reef edge. Although population densities of these species are low, the habitats can cover large areas and, therefore, this behaviour may have important demographic consequences. Since spawning migrations may also mean that individuals move either into, or out of, marine reserves, documenting the scale and drivers of this behaviour is a pressing research topic.

FAQ 4. Is It Possible to Increase Parrotfish Recruitment on Reefs?

As might be expected for a species that spawns daily, parrotfish settlement to Caribbean reefs occurs throughout the year, although in some locations there can be a summer peak (e.g. in Florida, Paddack and Sponaugle 2008). Settlement may also be higher around new moons (Tolimieri 1998a) and, like many species, settlement rates vary annually (Paddack and Sponaugle 2008). Settlement densities are also affected by microhabitat availability, with species such as *Sp. viride* having higher densities on the coral *Porites porites*, a pattern that appears to reflect post-settlement survival rather than larval settlement preferences (Tolimieri 1998a). Since survival of recently settled individuals represents a particularly important bottleneck in the demographics of many reef species (Almany and Webster 2006), an important consideration is whether the local breeding biomass of parrotfishes drives a commensurate increase in local settler density (a stock-recruitment relationship). Reef fish stock-recruitment relationships have proven extremely difficult to quantify (Haddon 2011), but are important to identify because protecting adult stocks through the cessation of fishing could potentially lead to a positive feedback: the increased biomass of adults could increase settlement rates and consequently further recovery of the local breeding population. Perhaps the clearest attempt to identify a stock-recruitment relationship in parrotfishes has occurred in Bermuda, where a ban on fishing traps led to an increased biomass of adults that was monitored over a nine-year period (O'Farrell et al. 2015). In contrast to expectations given that Bermuda is a largely demographically closed system, there was no increase in juvenile density as the adult population increased, and this appeared to be linked to an increase in the abundance of the meso-predator *Aulostomus maculatus*. Since large predatory fishes were still being fished in Bermuda, it appeared that *Aulostomus maculatus*, which is rarely caught by fishers, benefited from low predation pressure and an increased abundance of food. Their increased abundance seemed to have limited any potential stock-recruitment relationship. This finding has important implications for management, because the benefits of larval parrotfish spillover from marine reserves may be lost in non-protected areas if meso-predators can respond unchecked to prey enrichment because of the absence of large predators to control their biomass.

FAQ 5. How Important is the Protection of Mangroves and Seagrass Beds for Parrotfishes?

Many reef fish species recruit into 'nursery habitats', rather than directly into their preferred adult habitat, to benefit from lower predation rates or increased food availability. Consequently, a nursery habitat can be defined as an area that supports greater contributions per unit area to the adult population (Beck et al. 2001, see also Adams et al. 2006, Dahlgren et al. 2006). Mangrove stands and seagrass beds are frequently cited as nurseries within tropical marine seascapes, but other shallow habitats such as algal beds can also be important (Nagelkerken et al. 2000). Evidence that Caribbean parrotfish populations are enriched by these habitats is multifaceted. Firstly, juvenile parrotfishes, particularly *Sc. iseri*, are frequently seen in mangrove and seagrass beds (Nagelkerken et al. 2001), which is consistent with the nursery habitat hypothesis. The presence of higher densities of parrotfishes on nearby reefs provides a stronger justification for categorising mangroves and seagrass beds as nurseries, and this has been demonstrated in a number of locations. For example, surveys at different distances from a bay containing mangroves and seagrass beds in Curaçao highlighted that densities of *Scarus coeruleus*, *Sc. guacamaia*, *Sc. iseri*, and *Sp. chrysopterum* were higher close to the bay (Nagelkerken et al. 2000). In contrast, *Sp. aurofrenatum*, *Sp. rubripinne*, and *Sp. viride* were classified as 'reef species' with all life stages, including juveniles, normally found on the reef and not appearing to use nursery areas. However, for reef species such as *Sp. viride*, shallow water habitats such as back reefs and patch reefs represent important sites for juvenile fish (Tolimieri 1998a, c). Subsequent studies have suggested that juvenile parrotfishes are primarily using mangroves and seagrass beds because of increased food availability (Nagelkerken and van der Velde 2004, Verweij et al. 2006).

The enrichment of populations of nursery-using parrotfishes on nearby reefs has also been demonstrated at entire reef scales. Across six Caribbean islands, the abundance of *Sc. iseri* was significantly higher on reefs around islands supporting mangrove stands and provides further evidence for this species having a high dependence on mangrove and seagrass nurseries (Nagelkerken et al. 2002). Working within a single biogeographic region (Mesoamerican Barrier Reef) and controlling for reef area, Mumby et al. (2004) demonstrated an increase in the biomass of *Sc. iseri* of 42% in mangrove-rich reef systems compared to mangrove-scarce areas. Furthermore, *Sc. guacamaia* appeared to have a functional dependency on mangroves and had suffered local extinctions after mangrove removal (Mumby et al. 2004). This functional dependency has also been reported from Aruba, where juvenile *Sc. guacamaia* were only observed in mangroves while all adults were observed on reefs (Dorenbosch et al. 2006). For these nursery-using species, the benefits of nursery habitat availability can have greater effects on the abundances of smaller fish (<25 cm) than the cessation of fishing within marine reserves, although protection is more important for larger-bodied individuals (Nagelkerken et al. 2012).

Despite convincing correlative studies, there is still a need for studies that directly observe the ontogenetic migration of parrotfishes from nursery habitats to nearby reefs, which may be aided by increasingly sophisticated methods of tagging fishes. Such research is necessary to fully parameterise models of parrotfish population demographics, including the maximum distance to which nursery habitats enrich adult populations. Currently the best evidence is that mangroves affect fish assemblages up to distances of approximately 10 km (Dorenbosch et al. 2006, Mumby 2006b, Huijbers et al. 2013), but this is likely to vary among species and seascapes. Finally, there is a need to better understand how the enrichment of nursery-using species by nursery habitats affects other components of fish assemblages. For example, in seascapes where nursery habitats are extensive, the high abundance of nursery-using parrotfishes and piscivores reduces the abundance of other parrotfishes settling directly to reefs because of increased competition and predation, and affects benthic dynamics (Harborne et al. 2016).

FAQ 6. How Fast Do Parrotfishes Grow?

Growth rates are a key parameter for fisheries management, but data on the growth rates of Caribbean parrotfishes are limited. Most of the available data are for *Sp. viride*, which suggests growth in this species best fits von Bertalanffy growth equations (van Rooij et al. 1995, Choat et al. 2003, Paddack et al. 2009). Growth rates vary among size phases, with growth fastest in juveniles, slower in sexually inactive terminal phase males, and slowest in initial phase females and territorial, terminal phase males (van Rooij et al. 1995). These rates suggest that sexually inactive males trade higher growth rates for current reproductive activity in order to obtain a territory that will provide future high reproductive success (van Rooij et al. 1995). Consequently, *Sp. viride* can be categorised as a relatively short-lived species (maximum age ~12 years) with relatively consistent demographic parameters across the Caribbean, although data from Florida and elsewhere suggest that at some spatial scales there may be higher demographic plasticity (Choat et al. 2003, Paddack et al. 2009). Demographic models were recently created for the major Caribbean parrotfish species and tested against independent field data, although it would be advantageous to quantify regional variation in demographic rates in multiple species (Bozec et al. 2016), as has been carried out for *Sp. viride* (Choat et al. 2003).

FAQ 7. What are the Natural Predators of Parrotfishes?

The loss of parrotfishes caused by fishing and invasive lionfish is discussed in subsequent sections, so here we consider the demographic process of mortality caused by native predators. Although data are scarce, predation of parrotfishes can be assumed to decrease significantly with increasing body size, so that survival of new settlers on reefs is critical to replenishing adult populations (as for other species, Almany and Webster 2006). The list of parrotfish predators appears extensive, and they have been found in the stomach of piscivores such as small- and large-bodied groupers, jacks, snappers, and moray eels (Randall 1965, 1967). In addition, the trumpetfish *Aulostomus maculatus* elicits classic anti-predator responses by *Sc. iseri* (group formation) and *Sp. viride* (hiding in corals) (Wolf 1985). Over a time series of nine years in Bermuda following a reduction of parrotfish catches, an increased biomass of trumpetfish, but a stable density of juvenile parrotfishes despite an increased biomass of adults, also suggests this species is an important predator (O'Farrell et al. 2015, see FAQ 4). Interestingly, joining a school occurred less frequently when it contained fewer conspecifics (Wolf 1985), probably because of the increased predation risk for less-abundant species in groups (Almany and Webster 2004). On reefs where *Sp. viride* is relatively rare compared to *Sc. iseri*, such as near nursery habitats, this process may increase the mortality rates of non-nursery species even more than might be expected because of the increased biomass of nursery-using predators (Harborne et al. 2016).

The identity of parrotfish predators is likely to vary considerably with body size and life phase. Meso-predators, such as small-bodied groupers, are probably the most important predators of juvenile parrotfishes. For example, the abundance of recently settled fish, including numerous *Sc. iseri* and *Sp. viride*, were more abundant on patch reefs with higher densities of the large-bodied Nassau grouper, *Epinephelus striatus*, because it reduced the foraging of smaller *Cephalopholis* groupers (Stallings 2008). This experiment suggests that the indirect effects of Nassau groupers on the predators of juvenile parrotfishes are more important than any direct predation. Similarly, fishing of large-bodied carnivores in Belize was linked to increased densities and changed behaviours of meso-predators, and declines in populations of *Sp. viride* (Mumby et al. 2012). As parrotfishes increase in size, the number of potential predators decreases until only the largest piscivores, such as sharks, are capable of feeding on adult parrotfishes. Parrotfishes are the preferred food of juvenile lemon sharks, *Negaprion brevirostris* (Newman et al. 2010), and have also been found in the stomachs of nurse sharks (Randall 1967), but diet data for other sharks is scarce and mortality rates are very difficult to establish. However, constraining adult parrotfish mortality rates is important for population and resilience modelling because larger-bodied fishes have the largest contribution to grazing and bioerosion. Equally, currently poorly known changes in the behaviour of herbivores when threatened by predators may be more important to benthic dynamics than direct predation events, as demonstrated in both the Caribbean and Indo-Pacific (Madin et al. 2010, 2011, Rizzari et al. 2014, Catano et al. 2016).

A key consideration when establishing marine reserves to protect biodiversity, fisheries, and ecological processes is whether the direct benefits to parrotfishes of a cessation of fishing are greater than the negative effects of increased predation caused by higher abundances of piscivores. A study in one of the largest and oldest marine reserves in the Caribbean demonstrated that even in lightly fished systems, the reduction of parrotfish catches far outweighs increased predation rates, and will increase grazing intensity (Mumby et al. 2006b). This is primarily because large-bodied parrotfishes, which are key grazers, reach a size that dramatically reduces their risk of predation by large-bodied groupers that are a prime beneficiary of marine reserve establishment. However, increased predation inside the park appeared to reduce the mean size of smaller bodied parrotfishes, such as *Sc. iseri* and *Sp. aurofrenatum* (Mumby et al. 2006b). Despite this effect, the long-term impact of Caribbean no-take marine reserves is clearly to enhance grazing.

FAQ 8. What Role are Invasive Lionfish having on Parrotfish Populations?

Perhaps the highest profile invasive species on reefs is the introduction of the Indo-Pacific species *Pterois volitans* and *Pterois miles* (subsequently 'lionfish') into the Caribbean. The history of the invasion is reviewed in detail elsewhere (Côté et al. 2013), but in summary lionfish were recorded in Floridian waters in 1985, entered The Bahamas in 2004, and have colonised the majority of the western Atlantic, Caribbean Sea, and Gulf of Mexico. This rapid spread is driven by a range of factors including high fecundity of well-protected eggs, being habitat generalists, and the limited number of predators (Côté et al. 2013). The ubiquity and high densities of lionfish on many Caribbean reefs has led to concern about their effects on native fishes and invertebrates. These effects may be particularly acute because native species have not evolved anti-predation mechanisms when confronted by the unique hunting style of lionfish: lionfish are stalking predators that use a slow, hovering hunting style with pectoral fins spread out and angled forward (Green et al. 2011, Côté et al. 2013). Consequently parrotfishes may incorrectly assign lower threat levels to lionfish, as has been seen in gobies (Marsh-Hunkin et al. 2013).

A growing literature has demonstrated that the impacts of lionfish on reef fish assemblages are significant (Albins and Hixon 2008, Lesser and Slattery 2011, Green et al. 2012), and these effects include reductions of parrotfish populations. Parrotfishes have repeatedly been found in lionfish stomachs, including adult *Sc. iseri* and *Sp. viride* (Albins and Hixon 2008, Morris and Akins 2009), and juvenile *Sc. iseri*, *Sp. aurofrenatum*, and *Sp.*

viride (Green et al. 2011). This predation pressure reduced the recruitment rate of four out of five parrotfish species (*Cryptotomus roseus*, *Sparisoma atomarium*, *Sp. aurofrenatum*, and *Sp. viride*) settling on patch reefs, as part of an overall 79% reduction of fish recruitment caused by lionfish during a five-week experiment (Albins and Hixon 2008). *Sparisoma aurofrenatum* was also one of 42 small-bodied prey fishes whose biomass declined by 65% during a period of rapidly increasing lionfish abundance on a Bahamian reef (Green et al. 2012). Finally, the lionfish invasion appeared to lead to a local extinction of *Sp. atomarium* on a Bahamian mesophotic reef between 30 and 76 m (Lesser and Slattery 2011).

There are currently few data demonstrating how reductions in parrotfish populations because of lionfish predation translate into changes in grazing rate, and consequently the abundance of algae on reefs. This trophic cascade has been proposed as a potentially important consequence of the lionfish invasion (Albins and Hixon 2013), and there is some evidence that a lionfish-driven phase shift from coral- to algal-domination may have already occurred on some mesophotic reefs (Lesser and Slattery 2011). Lionfish predation particularly targets smaller parrotfishes, and the majority of grazing is typically undertaken by larger individuals. However, an increased mortality of parrotfish juveniles has the potential to reduce the abundance of larger fishes, but the full demographic impact of lionfish predation on prey species has not been documented. In addition, lionfish have sub-lethal effects on parrotfish grazing by altering their foraging behaviour and reducing bite rates (Eaton et al. 2016, Kindinger and Albins 2017). This combination of direct predation and non-consumptive effects on high-lionfish-density reefs in The Bahamas reduced algal removal by 66–80% (Kindinger and Albins 2017).

The Functional Role of Parrotfishes

FAQ 9. Parrotfishes Eat Coral, so Aren't They Bad for the Reef?

Of the Caribbean parrotfishes only species in the genus *Sparisoma* eat coral, and even then live coral comprises a small proportion of bites (<4%) (Bruggemann et al. 1994a). The main corallivores are *Sp. viride* and *Sp. aurofrenatum* (Miller and Hay 1998). The answer to whether these corallivores are bad for the reef depends on the habitat involved. On shallow reef flats and the shallower parts of some forereefs, the consumption of branching corals in the genus *Porites* by parrotfishes is profound, and can lead to local exclusion of this coral (Littler et al. 1989, Miller and Hay 1998). Even the massive *Porites astreoides*, which has a harder skeleton than branching forms, experiences heavy corallivory in this environment, though it is not excluded (Littler et al. 1989).

On forereef habitats, evidence of parrotfish corallivory is common, particularly on large massive species of the genus *Orbicella* (Bythell et al. 1993, Bruckner and Bruckner 1998, Rotjan and Lewis 2005), although the only coral that appears to be preferentially targeted is *Porites porites* (Roff et al. 2011, Burkepile 2012). However, bite lesions can heal rapidly leading to a rapid turnover of scars (Sánchez et al. 2004), with little apparent detrimental impact on the coral. Moreover, although the feeding behaviour of parrotfishes has been implicated in causing mortality in juvenile corals (Birkeland 1977, Box and Mumby 2007), the beneficial role of parrotfishes in removing macroalgae appears to be much more important for corals (Mumby 2009). Consequently, densities of juvenile corals are positively related to parrotfish density, biomass, or grazing in the Caribbean (Mumby et al. 2007b, Burkepile et al. 2013).

Thus, for most Caribbean coral reef environments, the net impact of parrotfishes on coral assemblages appears to be positive, although they might have a net negative influence on *Porites porites*. In addition to direct effects, an additional negative impact is that corallivory might potentially constrain the ability of *Porites porites* to take advantage of a loss of coral competitors as the cover of massive corals declines on some coral reefs (Roff et al. 2011). Of particular concern, however, is the role of corallivory when coral cover becomes low (Mumby 2009, Burkepile 2012); will predation overwhelm the capacity of corals to grow? There is mixed evidence available to address this question. Roff et al. (2011) found that the intensity of parrotfish corallivory across the Bahamas increased (although weakly) with an increase in coral density. Thus, if a decline in coral cover leads to lower average coral density then rates of corallivory might also decline. In contrast, Burkepile (2012) found that the frequency and intensity of corallivory increased at sites with low coral cover in Florida, but Florida appears to have unusually high levels of corallivory compared to the rest of the Caribbean.

FAQ 10. How much Evidence is there that Parrotfishes are Good for Reef Resilience?

In theory, grazing by parrotfishes can reduce the abundance of macroalgae and thick algal turfs and facilitate the recruitment, growth, and fecundity of corals (Mumby 2006a), which is a key mechanism promoting reef resilience following a disturbance (Mumby and Steneck 2008). Empirical evidence to support this hypothesis has been found by comparing the functioning of one of the Caribbean's oldest (60+ years), largest (450 km²), and most effective marine reserves (the Exuma Cays Land and Sea Park in the central Bahamas), to surrounding, unprotected reefs. Enforcement of park regulations since the mid-1980s has fostered a relatively intact fauna with abundant sharks and among the highest grouper biomasses in the entire region (Mumby et al. 2011). A cessation of fishing inside park boundaries means that parrotfish biomass is, on average, twice that outside the park, and the cover of macroalgae is four-fold lower (Mumby et al. 2006b). Density of coral juveniles show a simple linear positive increase with parrotfish grazing (Mumby et al. 2007b), and the trajectory of coral populations over time (2.5 years) was positive in the park but neutral to negative outside its boundaries (Mumby and Harborne 2010). These relationships were robust to other putative mechanisms, such as the possibility of there being natural variation in parrotfish abundance at the scale of the study site compared to controls, elevated coral larval supply to the reserve, changes in reef habitat complexity, differences in the density of damselfishes that can interfere with grazing behaviour, and densities of alternative herbivore groups including urchins or acanthurids. In addition, an experimental manipulation of parrotfish grazing evaluated the impact of fishing larger-bodied parrotfishes on the cover of macroalgae and the recruitment of corals to settlement plates (Steneck et al. 2014). The study was undertaken on two exposed forereefs in Belize, and used stainless-steel rods to prevent access of larger-bodied parrotfishes without the usual problems of caging effects on benthos. This 'removal' of large parrotfishes was sufficient to cause a large macroalgal phase shift and vastly reduce coral recruitment, highlighting the importance of adult fishes.

Other than those in the Bahamas and Belize, few studies have been able to track the relationship between a change in herbivory and consequent effects on algae and corals, but many studies have investigated individual components of this process. Mechanistically, one would first expect a negative relationship between parrotfish biomass or grazing and the cover of macroalgae. Secondly, a negative relationship would be expected between macroalgae and demographic and biological responses of corals. There is plenty of evidence for both. For example, negative relationships between total herbivore biomass or total parrotfish biomass and macroalgal cover have been found throughout the Caribbean (Williams and Polunin 2001, Kramer et al. 2003, Newman et al. 2006, Burkepile et al. 2013). These studies were mostly undertaken on forereef habitats where the major macroalgal species include *Lobophora variegata* and species of *Dictyota*. The exception appears to be shallow, eutrophic patch reefs inside Glovers Atoll, Belize (McClanahan et al. 2004) where parrotfishes are unable to influence macroalgal growth (McClanahan et al. 2001). A key point is that this habitat is very different to that of forereefs, and the dominant algae are erect and generally unpalatable, including the genera *Sargassum* and *Turbinaria*. A wealth of evidence exists on the competitive interactions between corals and macroalgae in the Caribbean, all of which shows negative impacts on coral, although this can be species and size-specific, with less importance in larger corals (Ferrari et al. 2012). Furthermore, macroalgae can negatively affect the growth and survivorship of juvenile corals (Box and Mumby 2007), coral growth rate (Lirman 2001, Ferrari et al. 2012), partial colony mortality (Lirman 2001, Nugues and Bak 2006), and coral fecundity (Foster et al. 2008).

In short, the evidence for the positive effects of parrotfishes on coral resilience is substantial for forereef environments, but impacts might be absent in environments where algal growth is strongly enhanced by high light and/or high nutrients, such as shallow patch reefs.

FAQ 11. Isn't There a Lot of Evidence of Reserves not Benefitting Corals even though Parrotfishes were Protected?

A few Caribbean studies have found no evidence of protecting parrotfishes improving coral health (Huntington et al. 2011, Toth et al. 2014), yet each has serious difficulties in interpretation. Huntingdon et al. (2011) compared coral cover in shallow patch reefs between 1998/9 and 2008/9 and asked whether those in a reserve had fared better: they had not. There are a number of problems with this study. First, there was no evidence that the reserves had any positive influence on the biomass of parrotfishes, perhaps because of poaching or the inhospitable nature of the profound macroalgal dominance throughout the study area. (as seen in the Pacific, Hoey and Bellwood 2011). If there was no trophic impact of the reserve then there is no reason to expect any cascading benefit to corals via a reduction in macroalgae. Arguably, the reserve was not functioning successfully for herbivores, and there was no clear mechanism by which the reserve could influence corals either directly or indirectly. Second, the reefs studied were impacted by two bleaching events and three hurricanes between census dates. The authors' study design was unable to resolve the impacts of either event, so it is perhaps not surprising that no systematic effect of reserves on coral cover was found. A more appropriate study design would have tracked the impact-recovery response of individual reefs and asked whether recovery rates were greater in those sites with the greatest herbivore biomass (Fig. 2).

A study of coral trajectories in Florida between 1998 and 2011 found no benefit of reserves on coral trajectories (which mostly declined), nor did reserves affect macroalgal cover (Toth et al. 2014). The authors conclude that reserves have no impact on the health of corals, contrary to other studies such as those described earlier from The Bahamas. Clearly, the reserves studied had no beneficial impact on coral, but the contrasts made with other studies are disingenuous and misleading. First, parrotfishes are protected from commercial exploitation in Florida and there is no evidence that reserves have any impact on herbivory at the study locations: even fished reefs are likely to have supported a parrotfish assemblage close to its carrying capacity. Indeed, the authors of this study did

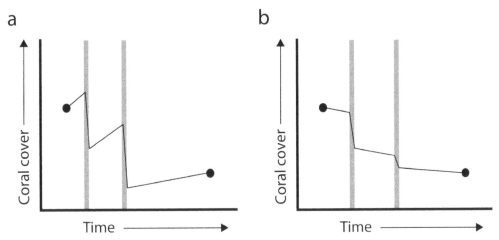

Fig. 2. A schematic representation of the difficulty in assessing coral cover trajectories (black lines) from only two sampling points (black circles) when the reef has experienced disturbance events (grey rectangles): (a) a resilient reef with high parrotfish biomass that experiences increases in coral cover between disturbances, and (b) a reef with low parrotfish biomass which experiences acute and chronic decreases in coral cover.

not even include herbivores in their study design, so we have to assume that herbivore levels were not systematically different between reserves and non-reserves. Under these circumstances, it is not surprising that reserve effects are absent on corals since the reserves have no mechanistic means of influencing corals. In contrast, reserves have a demonstrable benefit to parrotfish biomass where herbivores are the subject of fisheries (Hawkins et al. 2007, Valles and Oxenford 2014). It is under these circumstances that reserves have the potential to benefit coral populations. However, an important point made by Toth et al. (2014) is that coral populations can decline even when parrotfishes are not subjected to fishing. We will return to this issue in the next FAQ.

FAQ 12. If We Protect Parrotfishes, What is a Realistic Expectation for the Future Health of my Reef?

In the continued absence of the urchin *Diadema antillarum*, the current dominance – by biomass – of parrotfishes as the primary herbivore on Caribbean reefs is probably unprecedented. Importantly, although parrotfishes and *Diadema* competed when urchins were common (Carpenter 1988), it would be unrealistic to expect parrotfishes to expand their populations to completely fill the role vacated by urchins. After all, parrotfishes evolved in concert with other herbivores, including sea urchins. Consequently, following the mortality of *D. antillarum*, although densities of herbivorous fishes increased two- to four-fold across four reef zones in the U.S. Virgin Islands, their increased grazing intensity was not sufficient to halt an increase in algal cover and biomass (Carpenter 1990).

The functional importance of parrotfishes as a herbivore depends largely on the productivity potential of the benthos (Steneck and Dethier 1994). In highly productive coral environments exposed to high wave action, parrotfishes appear able to maintain between 30% and 40% of the reef in a grazed state of short algal turfs (Mumby 2006a). In contrast, the cover of macroalgae can become much lower where unexploited parrotfishes forage in low-productivity environments like leeward reefs (Renken et al. 2010). It follows that the response of corals to protection of parrotfishes will depend, in part, on the productivity of the benthos, which influences the frequency and intensity of competition with macroalgae. If productivity potential is low then parrotfishes are more likely to be effective in preventing macroalgal blooms, even when coral cover is low. This is likely to be a factor in the continued health of coral reefs in Bonaire, which is a leeward reef with high coral cover, high parrotfish biomass, and very little macroalgal cover (Steneck et al. 2007).

Since the positive influence of parrotfishes on corals acts through processes of recovery (enhancing recruitment and growth by reducing macroalgae), the net benefit needs to be weighed against opposing rates of coral mortality. First, if background rates of mortality are high, such as in an area with high prevalence of coral epizootics, then coral populations may show net decline even in the presence of parrotfish protection. This might explain the results of Toth et al. (2014) from Florida. Alternatively, even if rates of recruitment and growth exceed background mortality, the overall rate of recovery may be overtaken by frequent acute mortality events from hurricanes and coral bleaching. In this case, coral cover might show a long-term decline, but at least coral populations can continue to recruit and turnover between disturbances. A desirable aspect of such resilience, even in the face of net reductions in cover, is that it allows coral populations to continue evolving, thereby maintaining a window for adaptation.

Observed rates of coral recovery in the Caribbean are non-existent to low once cover falls below 20% (Connell 1997, Roff and Mumby 2012). Protection of parrotfishes in the Exuma Cays Land and Sea Park (The Bahamas) led to an annual increase of total coral cover of around 1% per year. This is low, but should be viewed in context. Coral cover on these reefs was already heavily depleted by the 1998 coral bleaching event and was only 7% at the beginning of the study. Moreover, it is not unusual for rates of coral recovery to follow a sigmoidal function, being slow when cover is low, and accelerating as adult coral populations increase (although examples are scare for Caribbean reefs, they are likely to follow patterns documented in the Indo-Pacific by Halford et al. 2004 and Gilmour et al. 2013). By protecting parrotfish populations, it is hoped that the currently slow increases in coral cover in the Bahamian park will also accelerate, but this will only occur in the absence of acute coral-mortality events.

In addition to increasing coral cover, an associated aim of parrotfish protection is to affect net carbonate production ('carbonate budgets'). Carbonate budget states, which are determined by the relative rates of carbonate production and erosion, represent an important tool for understanding the interactions between reef degradation and ecosystem services because they ultimately govern reef structural integrity and growth potential (Perry et al. 2008). Field-tested models of Caribbean population dynamics predict that protection of parrotfishes is vital to maintain positive carbonate budgets towards the end of this century, though this also requires significant action on greenhouse gas emissions (Kennedy et al. 2013). Note that these models include both the 'positive' effects of parrotfishes consuming algae and their 'negative' effects on coral carbonate budgets as a major source of bioerosion. In the absence of action on climate change, models predict that protection of parrotfishes still slow net reef decline by two decades or so, which may buy time for coral acclimation or adaptation to stress.

Parrotfish Fisheries Management

FAQ 13. How Important are Parrotfishes within Fisheries?

Fishing has had significant global impacts on herbivore assemblages, but large-bodied functional groups, such as scraping and excavating parrotfishes are particularly susceptible (Edwards et al. 2014, Debrot et al. 2008). Parrotfishes are not a primary target of Caribbean fishers, but are increasingly caught as more desirable species from higher trophic levels, such as grouper, become rarer (Mumby et al. 2012). Consequently, there is a clear correlation between fishing pressure and parrotfish abundance (Hawkins and Roberts 2004). Furthermore, even if not targeted directly, parrotfishes are frequently caught in traps primarily placed to catch species such as groupers and snappers. As they are designed to catch large-bodied fishes, traps also remove functionally important large-bodied parrotfishes, such as *Sc. vetula* and *Sp. viride* (Rakitin and Kramer 1996, Hawkins et al. 2007). Traps may be particularly effective in low complexity habitats where they provide refuges that are naturally scarce (Wolff et al. 1999). Large-bodied species, such as *Sc. guacamaia*, are also susceptible to spearfishing (Comeros-Raynal et al. 2012). Since many parrotfishes are caught by artisanal fishers and as bycatch, there are few reliable data for the weight of parrotfishes landed in the region.

FAQ 14. Are Marine Reserves Sufficient to Protect Parrotfishes?

No. If the only investment in parrotfish management constitutes protection in reserves, this will potentially leave harvested reefs with lower resilience. An important implication of this is that the structural complexity of harvested coral reefs will be more likely to decline than protected reefs (Bozec et al. 2015). Because many commercially important reef fish species prefer high complexity reef habitats (Bejarano et al. 2011), a loss of complexity will threaten future yield. Indeed, a model of the effects of lost habitat complexity on Caribbean reef food webs found that reef fisheries productivity could decline more than three-fold (Rogers et al. 2014). Management of parrotfishes throughout the seascape should provide a long-term benefit to reef fisheries. A national ban on parrotfish fishing has been implemented in some Caribbean countries, such as in Belize (Mumby et al. 2012) where compliance has been high (Cox et al. 2013), and there are growing hopes that stakeholders will accept similar restrictions elsewhere. Such species-specific regulations represent one of many additional management tools that will be required, along with marine reserves, to manage Caribbean reefs that may look and function very differently in the future (Rogers et al. 2015).

FAQ 15. If it's not Possible to Ban Parrotfish Harvesting, are there any Alternative Management Strategies?

One alternative management strategy that has been proposed is to review the degree to which parrotfish grazing is important for different reef habitats (Mumby 2016). Some habitats, like those with a coral-built framework – often referred to as *Orbicella* reef (formerly *Montastraea* reef) – appear to have the greatest dependence on parrotfishes for controlling algae. Algal populations in some other habitats, such as hardbottom habitats visually dominated by gorgonians, appear to be driven by physical rather than biological processes (Mumby 2016). In these gorgonian-dominated habitats wave-driven scour and dislodgement from resuspended sediments is likely to constrain successful algal colonisation and growth (Torres et al. 2001). Thus, one alternative management strategy is to protect parrotfishes on *Orbicella* reefs where they are functionally critical, but permit exploitation on extensive hardbottom habitats where benthic dynamics will be less affected (Mumby 2016).

    A recent study linked fisheries policies to the population dynamics of parrotfishes and their concomitant impacts on resilience of the ecosystem (Bozec et al. 2016). Bozec et al. (2016) found that even low harvest rates led to a large negative effect on ecosystem resilience. However, the adoption of two relatively simple management practices – a minimum size of 30 cm and ban on fish traps – led to not only a more productive fishery but better outcomes for reef health at a given harvest rate. However, to help mitigate losses of reef resilience it was also necessary to keep harvest rates (the proportion of fishable biomass extracted per annum) to less than 0.1.

FAQ 16. How Quickly do Parrotfish Recover if Fishing is Banned?

There are a number of examples where marine reserves have led to higher biomasses of parrotfishes on reefs compared to unprotected sites (e.g. Polunin and Roberts 1993, Chapman and Kramer 1999, Mumby et al. 2006b). However, there are few long-term data on parrotfish recovery rates after a cessation in fishing. One exception is in Bermuda, where in 1992 the government banned the use of fish traps, which were a major source of parrotfish exploitation. A nine-year data set demonstrated that adult biomasses of parrotfish species increased by a factor of 3.7, and reached this level after around six years, at which point biomass did not recover any further (O'Farrell et al. 2015). Sex ratios may also approach unfished values within 3-4 years (O'Farrell et al. 2016). Recovery trajectories will vary depending on whether initial conditions reflect light or heavy fishing pressure, but the Bermuda data suggest that parrotfish populations can recover relatively quickly following the establishment of protective measures.

FAQ 17. Which Species Should We Protect?

Larger-bodied parrotfishes are particularly susceptible to even modest levels of fishing pressure, as has been demonstrated in both the Caribbean (Debrot et al. 2008) and Pacific and Indian Oceans (Bellwood et al. 2012, Heenan et al. 2016). Consequently, all large-bodied Caribbean parrotfish species should have some form of management because it is particularly important to prevent a limited single-species fishery. The limited functional redundancy among Caribbean parrotfish species (Burkepile and Hay 2008, 2010, Adam et al. 2015b) and apparent weakness of interspecific interactions (Mumby and Wabnitz 2002) implies that there is limited capacity for the loss of one species to be compensated for by others. Therefore, if fishing reduces the abundance of one species it is unlikely that the biomass of other, unfished species would increase to compensate for the lost ecological function.

Despite the importance of large-bodied parrotfishes, most species are important for different reasons. Generally, species in the genus *Scarus* have higher size-specific bite rates than those of *Sparisoma*, making them functionally important in maintaining grazed algal turfs that are suitable for coral recruitment and growth (Bruggemann et al. 1994b, Mumby 2006a, Burkepile and Hay 2011). However, unlike species in the genus *Scarus*, sparisomatinines have a broader diet, and are the only reef parrotfishes that routinely consume several macroalgal species once they become established (Mumby 2006a). Sparisomatinines are therefore important for both preventing and constraining macroalgal blooms. Within each species, the rate of food consumption (bite size × bite rate) increases with body size, so larger-bodied fish have a disproportionately important impact on grazing (Bruggemann et al. 1994c, Mumby et al. 2006a, Hoey Chapter 6).

Lastly, although some of the largest parrotfish species, such as *Sc. coelestinus*, are protected in the US Caribbean, their densities are often so low that their functional relevance is questionable. That is not to say that protection is not important – indeed it might be vital to rebuild stocks – but it does not necessarily help restore the ecosystem process of grazing. Moreover, species like *Sc. coelestinus* appear to have low densities even in areas that had relatively light levels of parrotfish exploitation, such as Belize pre-2000, which implies that the rebuilding of stocks might only achieve modest densities of these species at best. However, the exploitation of large-bodied species such as *Sc. guacamaia* and *Sc. coelestinus* across the Caribbean, and removal of critical mangrove nurseries, means that their natural population levels and functional roles are poorly understood.

Conclusions

Parrotfishes have rarely featured as a model species for generic study, in contrast to the use of damselfish for population biology studies (e.g. Doherty and Fowler 1994), connectivity work with anemone fish (e.g. Almany et al. 2007), or examination of hybridisation in hamlets (e.g. Whiteman et al. 2007). The paucity of studies derives at least partly from the difficulty of working with non-site attached fishes, but also their high sensitivity to being handled means that manipulative studies have been fraught with difficulty. Consequently, although parrotfishes clearly have a critical functional role on Caribbean reefs, there are significant gaps in our knowledge of their basic biology. For example, the majority of studies have focused on *Sp. viride*, and demographic parameters for many other species are severely lacking, even though they have been estimated using a model-fitting approach. The incomplete answers to the FAQs on parrotfish management highlight the urgent need for us to solve these methodological issues and undertake further research. For example, additional data are critical to allow the construction of realistic population models that will provide new insights into the impacts of different fishing strategies on the process of grazing. Furthermore, understanding the impacts of climate change on the abundance and behaviour of herbivores is critical for understanding the stability and functioning of Caribbean reefs (Harborne et al. 2017).

Despite the many gaps in our knowledge, sufficient data exist to allow the incorporation of grazing into spatially explicit models of benthic dynamics, and consequently predict reef resilience on different reefs or under different disturbance regimes (e.g. Mumby 2006a, Mumby et al. 2014). These models provide important support to Caribbean reef managers that is currently unavailable in the Pacific, although similar tools are being developed (Ortiz et al. 2014). However, applying model outputs in real world scenarios relies on stakeholders fully understanding their derivation, limitations, and caveats. Therefore, we encourage everyone concerned with future reef health to frequently ask questions. As the adage says "The only stupid question is the one that is not asked".

Acknowledgments

We acknowledge the Australian Research Council for DECRA (DE120102459) and Laureate fellowships. This is contribution #39 of the Marine Education and Research Center in the Institute for Water and Environment at Florida International University. This manuscript was developed during a retreat to the Gold Coast, and we thank Jupiters Hotel & Casino for evening refreshments.

References Cited

Adam, T.C., D.E. Burkepile, B.I. Ruttenberg and M.J. Paddack. 2015a. Herbivory and the resilience of Caribbean coral reefs: knowledge gaps and implications for management. Mar. Ecol. Prog. Ser. 520: 1–20.

Adam, T.C., M. Kelley, B.I. Ruttenberg and D.E. Burkepile. 2015b. Resource partitioning along multiple niche axes drives functional diversity in parrotfishes on Caribbean coral reefs. Oecologia 179: 1173–1185.

Adams, A.J., C.P. Dahlgren, G.T. Kellison, M.S. Kendall, C.A. Layman, J.A. Ley, I. Nagelkerken and J.E. Serafy. 2006. Nursery function of tropical back-reef systems. Mar. Ecol. Prog. Ser. 318: 287–301.

Albins, M.A. and M.A. Hixon. 2008. Invasive Indo-Pacific lionfish *Pterois volitans* reduce recruitment of Atlantic coral-reef fishes. Mar. Ecol. Prog. Ser. 367: 233–238.

Albins, M.A. and M.A. Hixon. 2013. Worst case scenario: potential long-term effects of invasive predatory lionfish (*Pterois volitans*) on Atlantic and Caribbean coral-reef communities. Environ. Biol. Fishes 96: 1151–1157.

Almany, G.R. and M.S. Webster. 2004. Odd species out as predators reduce diversity of coral-reef fishes. Ecology 85: 2933–2937.

Almany, G.R. and M.S. Webster. 2006. The predation gauntlet: early post-settlement mortality in reef fishes. Coral Reefs 25: 19–22.

Almany, G.R., M.L. Berumen, S.R. Thorrold, S. Planes and G.P. Jones. 2007. Local replenishment of coral reef fish populations in a marine reserve. Science 316: 742–744.

Beck, M.W., K.L. Heck, K.W. Able, D.L. Childers, D.B. Eggleston, B.M. Gillanders, B. Halpern, C.G. Hays, K. Hoshino, T.J. Minello, R.J. Orth, P.F. Sheridan and M.R. Weinstein. 2001. The identification, conservation, and management of estuarine and marine nurseries for fish and invertebrates. Bioscience 51: 633–641.

Bejarano, S., P.J. Mumby and I. Sotheran. 2011. Predicting structural complexity of reefs and fish abundance using acoustic remote sensing (RoxAnn). Mar. Biol. 158: 489–504.

Bellwood, D.R. and J.H. Choat. 1990. A functional analysis of grazing in parrotfishes (family Scaridae): the ecological implications. Environ. Biol. Fishes 28: 189–214.

Bellwood, D.R., P.C. Wainwright, C.J. Fulton and A. Hoey. 2002. Assembly rules and functional groups at global biogeographical scales. Func. Ecol. 16: 557–562.

Bellwood, D.R., A.S. Hoey and J.H. Choat. 2003. Limited functional redundancy in high diversity systems: resilience and ecosystem function on coral reefs. Ecol. Lett. 6: 281–285.

Bellwood, D.R., A.S. Hoey and T.P. Hughes. 2012. Human activity selectively impacts the ecosystem roles of parrotfishes on coral reefs. Proc. R. Soc. B. 279: 1621–1629.

Bernardi, G., D.R. Robertson, K.E. Clifton and E. Azzurro. 2000. Molecular systematics, zoogeography, and evolutionary ecology of the Atlantic parrotfish genus *Sparisoma*. Mol. Phylogenet. Evol. 15: 292–300.

Birkeland, C. 1977. The importance of rate of biomass accumulation in early successional stages of benthic communities to the survival of coral recruits. pp. 15–21. *In*: D.L. Taylor (ed.). Proceedings, Third International Coral Reef Symposium. Volume 1: Biology. Rosenstiel School of Marine and Atmospheric Science, Miami, Florida.

Bonaldo, R.M., A.S. Hoey and D.R. Bellwood. 2014. The ecosystem role of parrotfishes on tropical reefs. Oceanogr. Mar. Biol. Annu. Rev. 52: 81–132.

Box, S.J. and P.J. Mumby. 2007. Effect of macroalgal competition on growth and survival of juvenile Caribbean corals. Mar. Ecol. Prog. Ser. 342: 139–149.

Bozec, Y.M., L. Yakob, S. Bejarano and P.J. Mumby. 2013. Reciprocal facilitation and non-linearity maintain habitat engineering on coral reefs. Oikos 122: 428–440.

Bozec, Y.M., L. Alvarez-Filip and P.J. Mumby. 2015. The dynamics of architectural complexity on coral reefs under climate change. Glob. Change Biol. 21: 223–235.

Bozec, Y. M., S. O'Farrell, J.H. Bruggemann, B.E. Luckhurst and P.J. Mumby. 2016. Tradeoffs between fisheries harvest and the resilience of coral reefs. Proc. Natl. Acad. Sci. U.S.A. 113: 4536–4541.

Bruckner, A.W. and R.J. Bruckner. 1998. Destruction of coral by *Sparisoma viride*. Coral Reefs 17: 350.

Bruggemann, J.H., M.J.H. van Oppen and A.M. Breeman. 1994a. Foraging by the stoplight parrotfish *Sparisoma viride*. I. Food selection in different, socially determined habitats. Mar. Ecol. Prog. Ser. 106: 41–55.

Bruggemann, J.H., M.W.M. Kuyper and A.M. Breeman. 1994b. Comparative analysis of foraging and habitat use by the sympatric Caribbean parrotfish *Scarus vetula* and *Sparisoma viride* (Scaridae). Mar. Ecol. Prog. Ser. 112: 51–66.

Bruggemann, J.H., J. Begeman, E.M. Bosma, P. Verburg and A.M. Breeman. 1994c. Foraging by the stoplight parrotfish *Sparisoma viride*. II. Intake and assimilation of food, protein and energy. Mar. Ecol. Prog. Ser. 106: 57–71.

Burkepile, D.E. 2012. Context-dependent corallivory by parrotfishes in a Caribbean reef ecosystem. Coral Reefs 31: 111–120.

Burkepile, D.E. and M.E. Hay. 2008. Herbivore species richness and feeding complementarity affect community structure and function on a coral reef. Proc. Natl. Acad. Sci. USA 105: 16201–16206.

Burkepile, D.E. and M.E. Hay. 2010. Impact of herbivore identity on algal succession and coral growth on a Caribbean reef. PLoS One 5: e8963.

Burkepile, D.E. and M.E. Hay. 2011. Feeding complementarity versus redundancy among herbivorous fishes on a Caribbean reef. Coral Reefs 30: 351–362.

Burkepile, D.E., J.E. Allgeier, A.A. Shantz, C.E. Pritchard, N.P. Lemoine, L.H. Bhatti and C.A. Layman. 2013. Nutrient supply from fishes facilitates macroalgae and suppresses corals in a Caribbean coral reef ecosystem. Sci Rep 3: 1493.

Bythell, J.C., E.H. Gladfelter and M. Bythell. 1993. Chronic and catastrophic natural mortality of three common Caribbean reef corals. Coral Reefs 12: 143–152.

Cardoso, S.C., M.C. Soares, H.A. Oxenford and I.M. Côté. 2009. Interspecific differences in foraging behaviour and functional role of Caribbean parrotfish. Mar. Biodiv. Rec. 2: e148.

Carpenter, R.C. 1988. Mass mortality of a Caribbean sea urchin: immediate effects on community metabolism and other herbivores. Proc. Natl. Acad. Sci. USA 85: 511–514.

Carpenter, R.C. 1990. Mass mortality of *Diadema antillarum*. II. Effects on population densities and grazing intensity of parrotfishes and surgeonfishes. Mar. Biol. 104: 79–86.

Catano, L.B., M.C. Rojas, R.J. Malossi, J.R. Peters, M.R. Heithaus, J.W. Fourqurean and D.E. Burkepile. 2016. Reefscapes of fear: predation risk and reef heterogeneity interact to shape herbivore foraging behaviour. J. Anim. Ecol. 85: 146–156.

Chapman, M.R. and D.L. Kramer. 1999. Gradients in coral reef fish density and size across the Barbados Marine Reserve boundary: effects of reserve protection and habitat characteristics. Mar. Ecol. Prog. Ser. 181: 81–96.

Choat, J.H., D.R. Robertson, J.L. Ackerman and J.M. Posada. 2003. An age-based demographic analysis of the Caribbean stoplight parrotfish *Sparisoma viride*. Mar. Ecol. Prog. Ser. 246: 265–277.

Choat, J.H., W.D. Robbins and K.D. Clements. 2004. The trophic status of herbivorous fishes on coral reefs II. Food processing modes and trophodynamics. Mar. Biol. 145: 445–454.

Choat, J.H., O.S. Klanten, L. Van Herwerden, D.R. Robertson and K.D. Clements. 2012. Patterns and processes in the evolutionary history of parrotfishes (Family Labridae). Biol. J. Linnean Soc. 107: 529–557.

Claydon, J. 2004. Spawning aggregations of coral reef fishes: Characteristics, hypotheses, threats and management. Oceanogr. Mar. Biol.Annu. Rev. 42: 265–301.

Clements, K.D., D. Raubenheimer and J.H. Choat. 2009. Nutritional ecology of marine herbivorous fishes: ten years on. Funct. Ecol. 23: 79–92.

Clements, K.D., D.P. German, J. Piché, A. Tribollet and J.H. Choat. 2017. Integrating ecological roles and trophic diversification on coral reefs: multiple lines of evidence identify parrotfishes as microphages. Biol. J. Linnean Soc. 120: 729–751.

Comeros-Raynal, M.T., J.H. Choat, B.A. Polidoro, K.D. Clements, R. Abesamis, M.T. Craig, M.E. Lazuardi, J. McIlwain, A. Muljadi, R.F. Myers, C.L. Nañola, S. Pardede, L.A. Rocha, B. Russell, J.C. Sanciangco, B. Stockwell, H. Harwell and K.E. Carpenter. 2012. The likelihood of extinction of iconic and dominant herbivores and detritivores of coral reefs: the parrotfishes and surgeonfishes. PLoS One 7: e39825.

Connell, J.H. 1997. Disturbance and recovery of coral assemblages. Coral Reefs 16: S101–S113.

Côté, I.M., S.J. Green and M.A. Hixon. 2013. Predatory fish invaders: Insights from Indo-Pacific lionfish in the western Atlantic and Caribbean. Biol. Cons. 164: 50–61.

Cox, C.E., C.D. Jones, J.P. Wares, K.D. Castillo, M.D. McField and J.F. Bruno. 2013. Genetic testing reveals some mislabeling but general compliance with a ban on herbivorous fish harvesting in Belize. Conserv. Lett. 6: 132–140.

Crossman, D.J., J.H. Choat, K.D. Clements, T. Hardy and J. McConochie. 2001. Detritus as food for grazing fishes on coral reefs. Limnol. Oceanogr. 46: 1596–1605.

Dahlgren, C.P., G.T. Kellison, A.J. Adams, B.M. Gillanders, M.S. Kendall, C.A. Layman, J.A. Ley, I. Nagelkerken and J.E. Serafy. 2006. Marine nurseries and effective juvenile habitats: concepts and applications. Mar. Ecol. Prog. Ser. 312: 291–295.

Debrot, D., J. Choat, J. Posada and D. Robertson. 2008. High densities of the large bodied parrotfishes (Scaridae) at two Venezuelan offshore reefs: comparison among four localities in the Caribbean. Proceedings of the 60th Gulf and Caribbean Fisheries Institute, Punta Cana, Dominican Republic: 335–337.

Doherty, P. and T. Fowler. 1994. An empirical test of recruitment limitation in a coral reef fish. Science 263: 935–939.

Dorenbosch, M., M.G.G. Grol, I. Nagelkerken and G. van der Velde. 2006. Seagrass beds and mangroves as potential nurseries for the threatened Indo-Pacific humphead wrasse, *Cheilinus undulatus* and Caribbean rainbow parrotfish, *Scarus guacamaia*. Biol. Cons. 129: 277–282.

Eaton, L., K.A. Sloman, R.W. Wilson, A.B. Gill and A.R. Harborne. 2016. Non-consumptive effects of native and invasive predators on juvenile Caribbean parrotfish. Env. Biol. Fish. 99: 499–508.

Edwards, C.B., A.M. Friedlander, A.G. Green, M.J. Hardt, E. Sala, H.P. Sweatman, I.D. Williams, B. Zgliczynski, S.A. Sandin and J.E. Smith. 2014. Global assessment of the status of coral reef herbivorous fishes: evidence for fishing effects. Proc. R. Soc. B 281: 20131835.

Ferrari, R., M. Gonzalez-Rivero and P.J. Mumby. 2012. Size matters in competition between corals and macroalgae. Mar. Ecol. Prog. Ser. 467: 77–88.

Floeter, S.R., C.E.L. Ferreira, A. Dominici-Arosemena and I.R. Zalmon. 2004. Latitudinal gradients in Atlantic reef fish communities: trophic structure and spatial use patterns. J. Fish Biol. 64: 1680–1699.

Floeter, S.R., L.A. Rocha, D.R. Robertson, J.C. Joyeux, W.F. Smith-Vaniz, P. Wirtz, A.J. Edwards, J.P. Barreiros, C.E.L. Ferreira, J.L. Gasparini, A. Brito, J.M. Falcón, B.W. Bowen and G. Bernardi. 2008. Atlantic reef fish biogeography and evolution. J. Biogeogr. 35: 22–47.

Foster, N.L., S.J. Box and P.J. Mumby. 2008. Competitive effects of macroalgae on the fecundity of the reef-building coral *Montastraea annularis*. Mar. Ecol. Prog. Ser. 367: 143–152.

Fulton, C.J., D.R. Bellwood and P.C. Wainwright. 2005. Wave energy and swimming performance shape coral reef fish assemblages. Proc. R. Soc. B 272: 827–832.

Gilmour, J.P., L.D. Smith, A.J. Heyward, A.H. Baird and M.S. Pratchett. 2013. Recovery of an isolated coral reef system following severe disturbance. Science 340: 69–71.

Green, S.J., J.L. Akins and I.M. Côté. 2011. Foraging behaviour and prey consumption in the Indo-Pacific lionfish on Bahamian coral reefs. Mar. Ecol. Prog. Ser. 433: 159–167.

Green, S.J., J.L. Akins, A. Maljković and I.M. Côté. 2012. Invasive lionfish drive Atlantic coral reef fish declines. PLoS One 7: e32596.

Haddon, M. 2011. Modelling and Quantitative Methods in Fisheries. Second Edition. CRC Press, Boca Raton, FL, USA.

Halford, A., A.J. Cheal, D. Ryan and D.M. Williams. 2004. Resilience to large-scale disturbance in coral and fish assemblages on the Great Barrier Reef. Ecology 85: 1892–1905.

Harborne, A.R., P.J. Mumby, F. Micheli, C.T. Perry, C.P. Dahlgren, K.E. Holmes and D.R. Brumbaugh. 2006. The functional value of Caribbean coral reef, seagrass and mangrove habitats to ecosystem processes. Adv. Mar. Biol. 50: 57–189.

Harborne, A.R., P.J. Mumby and R. Ferrari. 2012. The effectiveness of different meso-scale rugosity metrics for predicting intra-habitat variation in coral-reef fish assemblages. Environ. Biol. Fishes 94: 431–442.

Harborne, A.R., I. Nagelkerken, N.H. Wolff, Y.-M. Bozec, M. Dorenbosch, M.G.G. Grol and P.J. Mumby. 2016. Direct and indirect effects of nursery habitats on coral-reef fish assemblages, grazing pressure and benthic dynamics. Oikos 125: 957–967.

Harborne, A.R., A. Rogers, Y.-M. Bozec and P.J. Mumby. 2017. Multiple stressors and the functioning of coral reefs. Annu. Rev. Mar. Sci. 9: 445–468.

Hawkins, J.P. and C.M. Roberts. 2004. Effects of artisanal fishing on Caribbean coral reefs. Conserv. Biol. 18: 215–226.

Hawkins, J.P., C.M. Roberts, F.R. Gell and C. Dytham. 2007. Effects of trap fishing on reef fish communities. Aquat. Conserv.: Mar. Freshw. Ecosyst. 17: 111–132.

Heenan, A., A.S. Hoey, G.J. Williams and I.D. Williams. 2016. Natural bounds on herbivorous coral reef fishes. Proc. R. Soc. B-Biol. Sci. 283: 20161716

Hixon, M.A. 2011. 60 years of coral reef fish ecology: past, present, future. Bull. Mar. Sci. 87: 727–765.

Hixon, M.A., T.W. Anderson, K.L. Buch, D.W. Johnson, J.B. McLeod and C.D. Stallings. 2012. Density dependence and population regulation in marine fish: a large-scale, long-term field manipulation. Ecol. Monogr. 82: 467–489.

Hoey, A.S. and D.R. Bellwood. 2011. Suppression of herbivory by macroalgal density: a critical feedback on coral reefs? Ecol. Lett. 14: 267–273.

Huijbers, C.M., I. Nagelkerken, A.O. Debrot and E. Jongejans. 2013. Geographic coupling of juvenile and adult habitat shapes spatial population dynamics of a coral reef fish. Ecology 94: 1859–1870.

Huntington, B.E., M. Karnauskas and D. Lirman. 2011. Corals fail to recover at a Caribbean marine reserve despite ten years of reserve designation. Coral Reefs 30: 1077–1085.

Jackson, J., M. Donovan, K. Cramer and V. Lam (eds.). 2014. Status and Trends of Caribbean Coral Reefs: 1970–2012. Global Coral Reef Monitoring Network, IUCN, Gland, Switzerland.

Kennedy, E.V., C.T. Perry, P.R. Halloran, R. Iglesias-Prieto, C.H.L. Schönberg, M. Wisshak, A.U. Form, J.P. Carricart-Ganivet, M. Fine, C.M. Eakin and P.J. Mumby. 2013. Avoiding coral reef functional collapse requires local and global action. Curr. Biol. 23: 912–918.

Kindinger, T.L. and M.A. Albins. 2017. Consumptive and non-consumptive effects of an invasive marine predator on native coral-reef herbivores. Biol. Invasions 19: 131–146.

Kramer, P.A. 2003. Synthesis of coral reef health indicators for the western Atlantic: results of the AGRRA program (1997–2000). Atoll Res. Bull. 496: 1–55.

Kramer, P.A., K.W. Marks and T.L. Turnbull. 2003. Assessment of Andros Island Reef System, Bahamas (Part 2: Fishes). Atoll Res. Bull. 496: 100–122.

Lesser, M.P. and M. Slattery. 2011. Phase shift to algal dominated communities at mesophotic depths associated with lionfish (*Pterois volitans*) invasion on a Bahamian coral reef. Biol. Invasions 13: 1855–1868.

Lessios, H.A. 1988. Mass mortality of *Diadema antillarum* in the Caribbean: what have we learned? Annu. Rev. Ecol. Syst. 19: 371–393.

Lewis, S.M. and P.C. Wainwright. 1985. Herbivore abundance and grazing intensity on a Caribbean coral reef. J. Exp. Mar. Biol. Ecol. 87: 215–228.

Lirman, D. 2001. Competition between macroalgae and corals: effects of herbivore exclusion and increased algal biomass on coral survivorship and growth. Coral Reefs 19: 392–399.

Littler, M.M., P.R. Taylor and D.S. Littler. 1989. Complex interactions in the control of coral zonation on a Caribbean reef flat. Oecologia 80: 331–340.

Lobel, P.S. 1981. Trophic biology of herbivorous reef fishes: alimentary pH and digestive capabilities. J. Fish Biol. 19: 365–397.

Madin, E.M.P., S.D. Gaines and R.R. Warner. 2010. Field evidence for pervasive indirect effects of fishing on prey foraging behavior. Ecology 91: 3563–3571.

Madin, E.M.P., J.S. Madin and D.J. Booth. 2011. Landscape of fear visible from space. Sci Rep 1: 14.

Marsh-Hunkin, K.E., D.J. Gochfeld and M. Slattery. 2013. Antipredator responses to invasive lionfish, *Pterois volitans*: interspecific differences in cue utilization by two coral reef gobies. Mar. Biol. 160: 1029–1040.

McClanahan, T.R., M. McField, M. Huitric, K. Bergman, E. Sala, M. Nyström, I. Nordemar, T. Elfwing and N.A. Muthiga. 2001. Responses of algae, corals and fish to the reduction of macroalgae in fished and unfished patch reefs of Glovers Reef Atoll, Belize. Coral Reefs 19: 367–379.

McClanahan, T.R., E. Sala and P.J. Mumby. 2004. Phosphorus and nitrogen enrichment do not enhance brown frondose "macroalgae". Mar. Pollut. Bull. 48: 196–199.

McCook, L.J., J. Jompa and G. Diaz-Pulido. 2001. Competition between corals and algae on coral reefs: a review of evidence and mechanisms. Coral Reefs 19: 400–417.

Miller, M.W. and M.E. Hay. 1998. Effects of fish predation and seaweed competition on the survival and growth of corals. Oecologia 113: 231–238.

Morris, J.A. and J.L. Akins. 2009. Feeding ecology of invasive lionfish (*Pterois volitans*) in the Bahamian archipelago. Environ. Biol. Fishes 86: 389–398.

Mouillot, D., S. Villéger, V. Parravicini, M. Kulbicki, J.E. Arias-González, M. Bender, P. Chabanet, S.R. Floeter, A. Friedlander, L. Vigliola and D.R. Bellwood. 2014. Functional over-redundancy and high functional vulnerability in global fish faunas on tropical reefs. Proc. Natl. Acad. Sci. USA 111: 13757–13762.

Mumby, P.J. 2006a. The impact of exploiting grazers (Scaridae) on the dynamics of Caribbean coral reefs. Ecol. Appl. 16: 747–769.

Mumby, P.J. 2006b. Connectivity of reef fish between mangroves and coral reefs: algorithms for the design of marine reserves at seascape scales. Biol. Cons. 128: 215–222.

Mumby, P.J. 2009. Herbivory versus corallivory: are parrotfish good or bad for Caribbean coral reefs? Coral Reefs 28: 683–690.

Mumby, P.J. 2016. Stratifying herbivore fisheries by habitat to avoid ecosystem overfishing of coral reefs. Fish. Fish. 17: 266–278.

Mumby, P.J. and C.C.C. Wabnitz. 2002. Spatial patterns of aggression, territory size, and harem size in five sympatric Caribbean parrotfish species. Environ. Biol. Fishes 63: 265–279.

Mumby, P.J. and R.S. Steneck. 2008. Coral reef management and conservation in light of rapidly evolving ecological paradigms. Trends Ecol. Evol. 23: 555–563.

Mumby, P.J. and A.R. Harborne. 2010. Marine reserves enhance the recovery of corals on Caribbean reefs. PLoS One 5: e8657. doi: 8610.1371/journal.pone.0008657.

Mumby, P.J., A.J. Edwards, J.E. Arias-González, K.C. Lindeman, P.G. Blackwell, A. Gall, M.I. Gorczynska, A.R. Harborne, C.L. Pescod, H. Renken, C.C.C. Wabnitz and G. Llewellyn. 2004. Mangroves enhance the biomass of coral reef fish communities in the Caribbean. Nature 427: 533–536.

Mumby, P.J., J.D. Hedley, K. Zychaluk, A.R. Harborne and P.G. Blackwell. 2006a. Revisiting the catastrophic die-off of the urchin *Diadema antillarum* on Caribbean coral reefs: fresh insights on resilience from a simulation model. Ecol. Model. 196: 131–148.

Mumby, P.J., C.P. Dahlgren, A.R. Harborne, C.V. Kappel, F. Micheli, D.R. Brumbaugh, K.E. Holmes, J.M. Mendes, K. Broad, J.N. Sanchirico, K. Buch, S. Box, R.W. Stoffle and A.B. Gill. 2006b. Fishing, trophic cascades, and the process of grazing on coral reefs. Science 311: 98–101.

Mumby, P.J., A. Hastings and H.J. Edwards. 2007a. Thresholds and the resilience of Caribbean coral reefs. Nature 450: 98–101.

Mumby, P.J., A.R. Harborne, J. Williams, C.V. Kappel, D.R. Brumbaugh, F. Micheli, K.E. Holmes, C.P. Dahlgren, C.B. Paris and P.G. Blackwell. 2007b. Trophic cascade facilitates coral recruitment in a marine reserve. Proc. Natl. Acad. Sci. USA 104: 8362–8367.

Mumby, P.J., A.R. Harborne and D.R. Brumbaugh. 2011. Grouper as a natural biocontrol of invasive lionfish. PLoS One 6: e21510. doi:21510.21371/journal.pone.0021510.

Mumby, P.J., R.S. Steneck, A.J. Edwards, R. Ferrari, R. Coleman, A.R. Harborne and J.P. Gibson. 2012. Fishing down a Caribbean food web relaxes trophic cascades. Mar. Ecol. Prog. Ser. 445: 13–24.

Mumby, P.J., N.H. Wolff, Y.–M. Bozec, I. Chollett and P. Halloran. 2014. Operationalizing the resilience of coral reefs in an era of climate change. Conserv. Lett. 7: 176–187.

Nagelkerken, I. and G. van der Velde. 2004. Are Caribbean mangroves important feeding grounds for juvenile reef fish from adjacent seagrass beds? Mar. Ecol. Prog. Ser. 274: 143–151.

Nagelkerken, I., M. Dorenbosch, W.C.E.P. Verberk, E. Cocheret de la Morinière and G. van der Velde. 2000. Importance of shallow-water biotopes of a Caribbean bay for juvenile coral reef fishes: patterns in biotope association, community structure and spatial distribution. Mar. Ecol. Prog. Ser. 202: 175–192.

Nagelkerken, I., S. Kleijnen, T. Klop, R.A.C.J. van den Brand, E. Cocheret de la Morinière and G. van der Velde. 2001. Dependence of Caribbean reef fishes on mangroves and seagrass beds as nursery habitats: a comparison of fish faunas between bays with and without mangroves/seagrass beds. Mar. Ecol. Prog. Ser. 214: 225–235.

Nagelkerken, I., C.M. Roberts, G. van der Velde, M. Dorenbosch, M.C. van Riel, E. Cocheret de la Morinière and P.H. Nienhuis. 2002. How important are mangroves and seagrass beds for coral-reef fish? The nursery hypothesis tested on an island scale. Mar. Ecol. Prog. Ser. 244: 299–305.

Nagelkerken, I., M.G.G. Grol and P.J. Mumby. 2012. Effects of marine reserves versus nursery habitat availability on structure of reef fish communities. PLoS One 7: e36906.

Nemeth, R.S. 2012. Ecosystem aspects of species that aggregate to spawn. pp. 21–55. *In*: Y. Sadovy de Mitcheson and P.L. Colin (eds.). Reef Fish Spawning Aggregations: Biology, Research and Management. Springer, Dordrecht, Netherlands.

Nemeth, M. and R. Appeldoorn. 2009. The distribution of herbivorous coral reef fishes within fore-reef habitats: the role of depth, light and rugosity. Caribb. J. Sci. 45: 247–253.

Newman, M.J.H., G.A. Paredes, E. Sala and J.B.C. Jackson. 2006. Structure of Caribbean coral reef communities across a large gradient of fish biomass. Ecol. Lett. 9: 1216–1227.

Newman, S.P., R.D. Handy and S.H. Gruber. 2010. Diet and prey preference of juvenile lemon sharks *Negaprion brevirostris*. Mar. Ecol. Prog. Ser. 398: 221–234.

Nugues, M.M. and R.P.M. Bak. 2006. Differential competitive abilities between Caribbean coral species and a brown alga: a year of experiments and a long-term perspective. Mar. Ecol. Prog. Ser. 315: 75–86.

O'Farrell, S., A.R. Harborne, Y.M. Bozec, B.E. Luckhurst and P.J. Mumby. 2015. Protection of functionally important parrotfishes increases their biomass but fails to deliver enhanced recruitment. Mar. Ecol. Prog. Ser. 522: 245–254.

O'Farrell, S., B.E. Luckhurst, S.J. Box and P.J. Mumby. 2016. Parrotfish sex ratios recover rapidly in Bermuda following a fishing ban. Coral Reefs 35: 421–425.

Ogden, J.C. and N.S. Buckman. 1973. Movements, foraging groups, and diurnal migrations of the striped parrotfish *Scarus croicensis* Bloch (Scaridae). Ecology 54: 589–596.

Ortiz, J.C., Y.-M. Bozec, N.H. Wolff, C. Doropoulos and P.J. Mumby. 2014. Global disparity in the ecological benefits of reducing carbon emissions for coral reefs. Nat. Clim. Chang. 4: 190–194.

Paddack, M.J. and S. Sponaugle. 2008. Recruitment and habitat selection of newly settled *Sparisoma viride* to reefs with low coral cover. Mar. Ecol. Prog. Ser. 369: 205–212.

Paddack, M.J., S. Sponaugle and R.K. Cowen. 2009. Small-scale demographic variation in the stoplight parrotfish *Sparisoma viride*. J. Fish Biol. 75: 2509–2526.

Perry, C.T., T. Spencer and P.S. Kench. 2008. Carbonate budgets and reef production states: a geomorphic perspective on the ecological phase-shift concept. Coral Reefs 27: 853–866.

Polunin, N.V.C. and C.M. Roberts. 1993. Greater biomass and value of target coral-reef fishes in two small Caribbean marine reserves. Mar. Ecol. Prog. Ser. 100: 167–176.

Rakitin, A. and D.L. Kramer. 1996. Effect of a marine reserve on the distribution of coral reef fishes in Barbados. Mar. Ecol. Prog. Ser. 131: 97–113.

Randall, J.E. 1965. Food habits of the Nassau grouper (*Epinephelus striatus*). Proc. Assoc. Is. Mar. Lab. Carib. 6: 13–16.

Randall, J.E. 1967. Food habitats of reef fishes of the West Indies. Stud. Trop. Oceanogr. 5: 665–847.

Renken, H., P.J. Mumby, I. Matsikis and H.J. Edwards. 2010. Effects of physical environmental conditions on the patch dynamics of *Dictyota pulchella* and *Lobophora variegata* on Caribbean coral reefs. Mar. Ecol. Prog. Ser. 403: 63–74.

Rizzari, J.R., A.J. Frisch, A.S. Hoey and M.I. McCormick. 2014. Not worth the risk: apex predators suppress herbivory on coral reefs. Oikos 123: 829–836.

Robertson, D.R., F. Karg, R.L. de Moura, B.C. Victor and G. Bernardi. 2006. Mechanisms of speciation and faunal enrichment in Atlantic parrotfishes. Mol. Phylogenet. Evol. 40: 795–807.

Roff, G. and P.J. Mumby. 2012. Global disparity in the resilience of coral reefs. Trends Ecol. Evol. 27: 404–413.

Roff, G., M.H. Ledlie, J.C. Ortiz and P.J. Mumby. 2011. Spatial patterns of parrotfish corallivory in the Caribbean: the importance of coral taxa, density and size. PLoS One 6: e29133.

Roff, G., C.D. Doropoulos, G. Mereb, P.J. Mumby. 2017. Unprecedented mass spawning aggregation of the giant bumphead parrotfish (*Bolbometopon muricatum*) in Palau, Micronesia. J. Fish Biol. doi: 10.1111/jfb13340.

Rogers, A., J.L. Blanchard and P.J. Mumby. 2014. Vulnerability of coral reef fisheries to a loss of structural complexity. Curr. Biol. 24: 1000–1005.

Rogers, A., A.R. Harborne, C.J. Brown, Y.M. Bozec, C. Castro, I. Chollett, K. Hock, C.A. Knowland, A. Marshell, J.C. Ortiz, T. Razak, G. Roff, J. Samper-Villarreal, M.I. Saunders, N.H. Wolff and P.J.

Mumby. 2015. Anticipative management for coral reef ecosystem services in the 21st century. Glob. Change Biol. 21: 504–514.

Rotjan, R.D. and S.M. Lewis. 2005. Selective predation by parrotfishes on the reef coral *Porites astreoides*. Mar. Ecol. Prog. Ser. 305: 193–201.

Sánchez, J.A., M.F. Gil, L.H. Chasqui and E.M. Alvarado. 2004. Grazing dynamics on a Caribbean reef-building coral. Coral Reefs 23: 578–583.

Stallings, C.D. 2008. Indirect effects of an exploited predator on recruitment of coral-reef fishes. Ecology 89: 2090–2095.

Steneck, R.S. and M.N. Dethier. 1994. A functional group approach to the structure of algal-dominated communities. Oikos 69: 476–498.

Steneck, R.S., P.J. Mumby and S.N. Arnold. 2007. A report on the status of the coral reefs of Bonaire in 2007 with results from monitoring 2003–2007. University of Maine, Maine.

Steneck, R.S., S.N. Arnold and P.J. Mumby. 2014. Experiment mimics fishing on parrotfish: insights on coral reef recovery and alternative attractors. Mar. Ecol. Prog. Ser. 506: 115–127.

Streelman, J.T., M. Alfaro, M.W. Westneat, D.R. Bellwood and S.A. Karl. 2002. Evolutionary history of the parrotfishes: biogeography, ecomorphology, and comparative diversity. Evolution 56: 961–971.

Targett, N.M. and T.M. Arnold. 1998. Predicting the effects of brown algal phlorotannins on marine herbivores in tropical and temperate oceans. J. Phycol. 34: 195–205.

Tolimieri, N. 1998a. Effects of substrata, resident conspecifics and damselfish on the settlement and recruitment of the stoplight parrotfish, *Sparisoma viride*. Environ. Biol. Fishes 53: 393–404.

Tolimieri, N. 1998b. Contrasting effects of microhabitat use on large-scale adult abundance in two families of Caribbean reef fishes. Mar. Ecol. Prog. Ser. 167: 227–239.

Tolimieri, N. 1998c. The relationship among microhabitat characteristics, recruitment and adult abundance in the stoplight parrotfish, *Sparisoma viride*, at three spatial scales. Bull. Mar. Sci. 62: 253–268.

Torres, R., M. Chiappone, F. Geraldes, Y. Rodriguez and M. Vega. 2001. Sedimentation as an important environmental influence on Dominican Republic reefs. Bull. Mar. Sci. 69: 805–818.

Toth, L.T., R. van Woesik, T.J.T. Murdoch, S.R. Smith, J.C. Ogden, W.F. Precht and R.B. Aronson. 2014. Do no-take reserves benefit Florida's corals? 14 years of change and stasis in the Florida Keys National Marine Sanctuary. Coral Reefs 33: 565–577.

Tzadik, O.E. and R.S. Appeldoorn. 2013. Reef structure drives parrotfish species composition on shelf edge reefs in La Parguera, Puerto Rico. Cont. Shelf Res. 54: 14–23.

Valles, H. and H.A. Oxenford. 2014. Parrotfish size: A simple yet useful alternative indicator of fishing effects on Caribbean reefs? PLoS One 9: e86291.

van Rooij, J.M., F.J. Kroon and J.J. Videler. 1996. The social and mating system of the herbivorous reef fish *Sparisoma viride*: one-male versus multi-male groups. Environ. Biol. Fishes 47: 353–378.

van Rooij, J.M., J.H. Bruggemann, J.J. Videler and A.M. Breeman. 1995. Plastic growth of the herbivorous reef fish *Sparisoma viride*: field evidence for a trade-off between growth and reproduction. Mar. Ecol. Prog. Ser. 122: 93–105.

Verweij, M.C., I. Nagelkerken, D. de Graaff, M. Peeters, E.J. Bakker and G. van der Velde. 2006. Structure, food and shade attract juvenile coral reef fish to mangrove and seagrass habitats: a field experiment. Mar. Ecol. Prog. Ser. 306: 257–268.

Whiteman, E.A., I.M. Côté and J.D. Reynolds. 2007. Ecological differences between hamlet (*Hypoplectrus* : Serranidae) colour morphs: between-morph variation in diet. J. Fish Biol. 71: 235–244.

Williams, I.D. and N.V.C. Polunin. 2001. Large-scale associations between macroalgal cover and grazer biomass on mid-depth reefs in the Caribbean. Coral Reefs 19: 358–366.

Wolf, N.G. 1985. Odd fish abandon mixed-species groups when threatened. Behav. Ecol. Sociobiol. 17: 47–52.

Wolff, N., R. Grober-Dunsmore, C.S. Rogers and J. Beets. 1999. Management implications of fish trap effectiveness in adjacent coral reef and gorgonian habitats. Environ. Biol. Fishes 55: 81–90.

CHAPTER 17

Parrotfishes, Are we still Scraping the Surface? Emerging Topics and Future Research Directions

Andrew S. Hoey[1], Brett M. Taylor[2], Jessica Hoey[3] and Rebecca J. Fox[4,5]

[1] ARC Centre of Excellence for Coral Reef Studies, James Cook University, Townsville,
  Queensland 4811, Australia
  Email: Andrew.hoey1@jcu.edu.au
[2] Joint Institute for Marine and Atmospheric Research, University of Hawaii and
  NOAA Fisheries, Pacific Islands Fisheries Science Center, 1845 Wasp Boulevard, Building 176,
  Honolulu, Hawaii 96818, U.S.A.
  Email: brett.taylor@noaa.gov
[3] Great Barrier Reef Marine Park Authority, 2-68 Flinders Street, Townsville,
  Queensland 4810, Australia
  Email: Jessica.hoey@gbrmpa.gov.au
[4] School of Life Sciences, University of Technology Sydney, Ultimo, NSW 2007, Australia
[5] Division of Ecology and Evolution, Research School of Biology, The Australian National
  University, Canberra, ACT 2600, Australia
  Email: rebecca.fox-1@uts.edu.au

Introduction

Parrotfish (Scarinae, Labridae) are found on almost every coral reef of the world. It is this ubiquity, coupled with the uniqueness of their functional impact on the ecosystem, which makes them arguably one of the most important groups of fishes on coral reefs. With the morphological innovation of a jaw that has the power to bite through carbonates (Wainwright and Price, Chapter 2), and a pharyngeal apparatus that can grind corals and carbonate rocks into sand particles (Gobalet, Chapter 1), no other group of fishes is so inextricably linked to the structural dynamics of their ecosystem (Malella and Fox, Chapter 8). These innovations and unique ecological roles have stimulated much scientific interest in the parrotfishes. Indeed, the number of scientific publications in the past 50 years that have considered parrotfishes (697 publications) is almost double that of many other reef fish families (e.g., surgeonfishes: 326, rabbitfishes: 353, butterflyfishes: 384) over the same time period (Web of Science – June 2017). This body of work has provided a firm grounding in understanding the evolution, morphology, distribution and functional ecology of this group; however our understanding of several other topics is limited.

Parrotfishes are a monophyletic group with 100 species currently recognised (Cowman et al. 2009, Parenti and Randall 2011). Considered an independent family (i.e. Scaridae) for over 200 years (Rafinesque 1810), recent molecular evidence has shown that the parrotfishes are nested within the Labridae (Westneat and Alfaro 2005, Cowman et al. 2009). While no one questions that parrotfishes evolved from the wrasses, there is debate regarding the taxonomic status of parrotfishes with some calling for parrotfishes to be recognized as an independent family (Randall and Parenti 2014, Randall - Foreword to this volume). Such calls have been based primarily on their specialized feeding morphology and unique interactions with reef environments.

The functional decoupling of pharyngeal and oral jaws within the Labroidei has freed the oral jaws from potential constraints involved with food processing and facilitated increased diversity in diet and feeding mode (e.g., Wainwright et al. 2004, Bellwood et al. 2006). Importantly, the pharyngeal jaw apparatus of parrotfishes has several modifications from the pharyngeal jaw system that are shared by other labrid fishes that allow them to grind or mill, as opposed to crush, ingested material (detailed description in Wainwright and Price Chapter 2). Together with these functional innovations of the pharyngeal jaws, the strong beak-like oral jaws that characterise the group enable parrotfishes to scrape or excavate the calcareous layers of the reef as they feed (Bonaldo et al. 2014, Malella and Fox Chapter 8, Bonaldo and Rotjan Chapter 9). The unique ability of parrotfish to feed in this way, and hence the number of ecological roles to which they contribute (i.e., grazing, erosion, coral predation, production, reworking and transport of sediments), their high abundance and feeding rates has led to wide acceptance of their importance to reef processes.

Although all parrotfishes share functional innovations of the oral and pharyngeal jaws, among species differences in the morphologies of feeding apparatus, feeding behaviour, and spatial distribution (Wainwright and Price Chapter 2, Hoey Chapter 6) affect the nature and intensity of the disturbance impact of individual species on the reef benthos, and the locations in which such impacts are realised (e.g., Bruggemann et al. 1994, Hoey and Bellwood 2008). This variation in functional impact is important and may modulate the contribution of individual parrotfish species to the functioning and resilience of reef ecosystems (Burkepile et al. Chapter 7). Parrotfishes are often classified into three main feeding modes, or functional groups (excavators, scrapers, and browsers) based on the osteology and myology of the jaws (Bellwood 1994). However, differences in body size and volume of material removed among excavating taxa (e.g., Bellwood et al. 2003), and feeding rates among scraping taxa (Hoey Chapter 6) suggest these may need to be revised. Further, these functional groups often reflect their vulnerability to fishing (Bellwood et al. 2012) and habitat degradation (Emslie and Pratchett Chapter 14). Given these differences and the unique ecological role of some parrotfish species conservation efforts should look beyond the preservation of biological diversity and focus rather on the protection of key species and/or species groups.

Top-down versus Bottom-up Processes

The vast majority of studies to date have considered parrotfishes in terms of the disturbance impacts of their feeding on the biomass, composition and/or succession of benthic reef communities (e.g., Mumby et al. 2006, Burkepile and Hay 2008, Bellwood et al. 2012). This body of work has been motivated, at least in part, by dramatic changes in benthic reef communities following the large reductions in, or complete loss, of herbivores (including parrotfishes) in some locations (e.g., Hughes 1994, Rasher et al. 2013), or rapid increases in macroalgal biomass following the exclusion of herbivorous fishes from small areas of reef (e.g., Lewis 1986, Hughes et al. 2007). Collectively, these studies demonstrate that, under certain circumstances, parrotfishes can have a dramatic influence on the composition or standing biomass of algal assemblages (e.g., Mumby et al. 2006, Burkepile and Hay 2008), and/or the replenishment and potential recovery of coral populations (e.g., Trapon et al. 2013, Graham et al. 2015). Equally, benthic communities and reef condition may influence the dynamics of parrotfish populations; however the importance of such 'bottom-up' processes has received relatively little attention.

Feeding Ecology and Foraging Dynamics

The acquisition of nutrients through feeding is central to an individual's fitness, and the distribution of preferred nutritional resources can shape the distribution and growth of individuals and populations. Despite parrotfishes being one of the most studied groups of coral reef fishes, our understanding of their nutritional ecology has lagged behind that of other groups of reef fishes (Clements and Choat Chapter 3). Difficulties in identifying components of ingesta that have been ground into slurry by the pharyngeal apparatus, and the bias toward top-down process has resulted in studies of the feeding ecology of parrotfishes focusing primarily on where parrotfishes feed, not what they are targeting or ingesting. Although previously parrotfishes have been considered to be herbivores (e.g., Mumby et al. 2006), or detritivores (e.g., Choat et al. 2002), recent evidence based on feeding behaviour, digestive mechanisms, and biochemical analyses indicate they are microphages that target protein-rich epilithic, endolithic and epiphytic microscopic phototrophs, predominately cyanobacteria (Clements et al. 2017, Clements and Choat Chapter 3).

The abundance, distribution and growth of animals are often influenced by the availability of their preferred food (Krebs 1972). Although we are unaware of studies that have directly related the availability of epilithic, endolithic, and epiphytic cyanobacteria among reefs or reef habitats to the distribution or growth of parrotfishes, several recent studies have related the abundance and/or biomass to the availability of substrata covered by the epilithic algal matrix (EAM), the favoured feeding substrata of many scarinine parrotfishes (e.g., Adam et al. 2011, Gilmour et al. 2013, Russ et al. 2015). Rapid increases in parrotfish populations following coral mortality and increases in EAM-covered substrata suggest that parrotfishes may be food limited at higher levels of coral cover (Adam et al. 2011, Pratchett et al. 2011, Gilmour et al. 2013). This is counter to many previous studies that have reported the abundance of parrotfishes to be positively related to the topographic complexity provided by corals (reviewed by Graham and Nash 2013). Indeed, Heenan et al. (2016) suggested the non-linear response of scraping parrotfishes to topographic complexity across 33 central and western Pacific islands may represent a trade-off between topographic complexity and food availability. This presents a conundrum. Parrotfishes are widely accepted as being important in promoting coral-dominance, yet at high coral cover their populations may be suppressed (due to food limitation), and potentially unable to respond to large increases in algal-covered substrata following coral loss. It appears, therefore, that reefs with intermediate levels of coral cover may be optimal for parrotfish populations, and be more resilient to disturbance than those with higher coral cover.

Macroalgae, although often viewed as a consequence of reductions in populations of herbivorous fishes, is a natural and common component on many Atlantic reefs, as well as coastal and marginal reefs in the Indo-Pacific (Hoey et al. Chapter 12, Fox Chapter 13). Sparisomatine parrotfishes are often described as browsers of macroalgae and/or seagrass. However, Clements and Choat (Chapter 3) suggest they too are targeting protein-rich epiphytes, predominately cyanobacteria, rather than the macrophytes themselves. The only exception to this may be the Indo-Pacific sparisomatine *Leptoscarus vaigiensis* (Ohta and Tachihara 2004, Gullström et al. 2011). Further, few studies have considered how the abundance of macroalgae may influence patterns of habitat use or foraging of parrotfishes. These few studies indicate that most adult parrotfishes prefer areas devoid of macroalgae. For example, the abundance of *Sparisoma viride* increased on patch reefs following removal of macroalgae from patch reefs in Belize (McClanahan et al. 2000). Similarly, feeding by scarinine parrotfishes on the reef substrate was negatively related to biomass of *Sargassum* on an inshore reef of the Great Barrier Reef (GBR) (Hoey and Bellwood 2011, Fox Chapter 13). The influence of the abundance, composition, and canopy height of macroalgae on patterns of habitat use and foraging by parrotfishes, and the extent to which these effects extent beyond the boundaries of macroalgal beds is currently unknown.

Settlement Habitat and Seascape Structure

The settlement and juvenile habitats of the majority of parrotfish species are poorly understood, and is likely to be related to their small size at settlement (*Scarus*: 7-8 mm total length, Bellwood and Choat 1989), cryptic nature, and lack of identifying features, making in situ species identification difficult. For example, Bellwood and Choat (1989) provide descriptions of the juvenile colour patterns of 24 parrotfish species from the GBR, but only include broad qualitative descriptions of collection sites of juvenile fishes. Conversely, Green (1998) found recently-settled *Scarus* spp. associated with macroalgae in damselfish territories, but did not identify the individuals to species. Similarly, *Sp. viride* has been reported to settle to macroalgae in the Upper Florida Keys (Paddack and Sponaugle 2008, but see Tolimieri 1998), and juvenile parrotfishes are common within macroalgal beds on inshore reefs of the GBR (A. Hoey pers. obs.). The benefits of using these macroalgal habitats is unclear and warrants investigation, but may be related to the provision of food and/or shelter during this critical period.

For species that have distinct settlement and adult habitats, the spatial arrangement of these habitats can have important implications for the viability of populations. For example, juveniles of the Atlantic scarinine parrotfish *Scarus guacamaia* are found almost exclusively with mangrove habitats, while adults are associated with reefs (Dorenbosch et al. 2006). The proximity of juvenile and adult habitats has a major influence on local populations with abundance of reef-based adults decreasing exponentially with distance from mangroves (Claydon et al. 2015), and suffering local extinctions following mangrove loss (Mumby et al. 2004). Similarly, juveniles of the Indo-Pacific bumphead parrotfish *Bolbometopon muricatum* settle to branching *Acropora* on shallow lagoonal reefs close to mangroves, while adults typically occur on reefs further offshore (Hamilton et al. 2017). The loss of these juvenile habitats due to sediment input from logging operations in the Solomon Islands has markedly reduced juvenile abundance, with likely reductions in the replenishment of adult populations on offshore reefs. The reliance of branching corals on inshore reefs as a settlement/juvenile habitat may explain the distribution of adult *B. muricatum* in other regions. For example, *B. muricatum* are most abundant on outer-shelf reefs in the northern GBR, and mid-shelf reefs in the central GBR (Hoey and Bellwood 2008, Wismer et al. 2009). These differences coincide with a widening of the continental shelf, and hence proximity to shallow coastal reefs close to mangroves, in the central GBR. Finally, relationships between reef and island geomorphology and the biomass and structure of parrotfish assemblages have been suggested to be due to the associated broad-scale habitat diversity (Taylor et al. 2015, Heenan et al. 2016), further highlighting the potential importance of seascape structure in shaping parrotfish populations.

Parrotfishes are largely viewed in terms of the impact of their feeding on benthic communities. There is a clear need to change our thinking to consider how bottom-up processes may shape parrotfish populations and assemblages. This will require a shift from the current focus on the consequences of feeding to a holistic approach to understanding both the causes and consequences of parrotfish feeding, how changes in reef condition may influence foraging decisions, and an understanding of ontogenetic shifts in habitat requirements. Such detail is critical if we are to predict the likely short- and long-term effects of environmental disturbance on parrotfish assemblages, the effects of fisheries protection on benthic communities, or the effects of climate change on parrotfish-reef interactions.

Physiology and Thermal Tolerances in a Warming World

Parrotfishes are ectotherms, and as such their rates of biochemical and cellular processes are largely governed by environmental temperature (Pörtner and Farrell 2008). Any increase in environmental temperature within an organism's thermal tolerance range will, therefore, increase metabolic demands. These increased demands may be met through increased intake of energy, reduced energy expenditure, or both. For tropical ectotherms, such as parrotfish, theory predicts that they are already likely to be living close to their thermal optima and possess limited physiological flexibility because they only experience small temperature variations over an annual cycle (Tewksbury et al. 2008). Despite the importance of parrotfish to the health of reef ecosystems and the ongoing threat of climate change, our empirical understanding of the physiological responses of parrotfishes to environmental temperatures is currently extremely limited. There are, to our knowledge, no published laboratory studies examining the thermal sensitivity (in terms of their existing thermal reaction norms) of adult parrotfishes, nor has the potential for parrotfish to acclimate, either developmentally or transgenerationally, to projected increases in ocean temperatures been investigated. Although seasonal and latitudinal comparisons of the feeding rate of parrotfishes suggest that energy demand is greater at higher temperatures (Hoey Chapter 6), our understanding of the physiological responses of parrotfishes to changing temperature is limited to inferences drawn from other taxa. Increasing water temperature has been shown to increase the resting oxygen consumption and decrease the aerobic scope of small, site-attached coral reef fishes (reviewed in Hoey et al. 2016). However, not only has much of the work in this area tended to ignore the larger, more mobile reef fish species (see Johansen et al. 2014, 2015, Clark et al. 2017 for exceptions), but our understanding of the response of reef fishes to temperature is currently based just on laboratory testing of physiological performance. The interaction between behaviour and physiology in field-active individuals and the potential for larger, more mobile species such as parrotfish to use behaviour to buffer the effects of increasing water temperature has yet to be satisfactorily investigated. Future studies that can provide a complete and accurate investigation (*sensu* Hertz et al. 1993) of behavioural thermoregulation among mobile reef fish species will represent an exciting extension to our understanding of parrotfish ecology, and at the same time improve our ability to more accurately model the impacts of climate change on parrotfish populations (including whether they are likely to be able to continue to be exploited as a fisheries resource).

The Demographics of Sustainable Parrotfish Populations

There is still much to learn about the demography of parrotfishes and, equally so, there is much to learn through parrotfishes about broad-scale demographic processes. Almost every aspect of parrotfish biology and ecology can be considered complex and, in many cases, unique. Such features make them ideal for addressing a variety of questions regarding life history, community ecology, and fisheries sustainability. Examinations of age-based demography are uncovering scale-dependent patterns of life-history variation influenced by a wide range of biophysical and anthropogenic factors (Taylor et al. Chapter 4). Further development of the understanding of intra- and inter-specific life-history variation is imperative for improving fisheries sustainability as well as for predicting impacts of changing ocean conditions (e.g., temperature and benthic productivity) on parrotfish population dynamics.

There are now numerous examples of the effects of fishing on parrotfish assemblages and simple demographic traits such as body size consistently explain much of the differences among species in vulnerability to overexploitation (Hawkins and Roberts 2003, Clua and Legendre 2008, Taylor et al. 2014, Vallès and Oxenford 2014). Overall, the effects of fishery management strategies on parrotfishes and the types of management strategies available are variable (Harborne and Mumby Chapter 16). The most common type in tropical reefs environments is no-take marine reserves. There are many examples of positive effects of protection from fishing on parrotfish density and biomass (reviewed in Questel and Russ Chapter 14), but Harborne and Mumby (Chapter 16) warn that marine reserves alone are not sufficient for enhancing fisheries productivity or reef structural complexity through indirect effects. Ultimately, given their perceived functional role in coral reef ecosystems, there appears to be an emerging question of whether parrotfishes should be managed or outright protected.

From the demographic viewpoint, another area that deserves greater attention from researchers is the reproductive ecology of parrotfishes. Parrotfishes have an incredibly diverse array of reproductive strategies, including protogynous hermaphroditism, functional gonochorism, sexual dichromatism and dimorphism, as well as complex mating strategies, all of which vary intra- and inter-specifically. Their idiosyncratic and diverse reproductive nature makes parrotfishes a rewarding group for empirical evaluations of sex allocation strategies and the effects of biophysical factors in complex ecosystems.

Conclusions

In drawing up this list of emerging research directions, thoughts necessarily turn to what the future holds for parrotfish populations around the globe. The threats to parrotfish are numerous and severe: from the local effects of overfishing (e.g., Bellwood et al. 2012), to the twin evils of global warming and ocean acidification, and even extending out to land-based environmental issues such as the impact of logging on reef fish nursery habitats (Hamilton et al. 2017). The IUCN classifies parrotfish overall as facing a low risk of extinction globally, but individual species may face a high risk of extinction regionally, predominantly as a result of intense fishing pressure and habitat destruction at the local scale. This applies especially to the larger, functionally-important species such as *B. muricatum* (listed as Vulnerable, IUCN Red List, although not warranted by the US for listing under the Endangered Species Act) and *Sc. guacamaia* (Near Threatened, IUCN Red List), which are both attractive fishing targets and have experienced destruction of critical habitat. Even in parts of their range where fishing pressure is light, or where fisheries restrictions afford sufficient protection at the population level, parrotfish face challenges associated with the alteration of reef ecosystems caused by human-induced climate change. As we write, the Great Barrier Reef (GBR) has just experienced back-to-back bleaching events, including its most severe episode of coral bleaching ever recorded (Hughes et al. 2017a). Coming on the back of a severe Crown-of-Thorns starfish outbreak (2013-present) and with many reefs in the northern section of the GBR already having had coral cover dramatically reduced by the impacts of Cyclone Ita in 2014 and Cyclone Nathan in 2015, it is clear that coral reef ecosystems today are under greater pressure than ever before. Indeed, it is debatable whether reef habitats as we know them will be able to persist beyond the end of the current century (Hughes et al. 2017b). But how will the degradation of reef habitats impact on parrotfishes? We already know parrotfish are essential for healthy reef ecosystems (Jackson et al. 2014): but are healthy reefs essential for parrotfishes?

Unfortunately the question is not just a rhetorical one. With some predicting the end of reef habitats as we know them by the end of the century (Hughes et al. 2017b), asking whether parrotfish will be able to persist in the absence of coral reefs is a question that will require answering. Parrotfish evolved in tropical ocean waters prior to the mid-Miocene (Bellwood and Schultz 1991, Bernardi et al. 2000), with the earliest known fossil a *Calotomus* species (Bellwood and Schultz 1991). In terms of habitat associations, the phylogeny suggests that parrotfish have shifted over time from non-reef (seagrass) habitats to a more exclusive association with coral reefs, with the reef association being only a relatively recent one (Bellwood 1994, Bernardi et al. 2000, Streelman et al. 2002, Choat et al. 2012). It is argued by some that coral reefs have facilitated the diversification and speciation of the Labroidei, however, given the historical associations with non-reef habitats, it seems possible that parrotfish could persist in alternative environments, albeit at depauperate levels of species diversity, or potentially with a loss over evolutionary time of some of the oral jaw modifications that have evolved within the reef clade (Streelman et al. 2002). It is therefore possible that parrotfish of the future may devolve back into 'browsers' of seagrass.

The link between parrotfish diversification and reproductive ecology is not universally acknowledged (see Choat et al. 2012), however, there remains the potential for changes in habitat association to impact on mating and social behaviours within the tribe, with the potential for loss of social systems purported to have evolved within reef habitats (where territoriality and size advantage in males has led to the evolution of haremic mating systems, a sex ratio skewed towards females, as well as selecting for early maturation of primary males; Robertson and Warner 1978). Within seagrass populations, the lack of territoriality would likely see different sexual selection pressures, the evolution of alternative mating systems, alternative pressures on size dimorphism, and therefore different population dynamics, requiring different population management strategies. The future for parrotfishes has probably never been more uncertain, but with several hundred million people around the world dependent on these fishes as either a source of protein or a source of direct or indirect (via reef tourism) income, it is more important than ever for science to further our understanding of their biology and ecology.

References Cited

Adam, T.C., R.J. Schmitt, S.J. Holbrook, A.J. Brooks, P.J. Edmunds, R.C. Carpenter and G. Bernardi. 2011. Herbivory, connectivity, and ecosystem resilience: response of a coral reef to a large-scale perturbation. PLoS One 6: e23717.

Bellwood, D.R. 1994. A phylogenetic study of the parrotfishes family Scaridae (Pisces: Labroidei), with a revision of the genera. Rec. Austr. Mus. 20: 1–86.

Bellwood, D.R. and J.H. Choat. 1989. A description of the juvenile phase colour patterns of 24 parrotfish species (family Scaridae) from the Great Barrier Reef, Australia. Rec. Aust. Mus. 41: 1–41.

Bellwood, D.R. and O. Schultz. 1991. A review of the fossil record of the parrotfishes (Labroidei: Scaridae) with a description of a new *Calotomus* species from the Middle Miocene (Badenian) of Austria. Ann. Naturhist. Mus. Wien. 92: 55–71.

Bellwood, D.R., A.S. Hoey and J.H. Choat. 2003. Limited functional redundancy in high diversity systems: resilience and ecosystem function on coral reefs. Ecol. Lett. 6: 281–285.

Bellwood, D.R., P.C. Wainwright, C.J. Fulton and A.S. Hoey. 2006. Functional versatility supports coral reef biodiversity. Proc. R. Soc. B 273: 101–107.

Bellwood, D.R., A.S. Hoey and T.P. Hughes. 2012. Human activity selectively impacts the ecosystem roles of parrotfishes on coral reefs. Proc. R. Soc. B 271: 1621–1629.

Bernardi, G., D.R. Robertson, K.E. Clifton and E. Azzurro. 2000. Molecular systematics, zoogeography, and evolutionary ecology of the Atlantic parrotfish genus *Sparisoma*. Mol. Phylogenet. Evol. 15: 292–300.

Bonaldo, R.M., A.S. Hoey and D.R. Bellwood. 2014. The ecosystem roles of parrotfishes on tropical reefs. Oceanogr. Mar. Biol. Annu. Rev. 52: 81–132.

Bozec, Y.M., L. Yakob, S. Bejarano and P.J. Mumby. 2013. Reciprocal facilitation and non-linearity maintain habitat engineering on coral reefs. Oikos 122: 428–440.

Bruggemann, J.H., M.W.M. Kuyper and A.M. Breeman. 1994. Comparative analysis of foraging and habitat use by the sympatric Caribbean parrotfish *Scarus vetula* and *Sparisoma viride*. Mar. Ecol. Prog. Ser. 112: 51–66.

Burkepile, D.E. and M.E. Hay. 2008. Herbivore species richness and feeding complementarily affect community structure and function: the case for Caribbean reefs. Proc. Natl. Acad. Sci. USA. 105: 16201–16206.

Choat, J., K. Clements and W. Robbins. 2002. The trophic status of herbivorous fishes on coral reefs. Mar. Biol.140: 613–623.

Choat, J.H., O.S. Klanten, L. van Herwerden, D.R. Robertson and K.D. Clements. 2012. Patterns and processes in the evolutionary history of parrotfishes (Family Labridae). Biol. J. Linn. Soc. 107: 529–557.

Clark, T.D., V. Messmer, A.J. Tobin, A.S. Hoey and M.S. Pratchett. 2017. Rising temperatures may drive fishing-induced selection of low-performance phenotypes. Sci. Rep. 7: 40571

Claydon, J.A.B., M.C. Calosso, G.A. De Leo and R.B.J. Peachey. 2015. Spatial and demographic consequences of nursery-dependence in reef fishes: an empirical and simulation study. Mar. Ecol. Prog. Ser. 525: 171–183.

Clements, K.D., D.P. German, J. Piché, A. Tribollet and J.H. Choat. 2017. Integrating ecological roles and trophic diversification on coral reefs: multiple lines of evidence identify parrotfishes as microphages. Biol. J. Linn. Soc. 120: 729–751.

Clua, E. and P. Legendre. 2008. Shifting dominance among Scarid species on reefs representing a gradient of fishing pressure. Aquat. Living Resour. 21: 339–348.

Cowman, P.F., D.R. Bellwood and L. van Herwerden. 2009. Dating the evolutionary origins of wrasse lineages (Labridae) and the rise of trophic novelty on coral reefs. Mol. Phylogen. Evol. 52: 621–631.

Dorenbosch, M., M.G.G. Grol, I. Nagelkerken and G. van der Velde. 2006. Seagrass beds and mangroves as potential nurseries for the threatened Indo-Pacific humphead wrasse, *Cheilinus undulatus* and Caribbean rainbow parrotfish, *Scarus guacamaia*. Biol. Cons. 129: 277–282.

Gilmour, J.P., L.D. Smith, A.J. Heyward, A.H. Baird and M.S. Pratchett. 2013. Recovery of an isolated coral reef system following severe disturbance. Science 340: 69–71.

Graham, N.A.J. and K.L. Nash. 2013. The importance of structural complexity in coral reef ecosystems. Coral Reefs 32: 315–326.

Graham, N.A., S. Jennings, M.A. MacNeil, D. Mouillot and S.K. Wilson. 2015. Predicting climate-driven regime shifts versus rebound potential in coral reefs. Nature 518: 94–97.

Green, A.L. 1998. Spatio-temporal patterns of recruitment of labroid fishes (Pisces: Labridae and Scaridae) to damselfish territories. Environ. Biol. Fish. 51: 235–244.

Gullström, M., C. Berkström, M.C. Öhman, M. Bodin and M. Dahlberg. 2011. Scale-dependent patterns of variability of a grazing parrotfish (*Leptoscarus vaigiensis*) in a tropical seagrass-dominated seascape. Mar. Biol. 158: 1483

Hamilton, R.J., G.R. Almany, C.J. Brown, J. Pita, N.A. Peterson and J.H. Choat. 2017. Logging degrades nursery habitat for an iconic coral reef fish. Biol. Conserv. 210: 273–280.

Hawkins, J.P. and C.M. Roberts. 2003. Effects of fishing on sex-changing Caribbean parrotfishes. Biol. Conserv. 115: 213–226.

Heenan, A., A.S. Hoey, G.J. Williams and I.D. Williams. 2016. Natural bounds on herbivorous coral reef fishes. Proc. R. Soc. B 283: rspb20161717.

Hertz, P.E., R.B. Huey and R.D. Stevenson. 1993. Evaluating temperature regulation by field-active ectotherms: the fallacy of the inappropriate question. Am. Nat. 142: 796–818.

Hoey, A.S. and D.R. Bellwood. 2008. Cross-shelf variation in the role of parrotfishes on the Great Barrier Reef. Coral Reefs 27: 37–47.

Hoey, A.S. and D.R. Bellwood. 2011. Suppression of herbivory by macroalgal density: a critical feedback on coral reefs. Ecol. Lett. 14: 267–273.

Hoey, A.S., E. Howells, J.L. Johansen, J.P.A. Hobbs, V. Messmer, D.M. McCowan, S.K. Wilson and M.S. Pratchett 2016. Recent advances in understanding the effects of climate change on coral reefs. Diversity 8: 12.

Hughes, T.P. 1994. Catastrophes, phase shifts, and large-scale degradation of a Caribbean coral reef. Science 265: 1547–1551.

Hughes, T.P., M.J. Rodrigues, D.R. Bellwood, D. Ceccarelli, O. Hoegh-Guldberg, L. McCook, N. Moltchaniwskyj, M.S. Pratchett, R.S. Steneck and B. Willis. 2007. Phase shifts, herbivory, and the resilience of coral reefs to climate change. Curr. Biol. 17: 360–365.

Hughes, T.P., J.T. Kerry, M. Álvarez-Noriega, J.G. Álvarez-Romero, K.D. Anderson, A.H. Baird, A.H., R.C. Babcock, M. Beger, D.R. Bellwood, R. Berkelmans, T.C. Bridge, I.R. Butler, M. Byrne, N.E. Cantin, S. Comeau, S.R. Connolly, G.S. Cumming, S.J. Dalton, G. Diaz-Pulido, C.M. Eakin, W.F. Figueira, J.P. Gilmour, H.B. Harrison, S.F. Heron, A.S. Hoey, J.-P.A. Hobbs, M.O. Hoogenboom, E.V. Kennedy, C.-Y. Kuo, J.M. Lough, R.J. Lowe, G. Liu, M.T. McCulloch, H.A. Malcolm, M.J. McWilliam, J.M. Pandolfi, R.J. Pears, M.S. Pratchett, V. Schoepf, T. Simpson, W.J. Skirving, B. Sommer, G. Torda, D.R. Wachenfeld, B.L. Willis and S.K. Wilson. 2017a. Global warming and recurrent mass bleaching of corals. Nature 543: 373–377.

Hughes, T.P., M.L. Barnes, D.R. Bellwood, J.E. Cinner, G.S. Cumming, J.B. Jackson, J. Kleypas, I.A. van de Leemput, J.M. Lough, T.H. Morrison and S.R. Palumbi. 2017b. Coral reefs in the Anthropocene. Nature 546: 82–90.

Jackson, J., M. Donovan, K. Cramer and V. Lam. 2014. Status and trends of Caribbean coral reefs: 1970–2012. Global Coral Reef Monitoring Network.

Johansen, J.L., V. Messmer, D.J. Coker, A.S. Hoey and M.S. Pratchett. 2014. Increasing ocean temperatures reduce activity patterns of a large commercially important coral reef fish. Glob. Change Biol. 20: 1067–1074.

Johansen, J.L., M.S. Pratchett, V. Messmer, D.J. Coker, A.J. Tobin and A.S. Hoey. 2015. Large predatory coral trout species unlikely to meet increasing energetic demands in a warming ocean. Sci. Rep. 5: 13830.

Krebs, C.J. 1972. Ecology: the Experimental Analysis of Distribution and Abundance. Harper and Row, New York.

Lewis, S.M. 1986. The role of herbivorous fishes in the organization of a Caribbean reef community. Ecol. Monogr. 56: 183–200.

McClanahan, T.R., K. Bergman, M. Huitric, M. McField, T. Elfwing, M. Nyström and I. Nordemar. 2000. Response of fishes to algae reduction on Glovers Reef, Belize. Mar. Ecol. Prog. Ser. 206: 273–282.

Mumby, P.J., A.J. Edwards, J.E. Arias-González, K.C. Lindeman, P.G. Blackwell, A. Gall, M.I. Gorczynska, A.R. Harborne, C.L. Pescod, H. Renken, C.C.C. Wabnitz and G. Llewellyn. 2004. Mangroves enhance the biomass of coral reef fish communities in the Caribbean. Nature 427: 533–536.

Mumby, P.J., C.P. Dahlgren, A.R. Harborne, C.V. Kappel, F. Micheli, D.R. Brumbaugh, K.E. Holmes, J.M. Mendes, K. Borad, J.N. Sanchirico, K. Bunch, S. Box, R.W. Stoffle and A.B. Gill. 2006. Fishing, trophic cascades, and the process of grazing on coral reefs. Science 311: 98–101.

Ohta, I. and K. Tachihara. 2004. Larval development and food habits of the marbled parrotfish, *Leptoscarus vaigiensis*, associated with drifting algae. Ichthyol. Res. 51: 63–69.

Paddack, M.J. and S. Sponaugle. 2008. Recruitment and habitat selection of newly settled Sparisoma viride to reefs with low coral cover. Mar. Ecol. Prog. Ser. 369: 205–212.

Parenti, P. and J.E. Randall. 2011. Checklist of the species of the families Labridae and Scaridae: an update. Smithiana Bull. 13: 29–44.

Pörtner, H.O. and A.P. Farrell. 2008. Physiology and climate change. Science 322: 690–692.

Pratchett, M.S., A.S. Hoey, S.K. Wilson, V. Messmer and N.A. Graham. 2011. Changes in biodiversity and functioning of reef fish assemblages following coral bleaching and coral loss. Diversity 3: 424–452.

Rafinesque, C.S. 1810. Indice d'ittiologia siciliana ossia catalogo metodico dei nomi latini, italiani, e siciliani dei pesci che si rinvengono in Sicilia disposti secondo un metodo naturale e seguito da un'appendice che contiene la descrizione di alcuni nuovi pesci siciliani. Messina, 80 p.

Randall, J.E. and P. Parenti. 2014. Parrotfishes are not wrasses. Reef Encounter 29: 16–18.

Rasher, D.B., A.S. Hoey and M.E. Hay. 2013. Consumer diversity interacts with prey defenses to drive ecosystem function. Ecology 94: 1347–1358.

Robertson, D.R. and R.R. Warner. 1978. Sexual patterns in the labroid fishes of the western Caribbean. II. The parrotfishes (Scaridae). Smithson. Contrib. Zool. 255: 1–26.

Russ, G.R., S.A. Questel, J.R. Rizzari and A.C. Alcala. 2015. The parrotfish-coral relationship: refuting the ubiquity of a prevailing paradigm. Mar. Biol. 162: 2029–2045.

Streelman, J.T., M. Alfaro, M.W. Westneat, D.R. Bellwood and S.A. Karl. 2002. Evolutionary history of the parrotfishes: biogeography, ecomorphology, and comparative diversity. Evolution 58: 961–971.

Taylor, B.M., P. Houk, G.R. Russ and J.H. Choat. 2014. Life histories predict vulnerability to overexploitation in parrotfishes. Coral Reefs 33: 869–878.

Taylor, B.M., S.J. Lindfield and J.H. Choat. 2015. Hierarchical and scale-dependent effects of fishing pressure and environment on the structure and size distribution of parrotfish communities. Ecography 38: 520–530.

Tewksbury, J.J., R.B. Huey and C.A. Deutsch. 2008. Putting the heat on tropical animals. Science 320: 1296–1297.

Tolimieri, N. 1998. Effects of substrata, resident conspecifics and damselfish on the settlement and recruitment of the stoplight parrotfish, Sparisoma viride. Environ. Biol. Fish. 53: 393–404.

Trapon, M.L., M.S. Pratchett, A.S. Hoey and A.H. Baird. 2013. Influence of fish grazing and sedimentation on the early post-settlement survival of the tabular coral *Acropora cytherea*. Coral Reefs 32: 1051–1059.

Vallès, H. and H.A. Oxenford. 2014. Parrotfish size: A simple yet useful alternative indicator of fishing effects on Caribbean reefs? PLoS One 9: e86291.

Wainwright, P.C., D.R. Bellwood, M.W. Westneat, J.R. Grubich and A.S. Hoey. 2004. A functional morphospace for the skull of labrid fishes: patterns of diversity in a complex biomechanical system. Biol. J. Linnean Soc. 82: 1–25.

Westneat, M.W. and M.E. Alfaro. 2005. Phylogenetic relationships and evolutionary history of the reef fish family Labridae. Mol. Phylogen. Evol. 36: 370–390.

Wismer, S., A.S. Hoey and D.R. Bellwood. 2009. Cross-shelf benthic community structure on the Great Barrier Reef: relationships between macroalgal cover and herbivore biomass. Mar. Ecol. Prog. Ser. 376: 45–54.

Index

Milton Keynes UK
Ingram Content Group UK Ltd.
UKHW050455071024
449327UK00015B/392

9 780367 781408